1934 Official

SHORT WAVE
Radio Manual

Complete Experimenter's Set-Building and Servicing Guide

Featuring a complete directory of all 1934 shortwave receivers.
Loaded with projects and circuit diagrams.
Special new chapter on building transistor receivers.

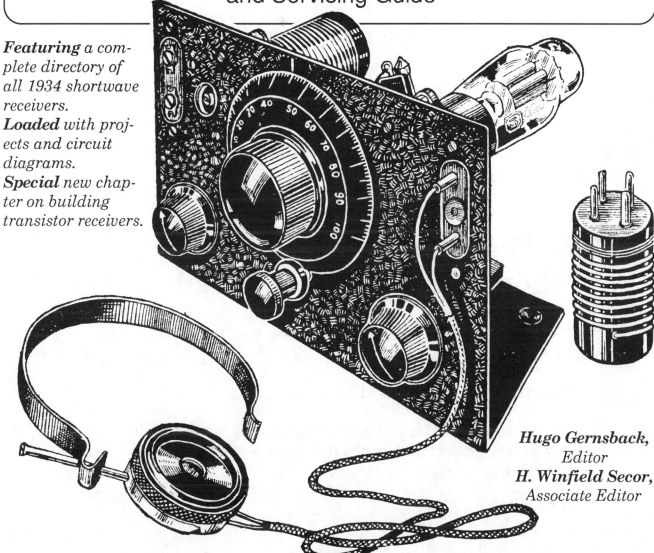

Hugo Gernsback,
Editor
H. Winfield Secor,
Associate Editor

reprinted by Lindsay Publications Inc

Autographed Edition

IT is my sincere wish that this, the first volume of the OFFICIAL SHORT WAVE RADIO MANUAL will afford you as much pleasure as I have had in compiling it.

Cordially yours,

H Gernsback

NEW YORK
Winter 1933-1934

1934 Official Short Wave Radio Manual

1 9 3 4
...Official...
SHORT WAVE
Radio Manual

COMPLETE
EXPERIMENTER'S SET - BUILDING and SERVICING GUIDE

Full Directory of all
SHORT WAVE RECEIVERS

HUGO GERNSBACK
Editor

H. WINFIELD SECOR
Associate Editor

Published by SHORT WAVE CRAFT

98 PARK PLACE
NEW YORK

INDEX

SERVICE SECTION

... *Introduction* ...

I HARDLY think that there can be any contradiction of the statement that short waves have finally "arrived," in every sense of the term.

While short waves have actually been with us since Heinrich Hertz in the late eighties discovered electro-magnetic radiation, the short-wave end of radio has not been exploited to a large scale until the past few years. The years 1932 and 1933, especially, were marked by a tremendous increase in short-wave radio; with some 8,500 short-wave stations, now actually transmitting, in every nook and corner of the entire world, it may be said that short waves have actually conquered the earth, taking it by storm.

The short-wave industry has already been recognized as such; while dozens of manufacturers are making either parts, short-wave sets, all-wave sets, or other incidentals which are comprised by the new industry.

Already, the short-wave art has become so great that it must be subdivided into many different branches. Up to this time, there has not been published a single work which covers the entire short-wave art in all its branches. The present volume, the first of a series, is intended to repair this shortcoming. In it we have endeavored to cover all branches in every way possible; so that the present and future student of short waves will not only look upon it as a historical work, but also use it for everyday purposes, wherever it may become necessary to do so.

Without trying to make this volume unwieldly, we have incorporated in it all the various branches of short-wave equipment, in such a manner as to cover the best that has appeared during the past few years.

We have tried our best to avoid duplication whenever this could be done, and to make the manual as live and up-to-date as it is possible to do so. We have had in mind particularly the newcomer to short waves—the student—as well as the research man who may wish to look for certain short-wave data.

We have not forgotten the radio service man either, and for him we have prepared an entire section showing practically every short-wave commercial set which has been manufactured since the inception of the art.

It is intended to bring out a similar volume once a year; in other words, a Year Book of Short Waves, to keep pace with this growing art.

HUGO GERNSBACK

New York, N. Y.
Winter, 1933-1934.

SHORT-WAVE RECEIVERS

4-Tube "MASTER COMPOSITE"
By Clifford E. Denton

● THREE thousand S. W. fans cannot be wrong. Results prove it, because tests on the receiver which was voted ideal by so many sure does the job 100 per cent. Stations all over the world at loud speaker volume, low back-ground noise, plus a very satisfactory degree of selectivity, are the result.

Permit the author to express his

thanks to the manufacturers of the components used. They have presented material for use in this set that will give maximum efficiency, highest gain and sturdy, long-life construction. There are other manufacturers making similar equipment that will give equal results but the manufacturers listed in the parts list are to be commended for

their vision and aid in developing equipment that will permit "real" short-wave reception.

Description of Receiver

Looking at the set from the front one will note the new National 7 inch *straight line* tuning scale. This is of great assistance in logging stations and makes for easy tuning.

Note that the coils are plugged in from the *front*—a real convenience for the fellow who wants to change from band to band quickly. It is possible to place the set in a metal case for further shielding and still be able to change coils without fuss or bother.

The tuning dial control is located at the center of the panel between the two coil *hand-hole* covers.

Starting from the left-hand side of the front panel and looking at the bottom row of knobs, we find that the first knob is the antenna potentiometer. This is used as a *volume control* when the set is used with a conventional antenna. When a doublet antenna or a Lynch *transposed* lead-in system is used, then the leads from the potentiometer are disconnected so that the primary winding of the antenna coil is not grounded. This permits *balanced* input conditions.

The second knob from the left controls the antenna compensating condenser and it "works like a charm."

The third knob controls the switching of the phones into and out of the circuit. When the phones are in use no signal will come from the speaker and when the speaker is in use, no signal will come from the phones. The

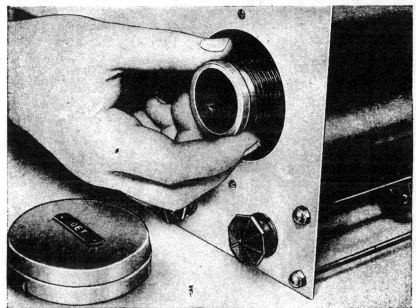

At two a. m. in the morn'—and your arm is tired perhaps, from lifting the lid of a receiver cabinet to change plug-in coils, you will give thanks for this method of changing plug-in coils, incorporated in the "Master Composite" receiver here described.

The "Master Composite" short-wave receiver, incorporating all major features is here described and illustrated. The set was designed and constructed under the supervision of Clifford E. Denton, well-known short-wave engineer, who has described so many good S-W receivers

Features of the present set are—only 4 tubes and therefore economical operation; together with the latest method of changing plug-in coils from the front of the set; a non-detuning regeneration control, and a powerful A.F. output stage, using the new 59 tube.

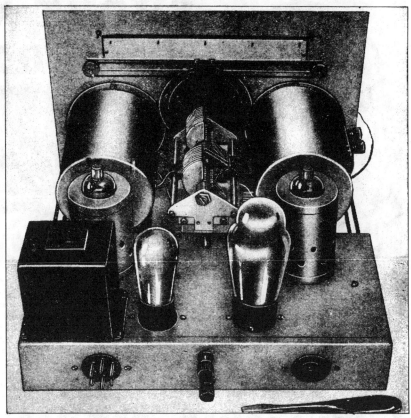

Photo at right shows rear view of the 4-tube "Master Composite" S-W Receiver, showing the shield cans, inside of which are the plug-in coils. National 6-pin coils with the R-39 low-loss forms are used.

The Ideal Receiver For You

Bottom view of the "Master Composite" receiver, showing the neat arrangement of the various resistors, R.F. chokes and by-pass condensers.

phones are permanently connected to the two binding posts provided on the rear of the chassis.

Regeneration in the detector stage is controlled by the remaining or fourth knob and is very smooth in action. The potentiometer of 50,000 ohms gives a *noiseless* variation of the voltage applied to the screen of the 58 type tube used as the detector. The action of the 58 type tube, as far as regeneration is concerned, is superior to the results obtained with a 57 type tube.

Of course, the two aluminum covers marked ANT and DET mark the placing of the openings through which the coils are inserted in their respective coil sockets. It is well to note at this time that the two coils used for any particular band are *identical in construction.*

When looking down on the top of the receiver it will be seen that the actual chassis is small in size.

The tuning condensers are mounted on the center line and are flanked by the two coil shields. Directly back of the coil shields are the two shields and tube sockets for the 58 type tubes. The shield to the left holds the R.F. amplifier tube and the shield to the right holds the detector tube.

The special detector coupling impedance is mounted to the rear of the detector tube and has the high inductance choke, coupling condenser and the grid coupling resistor mounted in the can. This method of coupling is more satisfactory than the standard resistance plate coupling so often used with screen-grid detectors. The main advantages are higher gain and smoother regeneration control.

On the right of the audio coupling

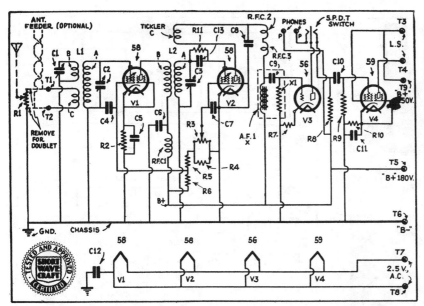

Schematic diagram of the "Master Composite" 4-Tube Short-Wave Receiver

phone binding posts are located in the center and they are flanked to the right (under the 59 output tube) by the plug and socket for the loud speaker connections. The remaining plug connection is for the *power* cable; five wires are all that are necessary for these connections even though the designer used a six-connector unit.

Note that the front panel is fastened to the chassis mechanically in five places, four by means of the threaded brass rods and the fifth by the brass collar used under the end-section of the tuning condenser.

The antenna volume control and the antenna compensating condenser are mounted on the front panel and the *phone-speaker* switch and the regeneration control are mounted on the chassis and are controlled from the front panel by means of the flexible couplings and the 5-inch long, ¼-inch diameter bakelite rods. This layout permits of easy wiring and much shorter leads, which improves the stability of the receiver and its general operation.

Examination of the receiver from underneath shows the layout of the various by-pass condensers and resistors.

Three .01 mf. mica condensers used as by-passes in the R.F. stage are mounted by a single through bolt near the R.F. tube socket. The two .25 mf. paper type condensers by-passing the plate of the R.F. stage and the screen of the detector are mounted one above the other, by means of small brass collars.

The small resistors, radio frequency

unit is the first audio stage; a 56 type tube is used here. The output of this tube is resistance-coupled to the power tube, which is a 59. This tube is more satisfactory than a 47 for one important reason—*less hum*. The 59 is a cathode type (heater) tube—which explains the hum reduction.

Space is provided on the rear of the chassis to the left for an *output transformer* if it is desired. The plate current of the 59 should not be allowed to flow through a pair of phones or a magnetic loud speaker as it will destroy their efficiency.

On the rear of the chassis, the

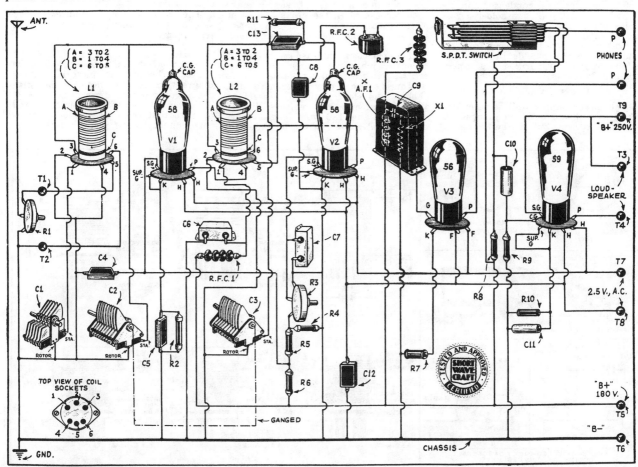

Picture wiring diagram, showing in A-B-C style just how to build the "Master Composite" short-wave receiver. You will find this set easy to build and also easy to tune—and speaking of a "hot" signal—wait till you hear it!

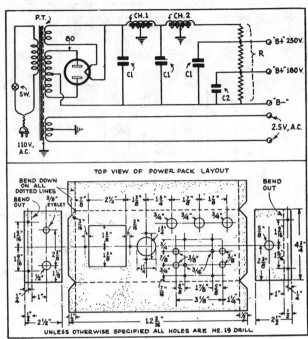

The two photos at left show the plate supply unit from the side and from the bottom. The diagrams at right show wiring diagram for the plate supply unit and lay-out of the metal subpanel.

chokes and mica condensers are held in position by the wiring. Therefore be sure to test every soldered connection for mechanical strength.

The detector grid condenser and the 5 megohm grid-leak are soldered into place before the coil socket of the detector tube is fastened into its shield.

Coil sockets are held away from the ends of the coil shields by means of 1-inch long collars drilled and tapped for a 6-32 thread.

Power Supply Unit

The power supply unit is simplicity itself and should offer no problems to the constructor, but a few words of description, together with the photographs, may prove helpful.

Many set-builders are not equipped to build the chassis. For that reason two possible sources of supply have been mentioned in the parts list. One of the chassis makers builds his chassis out of steel and the other uses aluminum.

If the chassis is obtained *ready-drilled*, then the job of assembly and wiring can be finished in "jig time."

Mount the power transformer, chokes, and electrolytic condensers on the chassis and the voltage divider (R), with the 180 volt by-pass condenser, under the chassis, as shown in the photographs here reproduced.

Place the rubber grommet in the hole on the side of the chassis; this will serve to prevent chaffing of the power cable.

The power switch is mounted on the side of the chassis near the power transformer.

Wiring can be done in less than 20 minutes after the soldering iron is heated. Follow the pictorial wiring diagram, if you are not familiar with the regular wiring diagrams.

Solder all connections carefully and be sure that there are no cold rosin-core connections if you want the best

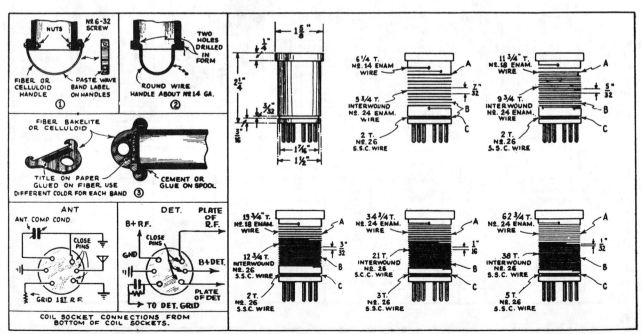

The two drawings at left show respectively how to make plug-in coil handles and also how to wire up the "coil" sockets. The data for winding the National type plug-in coils for the "Master Composite" 4-tube receiver are shown at right.

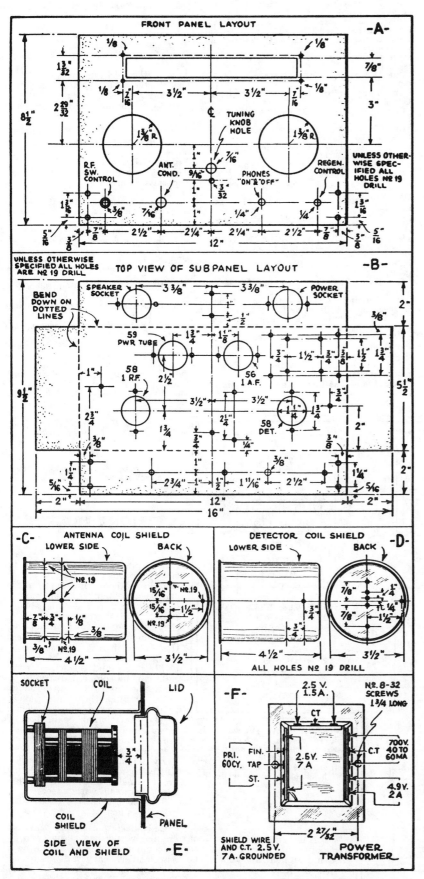

Details of chassis and other parts used in building the "Master Composite" Receiver.

results. Make sure that the solder flows *through* the joint thoroughly. The iron must be kept clean, well tinned and *hot!*

Tuning Frequency Range

Coils can be obtained ready wound to cover the wavelengths from 9 to 850 meters very efficiently. The following coils and the ranges covered were tested in SHORT WAVE CRAFT'S Composite Receiver and give the maximum of efficiency for their respective bands.

Range	Coil Number (National)
9 to 15 meters	No. 60
13.5 to 25 meters	No. 61
23 to 41 meters	No. 62
40 to 70 meters	No. 63
65 to 115 meters	No. 64
115 to 200 meters	No. 65

Those interested in band spread coils have a choice of the following coils:

Band	Coil Number
20 meter	No. 61A
40 meter	No. 63A
80 meter	No. 64A
160 meter	No. 65A

In the photographs showing the S-W-C "Composite" receiver the extra coils not in use are placed in a neat metal coil cabinet. This offers a convenient way of keeping the coils from being misplaced and preventing damage when not in use. The cabinet will hold 20 coils.

Antenna Recommendations

World-wide reception can be obtained under half-way decent conditions with an ordinary broadcast antenna at *loud speaker* volume. Nevertheless, it is recommended that one of the newer transposition lead-in types of antenna be employed, such as the system recommended by Arthur H. Lynch in the article "Reducing Noise on Short Wave Aerials," (See SHORT WAVE CRAFT for August, 1932, and "Good Antenna Design" in SHORT WAVE CRAFT for September, 1932).

Tests with this receiver in a suburban location shows marked improvement in the noise level and signal strength when the transposed lead-in was used with a 75 foot flat-top.

Speaker Considerations

Most every set builder has a dynamic speaker on hand and for that reason the choice of the loud speaker was left to the individual. Be sure that the loud speaker has a voice coil transformer with a primary that will match the tube; this is 7,000 ohms when used as it is in this set. The tube is used as a *pentode* for greater power sensitivity. Do not connect a magnetic type speaker to the output terminals of the set, without using an *output transformer*. Generally the magnetic speaker will not match the tube impedance and it may be damaged by the strong plate current flowing through the speaker windings.

Construction and Wiring Hints

So much has been written about the construction of radio receivers that little need be said at this time; study the photographs and the electrical circuit—then go ahead and build. Use care in the construction and see that everything is fastened firmly in place. Loose parts result in noise and noise is especially to be avoided in a short-wave receiver.

Make all connections in a direct manner. Do not have *loops* in the wiring and be sure that every lead from the tuned circuits of the radio frequency and detector are wired by connecting wires. *Do not depend on the chassis for common connections!* This will increase the stability of the receiver and result in smoother operation.

Conclusion

This receiver was designed in accord with the votes of readers of SHORT WAVE CRAFT and represents the majority opinion of the readers. Some of you may have wished that the set were a super-het and many may not like the method of regeneration control (for example) but the majority wins. So, here is YOUR set.

Parts List

One National Co. Type 9-SE 100 tuning condenser (Cap. 100mmf. each section). (C2, C3.)

Two National Co. Short Wave Chokes, Type 100, (RFC1, RFC2) (2.5 M. H.)

One National Co. Radio Frequency Choke, Type 90 (RFC3) (90 M.H.)

One National Co. Screen Grid Coupling Impedance Type S101 (AF1)

Two National Screen Grid Clips (V1, V2) type 24.

Two National Coils Sockets, Isolantite 6 prongs for National Coils (L1, L2)

Two National Isolantite Tube Sockets for 58 type tubes (V1, V2)

One National Co. Antenna Compensating Condenser, Type ST-50 (C1) (Cap. 59 mmf.)

One National Coil Cabinet (optional)

Two National Co. Tube Shields Type T58 (V1, V2)

One pair of the following National S.W. Coils, Nos. 61, 62, 63, 64, 65 (L1, L2).

One National "Full Vision" Tuning Dial, Type VKE

Three Micamold .01 mf. mica condensers (C4, C5, C12)

One Micamold .0001 mf. Mica condenser (C15)

One Micamold .00025 mf. mica condenser (C8)

One Flechtheim Tubular Condenser .01 mf. (C10)

One Flechtheim Electrolytic Condenser Type LT1000 (C11) 10 mf. 30 Vts.

Two Flechtheim .25 mf. Bypass condensers (C6, C7) Type GF25

One Acratest Wire Wound Resistor, 5 Watt, 7000 ohms Cat. No. 5900 (R6)

One Acratest Wire Wound Resistor, 5 Watt, 400 ohms Cat. No. 5900 (R10)

One Acratest Carbon Resistor, .5 Watt, 300 ohms Cat., No. 5860 (R2)

One Acratest Carbon Resistor, .5 Watt, 2000 ohms Cat. No. 3500 (R7)

One Lynch Mfg. Co. .5 Watt resistor .5 meg. (R9)

One Lynch Mfg. Co. .5 Watt resistor .1 meg. (R8)

One Lynch Mfg. Co. .5 Watt resistor 5. meg. (R11)

One Lynch Mfg. Co. 1. Watt resistor 2000 ohms. (R5)

One Lynch Mfg. Co. 1 Watt resistor 3000 ohms (R4)

One Frost 40 series potentiometer Cat. No. 6182 (R1) 3000 ohms.

One Frost 40 Series Potentiometer Cat. No. 6186 (R3) 50,000 ohms.

One Wafer Socket Type 59, 7 prong (V4)

One Wafer Socket 5 prong Type 56 (V3)

One Yaxley S.P.D.T. rotary jack switch (S)

One Wafer socket and male plug for speaker

One Eby Ant. Ground terminal strip (T1, T2)

One four-prong chassis mt'g plug and socket cable connector Type 7A-11 and 11A (T5, T6, T7, T8)

Two Eby Insulated Binding Posts (for phones) (P)

One Steel Chassis drilled and folded to specifications—Korrol Mfg. Co. or

One Aluminum Chassis drilled and folded to specifications Blan-the-Radio-Man

NOTE—The builder has a choice of chassis material

One Drilled panel. Aluminum panel is dipped and the steel panel is cadmium plated.

Two Blan—The-Radio-Man, special aluminum shields for the coils

Two Blan—The-Radio-Man "Hand-Hole" Covers for the coil openings

Two Blan—The-Radio-Man flexible couplings

Two 6 inch lengths of bakelite ¼ inch in diameter

Four small brown knobs

Four 5 inch lengths of 6/32 threaded brass rod

Two Raytheon 58 type tubes (R.C.A.)

One Raytheon 59 type tube (R.C.A.)

One Raytheon 56 type tube (R.C.A.)

Wire, soldering lugs, machine screws, etc.

Parts List of the Power Supply

One Jefferson Power transformer. Type 463-934. (P.T.) Federated Purchaser Cat. No. 2532

Two Jefferson Filter Chokes, Type SA2071 Federated Purchaser No. 2503 (CH1, CH2)

Three Flechtheim Electrolytic Condensers. 8 mf., 500 volts peak. Type JW800 (C1)

One Flechtheim Dry Electrolytic Condenser 8 mf., 500 peak volts Type KL800 (C2)

One Korrol Mfg. Co. Steel, cadmium plated chassis drilled and welded as drawings

Aluminum chassis by Blan, The-Radio-Man

One Federated Purchaser power switch No. 4112 (S)

One Federated Voltage Divider Cat. No. 3915. 25,000 ohms with taps (R)

One Five Wire Cable (Use No. 16 wire in cable if possible)

One Rubber Grommet

One Four prong wafer socket. Marked 280 (80)

One Raytheon 80 rectifier tube (R.C.A.)

My Idea of A Band Switch

By ANTHONY HOLTGREFE

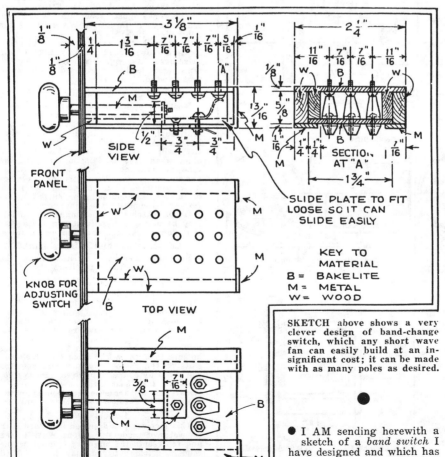

SIDE VIEW

FRONT PANEL

KNOB FOR ADJUSTING SWITCH

TOP VIEW

BOTTOM VIEW

SECTION AT "A"

SLIDE PLATE TO FIT LOOSE SO IT CAN SLIDE EASILY

KEY TO MATERIAL
B = BAKELITE
M = METAL
W = WOOD

SKETCH above shows a very clever design of band-change switch, which any short wave fan can easily build at an insignificant cost; it can be made with as many poles as desired.

●

● I AM sending herewith a sketch of a *band switch* I have designed and which has worked in a very satisfactory manner.

By looking over the sketch you will find that the switch

can be made very easily, with practically no cost, by simply using all the old material that every radio or short-wave "bug" has on hand.

Now for a few details about the switch. First you will notice that the bolts used for contact points have no nuts on them. By drilling the holes in the bakelite a little smaller, the bolt can be turned in to fit snugly and the wires from the coils can be soldered direct to same. By doing this I have saved on space.

By drilling the front panel for two wood screws, one on each side of the shaft it can be held in place by running these screws through the front panel into the front wood piece on the switch.

By marking the shaft at each setting with an awl, at the face of the front panel, the correct setting can be determined with ease.

This switch can be used with the "Best" short-wave converter

This design of short wave band-change switch has many good points to commend it. For one thing, it is a relatively easy matter to adjust the springs, made of phosphor bronze or German silver, so that they will have even tension at all positions of the switch.

The DENTON "ECONOMY THREE"

By CLIFFORD E. DENTON

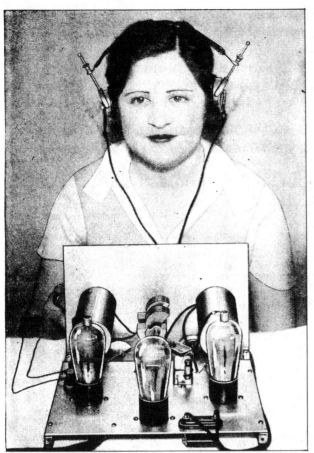

Here's a 3-tube short-wave receiver that you will derive a lot of pleasure from—it is up to the minute, with all controls mounted on the front panel, while switches change the wave bands.

Another view of the Denton "Economy Three" which utilizes the latest discovery in an efficient inductance-switching system for changing the wave-bands, without resorting to plug-in coils.

● THREE tube short wave receivers are very popular today. The results that can be obtained with a properly designed and constructed set are the reason for their popularity.

Most set builders use one stage of radio frequency amplification and a high-gain detector tube. This is the most satisfactory method of radio frequency amplification and detection.

If more than one radio frequency stage is used, then the problem of adequate shielding runs the cost of the set up so far that the average pocket-book will not stand the strain. With ordinary equipment now available tuned radio frequency below 20 meters will not offer high enough amplification to justify the cost.

A screen grid detector is used in the "*Economy-3*" because it offers the maximum sensitivity, coupled with smooth regeneration control, for the minimum cost.

As it is wise to have an audio stage and the cost of the additional parts for its construction is so little, a "high gain" pentode type tube is used, the output of which can be connected to a loud speaker or used in conjunction with phones, as desired. For the best results with this tube, use *high impedance* phones so that greater signal strength may be developed. If possible use an output-matching transformer for this purpose.

Circuit Description

In the outward appearance of the **circuit** diagram the receiver will have

Parts List

One Hammarlund Tuning Condenser Type MCD-140-M (140 mmf. cap. per section). (C2, C6)
One National Tuning Dial 4" Type VBD
One National R.F. Choke Type 100 (RFC)
Three By-pass Condensers .25 mf. (C4, C5, C8)
One Tubular Condenser .003 mf. (C11)
One Tubular Condenser .00025 mf. (C9)
One Tubular Condenser .015 mf. (C10)
One 50,000 ohm potentiometer with power switch (R3, SW3)
One Acratest Midget Condenser 25 mmf. capacity (C3)
Three Acratest 4 prong sockets
One Acratest 20,000 ohm, .5 watt resistor, (C2, C6)
One Acratest 1 meg. .5 watt resistor (R5)
One Acratest .25 meg., .5 watt resistor (R4)
One Acratest 5 meg. resistor .5 watt (R1)
One Acratest .0001 mf. mica Condenser (C7)
Two Acratest Tube shields, type 7268 (for the coils)
Two 1" diameter bakelite tubing 3¼ inches long
One ¼ pound spool No. 30 D.S.C. wire
One ¼ pound spool No. 22 D.S.C. wire
Two Acratest Selector switches (SW1, SW2)
One Eby Type 17 moulded twin-jack for phone or speaker connections (3, 4)
One Eby Antenna-Ground Type 22s Molded twin-posts (1, 2)
One Blan aluminum panel and chassis
One pair of Blan Brackets
Acratest grid-leak clips, Type 3892
Two Acratest 1" diam. black knobs
One Acratest 6 wire cable. (5, 6, 7, 8, 9)
One Type 33 Output Pentode (33) Triad
One Type 34 R.F. Variable-Mu Tube (34) Triad
One Type 32 S.G. R.F. Tube (32) Triad

a familiar look. After all, the old time "tried and found good" are the best. The problem is simply one of obtaining the most for the capital involved.

The coils are home-made, in fact the coils cannot be bought ready made for this set at all. The set constructor will have to follow the coil-winding directions as given in Fig. 2. There is no great job in winding these coils and it can be done in an hour; simply follow the directions carefully.

By the use of the two taps it is possible to cover three wave-length ranges. The first band will extend from 40-85 meters, the second band from 25-55 meters and the lowest band from 15-30 meters. Due to variations in the windings of the tuning coils when done by hand, there will be some differences in the bands covered, as well as differences in tuning condenser settings, due to change in constants and electrical values of the antenna and the detector coil. Experience in winding several sets of these coils indicates that this variation will not be so great that it cannot be corrected very simply.

One way to bring the antenna and detector coil circuits into alignment, so that resonance will be obtained with the single control dial tuning arrangement, is to spread the turns of windings L-1 and L-2 by prying them apart until repeated tests show that resonance is established fairly evenly over all of the bands. If there is a very great difference in the tuning range, due perhaps to differences in the tube shields used as coil shields, then it may

This 3-tube short-wave receiver does away with the necessity of removing and replacing plug-in coils; it employs instead a newly devised switching system, whereby the most used bands, the 20, 40 and 80 meter, can be tuned in by merely turning a pair of switches.

No Plug-in Coils—All Common Wave-Bands Tuned in by Switches Using New Circuit

be necessary to remove as much as a single turn on L-3. Due tests of the set will enable the constructor to judge the advisability of any coil correction.

After these corrections have been made, it is wise to apply small amounts of collodion, or some other high-grade insulating material, which will hold the windings.

The signals are fed into the receiver by means of the antenna connection to the home-made antenna series condenser. This home-made antenna series condenser is familiar to all. It consists of a simple two-inch long piece of bus-bar with No. 18 push-back wire wound around it in the form of a spiral spring and simply moved back and forth to increase or decrease the capacity in the antenna circuit.

Wave selection is given by means of the two switches S.W.1 and S.W.2. The input circuit to the first radio frequency tube has a tuned circuit, being tuned by condenser C-2, which is ganged to condenser C-6, tuning the

detector circuit. This condenser-leak method of detection is employed and C-7 and R-1 form an exit to this circuit. The return of the grid in the detector tube is directly to the chassis.

A typical regeneration circuit is employed,—control of screen voltage being the method whereby oscillation is controlled. The plate winding or feedback coil is divided into three equal windings of six turns each, interposed between the No. 22 tuning coil winding, so that satisfactory regeneration control can be had on all the bands with the switch in any position. Of course, with the switches S.W.1 and S.W.2 at the bottom of L-1 in both the input circuit and output circuit of the 34 R.F. tube, the receiver will tune to the shortest wavelength band. By moving the switch down to the tap N.L.2 this will tune to the next shortest band, and by moving still further so that the moving arm of the switch rests on an open contact, indicated near L-3, the complete coil is in the circuit and the highest wave bands will be covered.

The plate circuit of the 32 type tube which is used as the detector is bypassed to ground by means of condenser C-9 and the radio frequency choke. The detector is resistance coupled into the grid circuit of the 33 type tube, which has given excellent results, especially as far as power sensitivity is concerned. C-11 is used to equalize variation in the Pentode plate load impedance.

R-6 is mentioned, although it is not shown in the actual set. This is a 6 ohm rheostat used in series with the "A" supply when two dry cells are used in series, it being necessary to drop one volt in the rheostat to supply the two volts normal to the tubes in the receiver. If a 6 volt storage battery is to be used, an additional resistance must be inserted in the circuit so that the current to the filaments will not exceed the rated value.

In many cases sets of this type are equipped with Eveready Air-Cell batteries, which will give steady performance

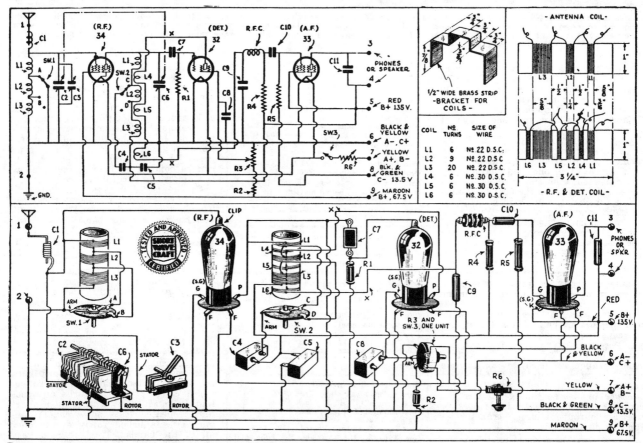

Both schematic and picturized wiring diagrams are shown above for those interested in building the Denton "Economy Three" receiver. This set employs three battery tubes and switch-type inductances instead of plug-in coils, to cover the most used bands, the 20, 40, and 80 meter bands.

for more than a year of steady use.

In some instances, when making tests in different locations, it was noted that improved results could be obtained with the use of tube shields on the first R.F. and detector tubes. If the constructor wishes to go to the expense, it would be a very good idea to include the two additional tube shields on the 34 and 32 type tubes.

Experiments with the 34 tube in the detector socket have been quite satisfactory, resulting in very smooth control of regeneration, but with an apparent falling off the sensitivity of the receiver, although the difference in sensitivity is not enough to cause the person testing the set to say that the 34 tube could not be used in this position with satisfaction.

Operation

Place the 34 type tube in the R.F. socket, a 32 type tube in the detector and 33 tube in the output tube socket. Connect the ear-phones, connect up the battery cable to the battery supply, adjust the filament volts to exactly 2 volts, and leave this at this point. The fact that these tubes work best at 2 volts from an electrical standpoint also means that they will last the longest at 2 volts from the standpoint of life. Therefore, always keep tubes in this 2 volt class at the rated 2 volts for satisfactory long-life operation.

Set S.W.1 and S.W.2 to the same identical tap, so that both coils can be tuned to resonance. Turn the regeneration control, which is mounted on the left-hand side of the panel, to the right and see if the detector tube will oscillate. If the detector tube will not oscillate, reverse the lead marked "X" in Fig. 1. If the tube oscillates, slowly turn the tuning dial which drives condenser C-2 and C-6. Keep the regeneration controls in position for maximum sensitivity. Vary C-3 for the maximum signal, as this condenser is used to enable condenser C-2 and C-6 to track throughout the band. If the condenser C-2 and C-6 does not track satisfactorily by means of condenser C-3, it will then be necessary to change the coils slightly so

that the tuning characteristics of the coils will coincide. If the 32 tube regenerates too quickly and not smoothly, vary resistance R-1 and reduce the capacity of condenser C-9, generally testing the value between .0001 and .00025 for C-9 and start at 5 megohms for R-1 and work down to 1 meg. if necessary. Variations between condenser C-9 and resistor R-1 are to be so adjusted that the regeneration control R-3 is about between ¼ and ½ way over to the right for maximum control of regeneration. An increase in size of R-2 or variation in R-2 will change the portion of the operating curve of R-3 for the maximum convenience.

Most experimenters are familiar with the problems to be encountered at this point and there should be no difficulty with a few hours of final adjustment with this receiver to obtain the smoothest operation possible. No set of this kind can be thrown together, of course, and have real satisfactory regeneration control right off the bat without a great deal of luck. Generally it takes time to get a set working smoothly.

Construction Pointers

There is little to be said about the construction, as the photographs clearly show the placement of the parts as well as most of the wires in such a manner that the set should go together with very little trouble.

Some mention should be made of the method of supporting the coils within their shields. This is done in a simple manner and although several methods were tried out this works out the best.

Cut two pieces of brass or thin aluminum as shown in Fig. 3 and bend into shape. Then spread a thin coat of PDQ Plastic Metallic Solder on the metal surfaces that contact the inner wall of the tube shield base; allow this coating to harden for a short time and then place a greater amount of the solder on the inside of the shield socket and allow the bracket to rest in place for as long a time as possible. It is a good idea to let the solder harden all night, if possible.

Of course the tap-switch and the tube shield socket have to be fastened to the panel before this can be done. If the hardening process is left for 24 hours the results will be perfect. No heat is necessary when using this solder.

Fasten everything in place. Those parts mounted by means of pig-tail connections should have good mechanical support without the aid of soldering. The best operation, when it comes to short waves, occurs when the set is free from noise. Noisy sets will ruin reception and most noise comes from loose connections.

Tighten up all nuts and bolts used to hold the chassis in place. Loose nuts and bolts here will cause noise.

Use care in soldering to the switch connections and do not permit soldering flux to drop down between the contacts, as this will cause losses; this is a common fault of constructors when soldering to tap switches. R2 is twenty thousand ohm resistor.

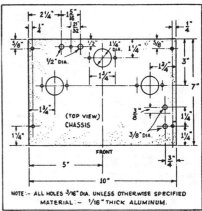

Top view of chassis.

A Receiver That Laughs At STATIC

By JOHN L. REINARTZ

This unusual circuit devised by Mr. Reinartz reduces the interference from non-tunable sources, such as static and line disturbances; it also permits two signals of different frequencies to be received simultaneously through the same amplifier system.

Fig. 2—Front-panel view of the receiver that "laughs at static" and other interference.

● TO circumvent the disturbance caused by static and other uncontrollable interference in the reception of radio has been the goal of many experimenters. A number of devices have been advocated that should help to reduce such unwanted interference. To date, however, no device is doing the job consistently or well.

For several years the writer has had in use a circuit that does reduce the interference from non-tunable interference such as static and line disturbances. It also has the advantage of allowing

two signals of different frequencies to be received at the same time through the same amplifier system; that is, one can listen to a signal in two of the amateur bands at the same time and hear both, or if desired the same frequency can be tuned to in each half of the two parts of this receiver, and advantage can then be taken of its capability to balance out such interference as static and other non-tunable interferences.

The receiver had its inception during a study of methods to reduce non-tunable interference during reception of a signal, the idea being to so adjust the receiver that the non-tunable interference would be allowed to enter the two parts of the receiver and when again combined at the audio part of the system it would cancel out, while the signal which was desired would go through only one part of the tuning system and then through

the audio and be heard in the regular way, minus the interference which may have been present. The result is so good that many of you will wish to build such a receiver, the description of which follows:

The circuit used (Fig. 1) will be recognized by the old-timers as the one which the writer has used for the last ten years and has found no good reason to displace, especially on amateur frequencies. The only difference is that there are two of them, so connected that the audio system starts in one and ends in the other of the two tuning systems. It is through this connection that unwanted signals are cancelled out, or that two signals of different frequencies can be tuned to at the same time and heard through the audio system. One precaution which must be taken is to keep the two systems

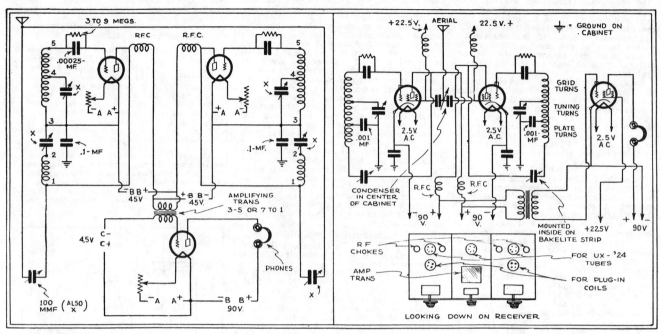

Fig. 1, at left, shows diagram of the Reinartz interference elimination receiving circuit for battery tubes; Fig. 3, at right, shows the same circuit adapted for use with A.C. tubes.

insulated from each other except through capacities. If we wish to house them in cabinets they must be insulated from each other. For the time being we will mount both parts of the system on a bakelite panel 7 x 14 inches in size, using a baseboard of the same size to mount the amplifier parts on at the same time. A connecting strip to connect our batteries to is mounted on the baseboard.

Figure 2A and B shows the panel and baseboard respectively. The large dials are the two tuning dials, while the two small ones are the regeneration control and the antenna coupling condensers respectively, one set for each tuning system.

We are going to use the type '30 tubes for detectors and the amplifier, as we have to use separate "B" batteries for all these tubes. It won't take as many as if we used the '31 or the '33 type tubes, although you may use them if you desire to. An amplifying transformer couples our detectors to the amplifier in such a way that energy from either detector energizes the amplifier. If the energies from the two detectors are equal and opposite at the same time, no signal will be transmitted by the amplifier; this is of course what we have been after.

Now that we have all the parts properly mounted we will make up a set of tuning coils. Starting with the 80-meter band we make two coils of 10 turns each, about 2 inches in diameter and of No. 16 wire. Then we make two more of 20 turns each with a tap at 10 turns, the diameter and wire size to be the same. The 10-turn coil connects between No. 1 and 2 and the 20-turn coil connects between Nos. 3, 4 and 5, the ends of the coils being to the right.

We are now ready to connect our "A" and "B" batteries. Each tube must have a separate "A" battery, so we connect two dry cells to each of our three tubes through a proper filament resistor and one 45-volt "B" battery to each detector, being certain that we make this important connection so that the negative of the "B" battery goes to the filament of one detector tube and the positive of the same "B" battery goes to the plate connection of the other tube through the radio frequency choke coil. The amplifier batteries are connected in the usual manner. Then we connect the amplifying transformer between the two positive connections of the two detector "B" batteries, the second connection going to the grid of the amplifier with a grid bias battery of minus 4.5 volts and a plate voltage through our phones of 90.

Now let's see what happens. Our antenna is connected through coupling condensers. We control our regeneration through the 100 mmf. regeneration condenser in each plate circuit and tune our two receivers by the two 100 mmf. tuning condensers. Let us tune one receiver to some signal in the amateur band and the other to a higher frequency signal in the same band. Detune them so that you obtain a 500-cycle tone frequency from one and an 800-cycle tone frequency from the other.

Let us now follow them through the system. The signal from the left receiver goes through the amplifying transformer to the plate of the tube in the left receiver and the signal from the right receiver goes through the amplifying transformer in the opposite direction to the plate of the tube in the right receiver. Both signals, because of the difference in tone frequency, will be impressed on the grid of the amplifying tube and we will hear them in the phones, one a 500-cycle signal and the other an 800-cycle signal. We will have lots of fun tuning to two different frequencies at the same time and listening to the conversations between two amateurs who are in contact with each other, doing this without the necessity of retuning for each signal, as we would have to do with our conventional tuners.

Now let us see what happens if we have some non-tunable interference. Suppose we set a buzzer going in the room. We tune to a signal with the left receiver and pull the tube out of the right receiver. The interference is very bad. Now we replace the tube in the right receiver and tune to some signal. We find that we can hear the signal which before was inaudible through the interference. To reduce the interference as much as possible, we carefully adjust our antenna condensers and regenerative feed back condensers until the interference is a minimum. The two tuning condensers are adjusted as close to the desired signal frequency as possible, one on one side of the carrier and one on the other side of the carrier, differing by an audio frequency which is passed on through the amplifying transformer to the amplifier.

Signals that could not be copied with just one of the receivers could be copied with no trouble at all when using both receivers and cancelling the interference at the amplifying transformer. It takes a little patience to make the proper adjustments, but the results are very worth while. To be able to listen to two frequencies in or out of the same amateur band makes listening-in even more desirable.

Many amateurs will find in this receiver a solution to their interference problems, and will build one for general all-around use. Such tubes as the '24 and '47 can be used, obviating "A" batteries. The circuit requirements for these tubes must, of course, be observed.

To reduce interference between circuits, the tuning coils should be turned at right angles to each other. When enclosed in a cabinet this precaution need not be followed. The coils themselves are so simple that no description is necessary. They can be made to suit conditions or can be found on plug-in forms to suit the maker. The radio frequency choke coils are wound with No. 30 wire on a 1-inch diameter cardboard tube 2 inches long and mounted on corks glued to the baseboard.

What to remember when building this receiver: Unless you use the separate heater type tubes you must use separate "A" batteries; you must use separate "B" batteries for any type of tubes; you must not ground the receivers directly, but through a .1 mf. condenser. You can ground the filament of the amplifier tube. You must be sure to make the plate connections of the two receivers just as shown; otherwise you undo what you have been trying to do. You can use as much amplification as you may desire. The connections for any additional amplification are conventional.

The present form of the receiver is as shown in the photo, each half of the receiver being in a section each side of the center section in which is the amplifier. This center section also has mounted on the front of the cabinet a two-section condenser by which the two halves are balanced to the aerial, the rotating part of the condenser being insulated from the cabinet. The two tuning condensers are insulated from direct ground connection with an insulating condenser mounted inside of the respective sections. This allows the tuning condensers to be mounted directly to the cabinet without insulating bushings. The amplifier tube is coupled to the two detector tubes through a regular amplifying transformer. Type '24 tubes are used throughout, and as they are of the separate heater type no separate "A" batteries are required, one filament transformer supplying all three tubes.

The number of turns for this range, 3,500 to 4,500 kc., is: plate turns, 15; tuning turns, 15; grid turns, 25; for other bands this is doubled or halved, the wire size being No. 24 B. & S. gauge. The circuit connections are as shown in the diagram.

A 3-Tube BAND SPREAD Loud Speaker Set

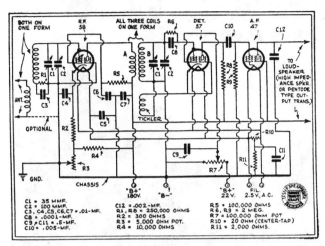

Schematic wiring diagram of the 3-tube band spread receiver here described. The two condensers C1 and C1 are ganged.

As far as operation is concerned, nothing more could be desired in the way of smoothness of control and over-all sensitivity of response. This "high gain" 3-tube circuit, employing pentodes of the latest type, makes the use of earphones unnecessary, as practically all received signals are strong enough to operate a "loud speaker" directly! Although designed primarily for the popular 20, 40, and 80 meter "amateur bands," Mr. Shuart's instrument is also a superlative receiver for "general" short-wave reception.

● IN designing the receiver here described, a number of items were taken into consideration. Among these was that the receiver complete, including coils and tubes but less power-supply unit, should not cost more than twenty dollars. In order to maintain this low cost the set had to be of such design that the chassis or foundation could be made at home with the usual tools found in the experimenter's workshop. The saving on this part of the set allowed the incorporation of many refinements that otherwise would have been impossible at this low cost.

The aluminum used for the base of the receiver is of 1/32nd inch stock. Material this thin would be far too weak if it were not for the particular design used in forming the base and the mounting of the two shields which form the compartments for the tuned R.F. and detector stages.

Metal Base and Shields

When bending the base and shields the only tools needed were two pieces of wood and two "C" clamps together

with a small vise. The "C" clamps can be purchased in any hardware store for about ten cents. It is advisable to have the aluminum cut to size when buying it rather than trying to cut it at home with a pair of tin snips. The cost of the aluminum cut to size as indicated in the drawing was only $1.25, not including the front panel. If care is taken in measuring and bending the material a very neat and sturdy chassis can be turned out.

Kind of Tubes Used

As can be seen, this set uses a type 58 as the tuned R.F. amplifier and a type 57 as detector, with a 47 as audio amplifier, a 56 can be used instead of the 47 if ear phones are to be used.

When using a stage of tuned R.F. ahead of an autodyne detector it is absolutely necessary to have some sort of R.F. gain control if overloading of the detector is to be eliminated. Therefore a type 58 is used in order to obtain control of volume by the cathode method. This type of control is very quiet in operation and has very little

effect on the tuned circuit. As can be seen in the diagram, the grid circuit of the R.F. stage is decoupled by a 250,000 ohm resistor and a .01 mf. condenser. This helps to prevent the R.F. stage from detuning the detector and allows the full benefit of the shielding. Screen grid voltage is obtained with a 100,000 ohm resistor which adds to decoupling and eliminates one wire in the cable. Isolantite coil forms and sockets and tube sockets are used because of their low loss.

Coupling R.F. to Detector

A separate winding on the detector coil form provides the R.F. coupling between the R.F. stage and the detector. This is the most efficient means of coupling and should be used when ever high gain and stability are required. Bypass condensers are used freely but no R.F. chokes are shown because no benefit was derived from them.

The type 57 was chosen as the detector because it oscillated much better at the higher frequencies than any other type. The detector circuit is of

Diagram component values:

C1 = 35 MMF.
C2 = 100 MMF.
C3, C4, C5, C6, C7 = .01-MF.
C8 = .0001-MF.
C9, C11 = .5-MF.
C10 = .005-MF.

C12 = .002-MF.
R1, R8 = 250,000 OHMS
R2 = 500 OHMS
R3 = 5,000 OHM POT.
R4 = 10,000 OHMS

R5 = 100,000 OHMS
R6, R9 = 2 MEG.
R7 = 100,000 OHM POT
R10 = 20 OHM (CENTER-TAP)
R11 = 2,000 OHMS

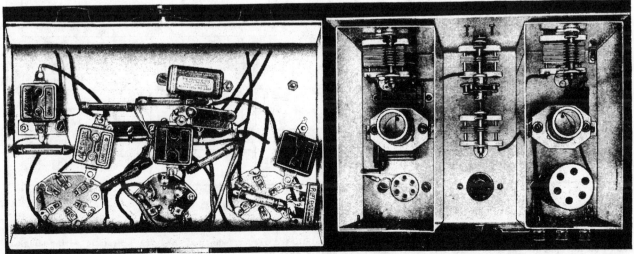

Bottom view of Mr. Shuart's 3-Tube Band Spread receiver is shown at the left, while we have an excellent top view of the receiver at the right.

Rome, Paris, and London on the loud speaker! Yes Sir, and with only 3 tubes! Thanks to the ingenious circuit here presented by Mr. Shuart, plus the utilization of three of the latest style tubes, this set provides unusually high amplification. This is one of the finest receivers we have ever had the pleasure of describing to our readers; moreover, it provides "band - spread" on any band. The receiver can be built for less than $20.00.

By GEORGE W. SHUART
W2AMN-W2CBC

the electron-coupled type. This allows the use of five prong coil forms and it is also more stable when oscillating. The feed-back coil is wound separate from the grid coil and is spaced about ¼ inch from it. This method is preferred rather than tapping the grid coil because it is not as critical and allows better control of regeneration.

There is no spacing between turns because distributed capacity was not an important item; in fact the opposite was desired. The total capacity across the grid coils is 135 mmf. when the condensers are tuned to maximum. The coils are wound so as to have high "C" (capacity) on any of the amateur bands. High "C" tends to make the circuit more selective and this is needed

in the amateur bands more so than for general short-wave reception.

The R.F. plate coil is threaded right along with the grid coil of the detector, beginning at the ground end. The wire for the R.F. plate coil, the tickler and the antenna coil is No. 32 silk covered. The grid coils are wound with No. 28 silk covered wire. The coil forms are Hammarlunds, intended for ultra short waves. These coils are small but provide ample room for the windings for all bands.

Regeneration Control Feature

Regeneration control is one of the most important items in any short wave receiver, and deserves careful consideration. To obtain smooth and noise-

less control of regeneration one must bear in mind that when regeneration is controlled by varying the screen grid voltage with a potentiometer, the voltage across the potentiometer should be only slightly higher than that required to make the tube oscillate and that the resistance of the potentiometer should be large enough to allow a very fine adjustment of this voltage. With the number of tickler turns designated in the coil table and 22 volts across the potentiometer, which is of 100,000 ohms resistance, it is possible to swing the regeneration control all over the scale without affecting the tuning of the detector circuit. It takes a swing of about one-half inch to bring the tube completely in and out of oscillation. From

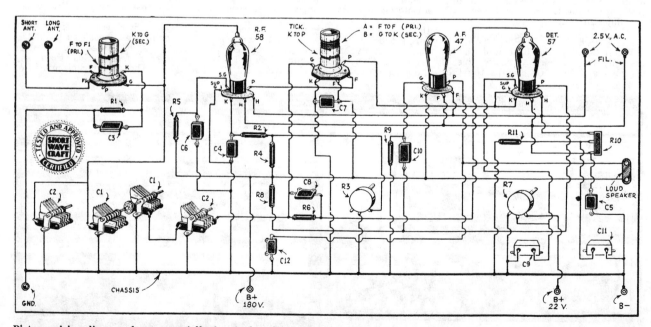

Picture wiring diagram drawn especially for the benefit of the uninitiated reader, who may find the schematic wiring diagram shown on the opposite page too difficult to understand. You should have no difficulty in following the wires connecting the various pieces of apparatus in this diagram.

this it can be seen that it is possible to get right on the edge of oscillation without a hair-splitting adjustment and receive phone signal with the same ease that C.W. signals are received. Then again, it takes a large change in line voltage to have any effect on reception, which means that one does not have to have one hand on the regeneration control all the while phone signals are being received.

47 Tube Used for Output

A type 47 tube is used in the audio stage because loudspeaker operation was desired. Of course, if one expects to use headphones it is suggested that a type 56 be used if the ears are valued at all. Resistance coupling is used in the audio stage because of its economy and the minimum of space it requires.

All necessary data are contained in the diagram for drilling and placing the parts, together with the plans for the construction of the chassis and the shields.

The two 35 mmf. midget tuning condensers are ganged with a flexible coupling. These condensers have to be mounted on posts in order to obtain height enough for the mounting of the dial. These posts can be made of ¼ inch copper tubing with machine screws run through them to hold the condensers in place.

The R.F. stage is mounted in the left-hand shield. The 100 mmf. tuning condenser and the R.F. volume control of this stage are mounted directly on the front of the shield itself. This layout is also used in the detector stage. The R.F. decoupling resistor and condenser are mounted on the bracket that supports the R.F. coil socket. The grid leak and condenser are mounted in the same fashion in the detector compartment.

The audio tube is mounted between the two shields and directly behind the two ganged 35 mmf. condensers. All parts are mounted on the chassis before the panel is attached, it only being necessary to drill five holes in the panel and mount it over the lock nuts holding the parts to the chassis. This allows a space the thickness of the nuts between the panel and the shields which is a form of double shielding and has been carried out extensively in this receiver.

It is necessary to have very nearly the same amount of capacity in the two tuned circuits of the receiver, if the controls are expected to track with any workable degree and be able to utilize the full tuning range of the two 100 mmf. tank condensers. This would be impossible, of course, if one of the tank condensers were set at a much higher degree of capacity than the other. Therefore, the number of turns stated in the table must be used if the two stages are expected to tune through the band without getting "out of resonance."

How to Tune Receiver

In tuning the receiver the two tank condensers are set at the approximate section of the band on which one desires to receive and the tuning done with the two smaller condensers which are controlled by the main tuning dial. This method allows band spreading at any frequency and still keeps the number of coils necessary to cover the usual short wave bands down to the ordinary amount. For ease in retuning to a previously recorded station, it is suggested that the two tank condensers be calibrated as to their frequency response.

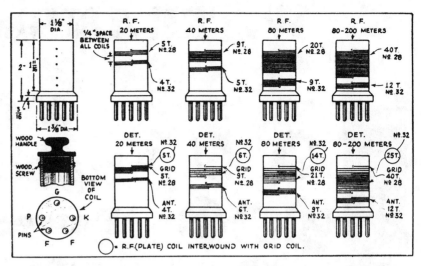

Coil Winding Data for Shuart 3-Tube Band-Spread Set.

In the receiver described it is usually impossible to use the full gain of the R.F. stage without overloading the detector so it is recommended that the volume control be set at about half way on in tuning in a station and then bringing the "volume level" up as desired. With the volume control set in this position the selectivity of the set is as good as the average short wave "super" and as to volume, well, any foreign short wave station that is received in this country *can be brought in on the loudspeaker, with ease.*

List of Parts

1—7"x10"x¹₁₆" aluminum panel
2—5½"x21½"x¹₁₆" sheet aluminum for shields
1—12"x10" aluminum for base
1—Type 58 tube Sylvania (R.C.A.)
1—Type 57 tube Sylvania (R.C.A.)
1—Type 47 tube Sylvania (R.C.A.)
6—Hammarlund coil forms (five prong) "small" Isolantite
2—Hammarlund five-prong sockets Isolantite
2—Hammarlund six-prong sockets Isolantite
2—Hammarlund 100 mmf. tuning condensers
2—Hammarlund 35 mmf. tuning condensers
1—Hammarlund flexible coupling
1—National type "B" dial
1—100,000 ohm Electrad potentiometer (Clarostat)
1—5,000 ohm Electrad potentiometer (Clarostat)
2—Aerovox .5 mf. bypass cond. (Polymet)
5—Aerovox .01 mf. fixed cond. (Polymet)
1—Aerovox .005 mf. fixed cond. (Polymet)
1—Aerovox .0001 mf. fixed cond. (Polymet)
1—Aerovox .00025 fixed cond. (Polymet)
2—Aerovox 250,000 ohm resistors (Lynch)
1—Aerovox 100,000 ohm resistor (Lynch)
1—Aerovox 15,000 ohm resistor (Lynch)
1—Aerovox 2,000 ohm resistor (Lynch)
1—Aerovox 300 ohm resistor (Lynch)
2—Aerovox 2 megohm resistors (Lynch)
1—Aerovox 20 ohm C.T. resistor (Clarostat)

Details of aluminum chassis used by Mr. Shuart in building his 3-Tube Band-Spread Receiver.

1—Eby five-prong socket
1—Hammarlund "Triple-grid" tube shield
1—Five-wire cable
1—Antenna binding post assembly
1—Speaker cord tip assembly.

THOSE HARMONICS AGAIN

MANY listeners who report the reception of short-wave signals from certain American stations are informed, much to their surprise, that these stations are not transmitting on the short waves at all! What they have picked up are "harmonics" of the regular broadcast waves.

Harmonics are secondary oscillations or vibrations that appear in any oscillating system. The harmonics you may hear on a S.W. receiver are of higher frequency (lower wavelength than the fundamental). Some types of radio transmitting circuits are more virulent producers of harmonics than others; even in the best systems, harmonics may be created as a result of the antenna's own radiation. Fortunately, most of the harmonics that appear in the short-wave bands are pretty weak, the strongest being heard between about 100 and 150 meters.

C. W. TRANSMITTERS

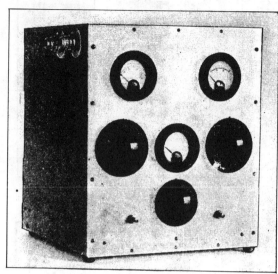

Front panel appearance of the extremely well designed 1-tube master oscillator-power amplifier transmitter, using a single UX-860 tube.

A ONE-TUBE
Master Oscillator
Power Amplifier
TRANSMITTER

By A. J. SPRIGGS, W3GJ

Mr. Spriggs, who holds an engineering position with Uncle Sam, here presents a fresh new design of improved transmitter for the short-wave experimenter. This transmitter, here illustrated in photograph and diagram, is an experimental type suited to amateur frequencies and based upon the use of a UX-860 tube

The original basic circuit is shown in Fig. 1. The filament, control grid, and screen grid of the tube are so connected as to form, with the associated apparatus, an oscillating circuit similar to the conventional Colpitts circuit. (The use of the Colpitts type of circuit is not strictly essential, however, as any of several standard oscillator circuits may be used.) The constants of the elements of this circuit determine the frequency generated, except for slight variations which may be caused by effects external to this circuit. The output or load circuit is connected between the plate and filament of the tube, and is carefully shielded to prevent electromagnetic or electrostatic coupling between the plate or output circuit and the frequency determining circuit. The purpose of the balancing condenser C is to permit neutralization of the small electrostatic coupling between the output

circuit and the frequency determining portion of the circuit, which would otherwise exist due to the effect of capacity between the shield grid and the plate of the tube.

Since there is no electromagnetic or electrostatic coupling between the frequency determining portion of the circuit and the output circuit, it is apparent that the radio frequency power developed in the output circuit is due to the production of pulses of the plate current corresponding to the variations of potential developed on the control and screen grids by oscillations in the frequency determining circuit. The means by which the oscillations in the frequency determining circuit result in the production of radio frequency power in the plate or output circuit has been called "electron coupling." The

Fig. 7 (left) shows rear view of Mr. Spriggs' MOPA transmitter; Fig. 10 (center) shows one of the plug-in inductances, and Fig. 6 (right) shows top view of transmitter.

over-all effect is similar to the action in a master oscillator-power amplifier transmitter, the action of the frequency determining portion of the circuit and the filament, control grid and screen grid of the tube being similar to the action of the master oscillator, and the action of the screen grid, plate, and output circuit being similar to the action of a power amplifier stage.

In Fig. 2 is shown a somewhat different circuit arrangement in which the necessity for the use of the balancing condenser shown in Fig. 1 is eliminated by making use of the shielding properties of the shield grid. In this arrangement the shield grid is kept at zero radio frequency potential, with respect to ground, and serves as an electrostatic shield between the plate and the frequency determining circuit. This arrangement necessitates the use of chokes in the filament leads, since the filament of the tube is at radio frequency potential with respect to ground. The basic principles of operation of this circuit are identical with the principles involved in the circuit of Fig. 1.

Frequency versus Voltage Changes

In connection with various tests which have been made of the operating characteristics of the circuit shown on Fig. 2, it has been found that under certain conditions variations of the screen and plate potentials have effects upon the generated frequency of opposite sign. By suitable choice of the operating voltages a condition can be established under which, with the screen and plate voltages supplied from a common source or from the same line, practically no change in frequency will result from variations of line voltage of as much as 25 per cent. The curves shown in Fig. 3 show the effects upon frequency of variations in screen and plate voltages using a UX-860 shield grid tube, in a circuit similar to that shown in Fig. 2. By choice of operating voltages for the screen and plate corresponding to points on the curves where the slopes are equal and opposite, it is apparent that variations of line supply voltage will necessarily have minimum effect upon the generated frequency.

In addition to the very desirable characteristics of the circuit arrangement of Fig. 2 as regards frequency stability under varying line supply variations, this type of circuit has other advantages as regards frequency stability. Tests show that with this circuit, changes in the output circuit or loading have only slight reaction upon the frequency as compared with standard types of circuits. The small magnitude of this reaction is to be expected, due to the fact that there is little or no electromagnetic or electrostatic coupling between the output circuit and the frequency determining circuit. The effect of tuning of the output circuit upon the generated frequency can be reduced to a practically negligible value, only a few cycles per million, by tuning the output circuit to the second harmonic of the frequency generated in the frequency determining circuit. When operating in this manner, the amount of power obtainable in the output circuit has been 25 to 50 per cent of the power obtained when working straight through on the fundamental. Changes in the ambient temperature cause variations in frequency of about the same magnitude as with conventional oscillator circuits. Changes of filament voltage cause little change in frequency, provided the filament emission is sufficient to produce saturation at the plate voltage used. In general, variations of filament voltage of 5 to 8 per cent above or below normal operating values will have negligible effect on the frequency generated.

Special Switches for Mercury Vapor Rectifiers

The circuit diagram of Fig. 4 and the photographs and sketches shown in Figs. 5 to 11, inclusive, show various features of an experimental transmitter for *amateur frequencies* based upon the use of a UX-860 tube with the Dow circuit. The general arrangement of parts and details of assembly are shown in the photographs of Figs. 5, 6 and 7, and the sketches of Figs. 8 and 9. The frame is constructed of sheet aluminum and angle brass, well bolted and banded together.

Particular attention is invited to the circuit arrangement of the two switches on the front of the transmitter panel (S1 and S2 of Fig. 4). In order to provide for necessary time delay in application of plate voltage to the mercury vapor rectifier tubes, the switches are so wired that the left-hand switch lights the filaments of the oscillator and rectifier tubes and the right-hand switch applies the plate voltage. The plate voltage cannot be applied unless the filaments are lighted. The necessity of providing a 15 to 30 second delay in applying voltage to the rectifier tubes after their filaments are lighted is a responsibility of the operator in manipulating the two switches. Between transmissions the plate voltage can be removed by operating the right-hand switch to provide a stand-by condition. Retaining the oscillator in a lighted condition results in less drift of frequency when keying is again resumed.

The inductance system used in the frequency determining circuit utilizes a novel method of providing for operation of the filament of

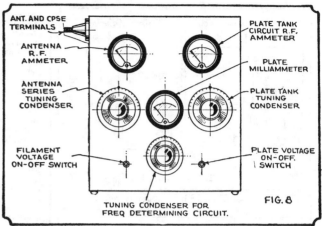

Fig. 8—Drawing above shows front view of Mr. Spriggs' single tube transmitter, which utilizes the famous Dow circuit. The names of the control dials and meters are given above.

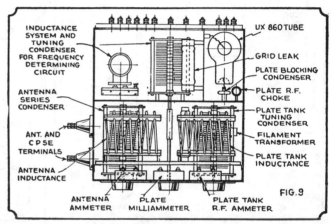

Fig. 9—Top view of the transmitter, showing position of UX-860 transmitter tube, together with location of various inductances and condensers.

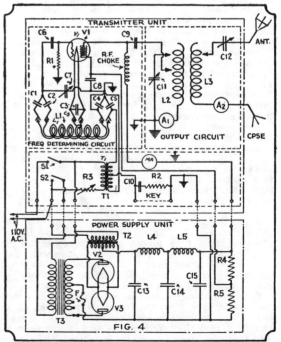

Fig. 4—Diagram above shows complete circuit of Mr. Spriggs' one-tube master oscillator - power amplifier transmitter, together with diagram of power supply unit. Dotted lines show shielding compartments.

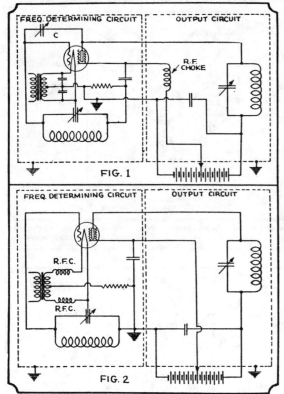

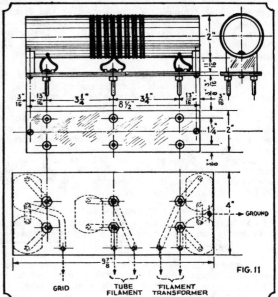

Left—Fig. 1 shows the original Dow circuit based upon the use of a four-element tube. The filament, control grid, and screen grid tube are so connected as to form, with the associated apparatus, an oscillating circuit similar to the Colpitts circuit.

Fig. 2, below at left, shows a somewhat different circuit from Fig. 1, and which eliminates the necessity for the balancing condenser by utilizing the shielding properties of the shield grid. Fig. 11, at right, shows details of plug-in inductances.

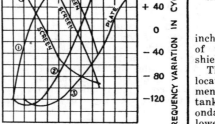

SCREEN VOLTAGE
360   400   440   480   500   560
PLATE VOLTAGE
800   900   1000   1100   1200   1300
① E_s AT 350 V. WHILE E_p VARIES.
② E_s AT 450 V. WHILE E_p VARIES.
③ E_s AT 500 V. WHILE E_p VARIES.
④ E_p AT 1000 V. WHILE E_s VARIES
⑤ E_p AT 1175 V. WHILE E_s VARIES
⑥ E_p AT 1250 V. WHILE E_s VARIES
INITIAL FREQUENCY SETTING
3992 KCS.   FIG.3

Fig. 3, at left—The curves here reproduced show the effects upon the frequency of variations in screen and plate voltages, using a UX-860 shield-grid tube, in a circuit similar to that shown in Fig. 2, above at left.

the tube at a radio frequency potential above ground potential, without the necessity of using radio frequency chokes in the filament leads by an arrangement similar to that shown in Fig. 2. Each coil is wound with two separate No. 14 enameled wires, wound side by side on a bakelite tube and arranged for plug-in connection in such a manner that the low voltage A.C. filament supply is fed through the two wires from the low potential end of the coil and taken off to the filament from the center of the coil. The two wires of the coil are by-passed at each end and in the middle in such a manner that, as regards radio frequency, the two wires are in parallel and equivalent in radio frequency effect to a single wire winding. The construction details of the coils and coil mounting are shown in Figs. 10 and 11. It has been found possible to cover the amateur 3.5, 7, and 14 megacycle bands by the use of only two coils for the frequency determining circuit, one of 22 turns tapped at the ninth turn, and one of 7 turns tapped at the third turn. For the lower frequencies, however, I recommend a coil of 19 turns tapped at the eighth turn instead of the 22-turn coil.

The radio frequency choke used in the plate circuit consists of a winding of No. 30 double-silk-covered magnet wire 4¾ inches long on a Victron tube 5¾ inches long and ¾ inch in diameter. No choke was found to be necessary in the grid leak circuit because of the high resistance of the leak itself, i.e., 20,000 ohms.

It will be noted that the tube is located in a separate compartment, shielded from both the frequency determining and the output circuit. This additional shielding is highly desirable to reduce to a minimum the possibility of electrostatic coupling between the plate or output circuit and the frequency determining circuit. Additional shielding to close completely the tube compartment or the frequency determining circuit compartment was found to be unnecessary.

Since the filament leads between the tube and the associated coil system are at radio frequency potential, it is desirable that there be wide separation between these leads and the shielding, that the length of the leads be as short as possible, and that the leads be rigidly supported. The filament leads in the set are of No. 10 solid enamel wire and are rigidly supported about 1 inch apart, well clear of all shielding on bakelite pedestals made for the purpose. The *isolantite* tube socket is mounted in such a manner as to provide about 1½

inches spacing between the base of the tube and the nearest shielding.

The filament transformer is located in the front compartment directly under the plate tank tuning condenser. The secondary leads run directly to the lower or grid end of the frequency determining coil system. In series with the primary of the filament are two resistors, one fixed and one variable. The fixed resistor is located in the front compartment just to the left of the filament transformer. (See Fig. 9.) The variable resistor is mounted on the underside of the subpanel and is not shown in any of the figures. The operating procedure includes an initial adjustment of the variable rheostat to obtain correct filament voltage, after which no further adjustments are undertaken except for excessive variations of line supply voltage. The differences in voltage drop resulting from changing of coils in the frequency determining circuit has been found to be so slight that no readjustments of the filament circuit have been considered.

The plate tank and antenna coupling coils are of conventional design, 3¼ inches in diameter, each of 10 turns, and wound of No. 8 solid wire with bakelite supporting strips. The antenna coil is arranged to slide back and forth to provide variable coupling. The antenna series tuning condenser is completely insulated from the shielding. By suitable shifting of the taps on these coils it has been found possible to operate satisfactorily on all three of the 3.5, 7 and 14 megacycle band. By operating the frequency determining circuit on 14 megacycles and tuning the output to the 28 megacycle band with a smaller output coil, good output can be obtained at 28 megacycles.

The variable condensers used are Cardwell .00045 mf., 3,000-volt transmitting condensers, and the condenser dials are Velvet Verniers. A condenser of somewhat larger spacing would be preferable for use in the plate tank circuit, as the condenser shown has a tendency to spark over on high power in the 3,500 kc. band, and also in the 7,000 kc. band if the antenna circuit happens to become detuned.

The meters I used are Westons. The 1.5-ampere antenna ammeter is not of sufficient capacity for use with a reasonable low resistance antenna. A 2.0 or 2.5 ampere meter would be preferable. The 15-ampere plate tank meter is adequate unless the tank circuit losses are substantially less than in the transmitter described. This meter can be dispensed with by using the dip of the plate milliammeter for tuning. The plate milliammeter is an 0-200 ma. Weston.

For normal operation with maximum output, the coil for the frequency determining circuit should be plugged in with the section of the coil with the larger number of turns in the grid circuit. Reversing this coil will reduce the input and output by a large percentage. No exact determination of the best ratio of turns in the grid to turns in shield grid section of this coil has been made, but it appears that about a 60-40 percentage is satisfactory.

The plate blocking condenser "C," Fig. 4, has a capacity of .000067 mf.

When first placing the equipment in operation the plate voltage applied should be not over about 1,500 volts. The tap on the voltage divider system which supplies voltage to the shield grid should be placed one-third to one-half the way from the low end; that is, the shield grid voltage should be somewhat less than half of the plate voltage. The plate current with the plate circuit out of tune is considerably greater than when it is in tune, although not high enough to be dangerous to the tube. The frequency determining circuit oscillates at practically the same frequency, regardless of the tuning of the plate circuit.

List of Parts

A1—Weston R.F. ammeter, 0-15 amps.
A2—Weston R.F. ammeter, 0-1.5 amps.
C1, C2, C3, C4, C5—.01 mf., 2,500-volt Sangamo fixed condensers.
C6—.0002 mf., 5,000-volt Sangamo fixed condenser.
C7—.00045 mf., 3,000-volt Cardwell variable condenser.
C8—.002 mf., 5,000-volt Sangamo fixed condenser.
C9—Two 5,000-volt Sangamo fixed condensers in series, .0001 and .0002 mf., respectively ; net total capacity, .000067 mf.
C10—½ mf., 1,500-volt G. E. Co. fixed condenser.
C11, C12—Same as C7.
C13, C14—1 mf., 3,500 Potter filter condensers.
C15—2 mf., 3,500 Potter filter condenser.
F—High voltage fuse rated at 2,000 volts, .75

amps.
L1—Frequency determining coil system.
L2—Plate tank circuit inductance.
L3—Antenna coupling inductance.
MA—Weston D.C. milliammeter, 0-200 milliamperes.
R1—Grid leak Ward Leonard resistor, 60-watt, 20,000 ohms.
R2—Ward Leonard 60-watt, 100-ohm resistor.
R3—Variable resistor, fixed Ward Leonard 60-watt, 300-ohm resistor in series with 20-ohm, 60-watt porcelain base variable resistor.
R4—Ward Leonard 200-watt, 14,000-ohm fixed resistor.
R5—Ward Leonard 200-watt, 11,000-ohm fixed resistor.
S1—Filament switch, Cutler Hammer.
S2—Plate power switch, Cutler Hammer.
T1—Thordarson 80 va., 12-volt filament transformer.
T2—Thordarson rectifier filament transformer for UX-866 tubes.
T3—Special home-made plate transformer, 0-1500-2000-2500 each side of center tap, 1,000 va.
V1—UX-860 screen grid, 75-watt transmitting tube.
V2, V3—UX-866 mercury vapor rectifier tubes.

The MONITOR—How to Build and Use It

By LEONARD VICTOR, W2DHN

It is possible to find out just where the transmitter is working in the band by tuning it in on the monitor and then tuning the monitor whistle in on the receiver. Obviously the procedure can be reversed for placing the transmitter in any desired part of the band.

The unit is simplicity itself and almost any changes necessary can be made to adapt the monitor to the parts available. All of the components should be solidly placed so that the monitor will hold calibration, i. e., return to the same frequency at a given point on the dial. This is a mechanical detail and should be given great consideration on the part of the constructor.

The monitor shown in the picture uses a 2-volt, 30 type tube, but any of the small size tubes, such as the 99

Front view of the monitor.

type, may be used, by supplying the proper filament voltage. A little experimentation will determine the best spot to place the monitor, both to give a healthy signal in the receiver and have a fairly loud note from the transmitter. The coils may have to be changed by the "cut and try" method in some cases so that adequate band coverage can be obtained.

If an extra pair of earphones is not available to use with the monitor at all times, a resistance having approximately the same value as the phones should be used in the monitor, when tuning it in on the receiver. This will minimize the unbalancing effect due to the changing of the voltage on the plate of the tube, which in turn shows up in the grid tuning circuit.

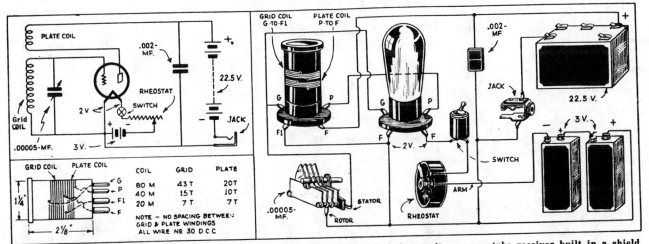

COIL	GRID	PLATE
80 M	43 T	20 T
40 M	15 T	10 T
20 M	7 T	7 T

NOTE — NO SPACING BETWEEN GRID & PLATE WINDINGS
ALL WIRE N² 30 D C C

Both schematic and picture wiring diagrams are given for the construction of the monitor—a one-tube receiver built in a shield box, so as not to pick up too strong a signal from the local transmitter. A 2-volt tube is used and no antenna is required.

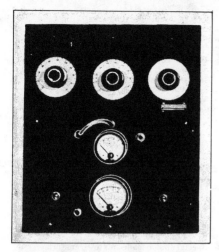

Front view of 2 tube portable transmitter.

Taking the "Headaches" out of Crystal Control

By George W. Shuart (W2AMN)

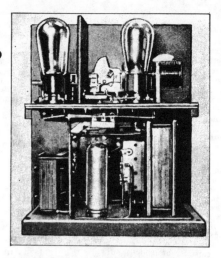

Rear view of 2 tube transmitter with crystal control.

● DUE to the popularity of crystal control among both amateur and commercial stations, many new circuits and tubes have been developed to increase the efficiency and stability of radio transmission. The principal tube adapted to crystal control was the type 47, which was originally intended as a pentode. This increased the efficiency and stability of crystal oscillatory circuits considerably, as most of us have discovered. However, the author has discovered a new use for this tube which has undoubtedly been overlooked.

The object of crystal control is obviously to obtain frequency stability. However, good frequency stability has only been obtained by the use of temperature controlled crystals, or by a very low-powered oscillator followed by numerous amplifier stages to obtain sufficient excitation for the amplifier feeding the antenna. It can readily be seen, therefore, that if the oscillator output were increased, a few amplifier stages could be eliminated. With the average circuits the application of 500 volts to the plate of the oscillator tube would undoubtedly wreck the crystal or cause serious heating which would result in *frequency drift*.

The Author's New Circuit Idea

For some time the author has been using a circuit in which 500 volts can be applied to the plate of the tube in the crystal stage and in which there is less heating and frequency drift than in the usual circuit using in the neighborhood of 150 volts.

It has been overlooked that the type 47 will operate very efficiently with the screen and control grids tied together and connected through a radio frequency choke to a filament center tap. This eliminates external grid bias of any kind, and holds the plate current to the order of 3 to 4 milliamperes at 500 volts. This causes the tube to have a very high amplification factor, which results in comparatively large output with a low input voltage. (As it creates a second harmonic only slightly weaker than the fundamental it is ideal when followed by a frequency multiplier.) With the grids tied together in this manner it is possible to drive the current to 80 ma. and still operate efficiently and with stability.

In the diagram is shown the type 47 in its role of oscillator and also class "B" amplifier. These tubes are both

A nifty 2 tube portable transmitter is here described, which employs crystal control in a greatly simplified circuit, utilizing '47 type tubes and an '82 mercury vapor rectifier.

operated from a common power supply, as the amplifier has no detectable effect on the oscillator, even when the amplifier is being keyed.

Oscillator Plate Tank Condenser

The plate tank tuning condenser of the oscillator stage should not exceed .00015 mf. in order to obtain full efficiency of the circuit, resulting in high LC ratio. The excitation lead to the amplifier should be tapped at approximately one-third of the distance from the low potential end of the oscillator tank. Extreme care should be taken, as this adjustment is very critical. The recommended capacity for the excitation coupling condenser should be in the order of .0001 mf. or less. Neutralizing for this type of amplifier should be done within the stage itself, as shown in the diagram. It has been found that the amplifier stage would not neutralize if the tuning condenser was shunted across the whole coil, a part of which is used in neutralizing. Therefore, the design of the amplifier tank coil and neutralizing coil should be carefully followed.

Details of Coil

The coils are wound with No. 14 B. & S. enameled wire on 2 inch forms, spaced approximately the diameter of the wire. This is accomplished by winding string along with the wire, then removing the string, and coating with clear Duco. The antenna coil (L1) has 12 turns, and is wound on the same form as the amplifier tank coil. Spacing between the two is one-half inch. The amplifier plate coil (L2) has 25 turns, 20 of which are shunted by

the amplified tuning condenser, the remaining five being the neutralizing coil. The oscillator coil (L3) has 20 turns and is wound in the same fashion as the above.

From the description it can be seen *that much apparatus can be eliminated from crystal controlled transmitters.* Therefore this would make an ideal portable. Following is a description of the portable built by the author.

Portable Transmitter

The complete outfit, including the oscillator amplifier portion and power supply, is inclosed in a case 14 inches high, 12 inches wide, and 7 inches deep. An aluminum shelf through the center separates the radio-frequency unit from its power supply. In the lower compartment is mounted T1 as in the diagram. The high voltage secondary of this transformer supplies 500 volts to the oscillator and amplifier, at approximately 150 ma. T2 is the filament transformer supling 2.5 volts to oscillator and amplifier filaments. However, a transformer with two separate windings could be used in order to key the amplifier at the filament center tap, if the builder does not wish to key the amplifier in the high voltage lead as shown in the diagram. The rectifier problem is nicely solved by the use of

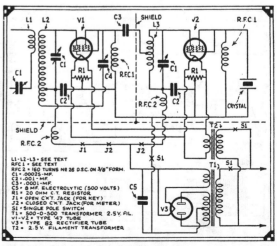

Wiring diagram of Mr. Shuart's 2 tube portable transmitter, featuring crystal control.

the type 82 tube, which is a full wave mercury vapor rectifier, capable of passing more than sufficient current for the set. In order to obtain a pure D.C. note an 8mf. electrolytic condenser shunted directly across the output of the rectifier tube is all that was required. This combination gives fine regulation.

Separating the oscillator and amplifier is an aluminum shield, through which only the excitation leads pass. All low potential wiring is done beneath the shelf. The center-tap filament resistors are mounted directly at the base of the tubes, and it has been found that filament by-pass condensers are not needed with these short leads. If a center tapped filament transformer is used, by-pass condensers must be located at the base of the tubes where the resistor is shown in the diagram.

Following the shield is the amplifier tank condenser, next to which is the antenna tuning condenser. As explained before the plate and antenna coils are both wound on the same form. These are mounted directly behind the two condensers. Behind the plate coil is mounted the amplifier tube. This is the layout used by the author and will give the reader a general idea of how these parts can be mounted.

The grid chokes which are of the lattice-wound variety commonly used in broadcast receivers are mounted as close as possible

to the tube case. The lead connecting the choke to the grid terminal should be as short as possible, the other end going to the aluminum shelf, which serves as a common connection for all negative leads. The neutralizing condenser is of the variety manufactured by Hammarlund, and mounted directly on the amplifier tuning condenser with a small bakelite post. This condenser is very small and can be easily mounted, with no danger of breaking down, due to its mica insulation. This condenser, of course, must be adjusted with an insulated screwdriver. Once adjusted this will need no further attention.

How Set Is Tuned

The tuning of this set is very simple. With every thing connected and checked carefully the builder should have no difficulty in getting the set to work. The filament voltage should be applied first, as these tubes take comparatively long to heat. Next is to plug in the 0-100 milliammeter into the jack provided for it in the plate lead of the oscillator tube. Now the plate voltage is applied to the entire set, providing the keying circuit of the amplifier tube is open. If the tube is not oscillating the plate current will be extremely low. (From 3 to 5 ma.) The plate turning condenser should be swung slowly from zero upward, watching the meter carefully. When the plate circuit approaches resonance with the crystal the plate current will rise, and as

resonance is passed the current will begin to decrease. Absolute resonance is the point at which the plate current is the highest. The value of the current will be between 60 and 80 ma., depending on the quality of the crystal. The adjustment of the oscillator should be left in this position.

Now to neutralize the amplifier. Under no consideration should the key of the amplifier be closed at this point, as it may damage the tube. Adjust the neutralizing condenser to full capacity, swing the amplifier tuning condenser gradually from zero upward until the point is reached, where there will be a dip in the plate current of the oscillator. At this point the capacity of the neutralizing condenser should be gradually reduced, and the amplifier tuning condenser swung gently back and forth till no dip occurs. When this point is reached the amplifier is sufficiently neutralized. The point at which the dip occurred is the approximate point of resonance, and the circuit should be left tuned as near as possible to that point. The plate circuit of the amplifier can now be closed, providing the milliammeter has been plugged into the jack provided for it in the circuit. The amplifier tuning condenser should be adjusted to a point where minimum plate current exists (approximately 20 ma.). The entire set is now properly tuned and ready to be coupled to the antenna.

Here we have a fine view of the complete 45 push-pull Beginner's Transmitter and power supply unit. Extra wave band coils are shown.

The "RT" Beginner's Transmitter

By GEORGE W. SHUART (W2AMN-W2CBC)

Here is just the transmitter the beginner is looking for—it employs two type 45 tubes as oscillators, which yield practically as great an output as the 10 type tubes, at far less cost. Coil data is given as well as specifications for building the oscillator and also the power supply.

● THE transmitter described in this article is primarily a low power, low cost outfit for the beginner. Almost any one aquainted with radio transmitters will admit that a push-pull arrangement is far superior to the single tube variety. There is really no extra cost in building a push-pull oscillator, the only added expense is in the additional tube required. And for this reason the author cannot see why anyone would build a transmitter and not incorporate a push-pull circuit. Push-pull circuits are much more efficient, this is proven by the fact that just about every ultra short wave transmitter is of this type. If this type of circuit is more efficient on the higher frequencies, surely it should perform more than satisfactorily on the lower frequency bands.

Uses 2-45's as Oscillators

The outfit shown in the photographs uses two of the type 45 tubes as oscil-

lators. This tube was used because of its low cost and the fact that it provides practically as much output as the regular 210, at one third the cost. The transmitter is divided into two sections namely, the radio frequency oscillator and power supply. The oscillator is mounted in bread-board fashion on a 7 by 15 inch plywood board which should be thoroughly dried and given a coat or two of shellac to prevent it from absorbing moisture in damp weather. The circuit is of the type using fixed-tune grid and tuned plate. All grid coils are wound on one inch bakelite tubing with fine wire, so that their natural frequency response is near the center of each amateur band. The frequency peak of this type of coil is rather broad and for this reason the entire band can be covered with the plate circuit, without the two circuits getting out of resonance. This is a desirable feature as any part of a given band can be worked on, with the plate

circuit being the only tuning control. To facilitate the changing of grid coils for the different bands, each coil is equipped with three small plugs that fit into sockets mounted on a strip of bakelite fastened to the baseboard. The construction of these coils can be clearly seen in the drawings, and the number of turns for each band is given in the coil table.

Plate Coils of Copper Tubing

Next we have the plate circuit. The plate coils for the 20, 40 and 80 meter bands are constructed of one-quarter inch copper tubing. The 160 meter coil is of a different type. This coil is made by winding No. 12 antenna wire (solid enameled) on a two and one half inch bakelite tube four and one half inches long. Wind 25 turns on this tube very tightly, and space the turns with string of approximately the same size as the diameter of the wire. This winding is then covered with clear lacquer or some such material that is a good insulator. The other coils are spaced so that they just fit on the two stand-off insulators which are four and one half inches apart. A 500 mmf. receiving type condenser is used to tune the plate circuit and is shunted directly across the plate coil causing both rotor and stator to be at high RF (radio frequency) potential. This will cause noticeable hand capacity effects to be present when the condenser is adjusted and allowance should be made for this. Be sure to use a knob or dial that does not have its set screw exposed where the body will come in contact with it, or else a nasty burn will be the result.

Simplification Achieved

From the diagram it can be seen that no grid or filament bypass condensers are used and the familiar "RF" choke is omitted from the plate circuit. No benefit was derived from their use and for that reason they were not used. However there is no law against them; should the builder wish to incorporate them in the set, a .0005 mf. bypass condenser in the plate voltage supply lead is all that was found beneficial.

The method of coupling the antenna to the out-put circuit of the transmitter may cause some to wonder whether or not there is something wrong with the diagram. This transmitter was built with a pair of antenna coupling coils, one coupled to each side of the plate tank and tuned with a variable condenser, just to see whether or not there was any difference in the out-put and the character of the signal. First one coil was eliminated and then both were taken out and the antenna connected directly to the plate coil with results being the same in either case.

Wiring the high-frequency part of the transmitter is done with regular hook-up wire and all connections well soldered, except the two leads connecting the plate coil to the tuning condenser; these leads are made of the same size copper tubing as the coils.

The power supply to operate the above transmitter delivers 400 volts at 150 mills (m.a.) for the plates of the tubes and 2.5 volts for the filaments. A type 83 mercury vapor rectifier is used because of its low voltage drop, which provides very good regulation. A filter consisting of a 30 henry iron core choke with a 2 mf. condenser on either side, produces very good "DC" signal from the transmitter.

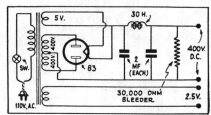

Power Supply unit diagram

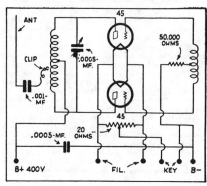

The diagram above shows the hook-up of the exceptionally few parts required in constructing the "RT" Beginner's Transmitter, which was especially designed by Mr. Shuart so as to embody the simplest possible construction at the lowest cost, without sacrificing quality and efficiency.

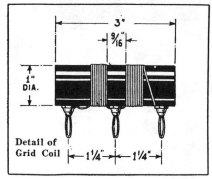

Detail of Grid Coil

Putting the Transmitter on the Air

Getting the transmitter on the air is no difficult task if a few pains are taken in adjustment. After all wiring has been checked carefully to make sure that no wrong connections have been made, connect the power supply to the oscillator unit, inserting a 0-100 milliammeter in the positive plate voltage lead. It is best to put a resistor in series with this lead also in order to start adjustments with a reduced plate voltage because if the plate circuit is not in resonance with the grid circuit the plate current will be very high and probably damage the tubes or meter. Do all tuning without the antenna connected. Insert all tubes allowing them to heat up sufficiently using the 80 meter coils as a starter, close the key and be prepared to tune the plate condenser *immediately* for lowest plate current as indicated by the meter. When this point has been reached the resistor can be removed from the plate circuit, allowing full plate voltage to be applied to the tubes. The plate current should now be in the order of about 50 milliamperes with no antenna load. All that remains is to check the frequency in the monitor, attach the antenna and we are "on the air."

The antenna suggested for use with this transmitter is the single wire feed Hertz, which was described in September's SHORT-WAVE CRAFT on page 311.

COIL DATA

Plate Coils

Band	Turns	Diameter
160	25	see text
80	12	2 ⅜ in.
40	6	2 ⅜ in.
20	4	2 ⅜ in.

The 80, 40, and 20 meter coils— ¼ in. copper tubing.

Grid Coils

Band	Turns	Size Wire
160	150	36 D.S.C.
80	78	36 D.S.C.
40	42	26 D.S.C.
20	16	26 D.S.C.

All grid coils wound on one-inch diameter bakelite tube, with no spacing between turns.

List of Parts for "Oscillator"

1 Set of coils—see coil table
4 Bakelite tubes 1 inch Dia. 3 inches long
12 Banana type coil plugs
3 sockets for coil plugs
1 50,000 ohm resistor, 5 watts or over (gridleak) Radio Trading Co.
2 4 prong isolantite sockets
1 .0005 mf. variable condenser, (receiving type.)
3 double binding post strip, (laminated) Radio Trading Co.
3 midget stand-off insulators
1 baseboard, 7x15 inches
2 type 45 tubes

List of Parts for "Power Supply"

1 power transformer—2.5, 5, 400-0-400 Radio Trading Co.
1 filter choke—30 henries, 150 milliamperes, Radio Trading Co.
2 filter condensers—2 mf. 1000 volt rating, Radio Trading Co.
1 4 prong socket
2 double binding post strips, (laminated) Radio Trading Co.
1 baseboard, 7x15 inches

A Pyrex Glass Lead-in for Ten Cents

● Passing a store window advertising Pyrex glass cups, about three and a half inches in diameter, on sale at five cents apiece and noting the resemblance to the regular bowl shaped radio antenna lead-ins sold at quite a high price, two of these were purchased to see if a cheap but good lead-in could not be made from them.

The principal thing to be done was to drill a hole in the bottom of each cup to take the assembly bolt. This was easily accomplished by cutting off a few inches of ⁵⁄₁₆ copper tube from an old automobile oil line, straightening and filling with carborundum powder ground from a hand bench grinder and mixed with turpentine, then placed in a drill chuck. Either a power or hand drill can be used as but little pressure is required.

A disc of sheet iron was cut of a diameter to fit near the bottom of the cup, with a five-sixteenths hole drilled in the center, to act as a guide for starting the hole, being removed later. A wood disc would do as well. Place a couple of teaspoonfuls of turpentine in the bottom of the Pyrex cup, set the guide disc in place and start the drill, held vertically of course. An annular groove the size of the copper tube will be formed in the glass due to the abrasive action of the carborundum powder; if the grinding action ceases to be noticed, file off the end of the tube, as this may have been worn round, place more abrasive in the tube and in the bottom of the cup and continue. Do not apply much pressure, especially when almost through the glass. When the drill comes through, reverse the cup and carefully grind through the other side.

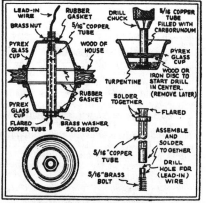

How to use Pyrex glass cups to make a first-class "lead-in."

Phone Transmitters

Front view of the C.W. (code) amateur transmitter, which is here described in detail by Mr. Victor, together with its power supply unit.

Amateur C. W. & Phone TRANSMITTER By Leonard Victor, W2DHN

As far as DX is concerned, during the week the set was under test (which was poor, owing to summer conditions) more than two dozen perfect contacts were made from New York, with England and Brazil as the furthest worked.

The outfit is divided into three parts, the tube transmitter, the power supply, and the antenna. This transmitter can be used with any four-prong tube, from a 30 with 90 volts of "B" battery up to a 10 with 500 volts from an A.C. power pack. For battery operation a 71A or an 01A with 180 volts will give very good results. The model illustrated uses a 45 tube for the oscillator and an 80 for the rectifier; because both of these are receiving type tubes which can be purchased very cheaply, but still will put out a very healthy, respectable signal.

Briefly, the transmitter operates as follows. The power transformer steps up the alternating current, which is rectified by the 80 tube and changed to a pulsating direct current. This current is then led through the filter, which consists of the filter choke coil and the two filter condensers.

The frequency on which a signal is sent is determined by the tank circuit L-1, C-1 (diagram). The grid circuit must be approximately in resonance with the plate circuit for feed-back and oscillation to take place.

It is possible to use a fixed grid coil for each amateur band, since one adjustment is adequate over a considerable range of the plate tank tuning.

The need for an antenna control is eliminated by use of a single-wire, untuned feeder connecting the plate circuit and the antenna. The antenna, or "flat-top" as it is called, is tuned to the desired transmission frequency by cutting it to a predetermined length.

TRANSMITTER SPECIFICATIONS

L1 and L2, plate and grid coils. Specifications given under the illustration of the coils.

C1, 350 mmf. (00035 m.f.) variable tank condenser. Any good receiving condenser will do.

C2, 2000 mmf. (.002 m.f.) fixed plate blocking condenser, mica, receiving type.

C3, 2 2000 mmf. (.002 m.f.) filament by-pass condensers. Same as C2.

C4, 250 mmf. (.00025 m.f.) grid by-pass condenser, mica, receiving type.

R.F.C., low resistance, high quality radio frequency choke, capable of passing 100 ma. The one used is a Hammarlund.

R1, 150 ohm center-tapped filament resistor, anything over 40 ohms can be used.

R2, 50,000 ohm, 5 watt wire-wound resistor.

S1, 4-prong socket to which filament, plate, and negative leads are lead. Cable from power supply plug into this socket.

CONSTRUCTING THE SET

There should be no difficulty about constructing the set, using the wiring diagram and the pictures shown with the story.

The base of the transmitter is a board 12 inches long by 7 inches wide. This should be sand-papered and given a coat of varnish so that its appearance is neat and workman-like. The board is mounted on four rubber or metal tacks, one in each corner, such as are used under the legs of chairs, to allow clearance for the two filament wires and prevent scratching of the table on which the transmitter is placed.

The grid coils are wound with No. 30 double cotton covered wire on regular coil forms. After the proper number of turns has been determined, the coils should be given a coat of collodion or clear varnish so they will retain their characteristics.

The plate coils are wound of ¼" diameter soft copper tubing. Wind the coil around a pipe or dry cell approximately 2⅜" diameter. Flatten the ends and drill them to fit over the machine screws on the stand-off insulators. The stand-offs behind the tank condenser should be mounted 5" apart between centers.

The lay-out of the apparatus is self-explanatory. The right-hand socket is for the grid coil and just behind is the grid leak and grid condenser (R2, C4).

The socket next to the tank condenser is for the tube. Behind the tube, the three condensers in a row are the two .002 mf. filament by-pass condensers and the plate blocking condensers, C2, C3. The plate lead of the tube is connected to one side of the condenser stator and the other side of the stator goes to the insulator at the extreme left. The right-hand stand-off is connected to the rotor of the condenser, the plate blocking condenser and the R.F. choke. The socket at the back is for the plug from the power supply cable. The Fahnestock clips at the right are for connection to the key. All the parts are wired with regular

Details of C.W. "Code" Transmitter

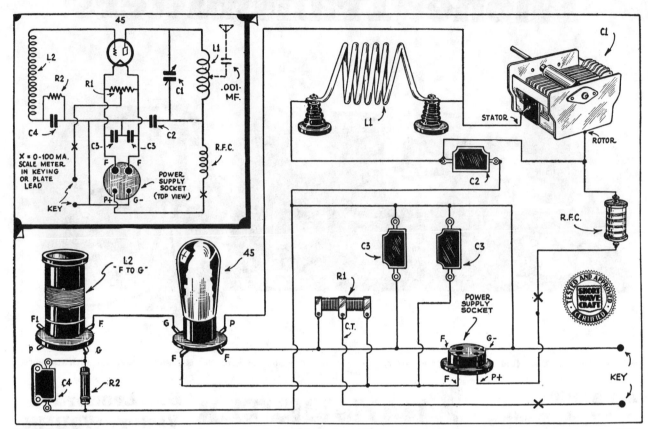

The diagrams above show in schematic and also physical form how to wire the simple lay-out of parts comprising the C.W. transmitter for amateur use, as here described by Mr. Victor. A 45 or a smaller tube may be used as the oscillator.

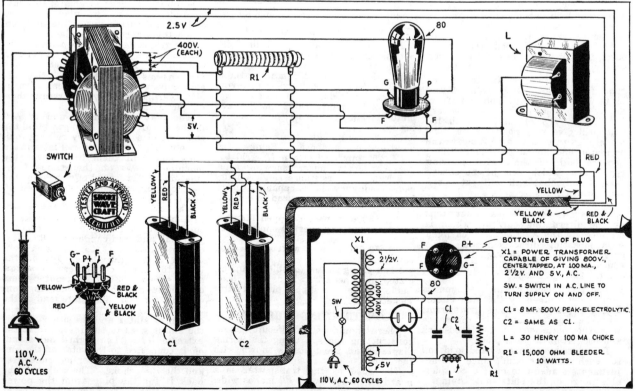

The "power-supply" unit, using an 80 type tube for the rectifier, is shown in diagrammatic form above. It supplies 350 volts D.C. for the plate and 2½ volts A.C. for the filament of the 45 transmitting tube.

KEY

Top view of the C.W. transmitter, the construction of which is fully described in the accompanying article by Mr. Victor. The set illustrated uses a 45 tube for the oscillator and an 80 for the rectifier.

square bus-bar, covered with spaghetti. The connections to the plate of the tube and the tank connections are made of very heavy bus-bar to carry the high current flowing in that circuit. The filament leads from the tube to the power supply socket are run under the board. The 0-100 milliameter should be connected in the *keying leads*. It is usually most convenient to have it right by the key.

The power supply is mounted on a varnished board, 12" x 9". It is a conventional unit of the same type as is used on receivers, supplying 350 volts direct current and 2½ volts A.C. for the filament of the 45 transmitting tube. A standard "brute force" filter is used, consisting of two 16-mf., 450 volt peak electrolytic condensers and a 30-henry, 150 milliampere choke coil. Any suitable arrangement adaptable to the parts at hand may be used for the power supply unit and the plate voltage may be anything available from the transformer from 250 volts up to 400.

A regular 80 tube is used as the rectifier. The filament and the high voltage plus and minus leads are connected to the cable that runs to the transmitter. A 10,000-ohm bleeder resistor is used between plus and minus of the high voltage to afford good regulation and prevent a chirping note.

TUNING UP

After the transmitter is wired and the antenna is up, the *monitor* is called into play. Suppose we intend to work in the 80-meter band, on 3,575 kilocycles. We have already cut our antenna according to the measurement in the table for that particular frequency. Put the 80-meter coils in the transmitter. Tune in the band on your *receiver* and set your *monitor* just inside the low frequency edge. Usually there are some stations operating right on the edge of the band, but always

to be on the safe side, unless your monitor is already accurately calibrated, set it inside the first ten stations you hear. Plug your earphones in on the monitor and then press the key on the transmitter. Turn the transmitter dial until the current goes down to its lowest point, about 30 (milliamperes). Somewhere in the portion of the dial where the transmitter draws *minimum* current, a whistle should be heard in the earphones, signifying that the set is operating O.K. The grid coils will probably be all right, but it may be necessary to add on or take off a few turns until the transmitter draws least current (without the antenna) at the low frequency end of the band.

Now clip the antenna on the tank coil, one turn for 20, two for 40 and five for 80 meters, from the "cold" or plate blocking condenser end.

As the dial is turned through the frequency of the antenna, there should be a sharp rise in plate current, as shown on the milliameter. The antenna should be clipped as far up to the 'hot end' of the tank coil as is possible, without spoiling the note, because for good reliable operation the note must be pure D.C. The transmitter will work only on the frequency to which the antenna is cut, hence if a frequency well in the band has been chosen, there will be no worry about *off frequency* operation. Always operate right on the peak of the antenna, although in case of interference a shift can be made up to 15 kc. each side of the peak. When doubling the antenna, say using an 80 meter antenna on 40 meters, the peak will be much broader. *Always keep the note pure d.c.*

This set has been built to operate on the three most popular amateur bands, 20, 40 and 80 meters. Twenty is the big daylight DX band. Forty is usually good for about fifteen hundred miles, although in the early morning it is possible to work Australia and New Zealand.

However, the best band for the fellow just breaking in the game is the 80-meter band. There are plenty of stations on all day and all night, and hundreds of good operators who will "pull up" and send slow for a newcomer. After experience has

been obtained on the 80 meter band, the outfit can be put down on 40 and 20.

ANTENNA DATA

Frequency	Antenna Length	Feeder Length from center of aerial
3550 KC	134 ft.	18' 8"
3600 KC	132 ft.	18' 5"
3700 KC	129 ft.	18'
3800 KC	125' 9"	17' 6"
3900 KC	122' 5"	17'

For 40 meter aerials multiply the frequency by 2 and divide the aerial and feeder placement lengths by 2. For 20 meters multiply and divide by four. The bands are as follows:

80 m.	3500	to 4000 kc.
40 m.	7000	to 7300 kc.
20 m.	14000	to 14,400 kc.

Use good insulators, measure carefully, and solder the feeder on tightly.

Power Supply Parts List

1 Acratest power transformer (T)
2 Eby socket 4 prong (S)
2 Dual 8-mf. Acratest electrolytic cond. (C)
1 30-henry, 150 ma. choke (S)
1 10,000-ohm Bleeder resistor 50 watts (R)
1 Wooden base board
1 4 wire cable
1 4 wire socket plug

Grid and Plate Coil Specifications

Grid Coils No. 30 d.c.c. wire	{ 80 m.—50 turns { 40 m.—15 turns { 20 m.— 6 turns	} Wound on } coil form— } 1¼" diam.
Plate Coils	{ 80 m.—14 turns { 40 m.— 6 turns { 20 m.— 4 turns	} ¼" copper } tubing

TRANSMITTER PARTS

1 set of grid coils (See specs.) L-2
1 set of plate coils (See specs.) L1
1 Acratest .00035 mf. tuning cond., C1
3 .002 mf. mica C2, C3
1 .00025 mf. mica C4
1 Hammarlund R.F. Choke R.F.C.
1 Acratest 150 ohm CT Resistor R1
1 Acratest 5 watt resistor 50,000 ohms R2
3 Eby Sockets
1 Wooden base board 7"x13"
1 0-100 Readrite milliammeter

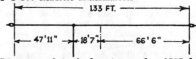

|← 133 FT. →|
|← 47'11" →|← 18'7" →|← 66'6" →|

Diagram of typical antenna for 3575 kc. Ant. of No. 14 solid enamelled copper feeder of heavy rubber covered lead-in wire. Use good insulators. Solder feeder on carefully. Feeder at right angles to flat top for first 30% of distance.

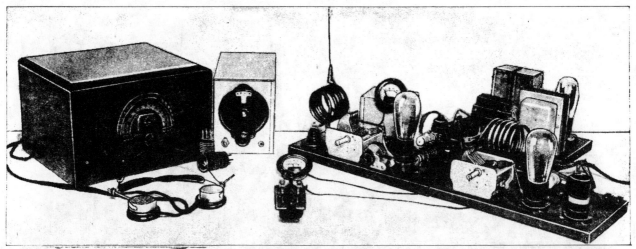

Complete short-wave station set-up, with S-W receiver at the left, together with the M.O.P.A. transmitter at the right.

Amateur Transmitters
How to Build, Install, and Operate Them

In this second article on Amateur Transmitters, Mr. Victor explains at length how to build the greatly desired *Radio Frequency Amplifier* for use with the oscillator.

Among other advantages gained by the use of an R.F. amplifier for transmitters are the steadier signal and the greater percentage of modulation which can be obtained.

By LEONARD VICTOR
W2DHN, W2DPT

● THERE are innumerable advantages to be gained by the use of a radio-frequency amplifier, among which only the major ones will be mentioned here. Uncle Sam, through the Federal Radio Commission, has enacted regulations to the effect that *phone* should not be used on a *self-excited* oscillator—hence the need for an *R.F. amplifier.* Modulating an oscillator directly causes a broad, mushy signal that eats up far too much space in our already over-crowded "ham" bands. Likewise modulating an oscillator on the 160 meter band will inevitably cause complaints from broadcast listeners in the neighborhood.

Oscillator modulation cannot be increased above 30% without bad distortion, whereas 100% modulation can be used on an amplifier. Since a ten watt, 100% modulated carrier is just as effective as a fifty watt, 30% modulated one, it will be seen that much less power gives far better results when an R.F. amplifier is employed.

An amplifier greatly increases stability, that is, gives a much steadier signal, and, usually, a better note than just a straight oscillator. Those who have used self-excited, self-controlled outfits have probably noticed that when the aerial moves in the wind, there is an annoying change in frequency. Likewise any change in plate voltage or load conditions brings about a corresponding change in the frequency of the transmitter. Using an amplifier, there will be no noticeable shift in the note, even during a heavy storm, and

the plate voltage may vary as much as 25% without detrimental results.

Just as when an amplifier is used on a receiver, greater power output is obtained. With a grid input from the tank of the oscillator of only 3 to 5 watts, outputs of the order of 15 to 25 watts can be obtained.

This particular arrangement of oscillator and amplifier works excellently on all bands and is admirably suited for future use as a *phone* transmitter.

As in the single-control transmitter, the buying cost is kept low, and the unit is very simple to build and operate.

During tests with a station in England, the power (plate volts times milliamperes) was gradually lowered until we were finally putting signals across the "big pond" with less than 10 watts input! That is, we were actually using less power to work 2,700 miles than would light a small electric light bulb!

Technical Description

A radio frequency amplifier is very much like an audio amplifier used in a receiver. It has a *grid input* circuit which gives it *push* or *excitation* as it is called from the preceding stage (in this case a self-excited, self-controlled oscillator).

A single tube of the 46 type is used; this tube is admirably suited to "ham" use because of its relatively high efficiency. It requires no *battery bias* for good operation, which eliminates one of the nuisances in *phone* work. Like-

wise, because of the very high *gain* of the tube, very little excitation is necessary. This allows running the oscillator greatly underloaded and produces a very steady note. This is likewise highly desirable for *phone* work, as with less excitation there will be less intercoupling between oscillator and amplifier.

The 46 has two grids between the filament and plate. These are tied (connected) together, making a very high impedance tube.

There is a tuned plate circuit which uses the same size coil and condenser as that employed for the oscillator, (described in first article, last month.) For *code* (C.W.) work it will be found better to use coils with a few more turns, and hence use less of the condenser capacity in tuning to the same frequency. The less condenser and the more coil used in a transmitter, the higher the efficiency. However, the less the condenser, the lower the dynamic stability. A happy medium has to be experimentally struck and is slightly different for phone or code.

The customary by-pass condensers are used in both the filament and plate circuits. Plate voltage is fed at the center of the tank coil as this is the "cold spot" of the transmitter.

Also there is a midget variable condenser, called the *neutralizing* condenser, which cancels the grid-to-plate capacity of the amplifier tube and prevents it oscillating of its own accord on a different frequency than the pre-

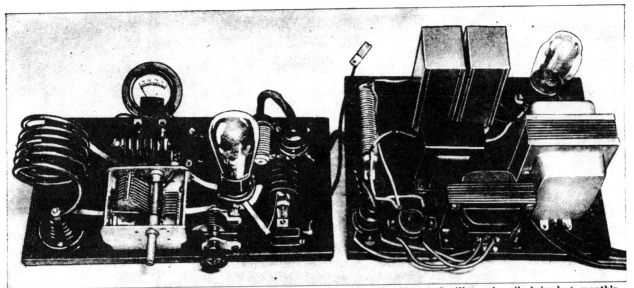

The neutralized amplifier and power-pack for use in conjunction with the Master Oscillator described in last month's article by Mr. Victor. This power amplifier uses a 46 tube and requires no external biasing battery.

ceding stage, which should not occur.

The R.F. chokes used are very heavy duty units which happened to be handy, but almost any good choke can be used. Test your chokes with a neon bulb. There should be no glow at the "cold end," but a pronounced glow at the tank end of the choke.

Chokes for transmitting purposes should be of low resistance and be capable of handling heavy current. The actual value of a choke can only be proven under test.

Care should always be taken to use good fixed condensers as this has a great deal to do with the final performance of the set.

Likewise the variable *tank* condenser should have very good insulation between the stator and rotor.

Building the Transmitter

Looking at the accompanying pictures and diagrams there should be no trouble experienced in building the unit. The baseboard is 7" x 13"; sandpaper and varnish this board for a neat appearance. The three condensers mounted in a row at the back of the board are the .002 mf. filament and plate by-pass condensers. The plug from the power cable plugs into the 4-prong socket at the rear right of the transmitter. What is normally the cathode connection on a 5-prong socket, and the grid connection, are soldered together for the two grids of the 46. Spacing between the stand-off insulators is five inches.

All tank connections are made with tinned copper braid, and the filament and choke connections are made of bus-bar covered with spaghetti insulation. Extreme care should be taken with the wiring, inasmuch as one poorly soldered connection will inexplicably spoil the operation of the transmitter.

Power Supply Details

A different power supply is used this month, because the current drain is much higher. Also, for voice (phone) work later it will be necessary to use two power supplies. The power transformer is a substantial unit supplying 400 volts at 150 mills (milliamperes). 5Z3, the successor to the 280, is used as a rectifier tube. This tube has very low voltage drop, and passes high current. A 4mf. electrolytic, a 30 henry 150 mill. (M.A.) choke, and an 8 mf. electrolytic condenser comprise the *filter* circuit. A 20,000 ohm, 50 watt resistor is used as the bleeder resistance.

The two power supply plugs, to go to the oscillator and amplifier have parallel plate and grid (really plus and minus) connections, but different filament supplies. A separate filament transformer is used to afford better regulation; if the filament windings were on the power transformer, their voltage would fluctuate with the load drawn from the high voltage circuit.

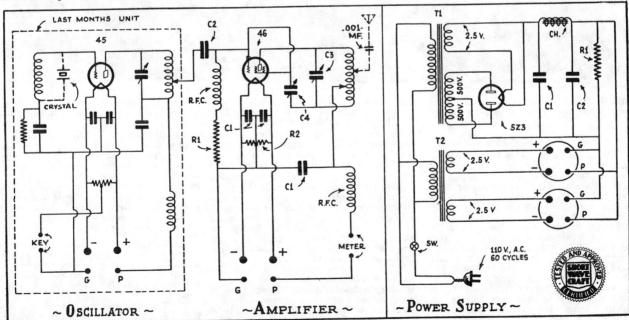

Diagram of Oscillator and Power Amplifier Circuit; also Power-Pack.

The same type of antenna, as was shown last month, namely matched impedance, single-wire voltage-feed, is used on the amplifier.

The *cold point*, however, is now the center of the coil, and the antenna is clipped towards the plate end from this point, until further advance spoils the note.

The oscillator should be carefully placed so it will receive no vibrations from anything in the room. Many an otherwise perfect set has put out a miserable note, *merely because the key was too near the oscillator and was shaking the tank coil!* Mount the oscillator board on a rubber kneeling pad, or rubber sponges.

Tuning Up

1. Clip the excitation lead from the .00025 mf. grid condenser about half-way up from the "cold" end of the oscillator tank, i.e. about two turns for 20, 3½ for 40, and 7 for 80. This excitation adjustment should be increased until an increase does not bring greater output. Never over-excite, as this will ruin the 46 tube.

2. Have the 46 in its socket but remove one of the leads from the Fahnestock clips, to cut off the high voltage. Now place a neon tube on the grid end of the amplifier tank and turn it until R.F. is shown by a reddish glow in the bulb. Then turn the *neutralizing* condenser C4 until the glow goes out. This point is usually found with the condenser almost fully opened. Check this by again tuning the amplifier tank and neutralizing any slight glow remaining. The monitor should be used as a final check of neutralization. With plate current off the P.A., (power amplifier) it should be possible to tune the P.A. tank condenser through resonance without making any appreciable change in the note from the oscillator as heard in the monitor.

3. Now connect the meter and apply voltage to the 46 tube. Tune the tank for minimum current.

4. The antenna is used in the same fashion as on the T.N.T. rig. However, since a more stable unit is supplying power, try for the highest possible output.

The transmitter is still keyed in the center tap of the oscillator, as the 46 amplifier does not draw current without excitation.

Will answer letters sent me in care of SHORT WAVE CRAFT, provided they contain *stamped, self-addressed envelopes.*

Amplifier Parts List

1 Eby socket 4 prong socket
1 Eby socket 5 prong socket (46)
3 Acratest mica condenser .002 mf. (C1)
1 Acratest mica condenser .00025 (C2)
1 Acratest 10,000 ohm 5 watt resistor (R1)
1 Acratest 50 ohm C.T. resistor (R2)
2 Acratest stand off insulators
1 Acratest .00035 mf. tuning cond. (C3)
2 Hammarlund R.F. chokes type CH500 (RFC)
1 Hammarlund .00005 mf. midget cond. (neutralizer) type MC-50-S (C4)
2 Fahnestock clips
1 Wooden baseboard 7" x 14"

Power Supply Parts List

1 Acratest power transformer, giving 500 volts either side of center tap (T1)
1 Acratest filament transformer (2.5 volts) (T2)
2 Acratest electrolytic filter condensers 4 mf. 1000 volts (C1, C2)
1 Acratest choke 150 M.A. 200 ohms D.C. (CH)
1 Acratest voltage divider 10,000 ohms 50 watt (R1)
2 Eby 4-prong socket
1 Wooden base-board 11½" x 9"

ADDING PHONE

In this photo we see the complete "Ham" transmitter, comprising the oscillator and amplifier described in the last two issues, also power supply unit in upper right-hand corner; in the upper left-hand corner MO indicates the new modulator unit, described this month, with hand mike "M."

How 100% Modulation Works

First, let us consider how an amplifier works when it is properly *grid-biased* for modulation (biased so the tube when tuned to minimum draws a very low plate current without antenna load). When the voltage on the plate of an amplifier is raised, the output is increased; conversely, when the voltage is lowered the output goes down. The function of the modulator is to vary the *instantaneous plate voltage* in exact proportion to the sound waves of the voice striking the microphone. To effect complete, or 100% modulation, it is necessary to vary the output from zero to twice the normal amplitude. Commonly the power output is varied by varying the voltage applied to the plate of the amplifier

tube that is being modulated. Likewise for 100% modulation the output of the modulator must be at least one-half the value of the input to the radio frequency amplifier for best results.

The plate input in watts is the plate voltage multiplied by the plate current in amperes; for example, 300 volts x 40 milliamperes (.04 amperes) equals 12 watts. The modulator is really an audio output power-amplifier, using the transmitter as a load resistance, instead of a loud-speaker.

The final requirement is that the load resistance be correct for the particular modulator used. Every tube is so designed that it is supposed to deliver its maximum power output to a certain value of load resistance. The particular amplifier used works into a load resistance of 8000 ohms, but we

divide this in two, without losing any appreciable power by coupling to the center tap of the audio output transformer and to the plate of one speech output tube instead of two. Hence we have an impedance of 4000 ohms, which works well into the 46 radio frequency amplifier. Here is a summary of the steps necessary to apply modulation to a transmitter:

1. Determine the power output in watts of the modulator (in this case 6 watts) by referring to the data charts supplied by the tube manufacturers, for the particular type tubes being used.

2. The plate power-input to the R.F. amplifier is twice this value; hence in our case it is 12 watts.

3. Ascertain the load resistance of

modulator and make sure it matches the R.F. amplifier. In order to find plate resistance of the R.F. amplifier divide the voltage by the current in milliamperes. Formula is $\frac{V}{MA}$. The load resistance of both must be approximately matched for maximum power transfer from modulator to R.F. amplifier.

Changes for 160 Meters

There are only a few slight changes to be made in the transmitter for 160 meter work. First, the coils: these are wound with bell wire, cotton covered, No. 18 or 20. The oscillator coil is wound with 30 turns, and the excitation tap is about one-third of the way up from the *cold*, or plate-blocking end of the coil. The exact turn must be determined experimentally, the idea being to keep it as near the *cold* end as possible, with the amplifier tube drawing only a few or no milliamperes, without the aerial connected. The amplifier coil has 35 turns, and is tapped at th center for the power supply clip. Both coils are wound on five-inch lengths of three-inch diameter tubing. The antenna coil is wound on a three-inch piece of the same diameter tubing, with 25 turns.

The Antenna

Next comes the antenna: For our purpose a very simple type of "sky-

In this installment of Mr. Victor's series describing Amateur Transmitters—How to Build, Operate and Install them, the theory and particularly the construction of a reasonably priced yet efficient "modulator" is described. Those interested in building an up-to-date "Ham" transmitter should study the previous articles which provide important data.

hook" is used. It should be a straight piece of wire, somewhere in the neighborhood of 150 feet in length, including the lead-in and ground lead. Ten feet more or less will make no real difference. The antenna is connected to the .00035 mf. aerial tuning condenser, the other side of which is connected to the 25 turn coil. The other end of the coil goes to *ground*. Try to get a good solid ground to a cold water pipe. This antenna ground arrangement is known as a Marconi system.

The antenna is coupled as follows: Tune the amplifier tank to the point where minimum current is drawn; then slowly turn the aerial condenser until there is a rise in current. Retune the tank for minimum again, which should be higher than before. Continue this until the set draws 40 mills (milliamperes) with the tank tuning at the minimum current point. This

is the proper load point, as the amplifier is now drawing 12 watts; if the voltage has been set at exactly 300.

The bias resistor on the 46 tube should be changed to 20,000 ohms, or better still, use a 45 volt battery as bias. Connect the plus of the bias battery to B minus on the power supply, and the B minus to the grid R.F. choke, in place of the resistor. If a battery is used, slightly better results will be obtained; connect a .002 mf. mica condenser between the plus and minus of the battery.

The Modulator

A modulator is really an audio frequency amplifier, such as is connected to any receiver after the detector. However, for transmitting purposes the amplifier must be capable of delivering 5 or 6 watts, which would be sufficient to run several large loud speakers. The particular modulator used is one which your author bought a short while ago for "public address" work. It is an excellent unit, very low in price, not running much over the ten dollar mark, including tubes. Likewise the type of tubes used are easily available and are very low in price. This particular amplifier is a manu-

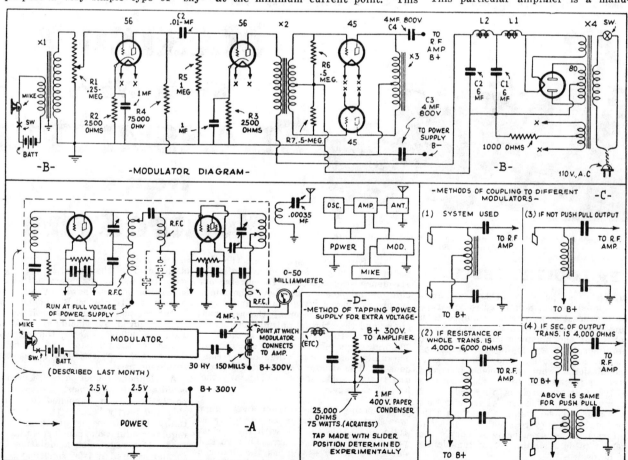

Fig. A shows block diagram with connection of modulator unit in the "Ham" transmitter set-up. Fig. B shows hook-up of the "modulator" unit here described. Fig. C shows possible methods of coupling a modulator to the R.F. stage. Fig. D, diagram showing method of tapping power supply to obtain exact voltage desired.

factured outfit, but there are many similar units available on the market at surprisingly low prices. For the fellow who prefers to "roll his own," a full description of the amplifier follows. It can be very easily built in "bread-board" form, and if the power supply used for the transmitter is good and husky, it might likewise run the amplifier. However, in most cases it will be advisable to build up a separate power supply, such as was described in the first article of the series, which appeared in the September issue.

The amplifier is conventional in design. A single-button microphone transformer is fed into a type 56 tube. This is the later model tube which replaces the 27 tube. If 27's are available they will work approximately as well. The first stage is resistance-coupled to another 56. This is transformer coupled to two 45 tubes working in push-pull. We couple to the primary side of the transformer with condensers. These are four microfarad units rated at 800 volts, although 600 volt units will stand up OK. All the resistors in the set are one-watt carbon units, except the 1000 ohm, 45 bias resistor, which is of 5 watts rating, wire-wound. The two 6 mf. filter condensers are rated at 600 volts. The filter chokes are rated at 30 henries, 150 M.A. each. A regular type 80 full-wave rectifier is used to supply the high voltage direct current. The *gain control* in the circuit of the first audio stage is a 250,000 ohm potentiometer. A *constant-current* choke is used in the plate lead of the RF amplifier. The rating of this choke is 30 henries at 150 milliamperes. Extreme care should be taken that a well-made, husky power transformer and modulation choke are used.

The microphone transformer has a primary input of 200 ohms, the same as the resistance of most good standard single-button microphones. The amount of battery used with the microphone depends on the

unit itself. The particular mike used is an RCA-Victor unit, which has been commonly selling for less than two dollars. Ordinarily, 3 volts are used, but for higher output, and to enable one to speak further from the mouthpiece, as much as twenty-two and one-half volts may be used without harming the "mike." A single-pole switch should be used to turn off the mike current "when listening to the other fellow."

There are several coupling arrangements shown in the diagrams, which will cover the problems arising with any type of amplifier used. Connect the transmitter power supply minus to ground. Likewise run the lead from the other side of the condenser that goes to the center-tap of the push-pull modulator output transformer, to *ground*. If a voltage divider (resistance) is used to get the exact 300 volts, be sure the tap is *by-passed* to ground by a one mf. 400 volt condenser.

Tuning Up!

The transmitter is set up with the 160 meter coils in it. Modulator, power supplies, (or supply, if only one is used for both oscillator and amplifier), and microphone are all hooked up. By means of the monitor, tune the oscillator to the part of the band in which operation is desired, tap the excitation coil one-third of the way from the "cold" end of the oscillator plate coil, and neutralize the amplifier according to instructions given in the previous issue. While neutralizing, the modulator should be *on*, but with the volume control turned all the way off. Likewise be sure that the antenna is *off* while neutralizing. Next tune the amplifier for minimum current, and adjust the antenna coupling until the plate meter reads 40 mills, (MA.), or whatever the proper value for the modulator is. As an example, if a ten watt modulator were used, the coupling would be adjusted until the plate current registered approximately

67 mills (M.A.) at 300 volts to obtain 20 watts of input power to the amplifier. With the antenna coupled, run the *gain* all the way up, and check with the monitor. At zero beat, the voice of a person talking into the microphone should be heard clearly and distinctly. If instructions and the rules set down have been scrupulously followed, and the proverbial grain of "horse-sense" has been used, there will be no trouble encountered. 73's and if you do strike any "snags," write to me and I will be glad to do what I can to help clear them up. However, please enclose a stamped, self-addressed envelope, as last month's mail ate quite a large hole in my pocketbook for stamps, not to mention stationery.

Parts List

1—Acratest microphone transf. X1.
1—Acratest push-pull input transf. X2.
1—Acratest push-pull output transf. X3.
1—Acratest power transformer 400-0-400 X 4
    5 V.—2 ½—2 ½.
3 30 henry 150 M.A. Acratest filter chokes
    (L1, L2, L3).
C1, 2—6 mf. 600 V paper or electrolytic condensers.
C2-1—Acratest .01 mf. bypass condenser.
2—5 prong sockets.
3—4 prong sockets.
C3, 1—4 mf. 800 volt condenser.
R1—250,000 ohm variable potentiometer, Acratest.
R2—2000 ohm resistor, Lynch (International).
R4, 1—75,000 ohm resistor, Lynch (International).
R5, 1—1 meg. ohm resistor, Lynch (International).
R6 & 7, 2—5 meg, resistors, Lynch (International).
1—1 watt carbon resistor—Acratest.
R8-1—1000 ohm 5 watt resistor.
R9-1—20 ohm CT (center tap) resistor.
(Note: The complete modulator as shown in the photograph and referred to by the author is manufactured by Federated Purchaser.)

The author with his transmitter constructed from an old neutrodyne receiver.

MAKING A
Short-Wave
TRANSMITTER
from a
Neutrodyne

By LOUIS F. LEUCK

◗ THIS is the story of how an antiquated 5-tube, 3-dial neutrodyne type of broadcast receiver was changed to a low-power amateur phone and code transmitter of the master oscillator - power amplifier type, employing two stages of speech amplification and 100 per cent modulation. From an obsolete, discarded receiver to one of the most modern types of transmitters surely is "reversed radio" in my opinion. The type of receiver that was used may be found in many an attic or purchased very reasonably. Sufficient information is given below to enable the reader to do a similar job of remodeling if he desires.

Reversing Receiver Into Transmitter

In making the change the wiring was altered, but practically all the parts of

the receiver were used except the detector coil and coil form. The two radio frequency stages became the *master oscillator* and *power amplifier*. The audio frequency system remained an audio system, the detector becoming the *first amplifier* and the original output tube being elevated to the position of *modulator*. The *antenna series tuning condenser* was originally the detector stage tuning condenser. The *master oscillator tube, coil and tuning condenser* was originally the first R.F. stage.

In making the conversion an attempt was first made to trace all wiring carefully and so do the job with the fewest possible changes of wiring. This worked out quite nicely as far as the filament wiring was concerned, and also fairly well for the balance of the changes in the audio frequency system. In fact, it is really important that the connections

to the audio frequency transformers remain poled as they were in the original set. If this is not observed there will be a tendency to "howl" at some audio frequency. If any such tendency exists in the completed transmitter it can usually be cured by placing a resistor of about 100,000 ohms across the secondaries of each of the transformers. If the tendency still persists, lower resistances should be used.

A telephone induction coil serves as a *modulation transformer*. This was used partly because it happened to be available and partly because there would actually have been too much gain if a regular modulation transformer had been used. If a regulation modulation transformer is used it is permissible to omit one of the stages of speech amplification, though the operator may have to raise his voice a bit above normal. An automobile ignition coil may be pressed into

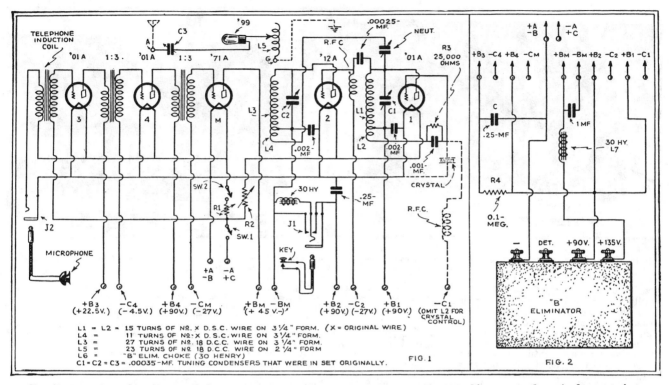

L1 = L2 = 15 TURNS OF Nº. X D.S.C. WIRE ON 3 1/4" FORM. (X = ORIGINAL WIRE.)
L4 = 11 TURNS OF Nº. X D.S.C. WIRE ON 3 1/4" FORM.
L3 = 27 TURNS OF Nº. 18 D.C.C. WIRE ON 3 1/4" FORM.
L5 = 23 TURNS OF Nº. 18 D.C.C. WIRE ON 2 1/4" FORM
L6 = "B" ELIM. CHOKE (30 HENRY)
C1 = C2 = C3 = .00035-MF. TUNING CONDENSERS THAT WERE IN SET ORIGINALLY.

FIG. 1

FIG. 2

Circuit above shows how Mr. Leuck hooked up the microphone and key connections to his revamped neutrodyne receiver, which then served him as a short-wave phone and code transmitter. Fig. 2, at right, shows connections of "B" eliminator.

service as a modulation transformer if no other is at hand, without introducing appreciable distortion.

All three of the tuning coil forms were removed and the detector coil form discarded. One of the others was then rewound with 30 turns, center-tapped, of the *original* wire. This became the plate and tickler coils of the *master oscillator* (L1 and L2 in Fig. 1). Twenty-seven turns of No. 18 D.C.C. wire were wound on another form, which serves as the plate coil of the power amplifier. Eleven turns of the original wire were wound on as the tickler. These are designated as L3 and L4 in the figure. The grid coils of this particular receiver each had 53 turns originally. If a receiver having different size coil forms and a different number of turns is to be remodeled, the correct number of turns can be arrived at by taking the same *ratio* of turns for the various coils as was done in the set described here. In order that the set may be properly neutralized, L3 and L4 must be wound in the same direction. Another way of saying it is that L3 and L4 are essentially a tapped winding. The connections of the master oscillator and power amplifier plate coils to their respective plates should be made in the same way. For example, the plates should be connected to the top ends of the coils in both cases. If this is done and both coils have been wound in the same direction there will be no trouble experienced in neutralizing. One of the original neutralizing condensers is satisfactory if it doesn't happen to be too small, as it was in this set.

Note that in a receiver the grid coil is tuned, while in a transmitter the plate coil is tuned. This makes it necessary to switch plate and grid connections on both of the R.F. tubes. That is one of

the reasons why it is best to just cut all wiring except filament leads away from these two tubes and their associated coils and rewire according to Fig. 1. This receiver happened to have two rheostats, two jacks and two switches and so use was made of all of them. All that is really necessary in this line is one filament switch (SW1) and a single fixed resistor which will handle five tubes. With the arrangement shown the Heising modulation system choke is short-circuited when the key is pushed in the jack for code work. A switch could be arranged to do this if keying impacts proved to be too noticeable with the choke in the circuit.

Tubes and Voltages

Three of the tubes are '01A's, the power amplifier is a '12A and the modulator is a '71A. With 135 volts on the plate of the modulator tube, 22½ on the first audio tube and 90 on the remainder, the whole outfit operates with rather high efficiency. A consideration of the rated output of the '71A and its plate resistance at 135 volts, and also that of the '12A at its operating voltage of 90, shows that conditions are just about right for 100 per cent modulation without overloading or overworking the '71A. The load resistance relations between the two tubes are just about ideal for the '71A to do its most effective work as a modulator. An additional 45-volt battery serves as a voltage booster for the modulator tube. Its positive is connected directly to the plate of the '71A and its negative is connected to the 90-volt plate lead to the '12A on the plate side of the Heising modulation choke. (A "B" eliminator filter choke is used as a Heising choke.) A "B" eliminator may be used as a source of plate supply if desired. This requires an additional 30-henry

choke and a couple of 0.25 to 1 mf. condensers. The connections are shown in Fig. 2.

This transmitter was intended for low-power work and for use in places where it is necessary to use battery power. It is easy on both the plate and filament batteries.

Tuning and Neutralizing

The setting of the master oscillator tuning condenser determines the wavelength. With the coils as given, the set will tune down to the "80-meter band" also, but will not be operating with "high C" which is desirable for stability of frequency. The first step in the tuning process should be to set the oscillator frequency within an amateur band. With the master oscillator tuning condenser set at 50 the wavelength is around 160 meters. This may be checked approximately by listening to other transmitting amateurs and comparing frequencies if no frequency meter is available.

The next step is *neutralization* and since this is something of a mystery until one has once successfully accomplished the feat, some pointers will be given. The reason for the elaborate row of Fahnestock clips along the rear of the transmitter now becomes apparent. Since the transmitter's power is low, it was found advisable to use the D.C. plate meter method of neutralization. In neutralizing the power amplifier with this method, its power should first be cut off by removing the 90-volt lead at the clip. A 25 ma. meter should then be connected temporarily in series with the 90-volt lead to the master oscillator. When oscillating properly without load, its plate current will be 6 to 8 ma. Next the plate circuit tuning condenser dial of the power amplifier should be rotated back and forth. When resonance with the master

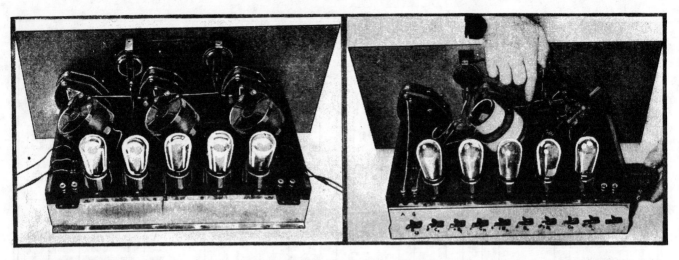

Two interesting photos of Mr. Leuck's short-wave phone and code transmitter constructed from a once-famous neutrodyne receiver, which he hauled down out of the attic.

oscillator is passed, the needle of the milliammeter will "kick." The neutralizing condenser should be adjusted until the "kick" is absent, or nearly so. Twirling of the dial back and forth should alternate with adjustments of the neutralizing condenser until the desired "no kick" position is found. The power amplifier is then neutralized. The milliammeter should then be changed to the plate lead of the power amplifier and its plate power applied. Its plate circuit must be tuned to the same frequency as that of the master oscillator. There will be a sharp downward dip of the milliammeter needle at this point.

The Antenna

Now we have arrived at the point where the radiating system should be connected. A wavelength of 160 meters requires an antenna approximately 120 feet in length. The ground should be as short and direct as convenient. If it has an appreciable length compared with that of the antenna, the antenna should be shortened just the length of the ground lead. If a counterpoise is used in place of an antenna, it should be the same length as given above for the antenna and may extend in any direction, but the opposite direction is preferable. The radiating system is tuned to the transmitter frequency by means of the antenna series tuning condenser. There will be a sharp increase of plate current to the power amplifier as the radiating system is tuned through resonance. If the plate current is high at all settings of the antenna series condenser the coupling is probably too close. This means that the number of turns in use in the antenna pick-up coil should be reduced. If connecting the radiating system and tuning it has no noticeable effect on the plate current to the power amplifier, the system is probably too long or too short.

No antenna current indicator is necessary. If one is desired the filament of a deactivated 199 tube in series with the antenna will serve. It will glow at just about normal brilliancy with the antenna current that this transmitter can supply. The silvery coating within the tube may be evaporated by holding the end of the bulb over a hot flame for a time and so making a "window" through which the filament may be more readily observed.

Does the set really work? Well, on the very first test a station thirty miles away was "worked," who gave a very fine report. Next a station fifty miles away was "worked" and a similar report received, and this was followed by a report from a station nearly a hundred miles away. Not bad for daylight work and the very first time on the air! The night range is much greater of course. A pleasant surprise was that duplex telephony was possible with this transmitter. On account of the low power it is possible to transmit and listen-in at the same time and in the same band.

Tube Data Chart

Tube No. in fig. 1	Type of Tube	Tube Use	Grid Bias	Plate Voltage	Plate Curr. mills
1	201A	M. Osc.	25,000 ohm g. 1.	90	12
2	112A	P. Amp.	—27	90	14
3	201A	1st A. Amp.	(—1)	22½	0.5
4	201A	2d A. Amp.	—4½	90	2.5
M	171A	Mod.	—27	135	17

With a completely battery operated set, all the above values except plate currents are predetermined. The plate currents to the master oscillator and the power amplifier are determined by adjustments and load. Antenna coupling, etc., should be varied until the above values are approximated. The above values are also helpful when a "B" eliminator is used

Transmitter Plate Supply from Ford Coils

● FOR the fellow who has no A.C. current at his elbow and has to rely on a bank of "B" batteries for the plate supply for his transmitter, a good way to obtain the current is to use the ordinary ignition coils taken from an old Model "T" Ford car.

By using two of these coils with 12 to 18 volts on the primary, from three to five hundred volts can be obtained.

I have been using two of these coils with 12 volts on the primary and have gotten fair "DX." The type of transmitter I use is a series-feed Hartley, but any other type may be used with the same results. In about four months of operation with these coils all but the 6th and 7th districts have been worked on the 80 meter band. I always get fine reports on signal strength and generally get the report that my signals are "pdc" and sometimes I get a report that my note is "xtal dc."

The vibrator on the coil must be made to vibrate at a higher frequency to get higher voltage. This is accomplished by cutting a piece of postal card large enough to be doubled and put between the vibrator and the magnet of the coil. The frequency then is adjusted by the little nut on the coil to a point where the vibrator has about

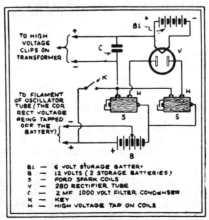

B1 — 6 VOLT STORAGE BATTERY
B — 12 VOLTS (2 STORAGE BATTERIES)
S — FORD SPARK COILS
V — 280 RECTIFIER TUBE
C — 2 MF 1000 VOLT FILTER CONDENSER
K — KEY
H — HIGH VOLTAGE TAP ON COILS

How to use two Ford spark (ignition) coils to obtain high voltage for plate supply of transmitter. The tube "V" rectifies the secondary voltage.

a 500 cycle note which is pleasing.

A separate battery must be used for the rectifier tube if it is one of the filament type, such as the 280, which is used at my station. A BH rectifier may be used, however, if desired.

If the filament supply for the oscillator tube of the transmitter is gotten from the same battery as the supply for the coils the center-tap connection on the filament leads to the oscillator will have to be taken loose, because this connection will already be made at the battery when it is connected

The keying is in one of the leads to the Ford Coils instead of at the transmitter proper as is generally the case.

The filter condenser is very essential and if it is left out an A.C. tone will result in the note of the transmitter. If it is found to hold a charge large enough to make the note of the transmitter have "tails" or a backwave on it, which can be told by listening on the monitor, a relay may be connected so that when the key is pressed, the lead to the plate of the oscillator tube will be completed and when the key is released it will break the circuit.

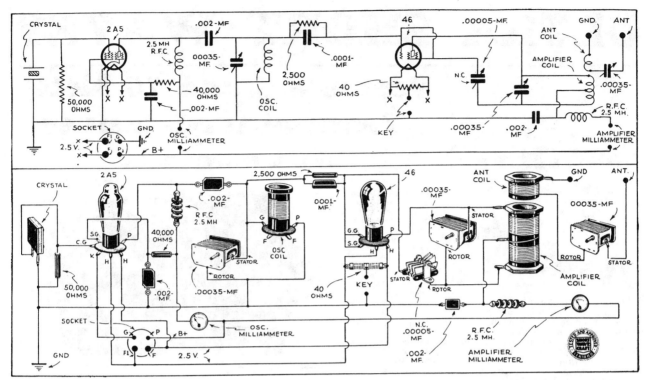

Both schematic and picture diagrams are given above, showing how to build the crystal-controlled transmitter here described at length by Mr. Victor.

HOW TO ADD CRYSTAL CONTROL

Even a self-excited oscillator is still O.K. provided pure D.C. power supply is used, and the frequency is carefully checked, but the safest thing to do is to get some other type of control that guarantees absolute stability and assures that the frequency is well within the band. *Crystal-control* is the answer to this problem. Provided a good quartz crystal is used, the transmitter will operate only on the frequency of the crystal, and thus forever eliminate all worry about being out of the band. Also with "xtal" it is much easier to get a pure note, even though little filter is used in the power supply and the rig is maladjusted. Try getting a good steady note with something "haywire" in a self-excited xmittr. 'Nuff sed!

Technical Description

The transmitter proper uses two tubes, a 2A5 tube as the oscillator and a type 46 tube as the amplifier. The 2A5 is one of the newer type tubes recently released. It is a pentode similar to the 47 tube but with an indirectly heated cathode. The efficiency with this tube is very high and voltages as high as 400 or even 450 can be used without straining the crystal. The 46 is used as an amplifier because it has several very good features. Firstly it requires very little excitation to produce high output, as it is a high mu tube. Likewise it needs no battery bias, which is a great saving, and eliminates one of the nuisances around an amateur station. The 46 is an excellent *doubler* tube, that is a tube to double the frequency of the xtal for operation on the higher bands, if it is ever desired.

Power inputs to a 46 can be as high as 30 or even 35 watts, especially when there is no worry about frequency stability, which is taken care of in this set-up by the xtal. This rig is designed for the 80 and 160 meter bands, which are the best bands for the fellow just getting up code speed, or wishing to do "message handling." The 160 meter code band extends from 1715 kilocycles to 1825 kc. Those that want to work both 80 and 160 meters without using more than one xtal should get one rated between 1755 and 1825 kc. Using a 160 meter xtal the 46 amplifier tube would be working as a doubler on 80 meters.

The Layout

The transmitter is mounted on a varnished board two foot by nine inches. The layout of the parts is exactly like the wiring diagram. From left to right the parts are: crystal holder, 2A5 tube, oscillator tank condenser, oscillator coil, 46 tube, amplifier tank condenser, amplifier and antenna coils, and antenna condenser.

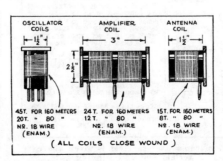

Behind the 2A5 tube are mounted the 40,000 ohm voltage dropping resistor, the oscillator R.F. choke, and the voltage dropping resistor by-pass condenser. In back of the oscillator coil is the excitation coupling condenser, and bias resistor for the 46. Plug-in jacks are used for the amplifier because they make a very neat arrangement and facilitate changing bands quickly. Behind the amplifier tank condenser is mounted the R.F. choke and by-pass condenser for that circuit. Along the back edge of the set the parts are as follows, reading from left to right: two binding posts for oscillator milliammeter, four-prong plug for power-supply cable, two binding posts for amplifier milliammeter and the two binding posts for *aerial* and *ground*. Filament, plate and ground leads are run under the board to give a neat appearance.

Parts

Receiving type parts are used throughout this transmitter, but care should be taken that they are of the best construction available. Make sure the variable condensers have good spacing and that all the fixed condensers are of the mica type, rated at least 400 volts. The R.F. chokes play a very important part in this set and should be of a type designed for transmitting use, although some short-wave receiving chokes work very well. The milliammeters need only be the cheap, "less-than-a-dollar" type. Both are 0-100 M.A. scale instruments. All the resistors are one-watt carbons, but be sure they are R.M.A. standard, as large quantities of poor resistors have recently been "dumped" on the market.

ULTRA SHORT WAVES

Portable 5 and 10 Meter Transmitter-Receiver

By L. L. HOTSENPILLER

Photo at left—portable 5 and 10 meter combination "transmitter-receiver" in actual operation in the field. A telescopic antenna is very desirable for the purpose. Plate and filament supply is readily obtained from batteries.

Photo at right shows top view of the portable 5 and 10 meter transmitter-receiver.

● IT is the purpose of this article to present a combined phone transmitter and receiver, to be used as a portable or as a complete "home station" working in the five and ten meter band. It is contained in a five by nine by six aluminum box and is readily set up for operation. In a favorable location a hundred thirty-five to one hundred eighty volts is sufficient B power to enable communication to be carried on over distances up to ten miles. When located on a high point, such as a mountain top, or when communicating with a plane much greater range can be expected however. To place in operation all that is necessary is to attach the antenna shown, apply proper plate and filament voltages for the tubes used. If operation is desired as a receiver the selector switch located on the front is turned to that position. To change to transmitting the selector switch is simply turned to that position.

Constructional Details

Insulation and careful layout are much more important in ultra-short wave work than in the customary short wave band. Failure of five and ten meter receivers and transmitters can often be traced directly to poor insulation in one of the component parts. All coil forms, condensers, and sockets, should be constructed of Isolantite or an equivalent material. The circuit shown consists of a No. 30 or No. 37 arranged in a series tuned, series-feed, Hartley circuit. When switched to the transmitting position, the oscillator is plate modulated by a No. 33 or No. 38 pentode. When receiving, a coil (L4) is introduced in the plate circuit of the oscillator tube, together with (L3) these coils cause additional oscillations to occur at 100 kcs. thus producing super-regeneration. The pentode modulator is changed into an audio frequency amplifier which will give loud-speaker operation on most signals if desired.

Either the two volt No. 30 series or the six volt No. 37 series tubes may be used with practically no change in the wiring except the substitution of one five prong socket. If the portable is to be operated in an automobile or plane it is suggested that the six volt tube be used. Identical results will be had with either series. It is recommended that the new 45v. *midget* "B" batteries be used. Due to their long life and small size these batteries enable any portable to compete on even terms with a permanent station. Six of these batteries delivering 270 volts occupy the same space as one standard 45 volt battery.

The portable is built on a four and three-quarter by eight by two inch steel chassis. It slips into a five by nine by six aluminum box. The tuning condenser and selector switch is located on the front panel. The filament switch on the left side of the box with the headphone and microphone jacks on the right side. Battery connections terminate at a six prong socket at the rear of the chassis. The socket for the No. 38 or No. 33 pentode must be held five-eighths of an inch below the chassis to allow clearance for the top of the tube. The socket for the coil must be supported one-fourth inch above the chassis.

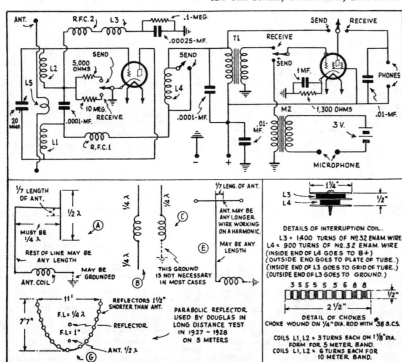

Schematic wiring diagram for the combination transmitter-receiver is given above, together with details of different styles of aerials and coil winding data.

The antenna coil is supported by two, one-inch isolantite insulators, located on the back of the box, the ends of the coils passing through the base of the insulators.

The interruption coil (L3) (L4) is located to the right of the selector switch when looking at the bottom of the chassis. The size of grid leak used on the interruption coil is fairly critical and successful super-regeneration depends on this resistance and the choke r.f.c. No. 2. The by-pass condenser shunted across the phones is essential before oscillation will occur and should not be left out.

If the receiver does not oscillate over the entire dial the fault probably lies in the r.f. choke No. 1. By removing a few turns from the choke and observing the change in the *dead-spot* it can be determined whether more or less turns are needed on the choke.

Operation

As stated previously the main factor in the ultra-short wave transmission is the location of the station. Often the signal that cannot be heard, or may be poorly heard at twenty foot elevation will be heard R7-R8 at 30 or 40 foot elevation. Due to the small physical size of a half-wave 5 meter antenna, it is very easy to erect a 30 or 40 foot mast, suitably guyed, and thus produce a satisfactory transmitting and receiving antenna at a reasonable cost. For the home station where the transmitter is probably located in the basement, the arrangement shown in A (Fig. 2) is probably the best. All feeder and transmission lines should be supported away from objects such as metal guttering, trees, and buildings. Right angle bends in the antenna should be avoided. The antenna itself should be constructed of new No. 12 or 14 enameled wire.

Antenna Systems

A number of practical antenna systems are shown in Fig. 2. However there are any number of other systems that work quite as well, although some are more difficult to tune. Any antenna that can be used in any of the amateur bands can by proper tuning be used in the 5 or 10 meter bands. For portable operation system (B) (C) are suggested. The parabolic reflector shown in (G) is highly recommended when conditions permit its use.

Ten meter antennas are shown in Fig. 2 and are very similar to those used on 5. The same attention must be given insulation and clearance of all wires. If transmission is desired to a fixed point, the signal may be increased considerably by the addition of a few reflector wires as shown in Fig. 2, producing a sharp beam in the direction the system is pointed.

To determine the length in feet of a half-wave antenna multiply the wave length desired by 1.56 except if operation is desired on five and one-half meters the length in this case would be $(5.5) \times (1.56) = (8.58 \text{ feet})$.

System A is suitable for fixed location. Systems (B) (C) are more desirable when the transmitter can be located at the antenna as in portable use.

System (E) is used where an existing transmitter antenna is already in place. It may be operated very successfully in the 5 or 10 meter bands by simply operating it on the proper harmonic. A parabolic reflector (6) for use where transmission is wanted in a certain direction. It is understood that systems A-B-C-E may be either horizontal or vertical also that any antenna that has an harmonic following in the five meter band may be used. Its length may be 8'-16'-32'-64'-138'. The type of feeder, of course, depends on the individual location. The transmission line shown in Fig. A-E are the simplest types of feeders.

List of Parts for Portable

2 .01 mf. by-pass condensers
2 .0001 mf. mica condensers
1 .00025 mf. mica condenser
1 5000 ohm resistor, Lynch, (International)
1 1300 ohm resistor, Lynch (International)
1 .1 megohm resistor, Lynch (International)
1 10 megohm resistor, Lynch, (International)
2 5-prong sockets, National or Hammarlund
1 Audio transformer, National
1 Single-button microphone transformer
1 100 k.c. interruption coil, (see coil table Fig. 3)
1 20 mmf. Hammarlund midget cond.
1 National type "A" vernier dial
2 speaker terminal strips, Eby
1 S.P.S.T. switch
1 2-point 4 gang switch
1 6-prong socket, Eby
1 6-prong plug, Eby
1 aluminum box—5x9x6 inches
1 steel chassis 4¾x8x2 inches
1 dial or condenser extension
2 1" insulators, National
2 ultra short-wave coil forms, Hammarlund 6-45 volt batteries
2 3 volt batteries
1 type 30 or 37 tube, Gold Seal (Arco, Van Dyke)
1 type 33 or 38 tube, Gold Seal (Arco, Van Dyke)

The "BEARCAT-3" 5-Meter Super-regenerative Receiver

By CLIFFORD E. DENTON

● CROWDED channels and the desire to explore the little known ultra high frequency regions has led to many interesting developments in the 5-meter band. Increased activity on the part of amateurs and other investigators has resulted in a great rush to start things in this band. Many interesting uses have been found for two-way intercommunication over short distances. For instance, two amateurs living in the same town or city will find that reliable transmission and reception can be carried on with a minimum of interference and this tends to relieve the congestion which exists on the lower frequency channels.

An example of how two-way conversations can be carried on is indicated in Fig. 1.

Stations A and B are located in the same city, say, New York, and stations C and D are located in some other town about 150 miles from A and B. Let station A transmit on the 80-meter band to station C. Station C listens to A and at the same time feeds the output of his 80-meter receiver into his 5-meter transmitter. Station D picks up the signal from C on 5 meters and transmits to station B on the 80-meter band. Thus, station A can talk to stations B, C and D at the same time. Note should be made of the fact that station A can converse with station B through stations C and D or direct on the 80-meter band.

A little thought will show that all parties can hear the remarks of any one station and can break in on the conversation without changing the adjustments of their receivers or shutting off their transmitters. This is indeed a very nice scheme and the beauty of it is that several fellows use it and commend it most highly.

Circuit Design

Three tubes are used in this design and the 6-volt automotive type has been selected as being the best for the purpose.

The detector tube, which is mounted directly in back of the tuning condenser, is one of the 37 type tubes. Note that the plate potential applied to this tube must pass through the resistance 13 and serves the dual purpose of controlling the regenerative action of the detector and limiting the amount of energy fed into the detector from the local oscillator.

The frequency of the signal fed to the detector is determined by the size of the coil 18 and the condenser 16.

The local oscillator, which supplies the quenching frequency, derives its power from the tap marked B-plus 67.5 and the proper operation of the set will depend on the obtaining of the proper value of this plate voltage. Voltages from 22.5 to 90 should be tried, as different tubes may have different characteristics as far as power output is concerned. Select the voltage giving the smoothest control of resistor 13. With the circuit as is, there is a compromise between the exact operation point for maximum sensitivity of detector action and proper voltage from the local oscillator.

The output of the detector is fed through the transformer to the grid of the pentode output tube. This raises the power sensitivity of the set as a whole and if a suitable coupling device is used to couple the output of the 38 to the reproducer, satisfactory quality will result.

Construction is a simple matter, as the parts are not numerous and there is plenty of room even though the chassis is very small. Drill and fold the chassis as per drawings. Many fans may want to purchase a finished chassis, which can be done.

Mount the tuning condenser in the center as shown in the photographs. It

A GOOD 5-meter receiver is in big demand just now, with hundreds of amateurs getting into operation with their 5-meter transmitters. Not only must the 5-meter receiver be selective, but it must also possess powerful amplification properties. The "Bearcat-3" possesses these qualities. Data are also given on the construction of the new antenna resonance coils which greatly increase the signal strength.

Max Pearlman listening to the mysteries of the 5-meter "ham" band as the waves roll in on the "Bearcat-3."

would be wise at this time to check up the drilling of the front panel, noting if the shaft of the tuning condenser lines up with the bushing of the tuning dial. The sockets can be secured in place, as well as the audio frequency transformer.

Most of the remaining parts, such as the resistors, can be held in place by the wiring. It would be wise to bolt the by-pass condensers to the under part of the chassis so as not to place too great a strain on the wiring.

The tuning dial, which is mounted on the front panel, can be locked to the condenser shaft and then the set can be wired.

Wiring

Little need be said as to the wiring. Do not use long leads in the detector circuit. There is a definite reason for using the type of socket for the detector —to insure short leads. Grid and plate leads must be as short and as clear from surrounding metal objects as possible. It is not necessary to use the same care with the balance of the set because the frequencies involved are much lower.

Coil Data

The specifications for coils 4 and 5 are given below:

	No. of turns	Wire size	Spacing
Coil 4	7	14	1/16-inch
Coil 5	7	14	1/16-inch

Coil 17.—Coil 17 consists of 650 turns No. 36 double silk covered wire, wound on a small bobbin ½-inch in diameter and closely coupled to the coil 18.

Coil 18.—The grid coil is number 18 and consists of 1,000 turns of the same size wire used on 17. This is wound in the same direction on the same bobbin and due to its small size can be bolted into place under the chassis.

Radio Frequency Choke No. 9.—This is a small choke and care should be used in building it. As the frequency range to which the receiver responds is very high, it is necessary that the distributed capacity of the winding be kept at a minimum. A satisfactory choke can be made by "jumble-winding" 30 turns of No. 36 double silk covered wire on a bobbin ½-inch in diameter.

A detail drawing is shown in Fig. 3 and should be studied carefully. Note that the coils are wound in the same direction and when they are mounted be sure that there is no change in the winding direction between X and Y.

These precautions should be exercised in the construction of the set. It seems that most builders have trouble making detectors oscillate. If the constructor builds his own coils as shown, then the only thing that will prevent the proper operation of the set will be defective tubes or "B" batteries reversed.

Keep all leads between coils and detector socket as short as possible.

Operation

The set is tuned to an incoming signal and the resistance controlling the plate voltage on the detector is varied for the best results.

The adjustment of the antenna series condenser is important and should be done with care. The band spread condenser (3) should then be adjusted so that the band required is spread over the tuning dial.

Vary the size of the oscillator tuning condenser (16) until the proper quenching frequency is obtained. This is important, as the sensitivity of the receiver will depend to a great extent on the frequency of the local oscillator. Use the frequency which gives the best results.

When the receiver is working right, there will be a loud rushing sound in the phones or loud speaker, and as the signal is tuned in, this rushing noise will disappear. When the incoming signal is weak, some of the rushing sound will remain in the back-ground.

Many builders of 5-meter receivers have not obtained the maximum results and then turned around and condemned the whole idea. It is more than likely that their antenna systems had something to do with it.

Mr. Dana Griffin of N. Y. City has built a

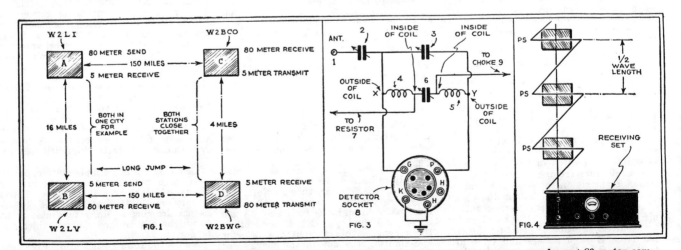

Fig. 1, at left, shows how 5-meter transmitters and receivers may be used in amateur stations to supplement 80-meter communication; Fig. 3 shows special part of the circuit in the super-regenerative receiver which requires accurate connections; Fig. 4, at right, shows positions for resonance wave coils along the receiving antenna.

device which permits the use of high vertical antennas for maximum pick-up and to develop the maximum signal voltage at the input of the receiver.

These units are called "phase shifters" and consist of a small coil and condenser capable of being tuned to the frequencies being received on the set. In general these circuits should be tuned to the center of the band on which the set is operating.

Figure 4 shows the voltage shift in the units after they have been tuned to the proper frequency in the band. It is a good idea to tune the "phase shifters" to the exact wavelength of the station being received.

The antenna can be as long as conditions permit. Run it straight up in the air, keep it free and clear from all obstructions. Place one of the phase shifters every 100 inches, starting 100 inches from the receiver antenna and ground posts. Use as many of these units as required and tune each one to the same frequency. This can best be done by building a small oscillator, calibrating it against some known 5-meter signal and then using this to adjust the phase shifters to the proper frequency.

Circuits for such an oscillator have been described in many of the past issues of SHORT WAVE CRAFT, so no further information should be necessary on this point. Many short-wave "bugs" have oscillators which will generate harmonics in the 5-meter band.

Some slight recalibration of the tuning condensers used in the phase shifters may prove necessary after they have been connected into the circuit of the antenna, the final adjustment being that of tuning for the maximum signal volume from some 5-meter transmitter.

Parts List

1 Antenna binding post (1).
2 Hammarlund equalizing condensers, .000035-mf. (2, 3).
1 Hammarlund midget condenser (6).
1 International Resistance Co., 1-watt, 2-meg. resistor (7).
1 Panel mount socket, 5-prong (8).
1 Radio frequency choke (9). See text for specifications.
1 Aerovox mica condenser, .001-mf. (10).
1 Medium ratio audio transformer (11).
1 Flechtheim by-pass condenser, .1-mf. (12).
1 Electrad 50,000-ohm potentiometer (13) with filament switch (28).
1 Flechtheim by-pass condenser, .1-mf. (14).
2 Wafer sockets, 5-prong (15, 19).
1 Mica condenser, .001-mf (16). See text.
1 By-pass condenser, .1-mf. or larger (20).
1 1,500-ohm resistor, 2 watts (21).
2 Output terminals (22, 23).
4 Binding posts (24, 25, 26, 27).
1 metal chassis and front panel.
1 Tuning dial.
1 Screen-grid clip.
Wire, etc.
Note—Coils 4, 5, 17 and 18 winding data included in text.

2 Eveready-Raytheon 37 tubes.
1 Eveready-Raytheon 38 tube.

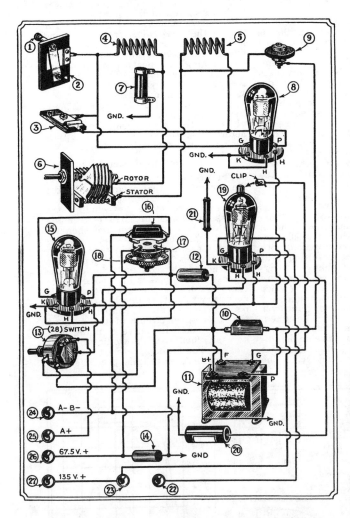

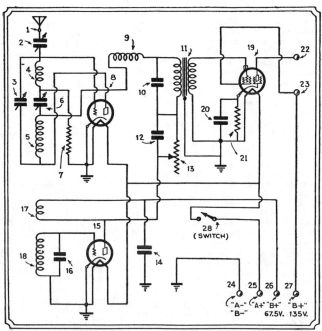

Above—Schematic diagram of 5-meter "Bearcat-3" receiver

Photo below shows the "Bearcat-3" 5-meter super-regenerative receiver here described by Mr. Denton. One of the new antenna resonance wave coils is shown just to the left of the receiver. This set uses but three tubes and works a speaker.

Left — Picture wiring diagram showing how easy it is to build the 5-meter "Bearcat-3" receiver. This receiver possesses high amplifying powers and good selectivity and uses but three tubes, it being possible to receive strong signals on a speaker.

Right—Details of antenna resonance coil and condenser.

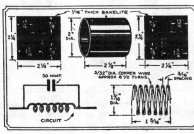

A 5-Meter S.W. Superheterodyne

By A. C. MATTHEWS,

Appearance of Mr. Matthews' 5-meter superheterodyne receiver which employs a single tuning dial; loud speaker appears in the background. The set uses seven tubes in all, including a rectifier.

1—Two gang variable condenser 100 mmf. cap. (Hammarlund) C1 and C2.
1—Midget variable condenser 18 mmf. cap. (Hammarlund) C3.
1—0.1 mf. 200 volt condenser.
2—Triple section 0.1 mf. 300 volt condensers.
1—0.001 mf. mica dielectric 200 volt condenser.
1—50 mmf. mica dielectric 300 volt condenser.
3—250 mmf. mica dielectric 300 volt condensers.
2—0.01 mf. paper dielectric 200 volt condensers.
1—0.02 mf. paper dielectric 200 volt condenser.
2—1.0 mf. paper dielectric 300 volt condensers.
2—2.0 mf. paper dielectric 200 volt condensers.
2—8.0 mf. dry electrolytic 500 volt condensers.
1—400 ohm ½ watt carbon resistor.
1—1500 ohm 1 watt carbon resistor.
1—5000 ohm ½ watt carbon resistor.
1—14,000 ohm 3 watt carbon resistor.
1—25,000 ohm ½ watt carbon resistor.
1—30,000 ohm ½ watt carbon resistor.
1—100,000 ohm ½ watt carbon resistor.
1—250,000 ohm ½ watt carbon resistor.
4—500,000 ohm ½ watt carbon resistors.
1—25,000 ohm 1 watt variable resistor with power switch.
1—200 turn universal wound coil, ½" form (Auto. Winding Co.).
1—85 mh. choke (Samson).
1—10 henry choke 35 ma. direct current.
1—20 henry choke 50 ma. direct current.
4—Six-prong sockets. Alden.
2—Four-prong sockets. Alden.
1—Five prong socket. Alden.
1—Power cord and plug.
1—Chassis (Blan—The Radio-Man).
1—Power transformer: Sec. Volts—2.5 volts c.t. 7.5 amps, 5.0 volts 2 amps, 700 volts c.t. 70 ma.
4—Tube shields for 58 type tubes.
1—Tube shield for 24 type tube.

● WITH the advent of actual broadcasting on the ultra short wave band between 43 and 80 megacycles, intensive receiver development has been taking place. The art has gradually progressed through the regenerative detector, super-regenerator stage, until at present the most satisfactory method is that of the *double-detector* or *superheterodyne*.

Tuned radio frequency amplification at such ultra short wavelengths is practically out of the question, since the low impedances encountered in the ordinary tuned circuit do not permit much amplification. Recent advances in tube design have resulted in decreased inter-electrode capacities. This is conjunction with the addition of an extra grid (R.F. pentode) has made it possible to realize some gain at very high frequencies, if extra precautions are taken in the circuit design. At its best, however, a tuned radio frequency receiver for these frequencies is complicated, due to the necessary design precautions that must be taken.

Superheterodyne

Briefly, the superheterodyne functions in the following manner. (Shown diagrammatically in Fig. 1.) The incoming signal frequency is mixed with a *local oscillator*. The resulting beat frequency, being lower than the original signal frequency, is therefore much easier to handle. The difference between the local oscillator and the signal frequency remains constant over the band for which the set is designed. Since the beat frequency remains constant, the design of a suitable amplifier having the desired characteristics is much easier than before. The

choice of the frequency difference between the oscillator and the incoming signal, however, is important and will be discussed further under the intermediate amplifier. Once the incoming signal has been transformed to a relatively low frequency, the design problem becomes simply that of a straight tuned radio frequency receiver with associated audio amplifier.

The development of this circuit for use in the *ultra high frequency* band has been rather slow. This has been due to the almost impossible task of maintaining the beating oscillator at a constant frequency. The success of the superheterodyne depends on the stability of this oscillator.

Having discussed the main difficulties to be experienced in the design of an ultra high frequency superheterodyne, we will now take up its design in a systematic manner.

First Detector—Mixer Circuit

The first detector circuit is tuned to the incoming signal frequency by the inductance L-2 and condenser C-1. The coils are made by winding the necessary number of turns (see table) or a one-half inch form and then removing the form. The wire size is rather large and this will tend to hold the coils in place. Pin jacks are soldered on the coil ends for convenience. This makes it possible to change coils in the event that it is necessary to shift to another frequency band. The oscillator is coupled through the screen-grid circuit of the 58 type tube, although inductive coupling may be used when a stable oscillator is employed. The author prefers the screen-grid method, since this precludes the

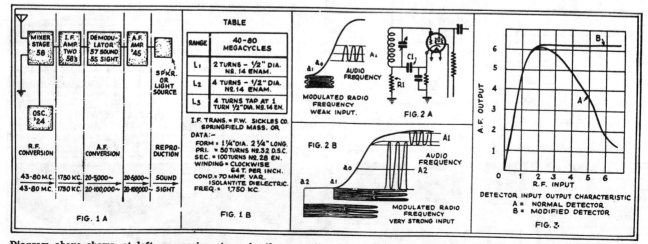

Diagram above shows, at left, successive stages in the reception of signals on a superheterodyne; coil winding data at Fig. 1B, while the graphs shown at Figs. 2 and 3 are used by the author in explaining the action of the receiver.

The 5-meter field is rapidly expanding. Many short wave "hams" operating in this field have undoubtedly found that one item badly needed was a good 5-meter receiver—one which would provide high sensitivity, suitable selectivity and sufficient volume to work a loud speaker. Mr. Matthews, author of the present article, is a prominent short-wave and television expert.

possibility of radiation through the antenna.

Oscillator

As has been said before, the oscillator is the heart of the ultra high frequency superheterodyne. The ordinary oscillator with inductive coupling, such as employed in the usual receiver, would be a complete failure in ultra high frequency work. The oscillator to be used must not only have a very high order of *frequency stability*, but also be capable of maintaining its intended frequency unaffected by the first detector circuit, with which it is connected. Frequency stability that is relatively impervious to changes in the supply voltage is necessary. The fact that its load circuit is subject to rather severe variations, since the oscillator is required to furnish power of a small order to the first detector, makes the oscillator requirements very severe to say the least. The degree of "pulling in" of the oscillator frequency with the tuning of the first detector unfortunately is greater as the frequency increases. In other words, the frequency stability of the oscillator decreases as the frequency increases. Therefore a combination that would be entirely adequate for broadcast reception would be entirely out of the question for ultra high frequency work. With the performance so dependent on a fixed frequency difference between the oscillator and the incoming signal frequency, it is easily seen that nothing but the most refined circuit design would be tolerable in this application. No doubt it is because of this fact that so much valuable time has been spent trying to improve on the straight regenerative and super-regenerative receivers.

Suppose we take a look at some commercial installation and see what precautions they take to maintain oscillator stability. Probably one of the best installations would be the trans-Atlantic receiver station of the R. C. A., at Rocky Point, L. I. In their diversity telephone receiving system used for picking up foreign broadcasts, they make use of a *buffer* or *coupling tube* between the oscillator output and the grid circuit of the first detector. This provides a high degree of oscillator independence but the additional tube makes for more complicated circuits and although it can be used for frequency doubling or tripling, it is hardly warranted in a receiver for Mr. General Public.

After having tried practically every type of oscillator circuit unsuccessfully, the *electron-coupled* oscillator was adopted. This oscillator, described by Lieutenant J. B. Dow in the December, 1931, *I. R. E. Proceedings*, has as good if not better all around *frequency stability* than the more complex oscillator-amplifier combination. The circuit employs a screen-grid tetrode; the cathode, control-grid and screen-grid forming the frequency generating circuits, while the plate is in the output circuit and is entirely independent of the oscillator frequency, since it is shielded by the screen-grid from the oscillator circuit proper. (The screen-grid is at ground potential, as far as radio frequency is concerned.) The coupling to the load circuit is therefore electronic rather than inductive or capacitive since the plate is effectively isolated by the screen grid. This reduces the interlocking effect between the oscillator and first detector tremendously and in no small measure

Top view of the 5-meter superheterodyne.

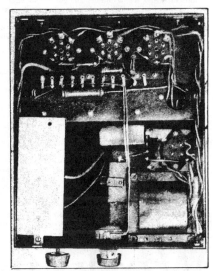

Bottom view of Mr. Matthews' 5-meter "super."

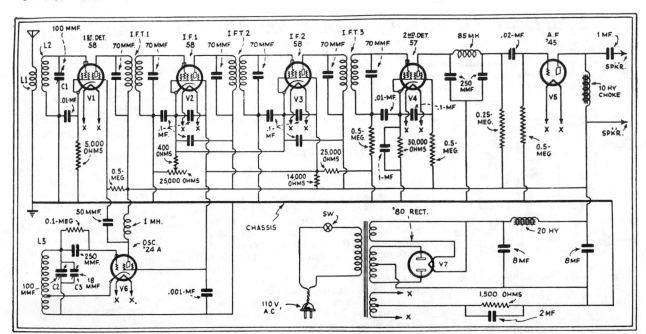

Here we have the complete wiring diagram of Mr. Matthews' superheterodyne designed for 5-meter reception. We believe that this is an ideal receiver for the average short wave fan interested in 5-meter reception, as it uses but seven tubes with rectifier.

contributes to the excellent frequency stability of the oscillator. The screen-grid being at ground potential (R.F.) necessitates operating the cathode above ground. This is completely satisfactory and when using uni-potential cathode type tubes having indirectly heated cathodes, no deleterious effects can be detected by having an R.F. potential difference between the cathode and heater. Although it might seem that the cathode-heater capacity might interfere with the satisfactory functioning of the circuit it compensates rather than incapacitates the frequency stability during the warming up period of the tube. A slight varying of the oscillator frequency with detector tuning has been noticed when using the fundamental of the oscillator; however, this can be eliminated by using the second harmonic of the oscillator to heterodyne with the incoming signal frequency to create the intermediate frequency beat.

Careful shielding of course is necessary if the oscillator is to be operated at full efficiency, since any coupling of the oscillator tuning circuit will defeat the excellent qualities of the system.

The coil data are given in table 1. The condenser C-2 determines the approximate frequency while the trimmer condenser C-3 acts as a vernier adjustment. Eventually when ultra high frequency super-heterodynes become as numerous as the regular broadcast variety, the receivers will then be truly *single control*. The vernier, however, is not a serious hardship to endure and without it the performance would surely suffer. The screen-grid voltage should be approximately 67 to 90 volts, the lower value being recommended for stability. The value of the gridleak should be 100,000 ohms for best operation.

Now that the degree of electrical stability far surpasses any other oscillator combination, it behooves the experimenter to exercise particular care in the mechanical construction to insure rigid mounting of the component parts which might affect the frequency stability. If ordinary precautions are taken in the construction of the oscillator, even the dyed-in-the-wool experimenter will witness a thrill at the stability of the electron-coupled oscillator.

Intermediate Frequency Amplifier

The choice of an *intermediate amplifier* is one of all importance, since the main characteristics of the receiver are obtained in this section. The intermediate frequency must be low enough so that sufficient gain can be realized with a good degree of selectivity. The frequency characteristic of course must also be considered, otherwise the quest for selectivity would result in undue attenuation of the high audio frequencies and poor quality would obviously result. However, there is another consideration to be taken into account in ultra high frequency work. Suppose the intermediate frequency was of the order of 400 kc. and the oscillator tuned to a frequency of 40,000 kc. It can readily be seen that a variation of only 0.01% in frequency would amount to so much that the resulting frequency would not be amplified by the highly selective intermediate stages.

Now, let us suppose an intermediate frequency of 1,750 kc. or thereabouts was chosen. The percentage allowable variation in the oscillator frequency could obviously be much greater, without affecting over-all performance.

The intermediate frequency finally adopted in this application was 1,750 kc. I.F. transformers may be purchased already built, or the experimenter may build his own. In the latter case the coils from a short-wave receiver covering this band will be satisfactory. Small Isolantite dielectric condensers may be substituted for the larger air condensers formerly used for tuning. The I.F. amplifier in reality is a fixed-tune radio-frequency amplifier and its design is not unlike any other R.F. amplifier covering this band. Such circuits are not so selective that the fidelity will be impaired by side-band attenuation. This applies only to sound reception; the requirements for television reception are somewhat more stringent.

Second Detector

The *second detector* for voice reception is of the orthodox *plate detection* variety. A 57 tube is employed in a circuit designed particularly to eliminate detector overloading. This scheme does not entirely eliminate detector overloading in the strict sense of the word, but it does greatly extend the usable range of inputs to the detector, without suffering an appreciable reduction in rectified output. This particularly applies to signals of low percentage modulation which heretofore have given the most trouble in detector circuits.

Figure 2 shows the essential circuit in its simplest form, together with a graphical explanation of the how-and-why of the improvement. C-1 and R-1 are chosen so as to have a time-constant of greater duration than the period of the lowest audio frequency to be reproduced, yet sufficiently short in duration to follow the variations in amplitude of the modulated carrier. C-1 must also be of such a value that it will have no effect upon the tuning.

Figure 2-A shows a typical grid-plate characteristic of a *power detector*. Point a-o represents normal bias with no signal applied. Upon the reception of a modulated signal (50% mod. shown) this point moves to a-1 and rectification takes place, giving the audio frequency component in the plate circuit as shown. Such a signal would result in the same output in either a straight bias detector or the modified circuit used here. Now let us consider a very strong signal which would normally overload the detector. With the normal circuit the effective grid bias is increased from a-o to a-1. This, however, is not sufficient to bring the envelope of the modulated wave on the *straight-line portion* of the tube characteristic, with the result that the audio output A-1 suffers severe distortion. Note also that the amplitude is greatly reduced.

Now consider the modified circuit. When grid current flows a voltage drop occurs across R-1, thus causing the bias point to shift to a-2. This results in a much greater amplitude than before, although with slight distortion. A close examination will show, however, that the distortion is more symmetrical and certainly less severe than that of A-1, without the decrease in amplitude experienced before. The voltage built up across R-1 can be returned to the I.F. grids through de-coupling resistors to effect further limitation on very strong signals.

Typical output curves with and without this circuit refinement are shown in Fig. 3.

Output Stage and Power Supply

The output stage of this receiver is left entirely up to the individual. The author prefers a single '45 tube for ordinary use. This can be used to drive a pair of 46s in push-push (class B) amplification, if the sound output is inadequate. The schematic diagram clearly shows the circuit and constants used, and needs no further explanation.

Working *the* 56 M. C. Band

Practical Operating Hints for the "5 meter" Boys

By HARRY D. HOOTON
W8BKV

THE necessity for improving our technique on the longer-wave amateur bands has more or less distracted attention from the "five-meter" band during the last few years.

Questions answered: Best tube to use; wiring 5-meter receiver; the 5-meter antenna and reflector; frequency measurement.

This band has not been entirely neglected, however, and a small group of experimenters have been doing fine work on it. It offers tremendous promise as a useful band for directive transmission and radio telephone work. As any development work in radio communication requires the cooperation of several experimenters, the primary purpose of this article is to stimulate the interest in this band, and to present some practical suggestions that may be helpful to the five-meter experimenter.

The 5-Meter Receiver

Let us consider the 5-meter receiver.

It will need special design and construction, but there is no reason for it to be a freak. The requirements are similar to a 20-meter receiver, for example, except that the leads must be shorter and the distributed capacity reduced before it will oscillate. Careful workmanship counts in the successful work on this band. The straight regenerative set will work on 56 Mc.; but it is not to be recommended because of the high background noise. The usual regenerative "hiss" is sometimes so loud that weak signals are drowned out entirely. A much better arrangement is to use a push-pull detector regenerative circuit. This type of receiver works very well on ultra high frequencies and has the added advantage of being quiet in operation; the noise mentioned above being extremely low or entirely absent. A push-pull radio-frequency stage may be added to this receiver if desired.

The Best Tube For 5 Meters

What will probably interest the ama-

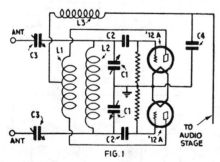

Fig. 1—The Push-Pull Receiver.

C1—Two section tuning condenser. About 20-m.m.f., each section.
C2—Grid condensers, 100-m.m.f.
C3—"Midget" variable condenser, 2 plates each.
C4—Not smaller than .005 mf.
L1—Tickler, to be determined experimentally.
L2—5 turns No. 20 D.C.C. wire, ¼" in diameter.
L3—R.F. choke, about 15 to 20 turns wound No. 38 on ¼" tube.

teur most are the constants for the five-meter receiving set. At W8BKV we had some difficulty in finding a suitable detector tube. Among the tubes we tried were the types '99, '01a, '24, '27, '12A and the '30; the most satisfactory were the '27 and the 12A, although the 01a and '30 types performed very well. A battery is used on the heater of the '27 instead of the A.C. filament supply.

The circuit for the push-pull receiver is shown in Fig. 1. The coils are self-supporting, being made of No. 20 D.C.C. wire and wound ¼-inch in diameter. It is necessary to use a coil of small diameter in order to reduce its field. Wind five turns on the grid coil if a 15 mmf. tuning condenser is used; if a larger ca-

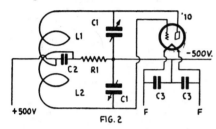

FIG. 2

Fig. 2—The Colpitts Transmitter.
C1—Two section transmitting condenser, 25-m.m.f., each section.
C2—Blocking condenser, .01-mf.
C3—Bypass condenser, .01-mf.
L1—1½ turns, 2¾" in diameter, copper tubing.
L2—Same as above.
R1—15,000 ohm transmitting grid leak.

pacity is used, use about four turns. The plate coil will have to be determined experimentally; as the correct number of turns is almost certain to vary with the different tubes; about five turns should start the receiver oscillating.

The tuning condenser is composed of two insulated stators and one rotor. The large "cut down" type of variable condenser should not be used; a remodeled "midget" tuning condenser of the proper capacity (about 15 to 20 mmf. each section) is desirable. Two small condensers of this type can be used, instead of the one two-gang midget, if desired. If a socket is used, solder the grid condenser (.0001 mf.) to its grid terminal, keeping the lead as short as possible; as this is very important. In some cases it is advantageous not to use a socket, but to solder directly to the terminals of the tube.

How to Wire the Receiver

The next step is to wire the set, getting leads as short as possible, and, at the same time, reducing the capacity between the various parts of the receiver. One stage of audio is usually enough, unless loud speaker operation is desired.

It is best to use separate batteries on the amplifier until the detector circuit has been adjusted, as this centers the trouble-hunting in the oscillating circuit. The detector will require a higher plate voltage than that commonly used on lower-frequency sets, because of the greater losses at five meters. About 90 volts should serve.

The 5-Meter Antenna

Almost any antenna will serve for reception of the five-meter signals, but the coupling must be less than that used on the longer waves. If too much coupling is used, the set will often become "cranky," and body capacity will be troublesome. An antenna is not especially necessary, on either receiver or transmitter, for distances up to five miles and possibly further. *We have heard our signal over twenty miles without any antenna on the receiver and with an eight-foot current-fed radiator on the transmitter.* The transmitting antenna was in the house at the time. The transmitting antenna shown in Fig. 3 is useful for receiving also; it gives quite a bit of gain over the plain type for use in receiving and, if it is used at both the transmitting and receiving stations, a high degree of efficiency is possible.

Transmitting on 5 Meters

Now with regard to the transmitter: the circuit shown in Fig. 2, is the split-coil Colpitts, which is especially fine for five meter work. The power for the plate and grid of the tube is fed at practically zero points of R.F. voltage, thus eliminating both plate and grid R.F. chokes which are very critical on the ultra short waves. It will be necessary to use a high-resistance grid leak, about 15,000 to 20,000 ohms.

The inductance is made of the usual ¼-inch copper pipe and is composed of 1½ turns each side of the blocking condenser, 2½ inches in diameter. The blocking condenser is .01-mf. The "tank" or tuning condenser is a two-gang type and should be very rigid, since a very small amount of vibration can do a great amount of tuning when the frequency is 56 megacycles! The leads and arrangement of the parts in the transmitter are not so critical as in the receiver; but it is best to get a good layout and as short a leads as possible. Any of the usual D.C. power supplies can be used at five meters, but for radiophone work it is best to use a battery filament supply; since the A.C. flowing through the filament seems to modulate the set in the same manner as "loop modulation".

Reflectors

Since the antenna is short, reflectors can be used easily. The procedure is to set a single wire, of the same length as the main antenna, at quarter-wave length distance behind the radiating wire coupled to the transmitter. This absorbs power and re-radiates it in the direction shown by the dotted arrow in Fig. 3. This is the simplest of the directive antennas and, as mentioned above, it can be used for receiving also. If a more highly directional result is desired, other wires can be supported vertically along a horizontal parabola, the main antenna being at the focus.

Of the plain type of antennas, the Zeppelin is probably the most efficient. It is desirable to use long feeders, say 5/4-wavelength, as the radiator can then be brought more into the clear. The

feeders can be spaced four or five inches with the usual spreaders. A radio-frequency ammeter should not be placed in series with either feeder, as it is possible to throw a zepp out of balance by doing so. The transmitter can be tuned to resonance by reading the maximum plate current. A small five-plate midget is placed in series with each feeder for tuning purposes. A five-meter zepp is 7 feet 10 inches long.

Frequency Measurement

In regard to frequency measurement: if an absorption meter is used it will be necessary to be very careful in tuning up, or the transmitter will be outside the band. An oscillating meter, such as the

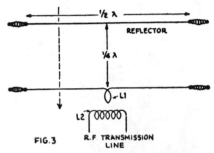

FIG. 3

Fig. 3—The Directive Antenna System.
Reflector: 7 feet 10 inches long.
Antenna: same length as reflector, including the single turn of the inductance.
L1—Single turn about 4 inches in diameter.
L2—One or two turns of wire.
Transmission Line: Can be twisted lamp cord 20 to 50 feet in length.

dynatron, is best and, as it is usually calibrated for the lower frequencies, it can be used for measuring the transmitter frequency on any of the amateur bands by multiplying the frequency by 2, 4, 8 or 16 according to the band in which it was calibrated.

A 5 Meter Super-Regenerator

(From Popular Wireless, London, England)
● RECENT experiments on 5 meters in England have brought forth a number of successful receivers covering this band. Included among these is the super-regenerative set shown in the illustration. It consists of two tubes, a detector and an oscillator for generating the variation frequency. The constructional details are as follows:

The two detector coils are each wound with 5 turns of No. 10 gauge copper wire (B&S) and are made by winding the wire on one-half inch bakelite rod, letting it slide off and pulling it out so that there is a space about one diameter between turns.

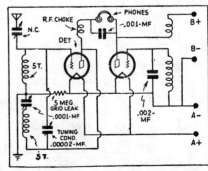

One of the latest 5-meter super-regenerative receiver circuits developed by English experimenters, is that shown above.

"5 and 10" Meter Receiver

● WITH the ten meter "amateur band" now made available for radiophone transmission, that is, the section from 28,000 to 28,500 kc., we can now expect to see great activity on this band and also a marked improvement in receiver and transmitter design.

Radio apparatus that will perform quite efficiently on the twenty meter band is liable to be entirely useless in the vicinity of ten meters. The requirements for a good ten meter receiver are stability, low background and set *noise level*, and adaptability to phone reception as well CW (telegraph code). The first thought naturally will be of the *superheterodyne*. This type of receiver if properly designed for the higher frequencies will no doubt prove to be by far the best.

But, on the other hand, the average superhet designed for general amateur use on the other bands may have a much higher noise-to-signal ratio than a well designed regenerative detector and one stage of audio combination. The author has, in many cases, seen the two tube set out-perform a seven or eight tube superhet; in fact the super fell down miserably on a signal that had a slight chirp or frequency change when being "keyed."

Tuned R.F. and Regen. Detector Preferred

After using both kinds of receivers for several months at the author's station, it was finally decided to build a stage of tuned radio frequency ahead of the detector in the straight regenerative set.

Various methods of coupling the R.F. (radio frequency) stage to the detector were tried and the old reliable

●

The 5 and 10 meter receiver designed and built by Mr. Shuart is here seen in actual operation. Among other signals, police calls on the new 8 and 10 meter systems were heard.

capacitive type of coupling was finally used, as it permitted less complication in circuit design and more effective coupling than that obtainable with the inductive method. The main objection to this system always has been that there was danger of the plate voltage of the R.F. tube leaking through the coupling condenser and getting to the grid of the detector tube, thus causing a failure of the set to function properly or else noisy reception. This was the case when the plate was attached directly to the grid coil, but gives no cause to worry when coupled through a condenser, because the grid condenser and coupling condensers are in series, which decreases this liability to practically zero. An alternative of course would be to use a low-capacity variable midget condenser or to construct a fixed air di-electric condenser. However, as stated before, if good mica condensers are used there will be no danger of any kind. The arrangement used in the receiver shown in the photographs was two 50 mmf. condensers in series, giving a total of around 25 mmf. and providing a third condenser between the plate of the R.F. tube and the grid of the detector tube.

The tubes used in the R.F. and detector stages are the type 57 and 58. The 58 being the R.F. amplifier and the 57 as regenerative detector, using the now famous *electron-coupled* circuit. These two stages are contained in the double-shield compartment mounted on the left-hand side of the base. Dimensions for constructing the shield and chassis are given in the drawings.

Super-Regeneration Added

An extra tube was added to the receiver to obtain *super-regeneration*, although this was not entirely necessary as very fine phone reception is obtained without it. The primary function of this addition is to enhance the reception of the very weak or broad

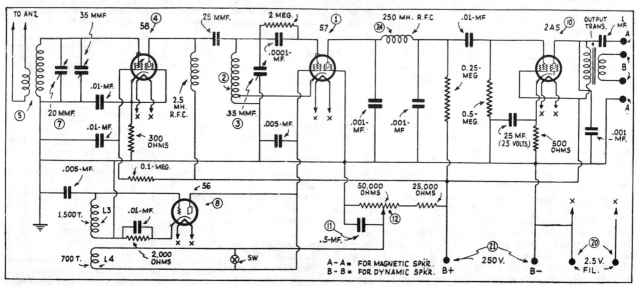

Above, we have the schematic wiring diagram of Mr. Shuart's 5 and 10 meter receiver.

modulated signals, such as those from new police broadcasting systems now operating on about eight and one half meters. These signals are so broad during modulation that it is impossible to receive them on a straight regenerative detector. However, when using super-regeneration the signal sounds first rate.

A type 56 is used as the generator of the *interruption frequency* oscillations, which produce the super-regenerative effect. The plate of the 56 is directly coupled to the screen-grid of the detector tube, the screen voltage to the detector tube and the plate voltage to the low-frequency oscillator being fed through L4 and controlled by the 50,000 ohm potentiometer. The voltage to both tubes is adjusted at the same time, providing *very smooth operation*. A 2A5 pentode is used as the output tube and is resistance-coupled to the detector; the output coupling is taken care of with a (single pentode to voice coil) transformer, working either as an output choke, for magnetic speaker or earphone operation, or for a dynamic speaker.

Bypass condensers were used quite freely in this receiver and are absolutely necessary at every point shown in the diagram, in order to obtain smooth and stable operation. This receiver will perform very nicely at frequencies as high as 60,000 k.c. and there is a decided gain present in the tuned R.F. stage, even at this frequency.

By GEORGE W. SHUART, W2AMN

All sorts of ideas arose in the author's mind as to what would be the best form of 5 and 10 meter receiver to build—after considerable experimenting, the receiver here described was finally evolved and it proved that it could "roll in the stations" in the 5 and 10 meter bands in excellent fashion! A tuned R.F. stage is used ahead of a regenerative detector, the detector being "electron-coupled." Super-regeneration is optional and is available at all times. A 2A5 pentode is used as the output tube. This set is the berries—no fooling!

Short Leads and Good Insulation Imperative!

As can be seen from the photographs the tube sockets are mounted above the base, not below, as is the usual practice. This was done so that all leads could be made as short as possible; if this were not done it would be impossible to get the set to perform on the five meter band. Remember: *short leads* and *good insulation* such as isolantite, are of utmost importance in ultra-high frequency receiver design.

Layout and placement of parts plays another most important part in this type of receiver. It is not advised that the builder should try to use any type of bread-board and panel arrangement, if good results are to be expected. An arrangement similar to the one used in this receiver should be used; it may be a trifle more expensive in the be-

ginning, but in the end it will pay higher dividends, as far as real results are concerned.

Antenna and Power Supply

Antennas are of prime importance in the reception of signals in either the 5 or 10 meter bands; in fact they spell the difference between the reception and non-reception of some of the weaker stations. About the best type is the vertical doublet, with each side measuring eight feet in length, and mounted as high from the ground as possible; for best results the feed line (lead-in) should be tuned. Such an arrangement is shown in the sketch, together with other types of antennas and their various lengths.

The power supply shown in the photographs was especially constructed to work on ultra high frequency receivers. To remove the main *hum* it was only necessary to use two 30 henry iron core chokes, with three 8 mf. electrolytic condensers. However on certain frequencies there was a decided hum, (this is usually termed *tun-*

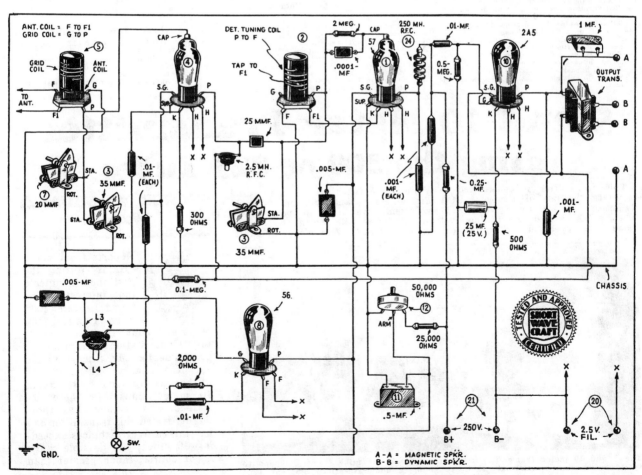

Picturized wiring diagram for the 5 and 10 meter receiver, which will make the construction of the set clear to even the uninitiated.

able hum) and about half a day was spent in trying to remove it; RF. chokes and by-pass condensers were tried in every part of the circuit, with no improvement at all. The power transformer used happened to have two extra filament windings that remained unused. These idle coils were finally sus-pected and one lead of each winding was *grounded* to the negative side of the circuit; sure enough the hum entirely disappeared, no trace of it could be found on any fre-quency. If you happen to be having trou-ble from *tunable hum*, watch all *unused* low voltage secondaries! To improve regula-tion a heavy-duty 20,000 ohm wire-wound resistor is connected across the output terminals of the high voltage.

The *power-supply* unit should be capable of furnishing no less than 250 volts under full load. This high voltage is necessary in order to obtain full gain of the tubes. Lower voltages will in all cases produce weaker signals on the speaker or phones, and may even cause the set to fail entirely on the 5 meter band!

Parts List of Receiver

1—Pentode output transformer. Acratest.
1—Chassis—see text and drawing for details.
6—4 prong coil forms; ultra-high frequency type; National.
2—4 prong isolantite sockets; National (Hammarlund).
2—6 prong isolantite sockets; National (Hammarlund).
1—6 prong wafer socket (laminated); Eby (Na-ald).
1—5 prong wafer socket (laminated); Eby (Na-ald).
2—35 mmf. variable tuning condensers; Hammarlund.
1—20 mmf. variable tuning condenser; Hammarlund.
1—Vernier dial; National, type B.
1—2.5 millihenry choke; National.
1—250 millihenry choke (universal wound).
1—50,000 ohm potentiometer; Acratest.
1—"Interruption Frequency" transformer, 700 turns pri. 1500 sec; Gross Radio.
3—.001 mf. mica fixed condensers. Flechtheim.
2—.005 mf. mica fixed condensers.
2—.00005 mf. mica fixed condensers (connected in series).
1—.0001 mf. mica fixed condenser.
1—.5 mf. bypass condenser.
4—.01 mf. bypass condensers (tubular).
1—25 mf. 25 volt electrolytic condenser; Acratest.

1—300 ohm 1 watt resistor, Lynch (International). Also following resistors.
1—500 ohm 1 watt resistor.
1—2,000 ohm 1 watt resistor.
1—25,000 ohm 1 watt resistor.
1—100,000 ohm 1 watt resistor.
1—250,000 ohm 1 watt resistor.
1—.5 megohm 1 watt resistor.
1—2 megohm 1 watt resistor.

Parts for "Power Supply"

1—Power transformer 325-0-325 plate, 2.5 fil, 5 v. R.T. Co.
2—30 henry, 60 milliampere chokes; Acratest.
3—8 mf. 500 V. electrolytic filter condensers; Acratest.
1—20,000 ohm bleeder resistor (20 watts rating).
1—4-prong wafer socket, Eby (Na-ald).

Coil Data and Receiver Chassis Dimensions.

5 and 10 Meter Transmitter
Using the 800 or 825 Tubes

This transmitter is constructed to facilitate the use of the new RCA Radiotron type 800 (or Sylvania 825) tubes, which have at this writing, just made their appearance on the market.

While these tubes are rated to stand around a thousand volts on the plate no attempt was made to operate them at this value, 650 being the highest voltage applied. With a thousand volts

GEORGE W. SHUART, W2AMN

Reports from many "Hams" during the past month indicate that interest in the 5 and 10 meter field is increasing by leaps and bounds. Several lead-ing "Ham" station operators have complimented Mr. Shuart on the excellent quality and steadiness of the wave radiated by the 5 and 10 meter transmitter here de-scribed. Mr. Shuart has thor-oughly tested this transmitter and has talked over distances exceed-ing 30 miles; the possible range is, of course, much greater than this.

The 5 and 10 meter transmitter ready for action; the mike shown happens to comprise an old "commercial" shell, enclosing a moderate-priced amateur type microphone unit.

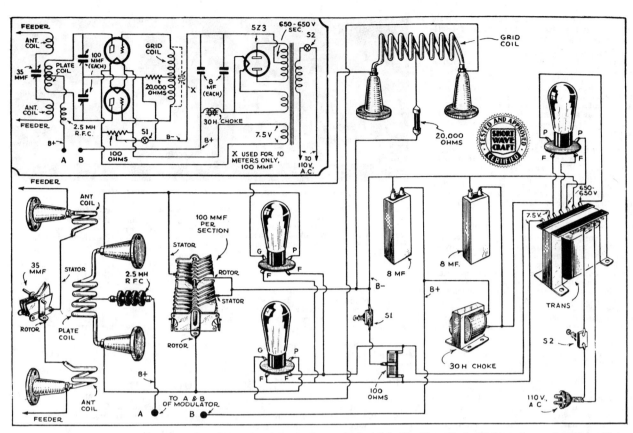

The schematic and picture diagrams above show how the various parts used in Mr. Shuart's 5 and 10 meter push-pull oscillator and power supply are connected together.

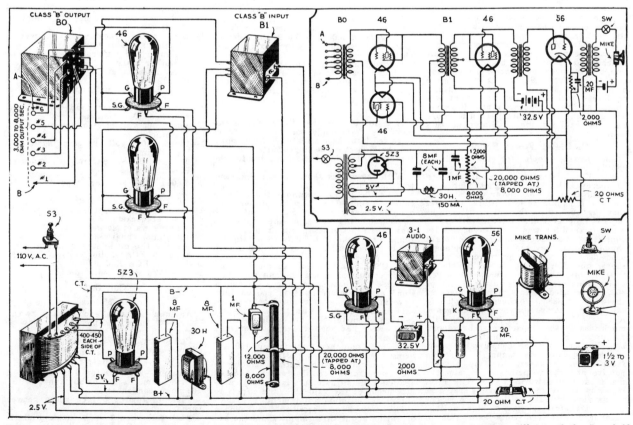

Here we have the schematic and physical wiring diagrams for the modulator unit used to excite the oscillator of the 5 and 10 meter push-pull transmitter.

on the plates of these tubes, it would be necessary to have a master-oscillator-amplifier arrangement to maintain frequency stability and it would be necessary to have a rather expensive *modulator* system.

With 650 volts on the plates of two of these tubes in a push-pull circuit, a very powerful and steady signal can be radiated, and can easily be modulated with a pair of 46's operated in class "B" push-pull. In the transmitter here described, every effort has been made to reduce losses, due to poor insulation, to an absolute minimum. Nothing but the highest grade parts with excellent insulation have been used. *Why use inferior insulation on a tube that has its grid and plate leads brought out on the glass envelope, for the sole purpose of improving their insulating qualities and to reduce the inter-element capacity.*

The Push-Pull Oscillator

The push-pull oscillator is mounted on a 7x10x2 inch aluminum chassis, in order that the filament and plate supply wiring could be isolated from the portions of the circuit carrying radio frequency currents. This isolation is really necessary if stable and efficient operation is to be obtained. **Remember, we are working on 60 megacycles, not 3.5.** And if there should be the slightest trace of R. F. in the filament circuit the chances of ruining the tubes are very great.

No filament by-pass condensers were used because without them, there was no trace of RF in the filament circuit. And the use of them provides a possible chance of the filament circuit being tuned to the frequency, or a harmonic of the frequency at which the transmitter is being operated. If no trouble of this sort is experienced without filament condensers *leave them out.* The switch shown in the filament center-tap lead is provided as a means of turning the oscillator on and off. If the switch were to be placed in the plate circuit, there would result a very heavy spark and in most cases it would continue to arc, due to the action caused by the modulation choke or the modulator output transformer secondary; at the center tap however there is only the slightest trace of a spark.

All inductances are provided with banana type plugs, in order that they can be changed easily. The stand-off insulators are equipped with jacks to accommodate these plugs. Be sure to use a good plug which makes a very *tight* contact, or losses will result at this point. A glance at the photographs and diagrams will give any further details and no more need be said of the oscillator at this point.

Oscillator Power Supply

The *power supply* for the oscillator should be capable of delivering around 650 to 675 volts at least one hundred milliamperes, although the highest plate current drawn by the oscillator was 90 mills. (M. A.) It should be well filtered in order that the quality of the voice will not be impaired by a "hum" on the carrier. A *brute force* filter was used because of its simplicity and effectiveness. Some might shudder when they see a 2 mf. condenser used directly across the mercury vapor tubes, but there is really no danger in this, because the tubes are operated well under their peak voltage rating. The type 866 can be used at this point in place of the 871's or 888's, whichever you wish to call them, that are shown in the diagram. A type 5Z3 can also be used here with much less "hash" (noise caused by vaporization of the mercury) in the receiver, but this tube will be considerably overloaded and long life cannot be expected from them. However, they are not expensive and the reduction in noise might make their use worth while. The filter choke is of the common variety, rated at 30 henries and 150 mills. (M. A.) The power transformer has a high voltage secondary with 650 volts each side of center-tap, which with the filter system and rectifier tubes used, gives around 650

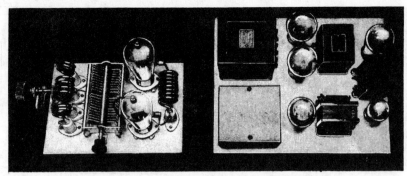

5 and 10 meter push-pull   oscillator and modulator,

volts at 90 mills (M. A.) pure D.C. This transformer also has a 2.5 volt winding for the rectifier tubes and two 7.5 volt windings, which are hooked in parallel to supply filament voltage for the oscillator tubes. So much for the oscillator *power supply.*

The Modulator

A good modulator unit is just as important as the oscillator, because it is the modulator that is responsible for the voice being super-imposed upon the R.F. carrier wave.

The most economical modulator is the class "B" type, which has become very popular with the "Hams." The one shown here uses a 56 class "A" speech amplifier transformer, coupled to a 46 Class "A" driver, which in turn furnishes excitation for the two 46's in class "B" push-pull. The plate voltages for the tubes are as follows—250 for the 56, 250 for the 46 "A" and about 450 for the 46's in Class "B." With these voltages on the various amplifier tubes it is possible to completely modulate the R.F. oscillator with 650 volts at 90 milliamperes input. Here also good equipment is necessary if full output is to be obtained from the tubes, with anything like BC (broadcast) quality.

The *output* transformer used in this modulator has a secondary with taps varying from 3,000 to 8,000 ohms impedance and provides a very flexible unit, which can be coupled to almost any output load efficiently and thus permit a maximum transfer of audio voltage. The method of calculating the load impedance of the oscillator is—*Plate voltage divided by plate current.*

It is to be remembered that the microphone is responsible for the quality of voice fed into the modulator. Use nothing but the highest grade single-button, or preferably a good double-button mike for the best quality.

With the values shown in the diagrams the plate current of the oscillator should be 60 mills (M. A.) without the antenna coupled, and 90 mills with the antenna coupled loosely to the plate "tank." Do not exceed this value, or the tubes will loose their activeness; higher plate currents also make it difficult to obtain 100 per cent modulation. If everything is connected properly, an indicating device such as a Christmas tree bulb or an R.F. ammeter connected in the antenna will show a decided increase in brilliancy (or deflection) as the microphone is spoken into. Careful adjustment is absolutely necessary if maximum results and a good percentage of modulation are to be obtained. The writer has seen an adjustment in a five meter transmitter so slight as to cause the frequency to shift only 100 KC, bring the signal strength from R3 up to R8. Cooperation with another amateur is very necessary, for his reports checked against your adjustments is the only definite proof of the value of your efforts to obtain an optimum adjustment.

The radiophone described has been in operation about three months and the author has received an R8 to R9 report from most every station contacted.

Parts List (Oscillator Circuit)

1—double-section Cardwell variable condenser 100 mmf. per section with micalex insulation (featherweight).
7—stand-off insulators with pin-jack receptacles (Johnson; Fleron).
1—35 mmf. midget tuning condenser, Hammarlund (National).
2—4-prong isolantite sockets, National (Hammarlund).
1—No. 100 National R.F. choke (2.5 mh.).
1—20,000 ohm grid-leak (10 watts or more).
1—100 ohm center-tap filament resistor.
1—single pole, single-throw snap switch.
1—set of coils—see coil table.
2—type 800 RCA Radiotron tubes (Sylvania 824).
1—aluminum chassis, 7"x10"x2" deep.

Oscillator "Power Supply" Parts

1—power transformer, 650-0-650, 2.5, 7.5 volt filament. R. T. Co.
1—30 henry 150 mil. filter choke. National.
2—2mmf. 1000 volt filter condensers, Flechtheim.
2—4 prong sockets, National.
2—type 871, 888 or 866 mercury vapor rectifier tubes. RCA Radiotron, (Arco).

Modulator Parts List

1—Microphone, Universal. (Single or double button.) Amplion; Lifetime; Miles; Mayo.)
1—Microphone transformer, Universal. (Single or double button type.)
1—3:1 ratio audio transformer. National (or other make).
1—Class B input audio transformer, National.
1—Class B output audio transformer National. (With tapped secondary.)
1—Power transformer 450-0-450, 2.5, 2.5 windings, 5V. National.
2—8mmf. 500 volt electrolytic condensers, Flechtheim.
1—1mmf. 400 volt by-pass condenser, Flechtheim.
1—20 mmf. 25 V., electrolytic condenser, Elechtheim.
1—20,000 ohm resistor, tapped at 8,000 ohms.
1—2 000 ohm 2 watt resistor, Lynch.
1—20 ohm center-tapped filament resistor.
4—5 wafer sockets, Eby.
1—4-prong wafer socket, Eby.
1—aluminum chassis 13"x10"x2".
Tubes for the modulator unit are:
3—type 46 RCA Radiotron.
1—type 56 RCA Radiotron.
1—type 5Z3 RCA Radiotron.

Coil Data

5 Meters

Antenna—2 turns each (make two).
Grid—8 turns, ⅛ inch space between turns.
Plate—4 turns. ¼ inch space between turns.

10 Meters

Antenna—same as for 5 meters.
Grid—same as for 5, but tuned with a 100 mmf. condenser.
Plate—6 turns, ⅛ inch space between turns.
*All coils are 1 inch inside diameter and wound with ⅛ inch diameter copper tubing.

The SHORT-WAVE
BEGINNER

Here we have a front view of the 1-tube "Twinplex" receiver developed by Mr. Worcester. 1 tube does the work of 2!

The "53"
1-Tube TWINPLEX

By J. A. WORCESTER, Jr.

● IT can be stated, without possible fear of contradiction, that the most popular short-wave receiver from the constructor's standpoint, at the present time, is a two tube affair consisting of a regenerative detector and one-stage audio amplifier. It is, of course, true that many home built receivers also include a stage of radio frequency amplification, either of the tuned or aperiodic variety, and possibly an additional stage of audio frequency amplification, as well; to provide sufficient volume for loudspeaker operation under favorable conditions. However, those fortunate enough to afford these more complicated receivers generally prefer to purchase one of the many excellent commercial receivers employing such circuits rather than to undertake the construction themselves; as the savings that can be effected thereby are generally not sufficient to justify such a procedure.

The average prospective constructors, becoming interested in short wave reception for the first time prefers as simple a receiver as possible consistent with satisfactory results. A one tube receiver is undoubtedly the ideal solution but unfortunately such a receiver of the conventional regenerative variety will not produce sufficient volume for satisfactory headphone reception. The writer has been interested for some time in designing a one tube receiver which would retain all the essential features of the conventional two tube

receiver and at the same time produce the simplification in wiring and apparatus effected by the single tube construction.

New 53 Tube Employed

The schematic wiring diagram of such a receiver is shown in Fig. 1. The tube employed is the new 53 which really consists of two tubes in one. This tube was designed as a Class B Twin amplifier but due to the comparatively large static plate current drawn, it can be readily adapted to detection and Class A amplification.

As an audio frequency amplifier this tube is very effective since its amplification factor is about 35. This permits an amplification approximating that of a

pentode without the latter's disadvantages of wiring complications and heavy plate current drain.

Looking at the back of the 1-tube "Twinplex," in which a single 53 type tube performs as both detector and A.F. amplifier.

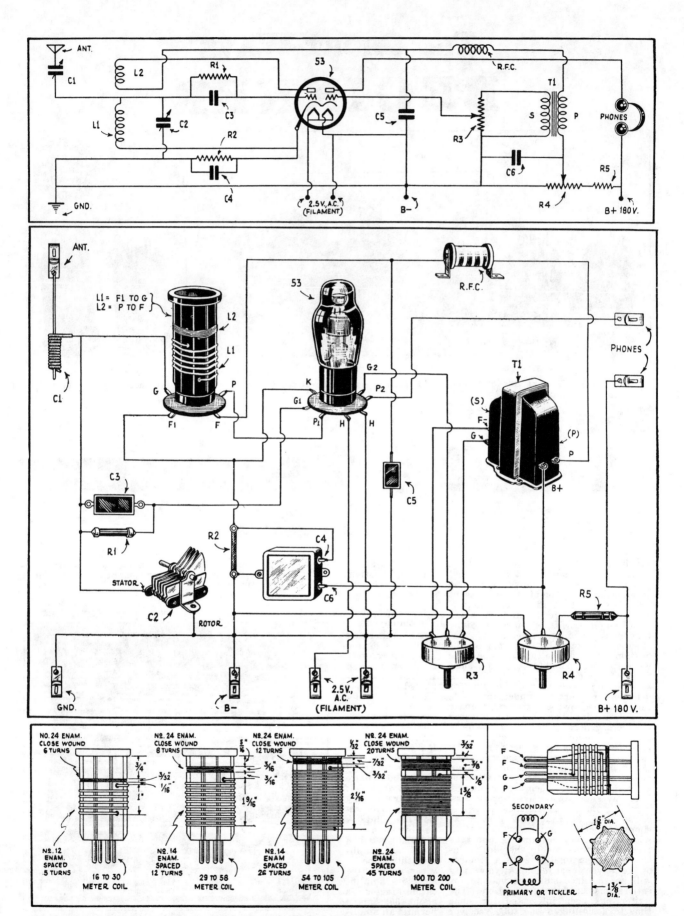

The drawings above show both the schematic and physical diagrams for the one-tube "Twinplex" receiver—in which a single 53 tube does double duty; that is, it performs both as a detector and as·an A. F. amplifier—true economy! Data for winding your own coils is given in the diagram above.

The use of this tube as a detector results in a substantially greater output than is possible from the usual low-mu triode, while maintaining the desirable characteristics of this type of regenerative detector; namely stable and foolproof operation and simplified construction. As is well known, a screen-grid detector is often rather tricky in operation, especially when regeneration is controlled by screen-grid voltage variation, which often proves somewhat confusing to a beginner.

Regeneration is controlled by varying the plate voltage by means of a 25,000 ohm potentiometer. Independent volume control is provided by a 200,000 ohm potentiometer across the audio frequency transformer secondary as the volume often becomes too great for comfortable headphone reception particularly on strong amateur and 49 meter broadcasting stations.

The tuning condenser has a capacity of 140 mmf. and is employed in conjunction with a set of short wave octo coils.

Plate Supply from Batteries or "B" Eliminator

It will be noted that a plate potential of 180 volts is required and this may be obtained either from dry batteries or a well filtered "B" supply. The heaters require 2½ volts A.C. which may be obtained from a suitable step down transformer.

The general layout of the various parts can be noted from the photographs. It will be seen that an aluminum panel is employed in conjunction with a wood baseboard. The panel is 6"x9"x1/16" and the baseboard 7"x9"x¾". The variable con-

denser along with the two potentiometers are mounted on the front panel while the remaining apparatus is mounted to the baseboard. External connections are made by means of Fahnestock clips mounted at the rear of the baseboard.

The antenna compensating condenser is made by connecting a piece of bus-bar wire to the antenna clip and bending upright as shown. The other electrode consists of about 15 turns of hook-up wire coiled around the bus bar. Adjustment is effected by moving the coil off of the wire until the desired coupling is obtained. For this reason, it is desirable not to wind the hook-up wire too tightly around the busbar or it will not be possible to slide the coil conveniently. The adjustment of this condenser is not critical and for normal operation can be left "all in." When "dead spots" produced by antenna absorption are encountered the coil can be moved off the busbar until the dead area is reduced to one or two dial divisions. As this results in decreased input it is advisable to increase this capacity when the "dead spot" area has been passed.

When wiring the set it is absolutely essential to ground one of the heater lines, as shown, if satisfactory operation is to be obtained.

Operation and Results Obtained

In operation, the set is exactly the same as the conventional two tube regenerative receiver and consequently it will not be necessary to go into detail regarding same. The results obtained during a week of testing have been exceedingly good. The for-

eign stations received during this period include EAQ, GSB, GSA, DJC, HKD and OXY. No listening was done during the daytime which accounts for the absence of 25 meter stations. The receiver is also very satisfactory for C.W. reception.

Parts Required

C₁—See text
C₂—Hammarlund "Midline" midget variable condenser—140 mmf., Type MC-140-M.
C₃—Molded mica condenser—.0001 mf.
C₄, C₆—.5-.5 mf. dual by-pass condenser.
C₅—.0005 mf. Molded mica condenser.
L₁, L₂—Set of short-wave Octo-Coils 16-200 meter.

RFC—Hammarlund isolantite R.F. choke, 8 millihenrys, Type CH-8.
R₁—3 meg. grid-leak; Lynch (International).
R₃—400 ohm wire-wound Resistor.
R₃—200,000 ohm potentiometer (Acratest)
R₄—25,000 ohm potentiometer (Acratest)
R₅—50,000 ohm resistor, Lynch (International).
T₁—Audio frequency transformer.
1—Alden 4 prong socket, type 481X.
1—Alden 7 prong socket, Type 487.
7—Fahnestock clips.
1—Type 53 Tube.
1—Roll hook-up wire.
1—National Type "B" Velvet-Vernier dial (0-100-0).
1—Aluminum panel 6"x9"x1/16".
1—Baseboard 9"x7"x¾".
1—Type 53 tube; Gold Seal, Arco, Van Dyke.

The "Tinymite" 1-Tuber Rolls 'em In

All ready for action—the "Tinymite" 1-tube S-W Receiver and phones.

Rear view of the "Tinymite."

● THE tickler switch of the *Tinymite* receiver rotates from the bottom to the top for higher waves, so that when the tickler switch is set on the third point from the bottom and the secondary switch is set on the third point from the top, they are on 160 meters and one should be able to hear amateurs, phone, police calls and airplanes. If the set refuses to oscillate on the lower waves, keep moving the tickler switch toward the top. If broadcast waves are desired, turn the secondary switch to the bottom point and leave tickler switch on the 160 meter or third point from the top, or move up to the top point.

The author has a confirmation on duplex speech between W3KL and Sandy Hook, N. J., and other "DX" records, which he can swear to if necessary. He has received (at Tionesta, Pa.) amateurs, phone, police calls, pilots in planes, plane stations and broadcast, all on a sixty-five foot aerial, one 45 "B" battery and 4½ volts "A" battery on a 99 tube, which seems to work best in the set. You might try a 30 type tube.

The author has heard police calls from Lansing, Pittsburgh, Cincinnati, Chicago; plane stations such as Toledo, Indianapolis, St. Louis, Cleveland and New York. The regeneration slide is critical on airplanes. If the set fails to oscillate pull the coils together, if too much, push apart, slowly turning the tuning condenser.

The aerial condenser is critical and seems to work good when set close; try adjusting it for different wavelengths. Give the set a critical test on a real aerial. If too much oscillation occurs shorten the aerial or adjust aerial condenser slowly.

Parts List for "Tinymite"

1—celluloid panel—bakelite may be used, 4⅜ x 5⅝.
L1—The secondary is a coil taken from old Crosley "pup." It was tapped at four evenly divided places.
L2 is the tickler, a coil taken from RCA old type superhet. It is tapped at four evenly divided points.
C1—.0001 mf. midget condenser, Hammarlund (National)
C2—vernier condenser, 10 to 20 mmf. capacity. —Hammarlund (National)
C3—.0001 mf. midget regeneration control condenser.

C4—midget vernier—2 plate condenser; about 10 to 20 mmf.—Hammarlund (National).
C5—80 mmf. trimmer (Hammarlund) aerial condenser.
C6—.0002 mf. grid condenser; .0001 or .00015 may be used instead.
R1—2½ megohm grid-leak. Lynch (International).
R2—10 ohm fixed rheostat.
—L. S. Hoover, Tionesta, Pa.

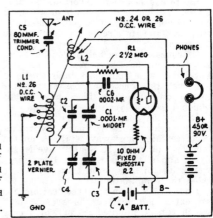

Hook-up for the "Tinymite."

The Short-Wave Megadyne

By HUGO GERNSBACK

The Megadyne receiver represents t h e culmination of n e a r l y ten years research work by Mr. Gernsback and we are sure that all short - wave "hounds" will be tickled pink with this ultra efficient a n d economical re- ceiver. T h e Megadyne 'uses but one tube and most stations c a n be brought in on the loud speaker with it.

The Short Wave Megadyne One- Tube Pentode Set is an adaptation of the author's Megadyne One-Tube Loudspeaker Set for broadcast re- ception. While the author does not claim that stations are received gen- erally with loud speaker strength, still we have actually listened to 'phone broadcasts of a great number of eastern amateurs, all of which were received with fair loud speaker strength; this, in New York City, where short-wave conditions are none too good. The author would be happy to hear from those who have built this really remarkable set.

● RECENTLY, after quite a good deal of experimenting, which covered a term of several months, I devised a new circuit which is termed the "Megadyne" circuit. (*Megas*, the Greek for "great"; *dyne*, Greek for "power.") The circuit originally was devised by the author for the broadcast band, and is unique in that it is actually possible to operate a loud speaker from a single tube.

To those interested in the broadcast receiver, I refer to the July 1932 of RADIO-CRAFT and my article entitled the *Megadyne One-Tube Pentode Loudspeaker Set.*

Incidentally, this simple set, which anyone can build, brought in stations from Pittsburgh, Cleveland, Springfield, Mass., Hartford, Conn., with good loud speaker strength without any trouble.

The circuit is, to the best of my knowl- edge, new in that the tube really works "backwards." In this circuit, I also make use of my old Interflex idea, where the crystal is inserted directly into the grid circuit of the tube.

Having had such astonishing results with this little set, I thought it worth- while to devise one for short waves also. Some modifications had to be made from the original broadcast set, and the short- wave set and its circuit are described in these pages.

Of course, I do not claim loud speaker strength on the Short Wave Megadyne, but I do claim that the signals that you get with this one-tube receiver are far more powerful than from any circuit which I have as yet tested or tried.

It may be possible in locations which are superior to New York City to re- ceive certain short-wave stations on the loud speaker with the Short Wave Mega- dyne, and if this is the case I certainly would like to hear from those who have tried the combination.

I have also given the specifications for an extra coil termed "broadcast coil," which can be plugged in, in which case the set becomes a one-tube loud speaker type for broadcast purposes. However,

the tuning of this broadcast coil is some- what broad, and the set cannot be used in the proximity of a powerful broadcast station. Out in the country, and away from stations, it will perform excellently.

The Circuit

The Short Wave Megadyne circuit dif- fers slightly from the original circuit of the Megadyne for *broadcast* reception. In the latter, regeneration was obtained by connecting the tickler coil in the con- trol grid circuit of the tube and coupling this coil to the input coil, which was

connected in the screen grid circuit. The screen grid in the Megadyne is used as a control grid.

In the Short Wave Megadyne, regen- eration is obtained ·in the conventional manner, with the tickler coil in the plate circuit. This gave better control of re- generation, which was found more criti- cal for short-wave reception when a num- ber of plug-in coils were used with fixed tickler windings than for broadcast re- ception with a single tuner with a vari- able tickler winding.

The elementary circuit is shown in Fig. 1. It employs a type '38 pentode tube (although any pentode or any screen-grid tube has been found satis- factory). The tube is connected as a

Rear chassis view of the short-wave Megadyne, as perfected by Mr. Gernsback.

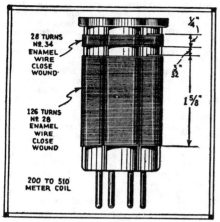

Above—Winding data for Octocoil, or similar shaped form, having sufficient inductance to cover the broadcast band with .00015-mf. condenser.

At right—Picture diagram which anyone can easily follow in building the short-wave Megadyne.

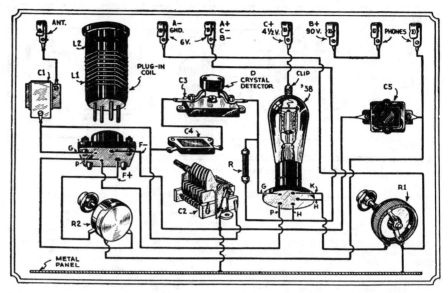

space-charge regenerative detector; that is, the control grid is connected to a plus potential of 6 to 10½ volts with respect to the negative side of the filament or cathode and the screen-grid is then used as a control grid. The plate potential may be 90 volts, although as low as 45 gives satisfactory results. These voltages may vary when other types of tubes are used.

A distinctive feature of this circuit is the use of a fixed crystal detector (D) connected in series with the (screen) grid of the tube, as shown; its use considerably improves the volume of the signals. It acts as a rectifier and therefore feeds a rectified or audio frequency signal to the input of the tube so that the tube serves the dual purpose of functioning as a regenerative R.F. amplifier and an audio amplifier. The method of controlling regeneration and the general method of tuning the circuits may be any of the several standard varieties now employed for short-wave work. With the use of plug-in coils with fixed coupling between plate and grid circuits a variable resistance or capacity regeneration control is essential. The former method is employed in this particular circuit, as it gives a more uniform control.

Construction of Set

As shown in the photographic illustrations, the set was built in a very simple manner in order to prove the efficiency of the circuit and also allow for making any changes in the wiring, if necessary, for experimental work. A metal panel 8 x 4¼ inches, mounted on a half-inch baseboard 8 x 6 inches, forms the framework of the chassis. The use of a metal panel is recommended, as it eliminates hand capacity.

On the panel are mounted a 10-ohm filament rheostat, R1, which also serves the purpose of opening the filament circuit when in the "off" position; a 400-ohm potentiometer, R2 (mounted at the left) which controls regeneration; and a .00014 mf. midget variable condenser, C2, for tuning the various wave bands. A vernier dial is used on the tuning condenser.

On the baseboard are mounted seven terminals, a home-made antenna series condenser, C1, a four- and five-prong socket, grid condenser, C4, crystal, shunt crystal condenser, C3, and bypass condenser, C5. The location of the various parts is clearly shown in the illustrations. The pig-tail grid leak, R, is mounted directly on the tube socket.

The antenna series condenser comprises a metal plate, 1½ x 1¼ inches, mounted flat on the board and held by the screw that fastens the antenna terminal. The other plate of the condenser is 1 inch wide and 2¼ inches long and is bent around the edge of the baseboard and held in place with a wood screw. Between the two plates is placed a sheet of mica. The upper plate tends to spring out and is held down with a screw, the turning of which varies the capacity.

The shunt condenser across the crystal detector was found necessary to act as a by-pass across the crystal and allow the R.F. currents to pass through to the grid of the tube and maintain oscillation. The lower the capacity required to maintain oscillation the better. Since this may vary with different coils in the circuit, an adjustable condenser having a maximum capacity of .001 mf. is employed.

Two metal contact springs from an old vacuum tube socket, mounted on the shunt crystal condenser as illustrated, are used to support and make connection to the crystal detector. This allows the detector to be quickly removed and replaced with a different one, or reversed. It has been found that the polarity of the detector is sometimes important.

The 400-ohm potentiometer that controls regeneration is connected directly across the tickler winding of the plug-in coil. The center connection of this potentiometer is connected to plus "B", and since it is at high potential, the shaft of the potentiometer must be insulated from the contact arm or any of the terminals. Some makes of potentiometers, such as the one used, have insulated shafts and can be mounted on grounded metal panels.

The complete wiring of the set is shown in Fig. 2. The various parts are lettered to correspond with the lettering in the other illustrations and in the list of parts, so that the builder will have

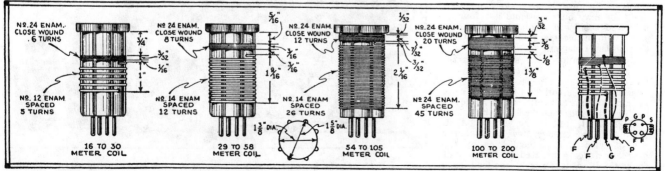

Coil winding data for the four principal short-wave bands, using Octocoils or similar shaped forms; round tubes can be used without much discrepancy in the wavelength covered.

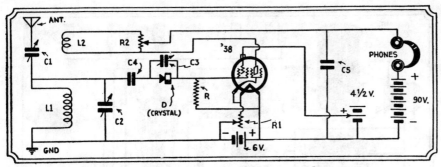

Schematic diagram of short wave megadyne.

little difficulty in assemblying and wiring the set.

The Coils

While any good set of standard plug-in coils may be employed, it is recommended that the builder wind his own either on tube-base forms or on regular plug-in forms. In winding your own coils the data given on page 395 of the April, 1932, issue of SHORT WAVE CRAFT may be closely adhered to. The socket connections, as shown in Fig. 2, are correct for these plug-in coils.

Operation

In operating the set, a coil covering any desired waveband should be inserted in the socket and the filament rheostat, R1, turned all the way on. After about 30 seconds the tube will warm up and the set will be ready for operation. With the regeneration control, R2, turned to one side the set should not oscil-

late. By turning this potentiometer slowly, the set should start to oscillate and then the tuning condenser, C2, can be turned until the heterodyne squeal of a station is heard. It may then be well to try reversing the crystal detector and leave it in the position that gives loudest response. For code reception the set should remain in this oscillating condition. For phone reception potentiometer, R2, should be turned until the set just stops oscillating and the station is heard loud and clear.

When using the small coils it will be found that the antenna condenser, C1, should be adjusted so as to have very little capacity; with the largest coils this condenser can be adjusted to maximum capacity. This adjustment may vary with different types of aerials and no set rules can be given.

The batteries used for operating this set depend upon the type of tube employed. Using a type '38, four 1½-volt dry cells connected in series furnish the filament current and two 45-volt units furnish the plate potential. A

4½-volt "C" battery is used for the space charge grid. The ingenious builder of course can make it for A.C. or D.C. operation if he desires, although the current consumption of the single tube is so low that we believe battery operation is more practical and less troublesome.

Those who operate this set will be amazed at the ease of control and lack of dead-spots in the tuning spectrum. The circuit goes gradually into oscillation.

In actual tests, stations all over the country could be picked up on any of the short-wave bands. This includes police signals from all over the United States, television signals, amateur phone and code; transatlantic phone conversations and, under favorable conditions, foreign broadcast stations. Canadian and U. S. A. short-wave broadcast stations came in with surprising volume—some of them loud enough to operate a loud speaker with fair volume.

List of Parts

1 set short-wave plug-in coils (Octocoils).
1 .00014 mf. Hammarlund midget condenser, C2.
1 .001 X-L-Radio adjustable condenser, C3.
1 .00025 mf. fixed condenser (Polymet), C4.
1 Home-made antenna condenser, C1.
1 .001 fixed condenser (Pilot), C5.
1 BMS fixed crystal detector (Brooklyn Metal Stamping Co.), D,
1 1-megohm grid-leak (Durham), R.
1 10-ohm filament rheostat (Carter), R1.
1 400-ohm Clarostat potentiometer with shaft insulated from terminals, R2.
1 Four-prong socket (Benjamin).
1 Five-prong socket (Benjamin).
7 Fahnestock binding posts.
1 Vernier dial, 0-100 (Kurz-Kasch).
1 baseboard, 6 x 8 x ⅝ inch plywood.
1 panel, aluminum, 4¼ x 8 inches.
1 '38 pentode, 6.3-volt type.

1-TUBER By CLIFFORD E. DENTON

● THIS article is not dedicated to the person who can build a short wave receiver but to the person who would like to but thinks that he would never get it together and if he did finish the construction that "it would'nt work" anyway.

Look at the picture diagram and see that little gadget marked 2; well, this unit has the effect of increasing and decreasing the length of wire that is called the aerial. The idea is to bring the electrical length of the aerial down to such a value that it will give the best results. How does it do it? Don't worry about that now. When an electrical current flows down the antenna into the set it is following the first law of electricity and follows the path of least resistance and continues through the coil marked 3.

Let's start with a simple one tube receiver; because, if we make a mistake, one tube may burn out and not four or five, and besides a minimum number of parts will be necessary for the construction of the receiver.

By what manner and means does the set work? Simple. Light your pipe and listen.

Passing through coil 3 as though it were running from a bill collector, it rushes to the ground and then back to the transmitter. Well, that is one way to tell it.

Now, as the energy flows through the coil of wire 3, it is held back by the magnetic force generated by the flow of energy. In other words, the coil says "let's talk this over." This halting or blocking action causes the energy to

build up to appreciable amounts, if the units called a condenser, 6 and 7, are so adjusted that their electrical length is the same as that of the transmitter. In fact, one would have perpetual motion right here if the coil and condenser were perfect. You see, the energy would chase around and around like a dog chasing its tail. See Fig. 1. When the dog gets tired he stops and the electrical losses soon stop the energy from going

around and that is a good thing too, as it permits more energy to flow into the circuit to merry-go-round awhile.

This energy causes a surprisingly large electrical force to appear between the control member of the vacuum tube called the *grid* and the heating element, called the *filament*. This force is greater in value than the original energy derived from the aerial.

Unit 8 stores this force as though it

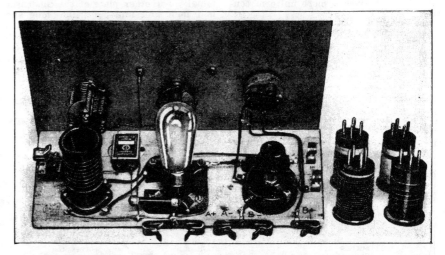

"Simplicity" is the keynote in this 1-Tube S-W Receiver. "Thrills by the carload" are yours at an insignificant cost.

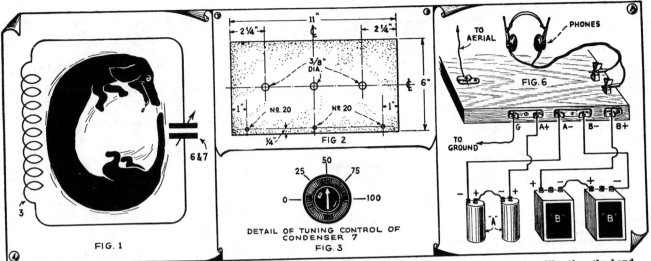

Fig. 1—Oscillations flow around a tuned circuit like a dog chasing its tail. Details for drilling panel, calibrating the band-spreader condenser dial and battery connections are shown above.

were a storage tank and delivers it to the control unit at regular intervals of time. The poor electrical path of unit 9, called a *grid-leak*, permits surplus energy to flow off to the ground.

This rise and fall in the value of force applied to the grid causes a symmetrical rise and fall in the energy drawn from the battery connected to the remaining element in the tube called the *plate*. This change in the amount of energy flowing in the phones causes the small diaphragm to vibrate, emitting speech, music, code signals, or "what have you."

Construction Remarks

The detailed drawing covering the drilling specifications of the front panel is shown at Fig. 2 with a special template for laying out the tuning control for the condenser 6. Condenser 6 is

known as a *tank condenser* and serves as a master control for the *band-spreading* tuning condenser 7. Units 6, 7, 12 and 20 are mounted on the panel. Each tuning dial has a template to aid in the drilling of the three holes necessary for mounting the dial.

The rest of the parts are mounted as shown in the pictures and the detailed picture drawing, Fig. 4.

Run the wires exactly as shown in Fig. 4, carefully soldering each lead as the work progresses. For those desiring to construct this set and who can follow a schematic diagram, Fig. 5 can be consulted. SOLDER EVERYTHING CAREFULLY AND CHECK OVER ALL CONNECTIONS!

Testing Your Receiver

After all of the connections are made the set can be tested. Connect the bat-

teries as shown in Fig. 6 and insert a type 30 tube in the socket 10. Turn the control knob on the left hand side of the front panel to the right and adjust the knob on the rheostat until the filament inside of the tube glows with a dim cherry light. If a voltmeter is available then adjust 11 until the meter reads two volts. This is the correct operating point for this type of tube.

Plug in the coil with the yellow band around the top and set the knob controlling condenser 6 to 25 on the scale. Turn the knob on the left hand side of the front panel full on. Turning the main tuning control under these conditions should bring forth a series of whistles and chirps and if there is any speech or music being broadcast at that time it will be badly distorted. After a signal has been picked up turn the

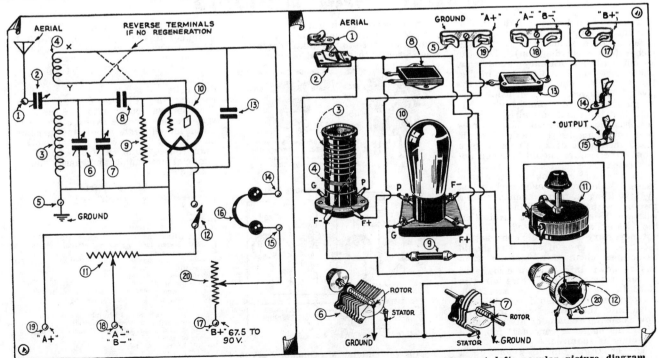

Regular schematic wiring diagram (Fig. 5) for the Denton 1-Tuber is shown above at left; popular picture diagram (Fig. 4) which anyone can follow, is shown at right.

knob of unit 20 slowly back until the signal clears up. Then readjust the master tuning control 7 for maximum volume.

Aerial and Ground

The aerial can be any length; use the same aerial that is connected to the receiver used for broadcast reception. Make sure that the connection to the ground is well made. Solder this connection if possible. Loose connections increase the noise that tends to swamp out the signal; the better the connections the less the noise and the more satisfactory the reception.

Troubles and Where to Find Them

Every radio set, no matter how simple, can refuse to work. The following information may be helpful:

If the Set Doesn't Work

"B" battery connections reversed.
Dry cells defective or worn out. ("A" battery.)
Wires left off in the construction.
Tube defective. (Have it tested.)
Antenna touching the side of the house and "grounded."
One of the windings of the coils "opened."

Tube Lights But Set Will Not Oscillate (Squeal)

"B" battery run down. Voltage too low.
Tube defective. (Tube can light and still be no good.)
Feed-back connections wrong. Reverse connections x and y, Figs. 4 and 5.
Open connection due to poor soldering or careless handling.
Improper adjustment of condenser 2.

Set Works but Signals are Weak

"B" voltage too low: increase the voltage by adding a new "B" battery in series.
Tube old or defective.
Adjust the small condenser 2 for maximum results.
Increase size of aerial.

Parts List

6 Fahnstock clips (1, 5, 14, 15, 17, 18, 19).
1 Pilot 4-prong socket for the plug-in coils.
1 Air-gap 4-prong socket for the type 30 detector tube (10).
1 Pilot 23-plate midget condenser (6).
1 Pilot 7-plate midget condenser (7).
1 Flechtheim midget condenser, .00025-mf. (8).
1 International Resistance Co. 3-megohm resistor (9).
1 Pacent 10-ohm rheostat (11).
1 Hammarlund equalizing condenser EC80 (2).

1 Alden Mfg. Co. set of short-wave coils (3. 4). 3 is the large winding and 4 is the small winding on all four coils.
1 Frost 100,000-ohm potentiometer with power switch (20, 12).
1 Aerovox mica condenser, .441-mf. (13).
1 Pair of good phones (16).
1 Wooden baseboard, 11 x 4½ x ½ inch thick.
1 Aluminum front panel, 6 x 11 x 1/16 inch thick.

Wood screws, solder, soldering lugs, wire, etc.

DATA ON ALDEN PLUG-IN COILS

	Number of turns		
(1)	4¾	6 Pitch No. 22 D.S.C.	Primary 4 turns No. 31 D.S.C.
(2)	10¾	12 Pitch No. 22 D.S.C.	Primary 6 turns No. 31 D.S.C.
(3)	22¾	16 Pitch No. 22 D.S.C.	Primary 7 turns No. 31 D.S.C.
(4)	51¾	40 Pitch No. 22 D.S.C.	Primary 15 turns No. 31 D.S.C.
(5)	68¾	Close wound No. 28 D.S.C.	Primary 28 turns No. 36 D.S.C.
(6)	131¾	Bank wound, 2 layers, No. 32 (Optional Litz)	Primary 32 turns No. 36 D.S.C.

WAVE BANDS:

(1) Blue—10 to 20; (2) Red—20 to 40; (3) Yellow—40 to 80; (4) Green—80 to 200; (5) White—200 to 350; (6) Orange—350 to 550.

D.S.C.—double silk covered. Pitch—turns per inch.

Use "sensitive" phones. Phones may be defective.

Poorly soldered connections. Resolder all connections, using a clean, hot iron.

These suggestions should help the fellow building his first short-wave receiver. If you have trouble, read these notes over again and see if you can be your own "trouble shooter."

Short-wave receivers of the more simple type have been described before but they have lacked several features that have been incorporated in this design.

Beginners have a habit of passing over any number of stations because they are used to tuning a broadcast receiver. A short-wave set should be tuned VERY closely, because if you don't you will miss them. Note that the main tuning condenser has but a few plates and is

equipped with a vernier dial for critical setting. The effect of tuning this small condenser is that of *spreading* the 100-degree tuning range over a small portion of the total range, which could be tuned by means of condenser 7, which is very large. Thus with a very simple set of coils the "tuner-in" can pick out that portion of the various tuning ranges that will give the maximum tuning control for accurate tuning. This feature alone will be of assistance to the beginner trying to find out about these short waves.

As experience crawls upon us, short-wave reception will lose some of its mystery and the builder of this simple receiver can look for bigger and better sets to build, but in the meantime let's try and build this one first.

2-Tube "Old Reliable"

● TO make the chassis of this instrument, procure an aluminum cooky sheet. These sheets are 12 x 15½ inches (cut as in diagram). This can be cut with an ordinary wood saw. Take piece No. 2 and bend as in Fig. 1. If no vise is handy bend it in the jaw of a drawer. Next drill all holes or punch out, then finish with file. Note that the small hole already in the pan has been enlarged, and is now used for bringing the cable through. The 7 inch piece of aluminum is for a front panel. Care should be taken not to scratch the surface.

To make the coils obtain three tube bases. Make a hole over each prong by first filing a groove then punching through with an awl. Wind the coils with No. 18 enameled wire. Note that the tickler coil is reversed. This is not necessary but it helps the set to oscillate. The ends of each wire should be scraped clean of all enamel, then pushed down into its own designated pin or prong. To hold these wires in place in the prongs, take a match, sharpen one end, then force this end down into the prong with the wire. Now snap the remainder of the match off. This makes a very neat job, and no solder is required.

The choke is jumble wound on a ½ inch core spool and consists of 150 turns of No. 24 wire. The 7 plate

antenna series condenser should have one stator plate removed. This will allow the set to oscillate correctly when used with an aerial of 100 feet.

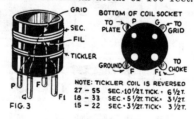

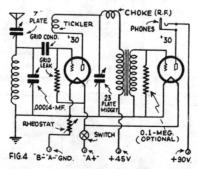

Details for building Mr. Simmons' 2-Tube Short-Wave Receiver, including wiring diagram.

The lead-in is brought in through a grommet in the right side of the set. The ground is connected directly to "A" minus on the "A" battery.

The filaments are lighted from a two volt battery or from a battery of two dry cells in series, with a ten ohm rheostat in "A" plus or "A" minus. This set was constructed to see if it was possible for the average person to build an efficient, small and dependable set at home with no tools other than those found in every home; of material obtainable anywhere; and at little cost.

List of Parts

1—Wearever cooky sheet 12x15½"
1—dial, 4 inch
2—knobs
1—filament switch
1—single circuit jack
1—Pilot (Hammarlund) 23 plate midget (Cap.-.0001 mf.)
1—Pilot (Hammarlund) 7 plate midget (Cap.-.00025 mf.)
1—tuning condenser 7 plate .00014 mf. (Hammarlund)
1—Sub-base transformer Stromberg Carlson 4-1; a 6-1 size will result in more volume.
2—wafer sockets (Na-ald)
1—Pilot socket (Na-ald)
1—Grid condenser .0001 mf. Aerovox (Polymet)
1—5 megohm grid-leak Aerovox (Lynch)
4—fibre washers
3—tube bases, some No. 18 wire plus hookup wire
1½ dozen nickel brass screws (not steel)
1—pair of good phones large size Brandes 2000 ohms.
2—type 30 tubes
2—"B" batteries 45 volt
1—"A" battery
1—Antenna system
1—ground
Coil data (see drawing)

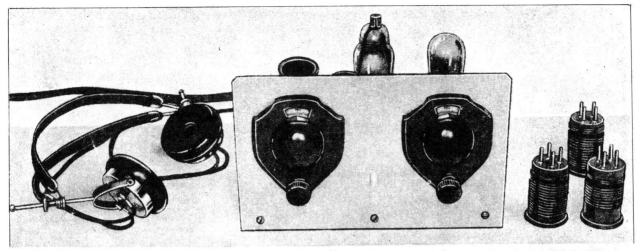

Looks like a very simple short-wave receiver to build, doesn't it? And it is, as you will agree, after reading the clearly written article by Mr. Shuart, well-known short-wave expert.

Rear view of the 110 Volt A.C. operated, 2-tube "Doerle" receiver. It provides world-wide reception as numerous tests have demonstrated.

The Famous
DOERLE
"2-Tuber"
Adapted
to A. C.
Operation

By GEORGE W. SHUART
(W2AMN-W2CBC)

● WITH all the fine reports from users of the famous "Doerle" receivers, the author decided to convert one of these receivers for A.C. operation using the new screen-grid pentode tubes. The results were so gratifying that it was decided to pass the information on to the readers of this magazine.

One of the latest models of this receiver was obtained for this purpose. This model uses two type 30, two-volt tubes; one as regenerative detector and another as transformer-coupled audio amplifier. The first operation is to remove all wiring, the two four-prong sockets for the two type 30 tubes, the filament rheostat, and the audio transformer; the four prong coil socket remains.

It might be well to mention at this point the list of parts necessary to do the job. They are as follows:

1—6-prong Wafer Socket
1—5-prong Wafer Socket
1—Screen-Grid Tube Shield, Type 50
1—2,000 Ohm Fixed Resistor, 1 watt

1—250,000 Ohm Fixed Resistor, 1 watt
1—2 Meg. Grid-Leak Type Resistor
2—.1 mf. By-pass Condenser
1—.005 mf. Fixed Condenser
1—1 mf. By-pass Condenser
1—Terminal Strip—5 lugs

The first of the above parts to be mounted are the two tube sockets. The six-prong socket is mounted in the center hole and the five-prong socket in the hole nearest the phone terminal strip.

Next mount the terminal strip with the five lugs on it in the center of the base on the under side. The one mf. by-pass condenser is mounted on the top side of the base in the position formerly occupied by the audio transformer. We are now ready to wire the set.

Hook-up "OK" for 2.5 or 6 Volt Tubes

Referring to the diagram it will be seen that the circuit is a straightforward regenerative one, with resistance-coupled audio amplifier stage and "throttle" (condenser) control of

regeneration. There are no changes in the circuit originally used in the Doerle receiver, other than those necessary to the use of the new type tubes. Either the 2.5 volt or the 6 volt tubes can be used in the new receiver, *with no change in the circuit* being necessary, the results being the same in either case. If the builder wishes to stick to batteries, and still have the benefit of the new type tubes with their high "gain," the use of the 6 volt tubes is recommended. In this case the detector should be the type 77, with a type 37 for the audio. This is very practicable as the set will operate on as low as 90 volts on the plates, although better results are obtained with from 135 to 180 applied to the tubes. A storage battery is used for filament supply for these tubes and lasts quite some time due to their low filament current rating.

Plate Supply
For 110 volt A.C. operation a

"power supply" is recommended; this should furnish 2.5 volts A.C. for the filaments, the *high-voltage* section supplying 180 volts, with a low voltage tap at 22 for the *screen*. This *screen voltage* is a very important point, as we are not controlling regeneration with a potentiometer in the screen-grid circuit, as is done in many other receivers. If this voltage is any higher than 22 volts the sensitivity of the receiver will be affected to a very great extent. Therefore one must remember, when the "throttle" condenser method of regeneration control is used with screen-grid tubes, the screen-grid voltage must be checked very carefully; otherwise poor results are liable to be experienced.

"B" Batteries May Be Used

If one wishes to use the 2.5 volt tubes and does not have on hand a regular power supply, a 2.5 volt filament transformer can be used to furnish the filament voltage with ordinary "B" batteries for the plate, (three, 45-volt batteries will operate the set very nicely and last for a long time, as the plate current of this set is in the order of 7 milliamperes. The foregoing paragraphs will give the builder an idea of just how flexible this set really is.

Wiring the set is a very easy task, and if the diagram is followed carefully no difficulty should be experienced in getting the set to "perk." All connections should be soldered with rosin-core solder and a *hot* and *well-tinned* iron. File the sides of the iron when they become corroded and retin by rubbing the hot iron in flux and solder. Rubbing it in sal-ammoniac or rosin and then appling solder is one of the old plumbers' tricks.)

Probably no other short-wave receiver of the 2-tube type has become so popular as the famous "Doerle."

Thanks to the use of the new type screen-grid pentode tubes, extreme increase in sensitivity is attained. Also the 6 volt D.C. tubes can be used, with no change in the circuit. Hundreds of S-W "fans" have requested data on how to rewire the Doerle receiver for 110 volt A.C. operation.— Well, Boys, here's how!

List of Parts

1—Antenna-Ground Terminal Strip
1—Phone Terminal Strip
1—Antenna Trimmer Cond. Cap. about 100 mmf.
1—5 wire cable
1—4 prong socket (Eby; Na-Ald; National; Hammarlund)
1—5 prong socket (Eby; Na-Ald; National; Hammarlund)
1—6 prong socket (Eby; Na-Ald; National; Hammarlund)
2—2 Meg. Resistors (Lynch)
1—250,000 Ohm Resistor (Lynch)
1—2,000 Ohm Resistor (Lynch)

1—.0001 mf. Mica Grid Condenser
2—.1 mf. By-pass Condenser
1—1 mf. By-pass Condenser
1—.005 mf. By-pass Condenser
1—Mounting Strip (5 lugs)
1—"Triple-Grid" Tube Shield
2—Hammarlund .00014 mf. Tuning Condenser
2—3-inch Vernier dials
1—Set of "Genwin" Plug-in Coils (15 to 200 meters)
1—57 or 77 tube (Triad)
1—56 or 37 tube (Triad)
1—Completely drilled A.C. Doerle chassis (Radio Trading Co.)

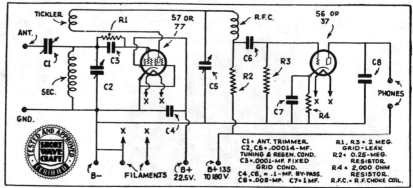

Here's the gratifyingly simple hook-up of the few parts used in constructing the A.C. operated "Doerle" set.

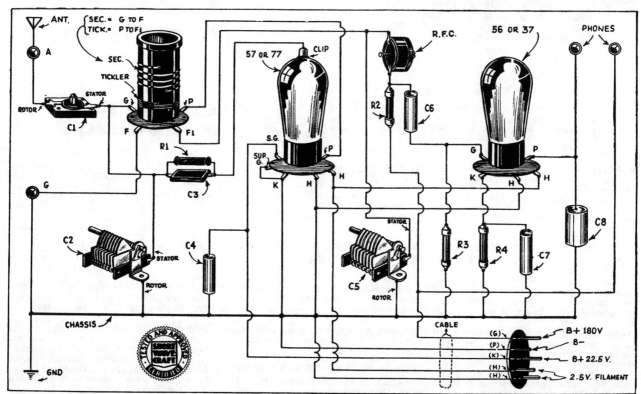

And in the event that you are not a dyed-in-the-wool short-wave "hound," who devours half a dozen R.F. chokes and a dozen plug-in coils for breakfast every morning, here's a "picturized diagram" which should make the construction of the A. C. Doerle a cinch!

The "OSCILLODYNE" 1-Tube WONDER SET

By J. A. WORCESTER, Jr.

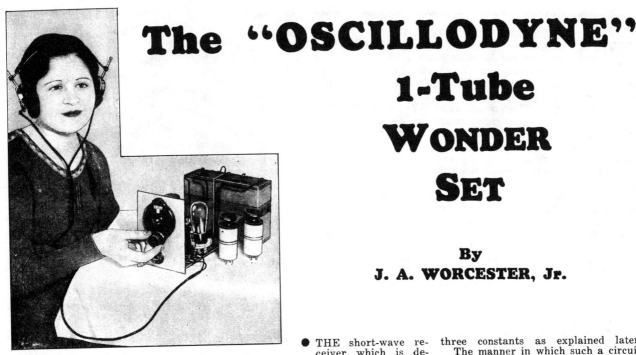

The 1-Tube "Oscillodyne" in actual operation.

● THE short-wave receiver which is described in this article depends for its operation on a principle which the writer believes is presented for the first time herewith. This receiver, while not presented as destined to replace existing methods of reception, is, nevertheless, in many respects the ideal receiver; particularly for the short-wave beginner or would-be beginner who is interested in obtaining the maximum "results per dollar" obtainable.

The fundamental circuit is shown in Fig. 1. A cursory examination will indicate that it is nothing more nor less than a simple oscillatory circuit. The feedback, however, is considerably greater than that required for the mere production of sustained oscillations, being of sufficient magnitude to produce *irregular oscillation*. This means that the oscillatory circuit is periodically rendered inoperative at a frequency dependent on the amount of feedback and on the value of the grid condenser and leak employed. In this receiver the oscillations are stopped and started at a super-audible frequency by proper selection of these three constants as explained later.

The manner in which such a circuit can be employed for the reception of radio frequency signals can be described as follows. In Fig. 1 is represented a high frequency disturbance of amplitude "A". If such a signal is present on the grid of the oscillator, this signal will build up as in Fig. 2B. In an ordinary oscillator, oscillations would build up to a value "B" (determined by the tube characteristics), as shown by the dotted lines of Fig. 2B. In this circuit, however, the feedback is too great to allow the electrons on the grid to leak off sufficiently fast to maintain a constant mean grid potential. The result is that the mean potential of the grid decreases, causing a corresponding decrease in the plate current as in Fig. 2C. As the plate current decreases the plate resistance increases, causing a decrease in the mutual conductance of the tube. Finally the plate current is reduced to a value "C" at which the mutual conductance is no longer sufficient to maintain oscillations and they die out as shown in Fig. 2B. The negative charge accumulated on the grid of the tube then leaks off at a rate determined by the time constant of the grid con-

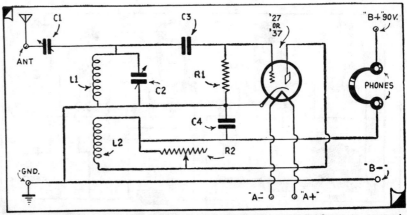

Schematic wiring diagram showing how to connect the few simple parts composing the "Oscillodyne."

Graphic diagrams employed by the author in connection with the text to explain the interesting action taking place in the "Oscillodyne."

A REALLY NEW CIRCUIT

WE are pleased to present to our readers an entirely new development in radio circuits.

Under the name of "The Oscillodyne," Mr. J. A. Worcester, Jr., has developed a fundamentally new circuit, and he describes the theory as well as the practical application in this article. This circuit, which is of the regenerative variety, acts like a super-regenerative set, although it does not belong in this class. Its sensitivity is tremendous.

The editor, in his home on Riverside Drive, New York, in a steel apartment building, was able to listen to amateurs in the Midwest on this simple one tube set, *using no aerial and no ground!*

With a ground alone, a number of Canadian stations were brought in, and with a short aerial of 40 feet length, many foreign stations were pulled in easily.

This circuit is certainly an epoch-making one which should find immediate acceptance by the entire radio fraternity. The circuit has the advantage that it is not tricky if good material and common sense are used.

The set was tested in different parts of the East, and it has been found that the results are satisfactory in practically every location.

In our own estimation, the Oscillodyne is one of the greatest recent developments in radio circuits, and the editors recommend it warmly to all readers.

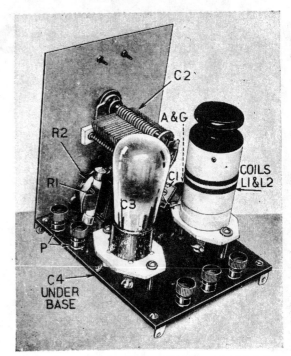

Rear view of the "oscillodyne," with parts labeled to correspond with those in the diagram.

---

denser and leak, whereupon the cycle repeats itself as shown.

A similar group of curves is shown in Fig. 3 for an initial disturbance having five times the amplitude of that in Fig. 2. The important thing to notice is that the average plate current ("D" Fig. 3C) is less than in the preceding case due to the greater number of "dips" the plate current makes during a given interval of time. Thus, it becomes obvious that a variation in the intensity of the signal applied to the grid results in a corresponding variation in the average plate current. Consequently, a modulated radio frequency signal will produce audible variations in the current flowing through the earphones in the plate circuit.

To sum up, it can be stated that the operation of this circuit depends on the fact that in an oscillatory circuit, prior to the establishment of sustained oscillations, *the time required* for an impulse to build up to a *given value* is proportional to the initial value of that impulse. This contrasts with the super-regenerative circuit in which

use is made of the fact that *the value* to which an impulse will build during a *given interval of time* is dependent on the initial value of that impulse.

Before leaving the theoretical side of the subject it might be advisable to point out that for proper operation of the circuit it is necessary that the oscillations in the grid circuit entirely die out during the period in which the charge is leaking off the grid. This is to enable the next train of oscillations to build up from the amplitude of the signal present on the grid at that time and not from the amplitude of the preceding train of oscillations which would otherwise be present. Thus it will be found that for satisfactory reception of broadcast frequencies the damping constant of the coil and condenser combination ($\varepsilon^{\frac{Rt}{2L}}$) is not large enough without adding considerable external resistance, which necessitates a corresponding increase in the feedback employed. The feedback cannot be increased indefinitely, however,

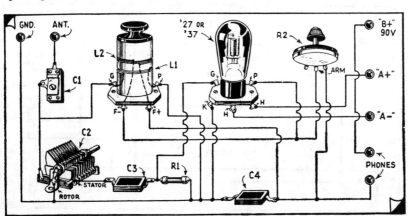

Picture wiring diagram for building the "Oscillodyne"; an "A-B-C" analysis of the set.

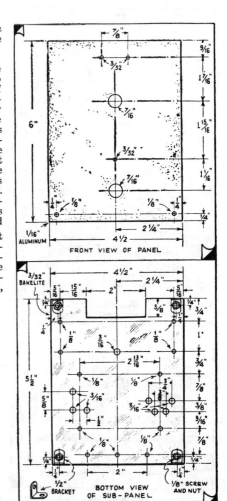

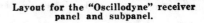

Layout for the "Oscillodyne" receiver panel and subpanel.

as it will be found that as soon as the natural frequency of the tickler coil becomes less than that of the tuned grid circuit, the plate load becomes capacitative and phase relations are no longer correct for oscillation.

It becomes evident, then, that as the frequency of the signals received is increased, enabling the use of smaller inductance coils, the damping constant, ($\varepsilon^{\frac{Rt}{2L}}$) increases, and operations of the circuit becomes more satisfactory. Henc it will be found that this circuit is particularly well adapted to short-wave reception (which is also true of the super-regenerative circuit and for the same reason.

How to Make the Simplest One-Tube Oscillodyne Set

In this article is described a simple one tube receiver employing the *oscillodyne* principle.

The schematic diagram for this receiver is shown in Fig. 4. The tube employed is a type 27 employing 2½ volts A.C., or a 37 using 6.3 volts D.C., on the heater and 90 volts plate potential supplied by a "B-eliminator," or battery. Other tubes such as the type 30, 56, 01A, 12A, etc., may be used if desired. The only change necessary is to supply the appropriate filament voltage for the tube selected. In general, screen grid tubes are not satisfactory in a one tube receiver due to the difficulty of matching the extremely high plate impedance of the tube to that of the earphone.

The plug-in coils employed are wound on tube bases. The specifications for the windings are given in the table accompanying this article. The turns of both windings are wound without spacing. It is essential that the two windings be wound in the same direction. This means that if the two inside terminals of the windings are connected together, the coil will appear like a continuous winding tapped near the center.

In regard to coil specifications, the following table is furnished for tube base coils wound with No. 36 D.S.C. wire and tuned with a 100 mmf. (.0001mf.) condenser. The first two coils may need a half turn adjustment one way or the other.

Approximate Wavelength (meters)	Sec.	Tickler
14- 25	4	6
23- 41	7	9
40- 85	14	12
83-125	23	23
120-200	36	36

About ⅛" separation between windings. It will obviously be necessary to extend the tube base forms if coils for the "broadcast band" are used. However, grid and plate windings of about 67 turns will tune from 200-360 meters and 105 turn windings will tune from 350-550 meters with the above condenser.

After the leads are soldered in the tube prongs, all superfluous solder should be carefully filed from the sides of the prongs to prevent damage to the coil socket when inserting. The windings should be so connected that the two outside leads go to the grid condenser and plate of the tube, respectively, while the two inner leads go to the cathode and phones respectively. If connections are not made in this manner *the tube will not oscillate!*

In order to provide exact coverage of the various frequency bands with suitable overlap at each end, it may be found desirable to vary the number of grid turns by a half turn or so for certain coils.

A suggested layout of parts is shown in the photographs. If other parts than the ones used are substituted it may be necessary to vary this layout somewhat. In wiring the receiver only nine leads are necessary and if these are carefully made no difficulty should be experienced from improper wiring. In preparing a lead to which several connections have to be made.

such as the ground connections, a much more convenient method of removing wax impregnated insulation than by scraping off with a knife, is to mash the insulation at the desired points with a pair of long-nosed pliers. The insulation can then be readily removed with the fingers. Soldered connections are not essential but should be made if possible.

The 50,000 ohm variable resistance should have an insulated shaft and bushing so that it can be directly mounted to the aluminum panel. Otherwise, it will be necessary to first mount the instrument to a strip of bakelite, which in turn is mounted to the panel. If this is done, the shaft hole should be large enough for proper clearance. A potentiometer can obviously be used for this purpose by employing only one of the two outside terminals.

In making connections to the variable condenser, the ungrounded terminal should be connected to the grid condenser so that the panel will be at ground potential.

Operating Notes

When ready to operate the receiver, the first thing to do, of course, is to make the various connections to the antenna, ground, "A" and "B" supply, and phones. The antenna compensating condenser, C1, should be set at close to its minimum value. The variable condenser should be set so that its plates are within about 15° of "all-in"; and coil No. 3 inserted in the coil socket. The variable resistance R2 should be set so that its maximum resistance is in the circuit.

The circuit is now tuned somewhere near the 80 meter amateur phone band. If the set is now turned on, a fairly loud high-pitched note should be heard in the earphones. The resistance R2 should now be decreased until this note becomes inaudible and a "hissing sound" is heard. If the variable condenser is now rotated slightly it should be possible to tune in an amateur phone transmitter. *When this is done the resistance R2 should be varied for best reception.* The antenna compensating condenser should now be set for maximum volume.

When using the 20 meter and 40 meter coils it will generally be found necessary to increase the resistance R2 to a greater value than required for the 80 meter coil. While this control is not nearly as critical as the regeneration control in a regenerative receiver, it is, nevertheless, necessary to exercise some skill in its manipulation before maximum results can be obtained.

It will generally be found that when foreign stations are to be received they will come in with nearly the same ease as locals; while when they are not to be received all the coaxing in the world will not bring them in. The absence of foreign stations on the dials can be attributed to a number of causes. In the first place, there may not be any broadcasting at the time the listening is being done; or the frequency band on which listening is being done may not be suitable for foreign reception at that particular time of day.

In general, it will be found that from daybreak to about 2 p. m. foreign reception is best on 14 to about 20 meters; from 2 p. m. to 9 p. m. on 20-35 meters, and from 9 p. m. to daybreak on 35-75 meters.

Even when listening at the right time to a foreign station that had been received regularly for days, it will often be found that the station has suddenly disappeared entirely only to reappear, just as suddenly, a week or so later. Experiences of this nature are very common on short waves and can only be attributed to the vagaries of short wave transmission.

Trouble Shooting

Difficulties encountered in getting the set functioning properly can be grouped in three classifications as follows:
1. Set refuses to operate.
2. Set oscillates but will not break into irregular oscillation

3. Set oscillates irregularly but does not function properly.

To determine whether the set is oscillating or not touch the terminal of the grid condenser that is *not* connected to the grid. If this results in a click in the phones the set is oscillating, and vice versa. If the set is not oscillating the first thing to determine is whether plate current is flowing. This can be determined by disconnecting one of the phone leads and making and breaking this connection by hand. If this results in corresponding loud clicks in the phones, plate current is flowing and the difficulty is elsewhere. If plate current is not flowing there is probably an "open circuit" in the plate or heater circuit. Make sure that the plate potential is not reversed; also that the coil is making contact with the socket and that the B— (minus) terminal is connected to the cathode. Also re-examine the plug-in coil to make certain that the connections have been made properly. Also make certain that the tube is not defective.

If the tube oscillates but does not break into irregular oscillation (high pitched note in earphones with R2 at maximum) make certain that the plate and filament voltages are correct. Also make sure that the tube is not faulty. Reduce the antenna compensating condenser to its minimum value or temporarily disconnect the antenna. If this procedure rectifies the trouble, the antenna condenser has too large a minimum capacity and a smaller one should be substituted. Rock the plug-in coil slightly to make sure there is not a *high-resistance* contact.

If the tube oscillates irregularly, but the set does not function properly, the trouble is probably with the tube or grid condenser and leak combination. If a new tube does not improve results, try a .00005 mf. grid condenser at C3 and experiment with different values of leak resistance from about one to seven megohms.

The editors of SHORT WAVE CRAFT had a special highly insulated model of the Oscillodyne built and this is the model shown in photographs herewith. Of course results can be obtained with a bread-board model, thrown together with odd parts, but, as in every piece of electrical apparatus—and particularly in the case of a sensitive radio receiving set such as the Oscillodyne, which is designed to realize the greatest possible strength of signal from one tube, it behooves us to thoroughly insulate evey part of the set to the best of our ability.

To that end, the coils were wound on Hammarlund *Isolantite* forms. As is well known, Isolantite is superior to ordinary Bakelite for use as an insulator in short wave and ultra short wave work. Next, Isolantite sockets were used for both the tube and the coil and all of the parts were mounted on a bakelite subpanel, to still further enhance the insulation.

Parts List For Building the Oscillodyne

1—Aluminum panel, 4½"x6"x⅟₁₆". Blan (Insuline Corp. of America.)
1—Bakelite subpanel, 4½"x5½"x3/32". Insuline Corp. of America.
1—50,000 ohm variable resistor, R2, Frost, (Clarostat).
1—Set of 4 pin plug-in coils wound on Hammarlund Isolantite forms 1½" dia., per specifications given in article.
1—Series antenna condenser, Cl, about 25 mmf. max., Hammarlund Compensator type condenser.
1—Variable tuning condenser, C2, .0001 mf., Hammarlund.
1—Grid condenser, C3, 100 mmf., or 50 mmf. Illini (Polymet)
1—Fixed resistor, R1, 3 megohms, Lynch.
1—Fixed condenser, C4, .0005 mf., mica type, Pilot or Flechtheim. (Polymet)
7—Binding posts, Eby.
1—3" midget National Velvet Vernier Dial, type BM.

Building the 2-Tube "Globe-Trotter"

The Cost Had to Be Low. The 2-Tube "Globe-Trotter" Is Complete, As Described, for Battery Operation. On Test It Brought in European and Other Distant Short-Wave Stations.

By ROBERT HERTZBERG, W2DJJ

● THE instrument illustrated on these pages is probably the simplest *complete* short-wave receiver that the beginner can build. The word *complete* is italicized because the set includes all the necessary "A" and "B" batteries right on the baseboard, along with the parts of the receiving circuit proper. A great many so-called "beginner" sets look simple and cheap because the batteries or other sources of power are kept separate from the receiver unit, yet these batteries may cost as much as and take up more room than the latter itself. The

writer recalls one such receiver which could be built for about four dollars and occupied as much space as an ordinary cigar box, but which required a fifteen-pound storage battery and four 45-volt "B" batteries.

Uses But Two Tubes

Using two tubes of the '30 type in a straightforward, "sure fire" regenerative hook-up, this little outfit is an *honest-to-goodness* receiver, and not a mere toy. It employs the minimum number of parts to make the circuit operative, with all fancy embellishments eliminated. The first tube is a regenerative detector, the second an audio frequency amplifier with transformer coupling. This combination is quite adequate for comfortable earphone reception. As a matter of fact, if it is adjusted carefully it will bring in most everything on the short waves worth hearing.

Since the two '30 tubes draw a total of only .12-ampere at two volts, the use of dry cells for filament supply is altogether practicable. Two No. 4 dry cells connected in series will last for several weeks of frequent listening. These batteries are only 2 inches square by 4 inches high—an extremely convenient size for radio purposes. They are much less bulky than ordinary No. 6 cells and much more satisfactory for continuous service than small "C" or flashlight batteries, which drop in voltage very quickly if more than a microscopic current is drawn from them. The chain stores sell the No. 4 batteries for ten cents—another feature in their favor.

Plate current is furnished by two 22½-volt "B" batteries of the smallest standard size, which measure 3⅜ x 2 x 2½ inches high. Along with the "A" batteries, these fit very nicely along the back of a wood baseboard 9 inches wide and 9½

Wiring diagram for the 2-Tube "Globe-Trotter"—a receiver that you will enjoy.

Left—Two views of the 2-Tube Short-Wave "Globe-Trotter" Receiver, designed and built by "Bob" Hertzberg. Photo at right shows set minus batteries and wiring.

inches deep, being held in place by a ten-cent web strap one inch wide.

Economical "B" Battery Used

No apology is offered for the use of three-element tubes instead of screen-grid tubes in this receiver, or for the use of a plate voltage as low as 45. The point is that screen-grid tubes require at least 135 volts of "B," which means a flock of expensive batteries, while three-element tubes work very sweetly on 45 volts and even less. Surely, screen-grid tubes would work better, but the cost and size of the set would be tripled! Such things are all purely relative.

Coil Data

There are numerous coils on the market that are designed to cover the 15 to 200 meter range with .0001-mf. tuning and regeneration condensers. A set sold by the Radio Trading Company, No. 1616, was used in the model receiver. These use four-prong form 2⅛ inches long and 1¼ inches in diameter, and are all wound with No. 24 double cotton covered wire. The grid winding in each case connects to the plate and right filament prongs in the base, the tickler to the grid and left filament. Grid and tickler coils are wound in the same direction and are separated about ⅛ inch. If you want to "wind your own," follow this dope:

Approximate Wavelength Range	Number of Turns	
	Grid Coil	Tickler Coil
15- 25 meters	6	7
25- 50 meters	12	8
50-100 meters	24	13
100-200 meters	54	20

Start the assembly work by mounting the two midget condensers in the vernier dials. Tighten the shafts of the condensers in the studs of the dials, and then fasten the latter upright in the position shown by means of brass angles 4 inches long and 5/16 inch wide. The K-K dials are fitted with convenient screws that make this construction possible. Now drill holes in the baseboard just under the threaded mounting feet of the condensers, so that when long 6-32 machine screws are passed through these holes they will go into the feet and prevent the condensers from turning when the dials are turned. Put two 6-32 nuts on each screw before turning them into the mounting feet; tighten one against the top of the baseboard and the other under the condenser foot. In this manner the condenser and dial assembly will be made quite rigid.

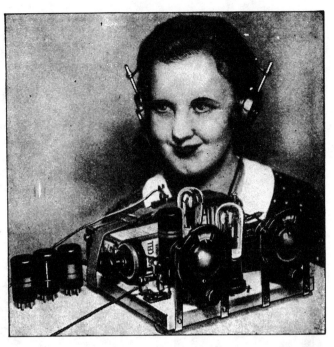

The "Globe-Trotter" in actual operation, with extra plug-in coils at the left.

LIST OF ESSENTIAL PARTS

In addition to the batteries and baseboard specified, the set uses the following parts:

2 Midget variable condensers, .0001-mf. (Hammarlund used in original model because they mount very easily.)
2 Self-supporting 3-inch vernier dials (Kurz-Kasch).
3 Four-prong tube sockets (Pilot).
1 Audio transformer, not less than 3:1 ratio, not more than 6:1. (An old Stromberg-Carlson was used because it was advisable; any other good make will do.)
1 .0001-mf. mica grid condenser (Aerovox).
1 3-megohm grid leak (Lynch).
1 Special (Blan) short-wave radio-frequency choke coil.
1 "Postage stamp" antenna condenser, about .0001-mf. (Hammarlund).
1 10-ohm filament rheostat. (This can be any make that has holes through the base or has other means for vertical mounting on the baseboard.)
4 Fahnstock spring binding posts for baseboard mounting.
1 Set of four plug-in coils, 15 to 200 meters.

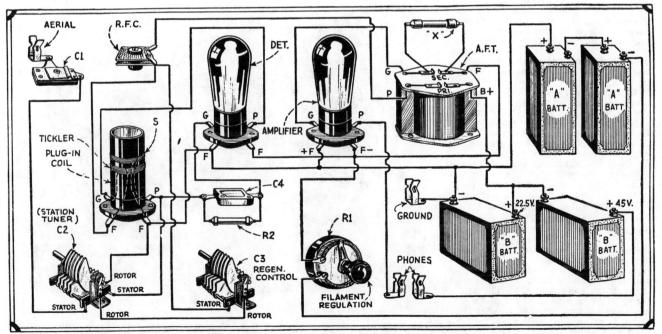

Even though you have never built a radio set, you can't help but go right with this 2-Tube "Globe-Trotter," as the "picture diagram" above shows you just how to wire it.

The PENTAFLEX

By J. A. WORCESTER, Jr.

"Two tubes for one"—that's what you actually achieve with this newest concoction of Mr. Worcester's, who is well-known to our readers for his "Oscillodyne" receivers. By utilizing the new 6A7 (or 2A7) pentagrid-converter tube in a reflex circuit, the author has indeed discovered a very remarkable combination. George Shuart, W2AMN, took a whirl at the dial and admitted that this was the "greatest" one-tuber yet.

Words fail us when it comes to telling about the results the editors obtained and which you can also obtain, without a doubt, with the "Pentaflex"—the latest brainchild of J. A. Worcester, Jr., originator of the Oscillodyne receivers

● THE receiver described in this article is so named because it utilizes the new 6A7 pentagrid converter tube in a reflex circuit. With this connection it is possible to obtain the equivalent of a two tube receiver employing a screen grid detector and one stage audio with only one tube; thus enabling an appreciable saving in space, equipment and power consumed.

In actual practice, this circuit has even proved superior in regard to volume to the conventional two tube circuit. This is probably due to the construction of the tube for its intended purpose whereby there exists a certain amount of electron coupling between the two circuits, thus producing a small amount of audio frequency feedback and a consequent increase in signal strength. The writer has tuned in GSA, Daventry, and DJC, Germany, every night for more than a week with this small set and although the volume appeared to fluctuate considerably from night to night, it was possible on all occasions to bring these stations in with sufficient volume to readily understand the announcement. On some occasions the volume was actually too great for comfortable earphone reception.

Description of the Circuit

Referring to the schematic wiring diagram, it will be noted that the input circuit is of a conventional nature. Inspection of the plate circuit, however, will reveal that the R. F. currents after passing through the tickler winding are by-passed to ground through the condenser C6. The audio frequency component of the plate current passes through the resistor R5, causing an audio frequency voltage drop across this resistor. This voltage is impressed on the first grid through the blocking condenser C5. The resistor R4 prevents a negative charge from accumulating and blocking the grid. The amplified currents flowing in the second grid circuit, which becomes the plate of the triode amplifier, pass through the earphones as shown.

Regeneration is controlled in the usual manner by varying the screen grid voltage with a 50,000 ohm potentiometer.

The resistor R2 provides bias for the triode grid, while the return of the control grid is made directly to the cathode.

The heater current is turned on and off by a double-pole, single-throw switch which also breaks the potentiometer return, thus preventing plate current drain through the potentiometer winding when set is not being used.

Batteries or A. C. Can be Used

Either a 6A7 or 2A7 tube may be used, which are identical except for the heater characteristics. The 6A7 requires a heater voltage of 6.3 volts at .3 amperes, while the 2A7 requires 2.5 volts at 1 ampere. Both tubes are designed for either A. C. or D. C. operation. If it is desired to use dry cells, it will probably be found more convenient to use the 6A7 with four dry cells connected in series, while if a 2½ volt A. C. source is available,

Yep! Only one tube to buy, a 6A7 or a 2A7 pentagrid-converter, plus a little simple wiring, and you have a sensitive receiver which has an output equivalent to "two" tubes—thanks to the reflexing.

the sturdy 2A7 tube can be used.

The front panel consists of a 5" x 7" piece of 14 gauge aluminum while the subpanel is formed by bending a 7"x5" sheet to a depth back of the panel of 3 inches and a height of 1 inch. The panel is fastened to the subpanel by three machine screws.

Mounted on the front panel are the Hammarlund variable condensers, the switch and the potentiometer.

Underneath the subpanel are mounted the bypass condensers, the .0005 mf. by-pass and the resistors. All of these parts are mounted by their pigtails and wherever it becomes necessary to expose any appreciable length of same, they are covered with spaghetti tubing for insulation purposes. The sockets are also mounted under the subpanel, although there is no objection to mounting them above if facilities for cutting the holes required are not available. It might be pointed out, in this connection, that there are two sizes of 7 prong sockets, having different pin circle diameters. The smaller size, having a .75" pin circle diameter, is the one that takes the 6A7 tube.

The twin binding post strip and the twin speaker jack assembly are mounted at the rear. Battery connections are made by connecting a five conductor cable directly to the required points.

The equalizing condenser and the grid condenser and leak are mounted directly on the Hammarlund variable condenser, as shown.

Coil Data

The coils can be obtained ready-wound, but specifications are furnished below for winding the coils on blank forms, if desired. The data furnished are for the manufactured Alden coils, but if these coils are constructed it is recommended that about fifty to a hundred per cent more tickler turns than those specified be employed. This is because this tube when used in this circuit requires more feedback than that required for the usual screen-grid tube.

Although all the manufactured coils will oscillate by decreasing the capacity of the antenna condenser sufficiently, the use of more tickler turns is recommended in that it permits closer antenna coupling with consequently increased input.

Coil specifications:

Coil Winding Data

Band Meters	Grid Coil Turns	Tickler Coil Turns	Space between 2 Coils
10- 20	4¾ T. No. 22 Wound 6 T. per inch	4 T. No. 31 Close wound	3/32"
20- 40	10¾ T. No. 22 Wound 12 T. per inch	6 T. No. 31 Close wound	3/16"
40- 80	22¾ T. No. 22 Wound 16 T. per inch	7 T. No. 31 Close wound	3/32"
80-200	51¾ T. No. 22 Wound 40 T. per inch	15 T. No. 31 Close wound	⅛"
200-350	68¾ T. No. 28 Close wound	28 T. No. 36 Close wound	⅛"
350-500	131¾ T. No. 32 Bank wound in 2 layers	32 T. No. 36 Close wound	⅛"

Data for Na-Ald coils form 1¼ inches dia. by 2⅛ inches long (4 pin).

Operating Features

The operation of this receiver is no different from the usual screen grid detector and one step with the following exception. As the feedback is increased to a point where oscillation begins, a series of regular clicks will sometimes be heard which vary slightly in frequency with the feedback employed. As these interfere with reception when receiving C. W., it is necessary to increase the feedback still further until the clicks stop. Outside of this one eccentricity at some positions of the tuning condenser, it will be found that the regeneration control is generally not as critical as in the usual regenerative receiver.

PARTS REQUIRED FOR THE "PENTAFLEX"

C1—Hammarlund Equalizer EC-35 (3-35 mmf.)
C2—Hammarlund 140 mmf. midget condenser (MC-140-M)
C3, C5—.5 mf. tubular condenser, 200 D. C. W. V.
C4—.01 mf. mica condenser, 200 D. C. W. V.
C6—.0001 mf. pigtail mica condenser
C7—.0005 mf. pigtail mica condenser

R1—Lynch 3 meg. metallized grid leak ½ watt
R2—400 ohm tubular wire-wound pigtail resistor
R3—50,000 ohm potentiometer
R4, R5—Lynch .25 meg. metallized resistor, ½ watt
L1, L2—Alden (Na-ald) plug-in coils (see text for details)
1—National type "BM 3" dial (0-100-0)
1—National grid-clip, type 24
1—D. P. S. T. switch
1—Eby twin binding post assembly (laminated)
1—Eby twin speaker jack assembly (laminated)
1—Eby laminated 7 prong socket, small (.75" pin circle diameter)
1—Hammarlund 4 prong isolantite socket (S-4)
1—6A7 or 2A7 tube
3—FT. 5-conductor cable
1—Roll Hook-up Wire
1—Blan Aluminum panel, 14 Ga, 5"x7"
1—Blan Aluminum panel, 14 Ga, 5"x5", bent to form 3"x5"x1" subpanel.

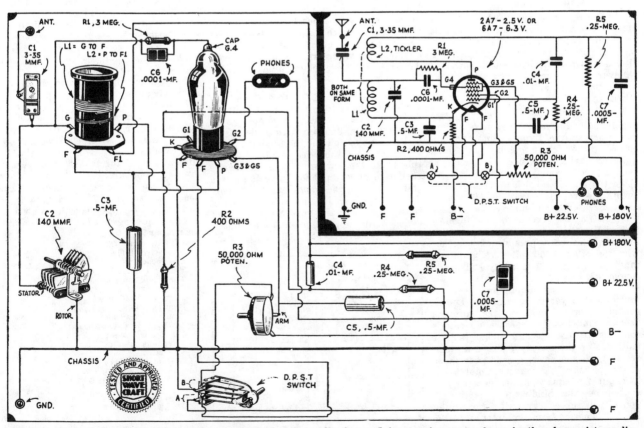

One tube does all the work—either a 6A7 or a 2A7, plus the really few and inexpensive parts shown in the above picture diagram. Anyone can build this "one-tuber," which really gives the same output as a "2-tube receiver," due to the reflex circuit used, as verified by Mr. Shuart, well known to our readers.

A 2-TUBE
Receiver
That
REACHES
the
12,500 Mile
MARK

By WALTER C. DOERLE

This low-priced head-phone receiver comprises a few well-chosen parts arranged in a well-tried circuit.

An easily built short wave receiver for the beginner, using but few parts of low cost. Note anti-capacity condenser controls.

WHAT the heck's" the idea of wasting power, of blasting out ear-drums, of going "bugs" with the performance of a costly short-wave receiver, when you can build a two-tube outfit that gets signals from the 12,500 mile meridan? Why, there is nothing to brag about when you "log" a bunch of stations with a powerful receiver; but listen to a man who "pets" a two-tube set, and then you get the "thrill of a lifetime".

Antenna and Ground

Now as to the antenna, a wire strung twice across the living room and anchored to the picture molding with small finishing nails, together with a good "water-pipe" ground connection, has enabled the author to pick up signals with such a receiver from stations 6,000 miles away, even on a hot summer day on the Pacific Coast (Oakland, California.) Say, fellows, if a well-insulated outside antenna had been possible of erection, why the other 6,500 miles of "no-man's land" would have been easily heard and conquered.

Time is moving along, and there is much ground yet to be gained. Let us consider for a moment the antenna "series condenser". For the operator's convenience, a seven-plate midget is quite suitable for the purpose; but in a small receiver of this price, a condenser made of two pieces of old condenser plates, cut to about 1½ square inches in area and spaced on the binding-post strip ⅛" apart, will serve very well for coupling the R.F. energy from the antenna to the oscillating circuit of the receiver.

Be sure that the post strip is of bake-

lite; as this is the cheapest, though not the best, insulation for the purpose. In some experiments made by the author, a home-made series condenser was mounted on ¼-inch plywood baseboard, but a surprise awaited—the signal intensity as heard in the phones was about three-quarters its value when the series condenser plates were mounted on the bakelite strip.

Since this type of receiver would undoubtedly call for home-made plug-in coils, because of their convenience, we follow up our diagram with a discussion of this type of coil for the oscillating circuit. To hold the wire in place on the tube-base, the author has found orange shellac to have small loss, and it gives a shiny finish to the form. As to the condensers for use in this receiver, select those that have the smallest amount of dielectric in supporting the stator plates.

Have you ever experimented with various values of grid condensers and leaks in the detector circuit? Well, get about twelve leaks (½ to 10 megs). and twelve different sizes of grid condensers (.006- to .0001-mf.), but first of all figure out the possible number of combinations.

Use a 5-megohm leak and .0001-mf. grid-condenser. These values will make the receiver very sensitive.

Now, in our discussion we are near the audio-frequency transformer and our eyes immediately behold an R.F. choke. Gee, what a mean thing for the temper; but, at any rate, 300 turns of No. 36 D.S.C., magnet wire, close-wound on a ½" wooden dowel, will choke the R.F. current out of the transformer primary, even at 20 meters.

As to the audio transformer, we can't boast for any type; but a good 5 to 1 ratio and a hefty type, will be good.

The following is a list of parts for the set proper:

1—Bakelite panel 7"x10";
1—Baseboard 9x11";
2—UX Sockets;
1—Tuning Condenser .00014-mf.
1—Throttle Condenser .00025-mf.;
2—Condenser Plates 1½" square;
7—Terminal Post-strip;
7—Binding Posts;
5—Megohm Grid-leak;
1—.0001-mf. Grid Condenser;
1—5:1 Transformer;
2—Telephone Binding Posts;
2—3" Dials;
1—20-Ohm Rheostat;
Hook-up wire, screws, etc.

COIL DATA

Range (meters)	Turns S	T
15-45	5	6
35-75	9	5
60-125	16	6

All coils are close-wound with No. 24 enamelled copper wire, and with no spacing between S and T.

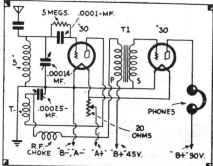

Circuit used by Mr. Doerle for the 2-tube "globe circler".

S-W Tuning Inductance Charts

The accompanying charts enable the Short Wave Fan to ascertain the inductance of a certain size coil without having to perform lengthy mathematical calculations. The graphs given are "direct-reading"—the inductance in microhenrys of a certain diameter coil being quickly and easily found.

By CLIFFORD E. DENTON

● SO many readers write in for "dope" on building coils for short wave reception that this article should be of interest to all.

The average experimenter selects a condenser and then starts to figure or guess the values, both *mechanical* and *electrical*, for the proper size coil to cover the various bands.

Now various sets of specifications can be found in every issue of SHORT WAVE CRAFT but in many cases when the coils are wound, certain factors creep in which seems to make the coils very unsatisfactory and the reader starts to think that the wrong "dope" has been given.

Let's review some of the fundamentals and perhaps this will give the prospective coil builder a better idea as to just what he is doing.

The unit of self-induction is called the *henry* and is defined as "a rate of current change of one ampere per second giving an induced potential of one volt." The henry, as a unit, is suitable for solution of problems in the audio and power-supply band of frequencies, but when *short-wave coils* are being designed a smaller unit is used

and these subdivisions are the *millihenry* and the *microhenry*, respectively equal to one-thousandth and one-millionth part of a henry. A seldom-used unit of inductance is the *centimeter* which is equal to one billionth part of a henry; it is the smallest subdivision. Note that the unit of inductance is also the unit of length.

The *farad* is the unit of capacity and is defined as a condenser which requires one coulomb of electricity to bring a potential difference of one volt to its plates. A coulomb is the quantity of electricity that passes through a circuit in one second when the flow is one ampere.

Again we find a unit that is too great for use in practical problems and circuits, so we subdivide farads into *microfarads* and *micro-microfarads*, respectively equal to one millionth and one million, millionth part of a farad. Here the base unit is the centimeter and one centimeter is equal to one nine-hundred-thousandth part of a microfarad. Thus the micro-microfarad and the centimeter are nearly the same in size, one centimeter being equal to 1.1 micro-microfarad. The unit

of capacity is also a unit of length.

Let's see; the capacity of a sphere is found to vary as its radius, and in the electrostatic system, a sphere with a radius of one centimeter has unit capacity. Thus, a condenser in the form of a sphere having a capacity of one farad would have a radius of 5,592,329 miles. The radius of the earth is approximately 650,000,000 centimeters; so its capacity should be about 700 microfarads. Say, this world of ours is not so large at that.

We will use microfarads and microhenries in the solution of our problems but it is interesting to note what would happen if we had circuits using henries and farads. Suppose we had a capacity of one farad connected to an inductance of one henry. It would take six seconds to complete one cycle; slow motion, more or less. The wavelength in this case would be 1,800,000,000 meters. Substituting a capacity of one microfarad would cause the circuit to oscillate 1000 times in six seconds and the wavelength would be 1,800,000 meters.

The combination of a coil (inductance) and a condenser (capacity) forms a *resonant circuit* and has the

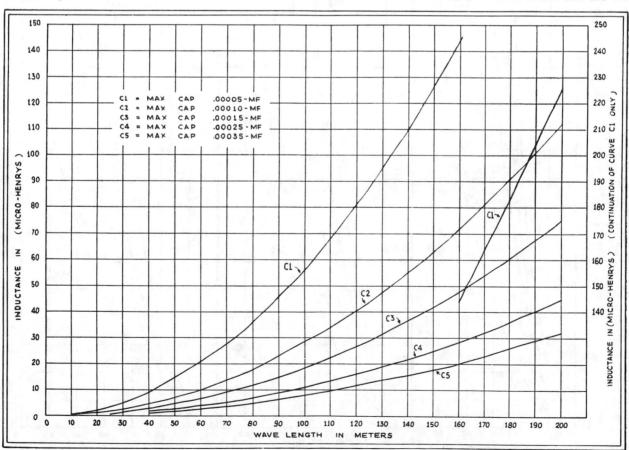

Chart showing relation between wavelength desired, condenser capacity and coil inductance in microhenrys.

ability to resonate large voltages and currents at some particular frequency. The frequency at which this phenomenon takes place depends upon the value of the capacity and inductance in the circuit. Thus,

$$f = \frac{159,200}{\sqrt{LC}}$$

Where f is the resonant frequency expressed in cycles,

L is the inductance in microhenries and C is the capacity in microfarads.

Instead of calculating the reader can refer to the LC Chart or Table of Fig. 1. This chart covers all wavelengths from one meter to two hundred meters. The chart is easy to use as the prospective coil-builder generally knows what type of tuning condenser he is going to use and its capacity. For example: A tuning condenser with a capacity of .00015 mf. is at hand. What must be the inductance of the coil to tune to a wavelength of 200 meters?

Referring to table 1 we find that the product of L times C corresponding to 200 meters is 11.26. Dividing this number by .00015 (the capacity of the tuning condenser in microfarads) we obtain 75,066 which is the inductance of the coil in centimeters. Now 1,000 centimeters equals 1 microhenry, so we must divide 75,066 by 1,000 to find the inductance expressed in the form of

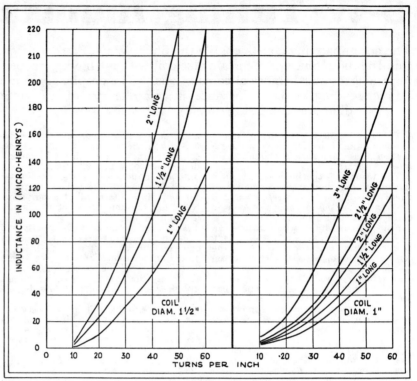

Additional "direct-reading" values of coil inductances with relation to the physical dimensions of the coil.

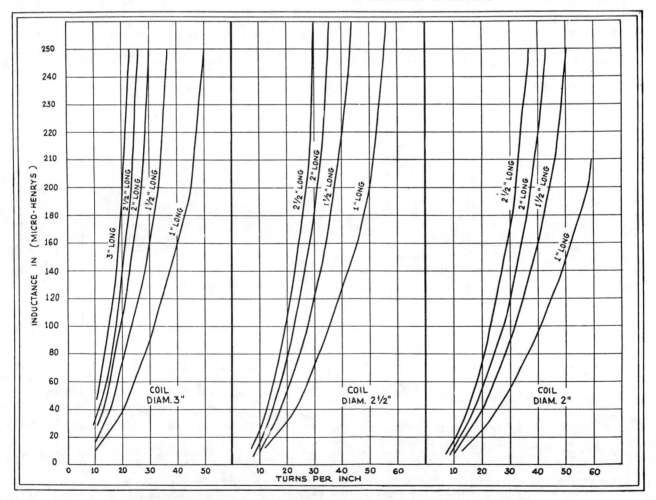

Graphs giving relations for various diameter coils and inductance in microhenrys.

λ Wave length meters	f Multiply values below by 1000	ω Multiply values below by 1000	CL C in uf (mf) L in cm
1	300000	1884000	0.0003
2	150000	942000	.0011
3	100000	628000	.0018
4	75000	471000	.0045
5	60000	377000	.0057
6	50000	314000	.0101
7	42900	269000	.0138
8	37500	235500	.0180
9	33330	209400	.0228
10	30000	188400	.0282
15	20000	125600	.0635
20	15000	94200	.1129
25	12000	75400	.1755
30	10000	62800	.2530
35	8570	53800	.3446
40	7500	47100	.450
45	6670	41900	.570
50	6000	37700	.704
55	5450	34220	.852
60	5000	31420	1.014
65	4620	28970	1.188
70	4290	26900	1.378
75	4000	25120	1.583
80	3750	23520	1.801
85	3529	22120	2.034
90	3333	20920	2.280
95	3158	19830	2.541
100	3000	18840	2.810
105	2857	17940	3.105
110	2727	17130	3.404
115	2609	16380	3.721
120	2500	15710	4.05
125	2400	15070	4.40
130	2308	14480	4.76
135	2222	13950	5.13
140	2144	13450	5.52
145	2069	12980	5.92
150	2000	12560	6.34
155	1935	12150	6.76
160	1875	11770	7.20
165	1818	11410	7.66
170	1765	11080	8.13
175	1714	10760	8.62
180	1667	10470	9.12
185	1622	10180	9.63
190	1579	9910	10.16
195	1538	9660	10.71
200	1500	9420	11.26

Fig. 1. "L C" Chart

MAGNET WIRE TABLE:
Turns Per Lineal Inch

Number, A. W. G. (B. & S.)	Kind of Insulation						
	S C	D C	T C	Asb	En	En & S C	En & S S
0000	2.14	2.10	2.07	2.06	.....	.....	
000	2.39	2.35	2.31	2.30	.....	.....	
00	2.68	2.63	2.57	2.56	.....	.....	
0	3.00	2.93	2.87	2.85	.....		
1	3.36	3.28	3.19	3.17	.....		
2	3.76	3.65	3.55	3.53	.....		
3	4.21	4.07	3.95	3.92	.....		
4	4.71	4.54	4.38	4.34	.....		
5	5.26	5.05	4.86	4.81	.....		
6	5.88	5.68	5.43	5.35	.....		
7	6.57	6.32	6.01	5.91	.....		
8	7.44	7.12	6.83	6.60	7.63	7.30	7.52
9	8.30	7.91	7.55	7.28	8.55	8.14	8.41
10	9.35	8.94	8.55	8.07	9.61	9.17	9.43
11	10.4	9.93	.....	.....	10.8	10.2	10.5
12	11.7	11.0	.....	.....	12.1	11.4	11.8
13	13.1	12.4	.....	.....	13.5	12.7	13.2
14	14.6	13.7	.....	.....	15.2	14.1	14.7
15	16.2	15.1	.....	.....	17.0	15.7	16.5
16	18.1	16.7	.....	.....	19.1	17.4	18.4
17	20.1	18.4	.....	.....	21.4	19.3	20.5
18	22.3	20.3	.....	.....	24.0	21.4	22.9
			SS	DS			
19	24.8	22.3	26.4	25.1	26.8	23.6	25.5
20	27.4	24.4	29.4	27.8	30.1	26.1	28.4
21	30.8	27.4	32.8	30.8	33.6	29.2	31.5
22	34.1	30.0	36.6	34.1	37.7	32.2	35.0
23	37.6	32.7	40.7	37.6	42.2	35.5	39.0
24	41.5	35.6	45.2	41.5	47.2	38.9	43.1
25	45.7	38.6	50.3	45.7	52.9	42.7	47.8
26	50.1	41.8	55.7	50.1	59.0	46.6	52.8
27	55.0	45.1	61.7	55.0	65.8	52.1	58.2
28	60.1	48.4	68.3	60.1	73.8	57.0	64.3
29	65.5	51.9	75.4	65.5	82.3	61.9	70.6
30	71.3	55.5	83.2	71.3	92.4	67.5	78.0
31	77.4	59.1	91.5	77.4	102.8	72.8	85.3
32	83.7	62.7	100.5	83.7	115.6	79.0	93.9
33	90.3	66.3	110.1	90.3	130.2	85.6	103.3
34	97.0	69.9	120.4	97.0	144.8	91.7	112.3
35	104.0	73.5	131.3	104.0	163.5	98.9	123.2
36	111.1	76.9	142.9	111.1	181.8	105.3	133.3
37	118.3	80.3	155.0	118.3	206.1	113.0	145.9
38	125.5	83.6	167.6	125.5	229.1	119.5	157.1
39	132.8	86.7	180.8	132.8	261.0	127.7	171.5
40	140.0	89.7	194.4	140.0	290.3	134.3	183.7

Courtesy John A. Roebling's Sons Co.

microhenries. Thus the inductance necessary to resonate a condenser with a capacity of .00015 mf. to 200 meters has a value of 75 microhenries.

In most cases the tuning of a *short-wave receiver* is accomplished by means of a condenser with moveable plates, although some set-builders have used tapped or variometer type inductances. This article will deal only with the standard method of condenser control for frequency selection.

The desire for high voltage gains at the short wavelengths will lead the coil builder to the choice of high values of inductance in conjunction with a tuning condenser. This brings several points to our attention which should be studied so that a compromise for good operation can be developed.

Modern short-wave tuning condensers of the better type have their electrical losses reduced to a minimum. Thus it becomes necessary that the efficiency of the tuning coil and its associated components be raised low losses at the frequencies to be received, it follows that the coils and the remaining components associated in the tuned circuits have their losses reduced to a minimum.

The selection of the tuning condenser capacity will depend on the range of frequencies to be received with a given coil. This presents a problem that every experimenter should be able to solve if headaches are to be avoided.

If a wide band of frequencies is to be covered it is necessary that the ratio between the maximum and minimum capacity values of the tuning condenser be made as great as possible. For example: condensers having a maximum capacity of .00014 mf. may have a minimum of .000007 mmf. This is a good condenser and the low minimum capacity value should not be misused by having the associated input circuit capacity of the tuned stage so high that the effective tuning capacity range is reduced. The effect of this shunting capa-

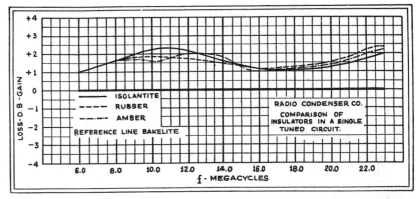

Fig. 2—Graph showing efficiency of different insulators at various frequencies.

to the highest degree. The reader will note that the losses in the various components are cumulative and unless care is exercised the losses will reach values that will nullify the efforts of the builder. An interesting graph showing the efficiency of various insulating materials used in condenser construction is shown in Fig. 2 and pictures the change in efficiency with the change in frequency. The base line for these tests has been defined by the efficiency of bakelite and the other materials have been judged as to the efficiency gain +, or loss —, as indicated.

Having selected a tuning condenser with

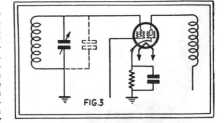

Fig. 3—Dotted lines represent the "lumped" capacity added by poorly designed circuit.

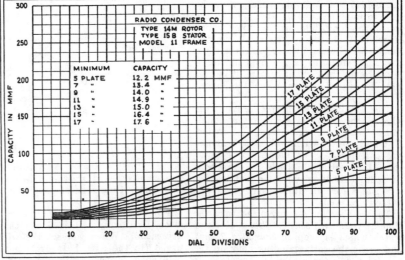

Fig. 4—Relation between capacity, dial divisions and condensers with different numbers of plates.

city is indicated in Fig. 3. Dotted lines represent the *lumped* circuit capacity shunted across the tuning condenser, thus limiting the *minimum* effective tuning range.

Figure 4 shows the capacity range plotted against dial divisions for condensers with varying numbers of plates. This chart is printed by courtesy of the Radio Condenser Co., Camden, N. J. The minimum has a fairly high value and this must be due to the construction. Heavy cast metal end-plates and "bathtub" construction will not permit the condenser designer to obtain low minimums. These minimum values of Fig. 4 are satisfactory for all practical purposes and will permit excellent band coverage.

To find the minimum wavelength to which a variable condenser and a coil will tune, multiply the inductance in microhenries by 1,000 and then by the capacity of the condenser in microfarads. This will give the "LC ratio."

Look this figure up in the chart of Fig. 1 and read off the wavelength in meters. For example: A coil with an inductance of 75 microhenries is tuned by a condenser with a minimum capacity of .000007 microfarads and the remaining circuit capacity is .000010 microfarads. What wavelength will the circuit tune to? 75 times 1,000

equals 75,000. Adding the two shunt capacities together gives us .000017 microfarads. .000017 times 75,000 equals 1.275. Refer to Fig. 1 and we find that the wavelength nearest this LC value is 70 meters. Note that most coil manufacturers specify that their coils for use with .00014 mf. condensers will tune from 80 to 200 meters. They figure, and rightly, that the builder will not have a condenser with a really low minimum and that the circuit capacities will be higher than the value used in the solution of the problem stated. If this coil is to tune to exactly 80 meters the lumped value of capacity at the minimum setting of the tuning condenser should be .000024 mf. If the minimum capacity of the tuning condenser is .000007 mf., then the circuit capacity must be the remainder or .000017 mf. Now we should be able to find the value of inductance to use with a given condenser to tune to a required wavelength and also know about what wave band or range can be covered.

How to Use the Charts.

Most coil builders want to have their coils designed for them, so the accompanying charts are given so that no mathematics are required at all. The best way to use these charts is as follows:

Ascertain the capacity of the tuning condenser that is to be used and then at the point of intersection between the wavelength desired and the tuning condenser curve, read the required inductance in microhenries.

Select the coil with a diameter suitable for use in the receiver in question. That is, the *physical size* of the coil. Having decided on the diameter of the coil form, try to use a ratio of length to diameter of *one-to-one*. If the diameter is one and one-half inches, then try and keep the length about the same. This refers to the *length* of the winding only.

Let a coil with an inductance of 100 microhenries be required; then, if the coil has a diameter of 2 inches and the winding is 2 inches long, a total of 54 turns will be required. With a winding length of two inches, a wire size should be selected that will wind 27 turns to the inch. This gives the necessary data for a coil without a lot of figuring.

The accuracy of these charts is close enough, for all practical coils being made by the "home-set" builder. Calculations involving ¼ turns would complicate the chart to such an extent that it would not be usable.

For the coils used on the very low wavelengths, the wire size will become larger, while the coils used above 80 meters will have comparatively fine wire.

International Time-Zone Chart and Converter

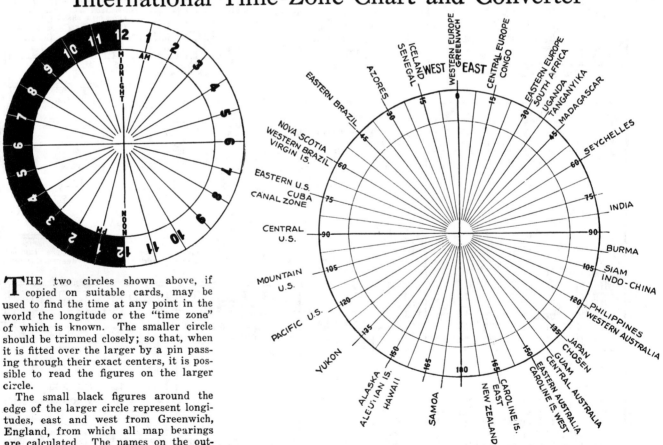

A Rotary International Time Chart.

THE two circles shown above, if copied on suitable cards, may be used to find the time at any point in the world the longitude or the "time zone" of which is known. The smaller circle should be trimmed closely; so that, when it is fitted over the larger by a pin passing through their exact centers, it is possible to read the figures on the larger circle.

The small black figures around the edge of the larger circle represent longitudes, east and west from Greenwich, England, from which all map bearings are calculated. The names on the outside of the circle indicate the countries lying in these longitudes; so that the position of a foreign station may be judged without reference to a map.

To find the hour at any place, take the time at your own station (use a clock or watch which is keeping *Standard Time*) and bring the corresponding hour on the rim of the inner circle to the place of your station, as marked on the larger card. The time at any place in the world, whose position is known, is then that on

the smaller card at the point opposite the proper time zone or longitude on the outer card.

For instance, you are in San Francisco, California. It is 3 o'clock in the afternoon. When you set the white figure 3 (in the dark half-circle indicating afternoon) opposite "Pacific U. S.—120" on the larger card, you will find that the black 11 is opposite "Western Europe—Greenwich" at the top of the card. That

is to say, it is 11 p. m. in England, France and Spain. This is also 1100 Greenwich Civil or Mean Time—which is the international time used by short-wave amateurs and others to standardize their schedules.

At the same time, you will find the black figure 8 (indicating morning) opposite "135-Japan, Chosen" on the larger card. That means it is 8:00 a. m. in Tokio, Japan.

SHORT WAVE POWER-PACKS

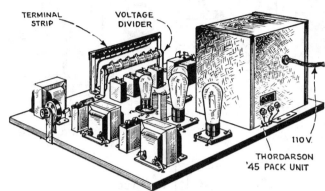

TERMINAL STRIP

VOLTAGE DIVIDER

110 V.

THORDARSON '45 PACK UNIT

Appearance of the finished power amplifier.

For the Short-Wave Fans who desire greater power output from the second audio stage, this power amplifier will be found useful. It employs two '45 tubes in the push-pull output stage and '27 first audio tube. This amplifier is to be operated on 110 volt, 60 cycles, A.C., circuit. The plate supply furnishes the "B" current for the tubes in the R.F. and detector stages. Hum is reduced to a minimum by liberal size chokes and transformers.

A POWER AMPLIFIER

POWER AMPLIFIERS suitable for use with short wave receivers have to be very carefully designed and balanced, otherwise there is liable to be an objectionable "hum" noticeable in the loud speaker and not every power amplifier is sufficiently stable to operate on short wave signals.

General Requirements

The two-stage, audio frequency, power amplifier here illustrated and described was constructed and tried out successfully, with practically no "hum" audible in the loud speaker and without audio frequency "howls" being set up. It is important to mention perhaps, in passing, that the amplifier was tested in connection with a Hammarlund short wave receiver, employing one stage of tuned R.F. ahead of the detector, with the usual throttle condenser control of the regeneration.

This amplifier is a good all-around piece of apparatus and can be used in conjunction with any broadcast receiver and also for amplifying phonograph pick-up signals, by connecting the output of the magnetic pick-up to the phonograph jack shunted across the input terminals of the first A.F. transformer. One of the most important points to watch out for in building any audio frequency amplifier, particularly those of the power type here described, is the proper positioning of the various transformers, choke coils, etc., so that the magnetic fields of the transformers do not interact on one another and thus constitute one of the frequent causes of an objectionable "hum" or other noise heard in the speaker as a "background" to the signal being received. It is therefore desirable that the inexperienced constructor follow the general layout of the apparatus comprising the amplifier.

The First Audio Stage

Looking at the wiring diagrams presented herewith the reader will see that there are two optional suggestions for building up the first audio stage, the first method involving the use of a Thordarson R260 (or its equivalent) A.F. transformer. The transformer used in

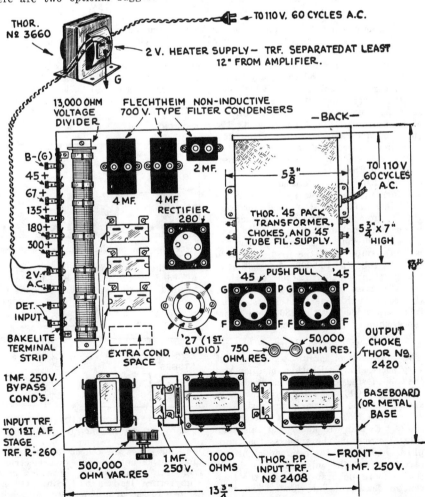

THOR. Nº 3660

TO 110 V. 60 CYCLES A.C.

2 V. HEATER SUPPLY – TRF. SEPARATED AT LEAST 12" FROM AMPLIFIER.

G

13,000 OHM VOLTAGE DIVIDER

FLECHTHEIM NON-INDUCTIVE 700 V. TYPE FILTER CONDENSERS

—BACK—

B-(G)
45 +
67 +
135 +
180 +
300 +

2 V. A.C.

DET. INPUT

BAKELITE TERMINAL STRIP

1 MF. 250 V. BYPASS COND'S.

INPUT TRF. TO 1ST. A.F. STAGE TRF. R-260

4 MF.   4 MF   2 MF.

RECTIFIER 280

EXTRA COND. SPACE

'27 (1ST. AUDIO)

500,000 OHM VAR. RES

1 MF. 250 V.

1000 OHMS

5 3/8"

TO 110 V 60 CYCLES A.C.

5 3/4" x 7" HIGH

THOR. '45 PACK TRANSFORMER, CHOKES, AND '45 TUBE FIL. SUPPLY.

'45   PUSH PULL   '45

G   P   P   G

F   F F   F

750 OHM. RES.

50,000 OHM RES.

THOR. P.P. INPUT TRF. Nº 2408

OUTPUT CHOKE THOR. Nº 2420

BASEBOARD (OR METAL BASE

—FRONT—
1 MF. 250 V.

13 3/4"

Plan view of power amplifier, showing exact position of the various parts.

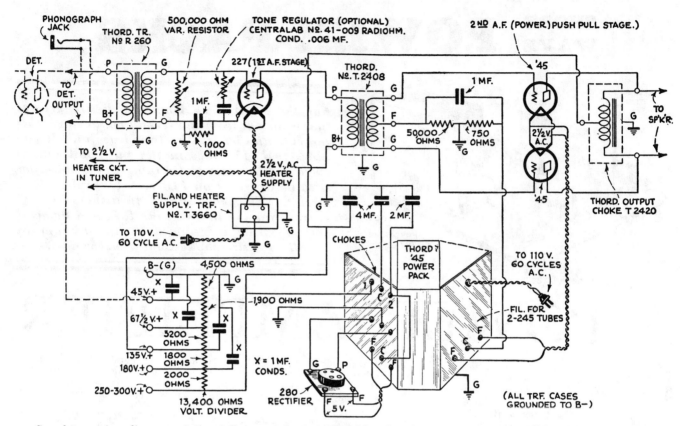

Complete wiring diagram of the A.C. operated power amplifier for short-wave reception, the amplifier using the Thordarson '45 "power compact" for the plate supply.

any case, should have a low ratio between the primary and secondary turns. The second method, and one which has received great favor at the hands of short wave enthusiasts, comprises an impedance or choke coil coupling as the optional diagram herewith delineates. The impedance used in tests by the writer was the Thordarson Autoformer, type R190, the detector plate lead being connected to the "P" terminal of the autoformer, the B plus feed wire to the "B" terminal and the grid terminal ("G") from the impedance connecting to one terminal of a .25 mf., fixed condenser (250 voltage rating). With impedance coupling of the first stage into the '27 tube, a 100,000 ohm potentiometer (Clarostat or other equivalent type), serves to balance the input to this tube. Grid bias for the first audio tube is provided by the 1800 ohm resistance, shunted by a 1 mf. condenser.

Tone Control Feature

A tone control circuit was tried out very successfully with this amplifier and the one tested comprised a fixed condenser of .006 mf., in series with a specially tapered, variable resistance (Centralab No. 41-009). It may be of interest to many to know that this tone control circuit is the same as that supplied on many of the commercial broadcast receivers, the only difference being that a more elaborate scale for the tone control feature is provided. The tone control regulator comprising the condenser and specially tapered resistance

in series, can also be shunted across the two grids of the '45 tubes in the second A.F. push-pull circuit.

A number of ground connections are

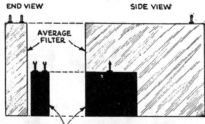

Note the remarkable saving in space afforded by the use of the Flechtheim thin-dielectric type filter condensers.

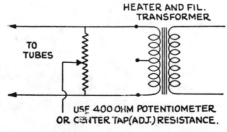

In some cases the center tap on transformers is not at the exact electrical center of the winding, in which case the return lead is best connected as shown to the arm of a 400 ohm potentiometer. This permits adjustment for exact balance.

indicated in the diagram and where the various transformers, condensers, etc., are not mounted on a metal sub-base, all of the ground connections indicated are joined to one piece of wire, not smaller than No. 14 B & S gauge, and of course, all joints should be soldered. The outside metal casings of all transformers and condensers should be connected to the common ground wire, so as to minimize all noises or hum in the reproduction at the loud speaker.

Second Audio Stage Is Push-Pull

For building up the push-pull power stage, which involves the use of two '45 tubes, Thordarson input and output transformers or chokes were utilized. The input transformer is a regular Thordarson T2408 push-pull type, with center-tapped secondary, while the output unit was a Thordarson center-tapped choke coil, type T2420.

The grid return circuit from the '45 power tubes has a 50,000 ohm and 750 ohm resistance connected in series with the center-tap terminal "C" of the filament transformer winding, supplying the 2½ volt A.C. to the '45 tubes. One of the Thordarson '45 compact push-pull amplifier plate supply units was employed, as the diagrams show, this unit containing two filament supply windings, the high voltage winding for the plate supply and also the two, high impedance choke coils for the main B supply filter.

A word of caution to those building an amplifier of this type, is to test out all transformers, choke coils and resis-

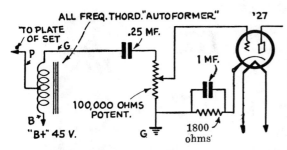

Instead of coupling the detector output through a regular transformer to the first audio tube, a Thordarson all-frequency "Autoformer" (impedance) may be used as per hook-up herewith.

Flechtheim 250 volt type, (250 being the working voltage). If you should see the plates of your rectifier tube get red hot, shut off the amplifier and start gunning for a short-circuited filter condenser.

De Forest tubes were used with very gratifying success in all of the stages of the amplifier and very satisfactory performance in amplifiers during the past year has proven that they do stand up and give quality as well as service.

Hints On Eliminating "Hum"

1—Ground all transformer and condenser cases.

2—Test grid return bias resistors for "continuity" and by-pass condensers for "short-circuits".

3—Center tap on filament transformers may not always be at exact electrical center; connect 400 ohm potentiometer across filament winding and join grid return lead to arm of potentiometer.

tances for electrical continuity. Most of these tests can be very well made with a milliammeter and a small B or C battery. If one of the resistances in the grid return circuit, such as the 50,000 ohm unit, should be open-circuited an objectionable hum would be heard in the loud speaker. All of the resistance-coils used in building this amplifier were of the baked enamel type made by the Ward Leonard Company and they have performed very satisfactorily indeed.

Source of Filament Supply

One of the important points about a good audio frequency amplifier is to see that not too many transformer windings are grouped together on one core. As a number of leading short wave experts have pointed out, it is better to have the filament supply transformers split up; so in this amplifier we find this condition. A Thordarson T-3660 filament supply transformer delivering 2.5 volts supplies the heater current for the R.F. and the detector tubes, as well as the first audio stage of the power amplifier. A separate filament transformer winding supplies the '45 tube current and a third separate filament supply winding furnishes the 5 volt current for the '80 rectifier tube. All of these points help to make a quiet operating amplifier and one of the leading radio engineers told the writer, that he never built any set, especially a power amplifier, unless he connected up the transformers on a "bread-board" and moved them around until the condition was found where a minimum hum was noticed in the loud speaker. Sometimes transformers have to be placed at right-angles or in other positions in order to prevent inter-action of their stray magnetic fields. It was found in the present case that in order to reduce the hum to the lowest possible limit that the heater supply transformer T-3660 had to be removed from the general layout of the amplifier and placed

at least 12 inches away from the other transformers and amplifier apparatus to prevent pick-up of the magnetic field.

Details of the Filter

Looking at the filter circuit for a moment we see that the two chokes are connected to the terminals 1, C, and 2 at the top of the left side of the Thordarson '45 compact. Three high voltage condensers of the Flechtheim extremely compact type were used, having capacities respectively of 4, 4 and 2 mf. The great

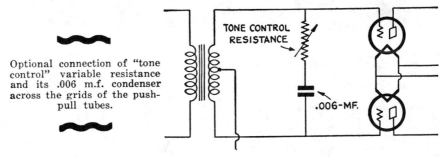

Optional connection of "tone control" variable resistance and its .006 m.f. condenser across the grids of the push-pull tubes.

saving in space afforded by use of these Flechtheim compact type condensers is shown in one of the diagrams herewith and they only occupy about ⅛ the volume of the average high voltage condenser supplied for filters of this type. In any case the condenser should have a working voltage of approximately 700.

The 13,400 ohm voltage divider resistance shown in the diagram is of the Ward Leonard baked enamel type and performs in very excellent fashion, without getting so hot that one can fry flapjacks on it, as some of these "19c special" resistances are wont to do. A potential of approximately 350 volts was measured with a Flechtheim voltmeter across the output terminals of the filter, or in other words across the end terminals of the voltage divider resistance. Each step or tap on the voltage divider is shunted by a 1 mf. condenser of the

4—Sometimes the A.C. supply plug, where it fits into the house service receptacle, has to be reversed.

List of Parts for Power Amplifier

1—Thordarson '45 power compact unit (Includes 2 chokes, high voltage plate winding, 5 volt fil. winding for rectifier, and 2.5 volt fil. winding for 2 '45 tubes.

1—Thordarson No. R260 input transformer.

1—Thordarson No. R260 push-pull transformer.

1—Thordarson No. T2420 push-pull output choke.

1—Thordarson No. T3660 filament-heater transformer.

2—4 mf. (700 volt working voltage) Flechtheim compact filter condensers.

1—2 mf. ditto.

7—1 mf. Flechtheim by-pass (250 working voltage) condensers.

1—13,400 ohm, Ward Leonard, voltage divider resistance.

1—50,000 ohm Ward Leonard resistance.

1—750 ohm Ward Leonard resistance.

1—1,000 ohm Ward Leonard resistance.

1—2 circuit (or other to suit builder's idea) jack for phonograph pick-up.

1—Baseboard (or metal sub-panel).

1—Terminal post strip—bakelite.

1—Set Terminal posts (X-L push posts used by author).

1 Coil No. 14 soft rubber covered wire for connecting apparatus.

1— .006 mf. condenser—tone control (Sangamo).

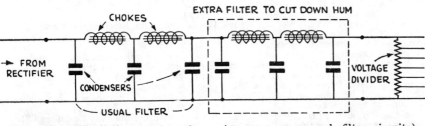

EXTRA FILTER TO CUT DOWN HUM

With some "B" eliminators (or when using some power-pack filter circuits) there is still an objectionable "hum"; the circuit herewith shows how to add an extra filter between the usual one and the voltage divider resistance. The chokes and condensers in the extra filter have values identical with those in the usual filter.

This Power Supply Unit
USES NEW 25-Z-5 TUBE
By CLIFFORD E. DENTON

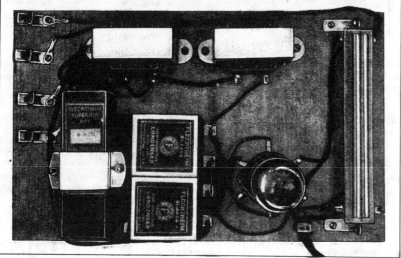

At left we have top view of the 25-Z-5 power supply unit as designed and built by Mr. Denton. No power transformer is required and among other features, voltage doubling can be obtained. Close-up view of the new 25-Z-5 rectifier tube is shown at right.

This very latest type S-W power supply unit designed by Mr. Denton utilizes the new "25-Z-5" tube, and does away with the cost of a power transformer. Voltage doubling is effected when the proper size condensers are used.

● EVERY day tube manufacturers are bringing forth new tubes that permit radical changes in the design of radio sets, audio amplifiers and power-supply units.

One of the many tubes announced is the 25-Z-5, which offers several inducements for the short-wave fan. It does away with the cost of a power transformer when used as the rectifier for small and medium size short-wave sets. It has satisfactory regulation and is not noisy, being of the high vacuum type. *Gas-filled* tubes have no place in the short-wave receiver power-pack.

The following characteristics covering this tube are given for reference, as every one is interested in having as much tube data on hand as possible.

25-Z-5 Tube

Heater Rating: Voltage, 25 volts; current, 0.3 ampere.

Operating Conditions and Characteristics: A.C. Voltage per Anode, 135 maximum RMS volts; D.C. Output Current, 100 maximum ma.; Voltage between cathode and heater, 300 maximum peak volts; Peak Plate Current, 300 maximum ma.

With the circuit as shown in Fig. 1, *voltage doubling* can be obtained, dependent on the values of condensers connected at C1. These condensers should be selected for their efficiency; low power losses and high internal resistance per microfarad will permit the best results.

When building the power supply unit study the chart in Fig. 2, which tells just what values of C1 must be used for a given result. For example, if 200 volts is required at 100 ma., then condensers C1 should have a value of 16 mf. each. Select the smallest value of capacity that will provide the required current at the proper operating voltage and that will be the only change from the specifications given.

The rest of the power supply unit should follow specifications as shown.

A radio frequency choke can be con-

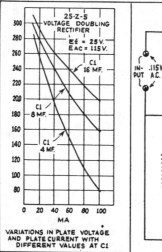

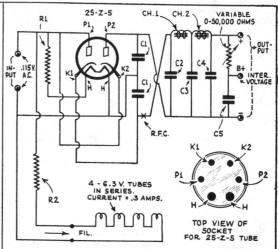

The graphs shown above, at left, tell the story of the variation in size of condensers used in conjunction with the 25-Z-5 rectifier. Wiring diagram for the complete rectifier is shown at the right. O-50,000 ohm resistor is R3.

nected in the filter circuit as indicated at X. This will help to reduce noise from the power line in many cases. If additional voltages other than the high voltage are to be obtained from the power supply unit, use a series feed circuit with a paper by-pass condenser at C5. The value of R3 will depend on the voltage required and the current drain in the circuit. In general this circuit will supply screen grid tubes, so the current through R2 will never be great.

The filament supply circuit as shown is for use with four 6.3 volt tubes in series with a total current drain of .3 ampere. If more tubes are used in this circuit it will be necessary to change the value of resistance at R2.

For those interested in using the filaments of the tubes in the radio set in parallel, it will be more economical to use a small separate filament supply transformer. As a great number of the short-wave sets in use today have four tubes, the resistor recommendations

given will be satisfactory. Remember this value of R2 is for use with the 6.3 volt tubes consuming .3 ampere; any other tubes will require different resistor specifications.

Parts List

Two Flechtheim By-pass Condensers, 4 mf., 250 volts (C1)—(Wego; Aerovox)

One Flechtheim Electrolytic Condenser 4 mf. Cardboard type (C2)—(Wego; Aerovox)

Two Flechtheim Electrolytic Condensers 8mf. Cardboard type (C3, C4)—(Wego; Aerovox)

One Eby six-prong socket

Two Federated Purchaser, No. 2505 Power Chokes (Ch1, Ch2)

Two Acratest Shielded Resistors, 310 ohms, 300 ma. (R1, R2)

One 25-Z-5 Rectifier (Sylvania)

One power cord and plug

One wooden baseboard, 7 by 11½ by ¾ inches

Wire, wood screws, etc.

Building

A

Shielded

Power Unit

for

Short Wave

Receivers

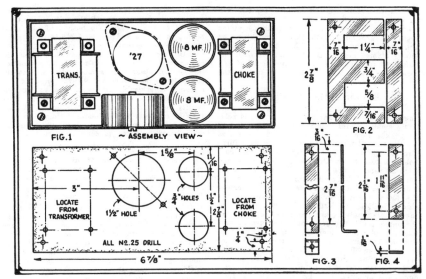

FIG.1 ~ ASSEMBLY VIEW ~

FIG. 2

FIG.3   FIG. 4

Drawing above shows details of metal shield, power supply unit and the filter system.

● At least half of the converter failures are due to the makeshift ways by which their power is taken from the receiver to which they are connected. By looking over the diagrams and drawings given on these pages the reader will see that this unit is very staple and inexpensive to make. While the shielding can is not required to make the power unit work it is put on to prevent any disturbances from interfering with the short wave tuning apparatus. Therefore this unit can be placed extremely close to the tuner or converter.

Some of these power packs take the high voltage for the rectifier plate directly from the A. C. line. This limits the rectified voltage to less than line voltage. At the same time this system requires the use of a filament transformer for the heaters. The transformer to be described will be found easy to construct and is properly designed for this job. By looking at the hookup diagram we find a pair of 2.5 volt heater supply leads that will take care of this end of the job. The plus lead will give a range of voltage from zero to 160 volts with plenty of current for all converter and short wave tuner needs. A knob on this voltage regulator is not needed as it will be set once to meet the need of the set and left.

Constructing Transformer

In Figure 6 we have a wood block; this will serve to wind the coil upon. Cut a strip of wrapping paper 1¼ inches wide and wrap this around the block to a thickness of nearly 1-16 inch. Put a machine screw through the hole and fasten it in the hand drill. This drill can now be clamped in the bench vise. Lay a strip of friction tape on each side of the paper form, leaving an inch project beyond each end of the block. Solder a lead wire to the end of the No. 29 B. & S. enameled wire and tape the joint. Twist this around the center screw in the block to anchor it and wind the first layer of wire over the paper and strips of tape. Before it winds to the end start the wire back over the first layer. With this second layer in place cover with a layer of thin waxed paper. Continue with two layers of wire and a layer of paper until 1,045 turns have been wound on. Now fold the ends of the tapes over the coil. They will hold the wire in place. Solder a lead on and cover the primary with three layers of the same paper used under the coil.

Now lay down four more strips of tape for the secondary binding. Solder a lead to the No. 36 B. & S. enameled wire and anchor as at the primary start. Wind about 400 turns on and cover with

waxed paper; continue thus for 1,700 turns. Do not try to wind this fine wire in even layers, simply wind back and forth. After the 1,700th turn solder a lead on and fold the tapes over the coil. Put on three layers of the wrapping paper. After which put down the tape strips as before.

Simple, clear directions for building a quiet plate and heater "supply unit" for satisfactory use with short-wave receivers has been rather scarce. The accompanying article gives details for "winding your own" power transformer; also the method of connecting the transformer, "home made" choke, condensers and rectifier tube. Keep this article for future reference.

Coil Is Finished

The No. 18 B. & S. wire is used next, and no leads are needed as the wire is strong enough. Wind on twenty-four turns and bind with the tapes. Cover with paper and wind on twenty-four turns of the No. 15 B. & S. wire. Now the coil is done; remove the block from the center and submerge the coil in melted wax or parafin. Do not have this too hot—just melt it. Figure 2 shows the kind of laminations to use. These are inexpensive to purchase (see note at end for this), or can be cut from lamination iron to the dimensions given. Using fifty of these laminations, place all the "E" pieces in the coil opening. Entering them alternately from one end, then from the other end. With this done, place the straight pieces at the open ends of each "E" lamination. A solid core is built up in this way.

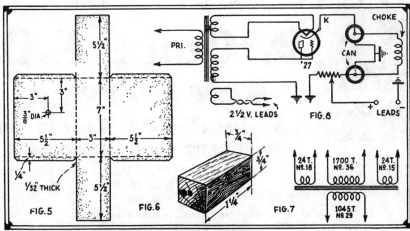

FIG.5   FIG.6   2½ V. LEADS   FIG.8   FIG.7   CHOKE   CAN   LEADS   PRI.

Here we see the various parts comprising the S-W "power unit" laid out on the base; also details of the transformer steel laminations.

Now with small wooden wedges tighten the laminations in the coil opening. Four of the brackets in Figure 3 and four clamping strips Figure 4 are made 1-16 inch thick brass or aluminum strip. Using No. 6-32x1¼ inch screws clamp these to the laminations with the strips on each side of the jointed ends of the core. The long brackets serve to mount the transformer to the base.

Buy or Make the Choke

A very satisfactory choke can be made, one having a high inductance, as the current is relatively small. Using a block of the same length but ¾x⅝ inch instead of ¾ inch square as for the transformer, cover with paper as before; also use the strips of tape for binding the coil. Solder a lead to the No. 36 B. & S. wire and wind at random 7,500 turns of this wire on the block. Put a lead on the end of the wire and bind the coil with the strips of tape. Remove the block and dip into the hot parafin. The core is built up of forty laminations, but all of the "E" laminations are put into the coil from the same end. The straight pieces are stacked up in a pile and clamped to the open end of the core by using four more of the strips shown in Figure 4. Brackets Figure 3 will be added for mounting the choke. Using a bought choke, select one of high inductance—the direct current resistance can be as much as 1,000 ohms.

A mounting base for the parts is shown, with dimensions, in Figure 1. Make this of 1-16-inch thick aluminum. The large hole is for the tube socket. The two ¾-inch holes mount the two electrolytic inverted condensers. Place the transformer and choke in position and mark for the holes from these. Drill holes for bringing the transformer leads and the choke leads under. The connections are very simple and are shown in the hookup diagram. The cathode terminal on the socket supplies the high voltage, rectified current to the choke. Tie both grid and plate terminals to one end of the 1,700-turn winding. Only one connection is to be made to the electrolytic condensers, this being shown as the center on each one. The can is shown grounded; this, as all other such symbols denote that the point is to be connected to the base if not already in assembly of part. Use a rubber-covered hookup wire for all connections. The twenty-four turn winding made up of number 18 wire connects to the heater terminals on the socket. Leave two long, twisted leads from the other heater winding to connect to set.

Shielding the Unit

A sheet of 1-32 inch aluminum 14x18 inch should next be cut as dimensioned in Figure 5. The hole is for mounting the voltage control resistance. Bend along the dotted lines, taking the flap edges first. A straight edge will be very useful if clamped to the bending line, then the aluminum may be easily formed around this edge. With four sides bent up rivet the flaps to the adjacent sides with small rivets or eyelets. Next drill four holes in the bottom to line up with those in the corners of the mounting base. In mounting the assembled and wired unit into the shield use number 6-32x1¼ inch screws with a 1 inch spacer under the base to give clearance to the parts under this base. If the variable resistance has its arm connected to the mounting stud this stud will have to be insulated from the shield by the use of fiber washers. A cover for the unit is made of the same material from which the shield was made. Using a piece 4x8 inch cut one-half inch squares out of the corners and bend the sides up so that the cover will slip over the shield tightly.

Testing

Using a '27 tube a test was made on the completed unit to determine the actual output at various current loads. A milliammeter and a high, variable resistor were placed in the lead of the unit for taking these measurements. The voltmeter was of the high resistance type and was put in circuit from the tap to the minus connection. After setting the external variable resistance so that the current read was 5 milliamperes (mils) the voltmeter gave the voltage as 175 volts. Another setting at 10 mils gave a voltage reading of 165 volts; at 14 mils this had dropped another 5 volts to 160. These readings will go to prove that this little unit is entirely capable of handling most of the converters and short wave tuners. The '27 tube used for rectification is very inexpensive as also are the two electrolytic condensers.

List of Material

Ninety Laminations (Type E-1-3 Allegheny Steel Co., Brackenridge, Pa.).

Two Electrolytic Condensers (8 mf.) Concourse, Aerovox, etc.

One 50,000-ohm variable resistance; wire wound.

One-third pound number 36 B&S enameled magnet wire (for transformer and choke).

One-eighth pound number 29 B&S enameled magnet wire.

Fifteen feet each of number 18 and number 15 enameled magnet wire.

One UY wafer type socket.

One 14x18x1/32 inch aluminum.

One 4x8x1/32 inch aluminum.

One 2⅞x6⅞x1/16 inch aluminum.

Courtesy of N. Y. Sun.

The right place for short-wave power-packs—away from the sets themselves. On the shelf just above the floor are (at the left) the factory-built power unit for the National SW-58, and (at the right) the "home-made" unit for the SW-3. A completely A.C. operated audio power-amplifier for the latter is mounted on a shelf above the tuners.

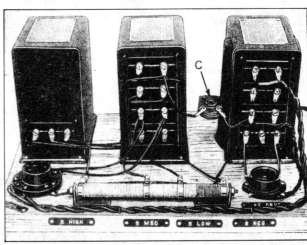

A complete A.C. power-pack for the S-W receiver. Note the R.F. choke "C" hanging in mid-air between the power transformer and the filter condenser. Socket on left for 5-wire cable and tube-base plug; socket on right for 80 rectifier. Any desired plate voltage up to 300 available.

A "QUIET" A.C. Power-Pack
for Small S-W Receivers
By ROBERT "BOB" HERTZBERG

A small, but "good," Power-Pack for use with small S. W. receivers is in great demand. Here's just the "job" you've been looking for.

● ONE of the earliest and most persistent bugaboos in the short-wave game is that concerning A. C. operation. It is kept alive by the many ex-broadcast experimenters who are finding new thrills in the short waves, and whose experience with all-electric short-wave set operation dates back to the days when adequate filters and properly designed cathode-heater tubes were unknown.

Now there is nothing especially mysterious or magical about A. C. power-packs, and it is not necessary to depart very far from ordinary practice in order to make a unit that will work *smoothly* and *quietly*. Instead of looking for queer circuit arrangements and phoney "hum balancers," the constructor should merely use good parts and a reliable hook-up, and above all else learn to keep the completed pack a respectable distance from the sensitive receiver itself!

Let that last statement sink in! In the writer's experience, which embraces part of the design work, in collaboration with Robert S. Kruse and David Grimes, on the original Pilot *Super-Wasp*, the first commercially successful short-wave receiver for all-electric operation, more trouble is caused by having the pack too close to the receiver, *than by any other single factor!* This remark should not be construed as a criticism of sets which have the A.C. power supply on the same chassis with the tuner; this sort of thing is all right if the set is designed from the start with the idea in mind and if the proper precautions are observed. When a man starts with a battery receiver and starts to "electrify" it, matters are altogether different.

Come to think of it, the manufacturer of some of the finest short-wave receivers on the market today insists on building the power-packs as entirely separate units, and supplies a long, heavy interconnecting cable with the advice that it be used *without being curled up.*

Built In One Evening

As an example of what can be done at little trouble and expense, the writer illustrates herewith a complete power-pack that he built in one evening, to go with a regular National SW-3, which is an orthodox short-wave receiver using one T.R.F. stage, regenerative detector and one audio stage. This set had been working on batteries for about a year, and with the demise of the "B's" it was decided to eliminate all batteries in favor of AC. The unit worked perfectly from the start, with only the barest trace of a hum in the earphones to indicate the nature of the power supply.

Two simple little tricks helped matters considerably. One was the use of an R. F. choke in the high voltage lead from the filament of the 80 rectifier to the filter system, to keep parasitic radio frequency currents generated by the rectifier tube out of the D.C. output; the other was merely the removal of the pack from the operating table and its placement on a shelf just above the floor. One of the accompanying photographs shows the receiver layout, with the SW-3 on the table and the pack underneath. The other unit on the shelf, incidentally, is the separate power-pack of a National SW-58. The writer usually keeps both sets running at the same time, and has a lot of fun listening to *both ends of phone conversations.*

Assembly On Wood Base-board

The various parts of the power-pack are spread out comfortably on a wooden base board measuring 14 by 8 by 1 inches. The power transformer, the filter condenser block and the filter choke block (reading from right to left) occupy the back section. In front of the transformers is a four-prong socket for the 80 rectifier tube (preferred over the 82 because of its

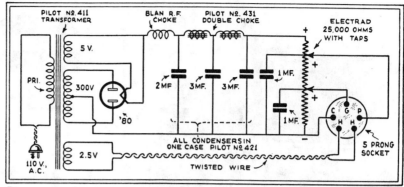

Diagram showing connections of various parts used in building the power-pack.

greater freedom from R. F. current). In front of the choke unit is a five-prong socket, at which the output wires terminate. Between the sockets is an Electrad 25,000 ohm wire-wound resistor with sliding taps.

The special Blan R. F. choke coil simply hangs in mid-air from its connecting wires between the transformer and the condenser block. This choke contributes noticeably to the quiet operation of the unit.

The wiring, as shown in the schematic diagram, is very simple. Note that heavy twisted wire (No. 14 flexible) is used between the 2½ volt filament terminals on the power transformer and the F posts of the five-prong socket. All other connections are made with No. 18 wire of the kind having *push-back* insulation.

Connection between the receiver and the power-pack is made through a flexible cable of five wires: two for filament, one for negative "B" and two for plus "B." If any particular set requires three different plate voltages, a six-prong socket may readily be used. The cable from the set terminates in a plug made from an old tube base. A neater plug designed for the purpose may be bought for a few cents.

Adjusting Output Voltages

By means of the sliding bands on the Electrad resistor, exactly the correct output voltages may be obtained. The values should be measured with a high resistance voltmeter. A fixed output resistor, or one having fixed taps, is absolutely worthless, as the voltages are never correct. If they are low, the set is weak and insensitive; if they are high, it is noisy and unstable.

Naturally the power units selected for a power-pack to fit a particular receiver must have sufficient capacity for the tube combination that is used. Don't work too close to the limit; the more margin you provide, the better. The units employed in the writer's pack are designed to supply a pair of 45's, in addition to a flock of the usual 24's and 27's, but the entire load imposed is less than 20 milliamperes, the tubes used in the SW-3 being two 35's and a 56. With this light drain the pack runs nice and cool, and there is no sign of the erratic behavior that indicates *"saturated" choke-coil cores* and *"overworked" rectifiers.*

Parts List

The following parts were used in the power-pack illustrated. The builder may use his discretion in making substitutions, depending on his own requirements. This pack will easily operate sets using up to six or seven tubes.

1—Power transformer, Pilot No. 411. 600 volt, center-tapped high voltage secondary; one 5-volt and two 2½ volt filament secondaries. (Franklin)

1—Filter condenser block, Pilot No. 421. One 2 mf., two 3 mf. filter sections, three 1 mf. by-pass sections.

1—Double choke unit, Pilot No. 431. Two 25 henry sections.

1—Blan special R. F. choke coil, uncased.

1—Electrad 25,000 ohm output resistor, with sliding taps. (R.T. Co. "Riteohm")

1—Four-prong socket for rectifier, Pilot (Alden)

1—Five-prong socket for connection cable. (Alden)

6—Special Blan black and white metal markers, as shown in photograph. These are very useful for labeling terminals, sockets, etc., in an unmistakable manner. Special labels will be made up to your special order.

1—Wooden baseboard, 14 by 8 by 1 inches.

1—Alden five-prong plug or old tube base. Wires, screws, odd hardware, etc.

A Simple Meter Stand

Most experimenters and constructors own only one or two meters, which they use for a variety of purposes. These instruments, being rather valuable, should be mounted in some way; so that they can not roll over on their edges and, possibly, off the table.

A simple and reliable support can be made from two small iron shelf brackets, such as the chain stores sell for a few cents each. One leg is cut short and its edge filed out to fit the curvature of the meter body. A small hole is drilled in it, to match the side mounting hole in the meter frame. Both legs are then bent over to form an angle of about 60 degrees with the longer legs, which provide a broad supporting base.

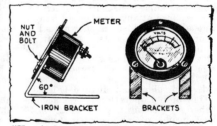

"Here's how" to support that portable meter at any desired angle.

USEFUL TIME CHART

ONE of the most useful gadgets the short-wave fan can have on his operating table is the Standard Time Conversion Chart sold by the U. S. Government. This is printed on heavy cardboard, 8 by 10½ inches, and costs only ten cents (coin or money order; *no stamps*). It takes all the trouble out of time conversion, because it is "direct-reading" and "foolproof".

Write to the Superintendent of Documents, Government Printing Office, Washington, D. C. Ask for Miscellaneous Publication No. 84.

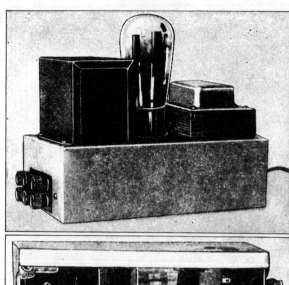

A Good 250 Volt POWER SUPPLY for Less Than $5.00

By ALBERT W. FRIEND, B.S., E.E., W8DSJ

A plate supply unit, well-filtered and furnishing not less than 250 volts D. C. is in great demand for operating the modern short-wave receivers fitted with the new type tubes. Mr. Friend provides us with the constructional data on a power-supply unit of this type.

Photos at left show external and internal appearance of 250 volt power-supply unit here described by Mr. Friend.

List of Material:

1 Chassis (bent as shown in Fig. 2; 20 gauge galvanized iron) ....$	.30
6 Binding Posts (Eby "Ace" W.R.S. Co. No. M13027) at 5c ea.	.30
Hardware—13 screws and nuts—1 angle bracket—4 ft. hook up wire—lamp cord and attachment cap—bakelite strip (panel)	.25
1 Transformer (Trutest) Pri. 110 volt, 60 cycle; Sec. (1) 700 V.C.T. 50 ma.; (2) 5 volts, 2 AMPS.; (3) 2.5 V.C.T. 5 AMPS. (W.R.S. Co. No. 4C1494)	1.25
1 Choke (Crosley) Double 30 henry. 80 ma. 150 ohms D.C. (W.R.S. Co. No. 4C1494)	.59
2 8mf. 500 V. Cardboard type dry electrolytic condensers (Trutest—W.R.S Co. No. D3348)......................	.86
1 2 mf. 500 V. Cartridge type dry electrolytic condenser (Trutest—W.R.S. Co. No. 2D3292......................	.33
2-0.002 mf. Mica condensers (Sangamo "Illini"—(W.R.S. Co. No. 4D4177)	.14
1 Resistor (Electrad Truvolt) 25,000 ohms 50 watts (4 taps) (W.R.S. Co. No. 4G7319)	.80
1 Socket (Eby wafer type—4 prong) W.R.S. Co. No. 4M13070)	.05
Total—(Parts for complete power supply) ....$	4.87

● EVERY short wave experimenter needs a good cheap power supply of small size. I have designed one which can be easily constructed for less than five dollars. It will give 250 volts of pure D.C. at a current drain of 50 milliamperes (more current at a lower voltage) as well as 2.5 volts A.C. at 5.0 amperes. The overall dimensions are only 8½ x 5¼ x 5¼ inches (without the tube).

The unit is very rugged, and the parts used may be easily obtained.

The Chassis

The first consideration is the chassis. It was constructed of No. 20 gauge galvanized sheet iron, which can be purchased at any tin shop. I bought mine already folded as shown in Fig. 2 for only thirty cents. All that remained to be done was to drill the holes and to cut slots along each fold from each end for a distance of one inch; bend and hammer in the end folds, and solder as indicated (Fig. 3).

The folding operation is best accomplished by placing a piece of 2" x 4" block inside of the chassis (with a squared edge at the folding point) and, after clamping the block and chassis in a vise, hammering the top deck ends down first. After relocating the block, hammer up the ends of the two flanges, and then the ends of the two sides. Solder the points indicated in Fig. 3 with acid core solder and wash off the excess flux to prevent corrosion.

The large holes in the top should be marked with a sharp tool and cut out with a chisel. The edges can be filed smooth to the marked lines. Most experimenters do not have means for drilling the larger round holes. By using curved or round files, a nearly perfect and very neat job can be done without large drills or special cutters.

The bakelite or fiber end panel of Fig. 4 can be cut from scrap found in almost any ham "junk box."

The binding posts will be found on some old battery set or amplifier. If desired they may be dispensed with and phone tip jacks or a six or seven prong socket may be used for making connections. The latter method will serve very well if all equipment is provided with cable connections. When using binding posts it is convenient to use spade tips on all connecting leads.

Voltage Divider and Choke

The voltage divider used is variable, by means of sliding bands, and allows any desired voltage combination to be obtained.

The Crosley double 30 henry choke is very compact, of ample rating, and neat in appearance.

Cardboard type (Trutest) dry electrolytic condensers are self healing, compact, of high rating, and very cheap.

The transformer used is a Trutest midget type which gives a high voltage and plenty of current for any ordinary short wave set or experimental layout.

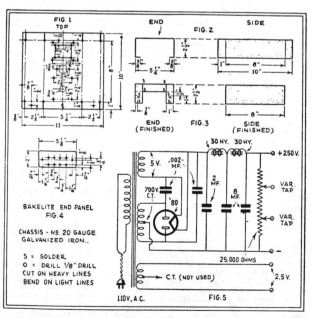

Wiring diagram and constructional details of chassis of 250 volt power-supply unit.

SPACE-WINDING S. W. COILS

The type of twine used to space the turns of wire is of the waxed variety, such as is used by electricians in tying several insulated wires together. Any other kind can be used but this type is preferred.

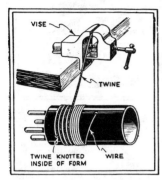

VISE
TWINE
TWINE KNOTTED INSIDE OF FORM
WIRE

Cut the proper length of wire to be wound, run one end through the first hole in the form, and solder to the prong in the usual way. Clamp the free end of the wire in a vise, and holding the form in both hands, stand away from the vise until the wire is taut. Put on the desired number of turns, close together, by turning the form toward the vise.

When all the turns are on, and the other end of the wire has been passed through the further hole in the form and soldered to its prong, take a piece of waxed twine somewhat longer than the wire, run one end through the hole at the start of the winding, and knot the end of it so it will not pull through. Clamp the free end in a vise, and wind it between the turns in the same way that the wire was put on.

It will slip easily between the turns of the wire, forcing them apart. The accompanying sketch shows how the coil will look when about half the turns have been spaced. It can be seen that when the remaining five turns of wire have been spaced by the twine, they will be moved toward the further hole where the end of the wire is secured.

When all the turns of twine have been put on, the end of the twine can be fastened to any convenient point on the form to hold it in place until any other windings are placed on the same form.

When the coil is entirely finished, the twine is carefully removed, and because the wire was put on fairly tight and the turns close together in the first place, it will be found that the twine, in forcing them apart, has strengthened their position on the form.

If waxed twine is used, it should always be removed before the coil is placed in operation. If this is not done, its dielectric quality will increase the distributed capacity of the coil and the original advantage in space winding is lost.—R. S. Peltier.

▼ ▼ ▼

STAND-OFF INSULATORS

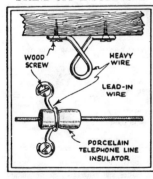

WOOD SCREW
HEAVY WIRE
LEAD-IN WIRE
PORCELAIN TELEPHONE LINE INSULATOR

A good aerial system must have a good insulation system. The lead-in must be kept away from the sides of buildings at all times. A good "stand-off" insulator may be made from an old telephone line insulator or porcelain insulator knobs. Take some heavy wire and make one turn around the grooved part of the insulator; then spread the legs apart and fasten to the sides of the building with wood screws. The lead-in may be then run through the hole. This provides a sturdy and inexpensive insulator for the short-wave set.—Y. H. Mori.

▼ ▼ ▼

TUBE-BASE COIL FORM

Here is a novel idea for making short-wave coil forms from old tube-bases. Each form is made, for instance, of two

SHORT WAVE KINKS

tube-bases cemented together in the manner shown. A handle is formed from a piece of wire soldered into two of the pins

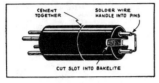

CEMENT TOGETHER
SOLDER WIRE HANDLE INTO PINS
CUT SLOT INTO BAKELITE

left on the upper base, the other two pins have been cut away to reduce dielectric losses.

▼ ▼ ▼

WINDING TRANSMITTER COILS

A few hints are given in the accompanying illustration on how to wind copper tubing and strip for transmitter inductances. Copper tubing may be wound "cold" around a cylindrical form, one end of the tube being held in a lathe chuck for example, if a lathe is available. Flat

COPPER TUBE
WOOD OR IRON
COPPER TUBE
END SET BETWEEN CHUCK JAWS OR IN SOME OTHER SLOT TO HOLD IT.
MADE TURNABLE BY HAND OR ELSE FIXED, AND WALK AROUND BLOCK WITH TUBE.
FLAT COPPER STRIP
FIBER OR RAWHIDE MALLET
HOLD IN LATHE CHUCK OR SLOT IN BLOCK
NAILS IN WOOD (OR PINS IN METAL, OR SCREWS IN PIPE AS GUIDE.)
EDGEWISE WOUND COPPER STRIP. (MAY BE ANNEALED OR BENT HOT.)

copper strip is wound around a form and the operation aided by means of a fibre mallet. Tubing may also be wound by walking around a stationary form with it. Copper strip may be "edge-wise" wound between nails driven into a wooden form as shown, (or pins or screws in a metal drum or piece of pipe).

▼ ▼ ▼

HANDY MIDGET CONDENSER

Double adjustment is provided in the midget condenser design here suggested. First, one of the circular plates is adjusted by means of the threaded rod and

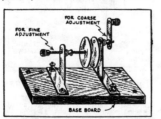

FOR COARSE ADJUSTMENT
FOR FINE ADJUSTMENT
BASE BOARD

check-nut shown at left of the condenser. The second plate, at the right, may be moved side-wise by means of the lever and insulated button.

▼ ▼ ▼

LEAD-IN INSULATOR

Many different types and styles of lead-in insulators have been designed and

sold on the market, and also described by various experimenters, but here is one that is simple to make and which can be made from parts usually available at slight cost. This lead-in insulator is made from the porcelain block obtainable from an old spark-plug. It is feasible, where thick walls are encountered, to use two of these, one on each side of the wall —Christian Jorgensen.

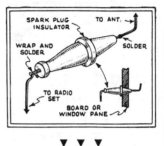

SPARK PLUG INSULATOR
TO ANT.
WRAP AND SOLDER
SOLDER
TO RADIO SET
BOARD OR WINDOW PANE

▼ ▼ ▼

TUBE SHIELD

The accompanying drawing shows one way to improvise tube shields—this idea calling for the use of a piece of cardboard mailing tube. The tube may be boiled in molten paraffin wax to render it non-hy-

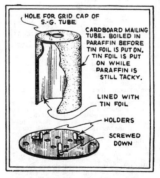

HOLE FOR GRID CAP OF S.-G. TUBE
CARDBOARD MAILING TUBE. BOILED IN PARAFFIN BEFORE TIN FOIL IS PUT ON. TIN FOIL IS PUT ON WHILE PARAFFIN IS STILL TACKY.
LINED WITH TIN FOIL
HOLDERS
SCREWED DOWN

groscopic and after that a layer of tin-foil is placed around the tube while the paraffin is still tacky. A tin-foil disc is cut out for the cap and this is lapped over the tin-foil on the cylinder.—Norman Harris.

▼ ▼ ▼

COIL FORMS FROM RECORDS

As the sketches reproduced herewith show, very good coil forms may be made from discarded phonograph records. The

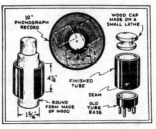

10" PHONOGRAPH RECORD
WOOD CAP MADE ON A SMALL LATHE
4¼"
FINISHED TUBE
SEAM
OLD TUBE BASE
ROUND FORM MADE OF WOOD
1⁵⁄₁₆"

record is heated over a gas flame until it is soft, when you will find it possible to cut out square pieces with a knife. While still warm, wrap the flat piece around a

suitable form and when it has cooled remove it; the ends may be glued if found necessary. The tube may then be glued on to a vacuum tube base and a wooden knob provided at the top, the knob being turned out on a lathe.—John Hengel.

▼ ▼ ▼

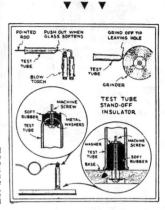

POINTED ROD
PUSH OUT WHEN GLASS SOFTENS
GRIND OFF TIP LEAVING HOLE
TEST TUBE
BLOW TORCH
TEST TUBE
GRINDER
SOFT RUBBER
MACHINE SCREW
METAL WASHERS
TEST TUBE
TEST TUBE STAND-OFF INSULATOR
TEST TUBE BASE
WASHER
MACHINE SCREW
SOFT RUBBER

TEST-TUBE STAND-OFF INSULATOR

Making the hole in the closed end of the test tube is the hardest part of the whole job. A small spot on the end is heated with a small sharp flame (a Bunsen burner and a small blow-pipe work out very well). While the glass is still soft a pointed rod is used to form a little tip as shown in the diagram. Now care must be taken to allow the test tube to cool very slowly because rapid cooling will cause it to crack. After it has cooled thoroughly the tip is carefully ground off on an ordinary bench grinder. A fine grade abrasive wheel is best for this work.

The rubber washers at the top of the test tube should be carved out slightly to give a better fit. The rubber stopper used in fastening the tube to the base should come down to the base to give a firm mounting. At neither end should the pressure used be greater than absolutely necessary, because remember that you are still working with glass. Use "Pyrex Glass" test tubes if possible, because they are much stronger mechanically and are not as likely to crack from heat.

Test tubes come in quite a range of sizes and various sized insulators can be made for different purposes. If well made these supports will add to the attractiveness of any job.—Joseph Kelar, W9GEC.

▼ ▼ ▼

PORTABLE BATTERY CASE

When the amateur "set-builder" takes his portable short-wave receiver with him on trips he frequently finds difficulty in setting up and connecting his "power supply." He has several dry batteries that must be disconnected and packed up every time he moves his location; here is an original stunt that will do away with all of this bother.

Take any kind of a box which is provided with a handle; the box that was used was of tin, rather deep and longer

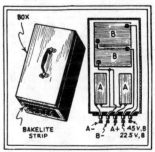

BOX
B
B
A
A
BAKELITE STRIP
A— B— A+ 45 V, B
22.5 V, B

than it was wide. A square opening was cut in one end and a bakelite panel was placed there; binding posts were set in this panel.

The batteries that are placed in it are the smallest types that can be used efficiently in the ordinary portable set. The "A" batteries of which there are two, are No. 4 dry cells and are only two inches square by four inches high. The "B" batteries, of which there are also two, are of the 22½ volt size; together they supply 45 volts. There are five posts, —A, —B, +A, 22.5 v. "B" and 45 v. "B." Of course, the same size batteries need not be used and more "B" batteries may be added. However, this makes a compact portable power supply and all connections can be made simply between the set and the batteries by means of the various binding posts.—Y. H. Mori.

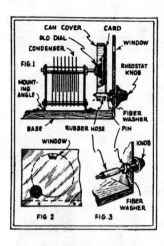

FIG. 1 — CAN COVER, CARD, OLD DIAL, CONDENSER, WINDOW, RHEOSTAT KNOB, FIBER WASHER, PIN, BASE, RUBBER HOSE, MOUNTING ANGLE

FIG 2 — WINDOW, KNOB, FIBER WASHER

FIG 3

SIMPLE VERNIER DIAL

As most everyone has been hit by the depression, these ideas ought to come in handy. A cover from a tin coffee can, an old flat type dial and a few odds and ends from the junk box will make a serviceable vernier dial, as shown in Fig. 1. The tuning control is mounted in a single ¼ inch hole drilled in the panel. A hole for the condenser shaft is drilled through the knob of the dial so that the latter may be held to the shaft by the usual set-screw. The can cover is then fastened to the dial by two short machine screws as indicated. A circular card is fastened to the can cover at front where it may be seen through a hole cut in the panel. The usual dial markings are filled in or the various call letters printed on the card. The small window may be covered by a celluloid cover and illuminated from near by a small flash-light bulb. The panel arrangement is shown in Fig. 2 and the drive-shaft in Fig. 3.—Arthur Buchtenhirch.

▼ ▼ ▼

NOVEL TUNING SYSTEM

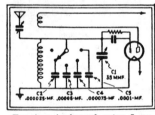

C1 35 MMF. C2 .000025-MF. C3 .00005-MF. C4 .000075-MF. C5 .0001-MF.

Here is a band-spread system I use, and find much superior over many types. When using a vernier dial on the small tuning condenser you can log stations and find them again at the same dial setting. With this system using a lumped capacity for band-centering, logging is impossible due to the fine tuning necessary on the lumped capacity to bring it to correct setting to match the vernier.

A tapped switch is used in conjunction with fixed condensers for the lumped capacity. This switch is mounted behind the panel, while the condensers are mounted directly on the switch. A bakelite rod and a coupling are used to extend it through the panel. Marks can be made on the panel to correspond with the condenser switch points.

The variable condenser should be of a good make, with a low "minimum" capacity to give a little "lap" on each setting. Not being able to get a .000025 mf. condenser two .00005 mf. were used in series to give the right capacity.—Marion Henley.

▼ ▼ ▼

ANTI-BODY CAPACITY KINK

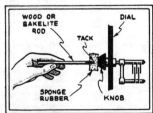

WOOD OR BAKELITE ROD, DIAL, TACK, SPONGE RUBBER, KNOB

For those who have constructed any of the very compact short-wave sets as described in Short Wave Craft and are still bothered by "body capacity," I offer the following solution.

In most cases, where space is at a premium, there is not sufficient room to mount permanent bakelite rods as exten-

sions of the condenser shafts. By mounting a small block of sponge rubber on the end of a wooden dowl or bakelite rod and applying it to the dial of the condenser, the set may be tuned just as easily. The rubber block gives a very good grip on the dial and will not slip in fine tuning.—A. F. Kennard.

▼ ▼ ▼

WIRE HOLDER

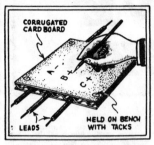

CORRUGATED CARD BOARD, A, B, C, HELD ON BENCH WITH TACKS, LEADS

The usual experimenter's table is littered with unsightly tangles of wires; often he cannot tell where the wires lead to. This can be remedied by this kink which I have used myself. Take an ordinary piece of corrugated cardboard and run your wires through the corrugations. Then tack the cardboard on your table and you will have a handy way of keeping track of the wires. You can write in "identification" numbers or words on the cardboard to indicate where each wire leads to or comes from.—Y. H. Mori.

▼ ▼ ▼

"STEPPING OUT" ON 160 METER PHONE

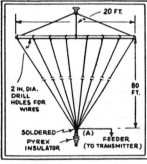

20 FT., 2 IN. DIA. DRILL HOLES FOR WIRES, 80 FT., SOLDERED, PYREX INSULATOR, (A), FEEDER (TO TRANSMITTER)

With the new amateur regulations restricting the use of 85 meters (3500 kc) for phone use, many hams are facing the problem of getting out with low power on the 160 meter fone band (1874 kc.)

The antenna system for use on 160 meters is really more critical than the average operator figures it is. My idea for the aerial described here is based on the broadcast type antenna, a large multi-wire flat-top to radiate all of the power produced by the oscillator. Broadcast transmitters either low or high power step out with the multi-wired system. Why shouldn't the ham fone do the same thing?

The flat-top consisting of eight, eighty foot wires made up in "kite shape," No. 16 gauge is best, but seven strand No. 26 will work just as well. The point end, A, to which the feeder connects should be toward the transmitter so as to make the feeder as short and direct to the tank as possible, and not over 25 feet long if possible. Good Pyrex or glass insulators should be used to reduce the leakage of the system.

The counterpoise is constructed on the same layout except it is thirty feet wide and approximately 7 feet high, while the flat-top should be placed as high as pos-

sible. It is not necessary to have the counterpoise directly under the flat-top but it is best to place it there if possible.

I have designed this antenna for my transmitter which is a Hartley circuit, one UX 210 as Oscillator, 1-250 modulator and two 227's as speech amplifiers. 95% of the stations "worked" with this aerial report steady signals and perfect modulation.

By ordering the antenna wire from some salvage house you can obtain the wire very cheaply. Be sure that all splices are soldered when making up the antenna, as this reduces the total resistance of the system. Approximately 1300 feet of antenna wire is used in the entire system.—W. T. Golson, W4AV; W4ZZA.

▼ ▼ ▼

IMPROVED DIAL KNOB

This idea is particularly applicable to the National BMD Velvet-Vernier dial but it can be adapted to other makes.

Desiring to increase the size of the knob, to give a better grip and take some of the "cramp" out of "cranking" from one end of the band to the other, and being unable to secure a larger knob to fit this shaft, I used the following method to accomplish the desired result.

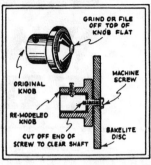

GRIND OR FILE OFF TOP OF KNOB FLAT, ORIGINAL KNOB, RE-MODELED KNOB, MACHINE SCREW, BAKELITE DISC, CUT OFF END OF SCREW TO CLEAR SHAFT

Grind or file the top of the knob flat. Drill a hole in the center and tap it for a 6/32 or 8/32 machine screw. Procure a disc about one-quarter of an inch thick and of the desired diameter (I cut mine from an old bakelite panel). Drill a hole in the center and tap it also. This disc is then fastened to the top of the knob with a machine screw just long enough to go through the end of the brass bushing and still clear the end of the shaft.

This arrangement makes a ship-shape job and certainly makes "dial-twisting" a pleasure.—R. E. Lauth.

▼ ▼ ▼

BATTERY CASING FOR COILS

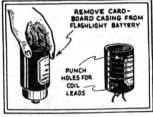

REMOVE CARDBOARD CASING FROM FLASHLIGHT BATTERY, PUNCH HOLES FOR COIL LEADS

When tube bases are not handy and you want to wind some coils for an experimental hook-up, a cardboard casing off of an old flashlight battery is the answer. Wrap your wires around the casing just as on any coil form. Punch holes in the casing to bring in the wires. This is not meant for use on a set but to experiment with.—Y. H. Mori.

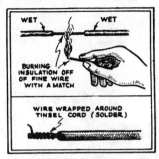

WET, WET, BURNING INSULATION OFF OF FINE WIRE WITH A MATCH, WIRE WRAPPED AROUND TINSEL CORD (SOLDER)

SOLDERING TINSEL

The fine tinsel should first be wrapped with a single strand of small copper wire for a distance of half an inch, then tin the wrapping. If tips are to be soldered on they should be half filled with solder. Heat the tip and then the protected end is deftly plunged into the tip's opening. Broken connections can be repaired the same way.

Removing Insulation From Fine Wire: After determining the length of insulation it is desired to remove, wet the wire several inches back from the place where the wire is to be burned off. The insulation is then burned off by the flame of a match and the wire is rubbed lightly with fine emery paper.—Joe Kocsorak.

▼ ▼ ▼

LENGTHENING TUBE BASES

Many times an experimenter will find that he cannot wind all the turns of a certain coil onto a tube base. This can be remedied by this useful kink. Wrap a strip of celluloid around the tube base so that the edges just overlap. Now, cement the celluloid to the base with some good glue. Also cement the edges of the celluloid together with some acetone. Holes can be bored in the celluloid form to bring the wires to the different prongs on the tube base.—Y. H. Mori.

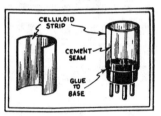

CELLULOID STRIP, CEMENT SEAM, GLUE TO BASE

▼ ▼ ▼

HOME-MADE QSL CARDS

LINOLEUM ¼ THICK, CUT TO 1/32 DEPTH, WOOD BLOCK, W8KAK, SPACE CHIPPED OUT, NAILED TO THIS, IRON, PRINTERS INK OR INKED CLOTH, GUIDE, HANDLE, 4" LONG ROLLER, WOOD

Materials needed: linoleum, cut to size of card and sandpapered smooth—on this the border and letters of any design are cut; some printers' ink or ordinary ink, and a roller, which is a wringer roller cut short. The cards may be bought from the post office, already stamped, at a cent a card. A guide, to hold the cards and to enable better printing, is also needed.

The border and call letters are first drawn on paper, then pasted on linoleum. A sharp knife cuts out the drawing and the space between is chipped out. This type is nailed to a wood block, size of a card. The ink is spread on an iron plate, over which the roller runs, then transferring the ink to the type. Care should be taken when drawing or else the print will read wrong. The small type may be done by hand or gotten some way; it's a good idea to have two types, one for the border and the other for call letters.—Matt J. Surofka.

CHEAP "CODE" SPEAKERS

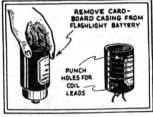

OLD EARPHONE, MEGAPHONE (A), ANGLE BRACKET, E, LEADS TO SET, WOOD BASE; OLD EARPHONE, COFFEE CAN, LEADS TO SET, SOLDER, NAIL, ANGLE BRACKET, WOOD BASE, PORCELAIN INSULATOR CLEAT (B)

The megaphone A can be made of thin tin or other metal or cardboard and is cemented with good strong glue on the earphone and is mounted at point E with glue. The angle bracket is bolted on the phone through holes that are to be drilled through the case.

At B, an old coffee can is used with a nail or metal rod three inches long soldered in the middle of the can on one end, and the other end to the earphone diaphragm. The can should be bolted on the base separated by a porcelain cleat. With a 2-tube set these give fair volume, considering their simplicity.—M. Hermess.

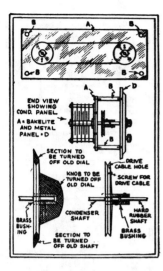

CONDENSER GANGING

The method used in ganging two condensers as shown in this drawing is nothing new. After many an attempt and equally as many failures, at cutting a bakelite disc for this purpose, I conceived this idea of using some of the odd dozen old dials in my junk-box. There are quite a few of the old dials (three- and four-inch) hard rubber or bakelite, which have a brass shaft bushing (note drawing).

I find it easy to make this disc or drum, as you might call it, by filing the knob off level with the end of this bushing. Then place each on a ¼ inch shaft; with the aid of a hand- or power-drill and a large file, we now proceed to turn down the knob until only the brass bushing remains. Then by placing both dials on one shaft, flat sides together, we are now ready to turn down the outer edges to the size disc we want, about 2¼"; by holding the file against both edges, presto, we find when done both disc are the same size! Now with a small three-cornered file, we cut the groove to accommodate the cord used to drive the condensers, both disc to have the same depth groove. Next, we drill the drive cable hole, and drill and tape hole for drive cable screw (note drawing). Then on only one of the brass bushings, we drill another set-screw hole near the outer edge, and tap for second set-screw; this we find serves nicely for a shaft coupling.—C. J. Fink.

▼ ▼ ▼

HEATER SUPPLY

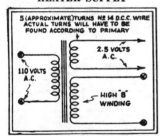

I was recently building a 4-tube A.C. receiver, but I lacked a filament transformer. I solved my problem by removing the power transformer from my power supply and winding five turns of No. 12 D.C.C. copper wire over the high voltage "B" winding. This gave me enough amperage to supply four 2½ volt tubes. This method can be utilized for different voltages by varying the number of turns.—Renaldo Karas.

▼ ▼ ▼

HOW TO USE TWO AERIALS

I have here two aerials. One is about ten feet long and the other about fifty feet long. When using the short one I have very good regeneration, but the signal strength is not as great as with the long aerial; with the long aerial the signal strength is appreciably greater but the regeneration is poor. By merely connecting both aerials to the set at one time, I obtained excellent results—a gain in signal strength and also better regeneration.—Allen D. Rickert, Jr.

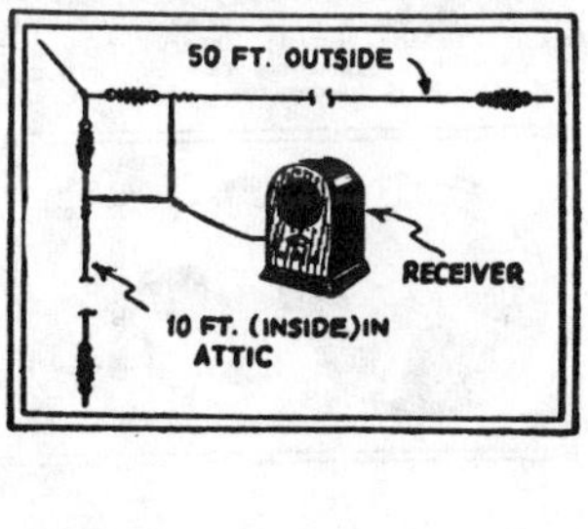

▼ ▼ ▼

PLUG-IN XMITTER COIL

Here is a sketch and description of a plug-in grid coil for transmitters of the push-pull and single-control type. Procure a machine screw large enough to fit the phone tips and solder within. Fit these into each end of the grid coil and solder the loose ends of wire of the coil

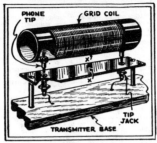

on to it, after fitting nuts into the machine screw. Then, as illustrated, these should fit into tip jacks. For push-pull an additional tip and tip jack is needed at the point marked "X".—James L. Paul.

▼ ▼ ▼

"MIKE" STAND

This "mike" stand may be made of scrap material that is usually found in any "workshop." The base of the stand pictured was taken from a "Crosley Musicone" speaker, but any sort of a base may be used. The arm supporting the cone was sawed off leaving a "stub" about one inch high. This was drilled with an 11/32 drill, and threaded for the 12" length of ⅜ rod on which the ring is mounted. The ring is made of a 24" length of ½" clock spring. Brass will be better, if available. The ends should be lapped 1", clamped together firmly, and

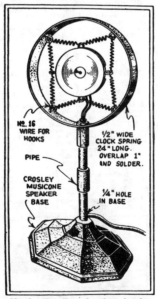

soldered. The rings for the microphone springs are made of No. 16 wire formed around a lead pencil. They should be attached to the large ring through small holes, drilled 5½" apart, soldered firmly on both sides, and smoothed down with a file. The overlapped portion is drilled with a ⅜" hole and the ring attached to the standard with two hexagon nuts. The cord is brought out through a ¼" hole in the pipe near the base. After assembling the stand, give it a coat of shellac or varnish. Before this finish dries, brush on bronze powder, covering thoroughly. After this dries hard, cover with brown paint (enamel or oil paint of any kind) and wipe off immediately with a cloth saturated in turpentine. This will give an antique bronze finish.—W. E. Carson.

▼ ▼ ▼

CUTTING DOWN A CONDENSER

Although midget condensers are relatively inexpensive, many experimenters still prefer to "cut down" standard receiving condensers when building S-W receivers. After much experimenting with all manner of variable condensers, using the Doerle "rig," it was found that as much as 25 per cent increase in volume could be obtained by copying transmitter variable condenser design and spacing

the plates, instead of cutting the rotor or stator plates to fit the capacity. The condenser to be altered should have washer spacers on both the stator and rotor mount-

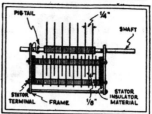

ings. Space the stator plates about ¼" apart, by the simple procedure of putting in twice the number of washers between plates. The same should apply to the rotor plates. The number of plates needed with the new spacing may be readily determined by experiment or formula. This method has another advantage in that the original frame retains its balance, and that scraping plates may be more easily avoided.—Carroll Moon.

▼ ▼ ▼

CELLOPHANE CONDENSER

Have you ever built a short-wave receiver and then have it not work? All because you did not have a midget condenser handy to tune the aerial. Or perhaps in making a very compact model, there was no place to mount one, without making it a cumbersome job. Try this method, which I have used successfully and see how it works.

Secure a piece of Cellophane off a cigarette package or a cigar and wrap it around the bare wire on the end of the aerial lead-in and then insert it in the antenna binding post. You will find that this method will save you lots of trouble, espe-

cially in experimental hook-ups. Capacity can be varied by the amount of Cellophane wrapped on the wire and also by tightening or loosening of the antenna binding post screw.—R. E. Thayer.

▼ ▼ ▼

ADAPTER PLUG

Here is how I make my adapter plugs: I got an A.C. plug at the Nickel & Dime store, and, as the blades of the plugs have holes in them, I soldered a

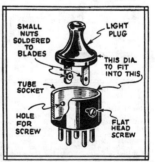

small nut on the inside of each blade and opposite the hole. Then, I drilled the tube-base for the bolts to go through the sides to the blades of the A.C. plug. The A.C. plug is slightly smaller than the tube-base, but if it is filed down close, this makes a very neat "adapter" plug.—Charles Cassell.

▼ ▼ ▼

KEY FROM CLOTHES-PIN

Here's one for you fellows who want an extra key to practice on or perhaps to take the place of a regular key until the pocket-book gets a little fatter. Secure a clothes-pin from your mother's clothes line; one of the clip variety with a spring.

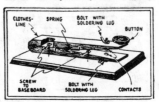

Mount the pin on a suitable base, put a button on top for the fingers to grip and insert a couple of screws for contacts; a very serviceable key results. This method can also be used for making push-buttons where appearances won't count against it.—R. E. Thayer.

▼ ▼ ▼

CHEAP SPRING CLIPS

On small 4½ v. "C" batteries there are usually two Fahnestock clips. Take a small wood chisel and knock the red substance from around the clip. With a little patience the clip can be removed. The clip will be bent as it was on the real battery. Put this bent part in a vise and clamp the vise together. With one or two

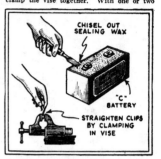

clamping movements the clip will be straightened. Old, dead "C" batteries and 45 v. "B" batteries will be given to you by radio shops, etc. In the long run this saves money, especially if Fahnestock clips are going to be used in a multi-stage transmitter. A drill will put a hole in the clip for fastening.—H. S. Harrison.

▼ ▼ ▼

USE OF VIBRATOR ARMS

Vibrator arms from old Model T Ford coils make good mountings for "one-hole" mounted radio apparatus. This is useful in dealing with experimental apparatus. This also may be used for holding panels to subpanels. If the hole is too small a reamer will help.—H. S. Harrison.

▼ ▼ ▼

"K.C." LABEL FOR COILS

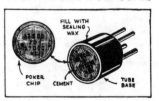

Here is my idea of a smart looking though simple "coil marker." Take an ordinary white poker chip, which is just a shade too large to be slipped into an old tube base, and file it to the proper size, at the same time roughening it. Apply ordinary household cement to the edge of the chip and inside edge of base and place the chip in position. The coil may then be marked as illustrated.—S. Ivan Rambo.

▼ ▼ ▼

AERIAL RIGGING

The idea of this double system is to get over the difficult and exasperating job which sometimes occurs if the hoist wire jumps the pulley wheel and jams

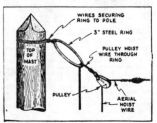

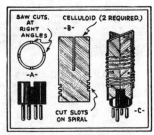

$5.00 PRIZE
SPACE-WOUND COILS

Many would like to make a space-wound coil with a high percentage of "air dielectric." This is easily accomplished with an old tube-base and a sheet of celluloid. For holding the coil, slots are cut in the tube base with a saw bent at right-angles, as shown in Fig. A. The coil form is made of two pieces of celluloid which are both alike as in Fig. B. The slots for fitting together and holding the wire can be cut with a coping saw. Care should be taken in making the slots for holding the wire. These slots should be made so the wire when wound will be in a spiral. A good glue for assembling and holding the wire in place can be made by dissolving celluloid in acetone until fairly thick. If the coil form is made of thin celluloid and is weak, small triangular pieces (braces) of celluloid can be glued on it. This same principle can be used in making plug-in coils with a base having a variable primary.—Jack Thorpe.

▼▼▼

IMPROVED "GROUND"

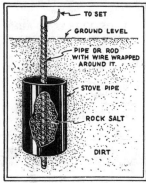

This "ground" works much better than an ordinary one. The following material is needed: An iron pipe or rod about 5 feet long, about 20 lbs., of coarse (rock) salt, a piece of No. 14 wire about 15 feet longer than is needed to reach from the "set" to the ground, and a section of stove-pipe. A hole is first dug in the ground big enough for the stove pipe to slip in. The insulation is scraped from about 15 feet of one end of the wire. This end is coiled around the rod and the rod is put in the pipe as shown in the illustration. Soil is then thrown in the hole and rock salt is mixed with it in the pipe. The hole is then filled with soil. The loose end of the wire is then connected to the set in the usual manner. A few holes punched in the stove pipe will increase the efficiency as more moisture is admitted. The stove pipe keeps the salt from washing away and the salt draws moisture.—Elbert Wehrheim.

▼▼▼

COIL HANDLE

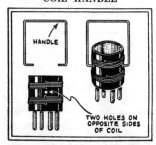

Here is a handle for quickly changing coils in a short-wave receiver; it consists of a piece of stiff wire bent in the shape shown in the diagram. It is inserted into 2 holes drilled in the coil forms and may be used for lifting or inserting the coils. To operate, the handle is compressed so that the two ends enter the two holes drilled on opposite sides of the coil; the coil is then inserted or removed. One handle can serve for all coils; it is especially useful for tube-base coils.—Clyde Preble.

R. F. CHOKE COVER

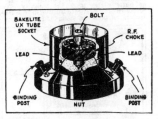

Remove all the binding posts and rods from a tube socket; then cut a piece of thin wood, fibre, etc., to fit on the inside of the socket. Place a small bolt through the center of the wood disc. The R.F. choke fits over this bolt and the nut is put on to keep it from slipping. Two of the binding posts are fastened on to the socket, where the choke leads are soldered.

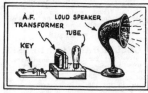

CODE PRACTICE FOR GROUP

Your February-March 1931 issue had a code practice outfit by Paul Skitzki, on page 388. The writer built this outfit and found it to be just what he wanted. But for "group" instruction of five people, the question of five pairs of head-sets and five keys was out of the question, so substituting a magnetic speaker and putting a 4½ volt "C" battery in the plate circuit with positive to plate, it was found that we could hear and understand the code ten feet away from the speaker. I am passing this on to you as it might help some more fans to get up a "code practice" club with a few of their friends.—J. B. Vepper.

▼▼▼

AERIAL TUNING CONDENSER

The accompanying sketch shows how I built an efficient series antenna condenser. The insulation is maintained at a high value by mounting the oppositely charged metal parts of the condenser on a bakelite or other equaly efficient base. The capacity of the condenser is varied by turning the insulating handle mounted on one

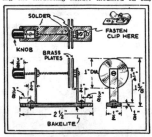

end of the threaded brass rod, the rod being threaded 8-32 or 10-32 pitch. The condenser members comprise two brass or other metal discs one inch in diameter, the thickness of the disc having no effect on the capacity. Not only will a condenser such as this be found extremely valuable to all short-wave operators, in helping to eliminate "dead-spots," but in some cases a set which will not oscillate at all can be made to do so by varying the capacity in the aerial circuit. If you have a fixed condenser in series with the antenna and your set is a little stubborn in one way or another, try this condenser and smile.—Joe Casalett.

▼▼▼

TICKLER REVERSAL

Being an experimenter I have several makes of plug-in short-wave coils and sometimes it becomes necessary to reverse tickler connections, so I hit upon the idea of placing a small D.P.D.T. switch close to the detector socket. Thereby I solved the problem of frequent soldering and resoldering of connections.—William S. Russ.

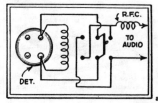

"PLUG-IN" CHOKES

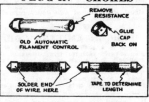

Most of us who have tinkered with battery sets have some of the automatic filament controls, contained in a glass tube, lying around, so why not put them to work as follows: Remove the metal end caps and take out the resistance element; then glue the metal end caps back on. You now have an R.F. choke form with winding space that will accommodate enough wire to choke in the 160 meter band, and all the others.—Grover E. Hall.

"LONG WAVE" ADAPTER

Here is a description of a "long wave" adapter for short-wave sets using plug-in coils. It consists of a variometer or variable tuning coil and an old tube-base. The tickler leads of the variometer go to the tickler prongs of the plug-in coil form

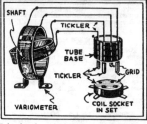

(tube base) and the same with the grid leads. The variometer is mounted on a small baseboard with a panel. The set I used it with was the "Globe Trotter" described in SHORT WAVE CRAFT, November 1932, page 400. In the first night's test many of the larger broadcast stations of eastern and central United States were logged.—Roy W. Neads.

▼▼▼

"PLUG-IN" CONDENSERS

I am a reader of your unique book, SHORT WAVE CRAFT, and am one of the many that really enjoys reading it and in doing so have accumulated much knowledge during the few hours I get to read it. I am enclosing herewith an idea which I think could be listed as a wrinkle either in short or long wave reception or transmission. Here is the dope: "Plug-in" fixed condensers made by means of electric light plugs and sockets. The condenser is secured to the plug by means of bus wire as illustrated.

In my case I flattened the wire up where the condensers are to go, so as to make the condenser easy to change to different values when experimenting. The sockets I used were of some white non-conductor (porcelain) which can be gotten from any electrical shop. This I believe is

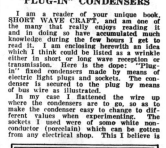

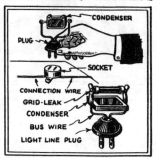

of most value to the experimenter, although also of value to the set-builder, whether he is a short wave or long wave "fan;" also the man building transmitters.—George Purnell.

▼▼▼

BURNT-OUT A. F. KINK

It is usually the primary coil of A. F. transformers which burns out, but they can be very satisfactorily fixed by connecting a 100,000 ohm resistor across the primary terminals and a .006 to .01 mf. condenser between the grid and plate terminals of the transformer. You can fix these transformers in a very short time by making some clips which may be mounted on the binding posts of the transformer, which will hold the resistor and condenser very nicely. These connections provide "resistance-capacity" coupling, with an "impedance leak" and will be found to give good

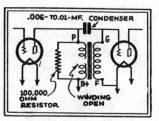

tone from even cheap transformers, but it will give slightly less volume. But why buy new ones when you can fix the old ones?—Alfred Oberstaedt.

▼▼▼

TICKLER INSIDE COIL

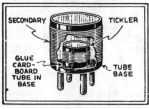

Tube-bases are often too short for both secondary and tickler windings. To overcome this the secondary is wound on the tube-base and the tickler is wound on a one inch cardboard or bakelite tube which is glued inside as shown. Connections are made to the prongs in the usual manner. —L. H. Wilson.

▼▼▼

SUBSTITUTE "MIKE"

I proceeded to build a low-power transmitter out of the junk box which contained many receiving set parts of varied sorts. I got along well and finally got "her" done. Then came the question of a "mike." What to use for the "mike?" The cheapest "mike" on the market at that time was well above five dollars, which was the one thing I didn't have. So instead of using the regular microphone transformer in the modulator, I substituted an ordinary audio transformer of "ancient vint-

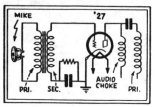

age," and with this I was able to use an old Vemco magnetic "speaker unit" with a little alteration as a "mike." To say the least, the results were excellent and much better than could be had from most carbon "mikes." The only alteration is the diaphragm; the old one is taken out, and a new one is made from the tin of a coffee can. Cut it out the same size and sandpaper down quite thin, replace, and it's ready. The output is good and strong and needs no pre-amplifier, but can be used with the conventional two-stage modulator. The quality is par-excellent. Here's my circuit in part—balance on request.—John Markovich.

▼▼▼

IMPROVISED VERNIER

Here is a quickly improvised and cheap substitute for the ordinary vernier dial. The materials used were one 10-inch phonograph record, one switch knob with shaft, one rubber washer, two metal washers, two nuts, and part of a connector of the type used to connect gang condenser shafts in broadcast receivers. The latter serves to connect the record to the condenser shaft. The knob assembly is shown in the sketch, a short length of thick-walled rubber tubing serving as a rubber washer, which makes firm contact with the rim of the record. If the set up is to be used for some time on a bakelite panel, a metal bearing should be provided for the shaft of the knob where it passes through the panel. A satisfactory scale may be laid out on heavy white paper and "shellacked" to the record; library paste and the so-called "household cement" will not hold after they become dry.—H. O. Ervin.

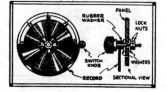

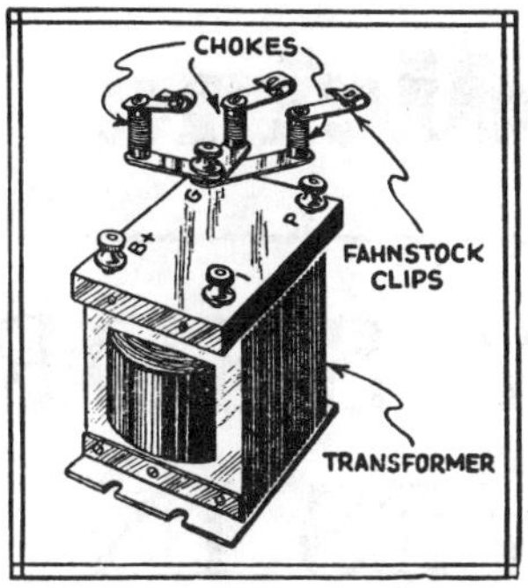

VARIABLE R.F. CHOKE

Above is a diagram showing a simple method of quickly changing the value of the radio frequency choke. Fahnestock clips are screwed to the top of half-inch wooden dowel rods and one end of the choke is fastened to the clip. A hole is drilled in each end of a strip of copper or brass. One end of the strip is bolted to the binding post of the transformer. Screw the other end to the dowel rod. The remaining end of the choke is fastened to the strip.—Ivan Ross.

▼▼▼

PLUG-IN COIL HANDLES

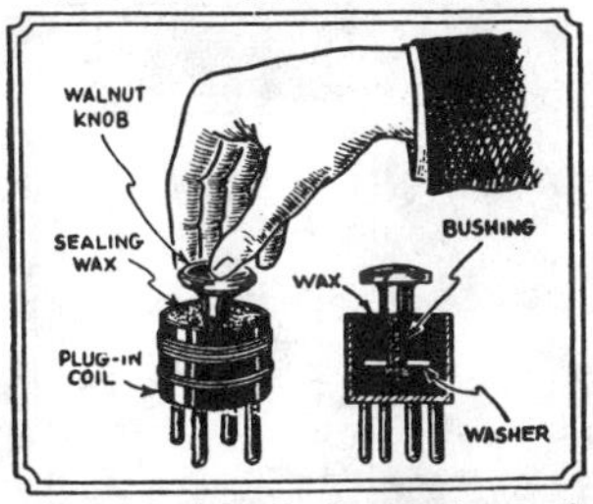

A very good handle for tube base coils can be made from a common walnut knob, the variety sold in the Five-and-Ten-Cent stores; a fiber washer and a brass bushing. Remove the nut on the knob and slip on the bushing, which should be about ½" long; next, the washer, and finally the nut. Tighten the whole thing and the handle is ready for use.

After the coil is made, fill up form with sealing wax and push the handle into the mass until the bottom of the knob is flush with the surface of the wax. Once the wax solidifies you have an ever-lasting coil "pull."—Thomas A. Blanchard.

▼▼▼

JOINING WIRE TO PRONGS

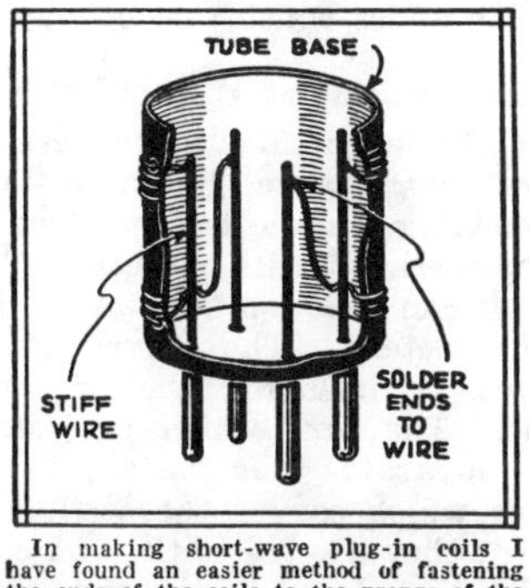

In making short-wave plug-in coils I have found an easier method of fastening the ends of the coils to the prongs of the coil form or tube base. For UX bases take four pieces of stiff wire (as tall as necessary) just large enough to fit tightly into the prongs of the form and solder the ends of the coils to the wire. For UY use five wires, etc.—Harry F. Sieber.

▼▼▼

"HUM" REMOVER

Here is a kink that will be very useful to those who use "A" eliminators for their receiver filament supply, which pass an A.C. hum, detectable in the speaker or

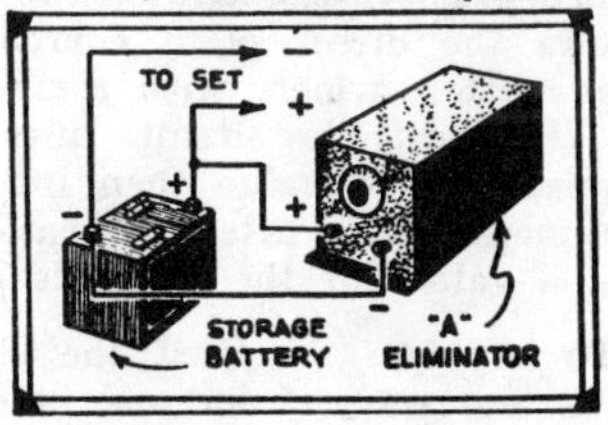

phones. I have an "A" eliminator which was extremely noisy, so I bought a storage battery, intending to use the eliminator as a charger. This took a lot of time and bother so I shunted the battery across the output of the "A" eliminator. This eliminated the A.C. hum completely. If this hum still persists take out the acids from the battery and fill it with condenser oil instead.—G. Zemanovich.

▼▼▼

SIMPLE "TEST" BATTERY

To construct this "everlasting battery" you sandwich the blotter between the metal plates. Drill two holes in the bakelite from each end; perform the same operation on the metal plates. The piece of bakelite I used was ⅞" wide and 2½" long. The plates are part of two variable condensers, one brass and the other aluminum. Any size plates of bakelite may be used; the blotter keeps the plates from touching one another. Instead of two clips or binding posts one may be used and a

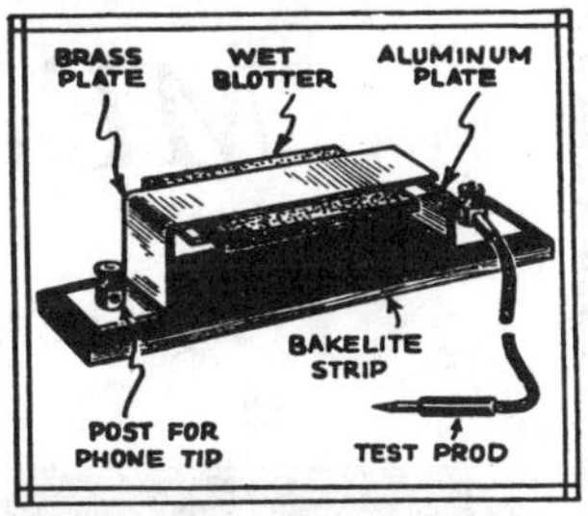

piece of flexible wire fastened under the head of one of the machine screws; something should be soldered to this end for a prod. The tip of the aforementioned phone cord is used for the other end of the test cord. To charge this battery simply immerse the blotter in water. If the blotter dries before you wish to use it, the process is simply repeated. It is not necessary to remove the blotter from between the plates to wet it. Wet the complete battery, and then wipe excess water from the plates and strip of bakelite, being sure the plates are so placed that they hold the blotter firmly between them.—James Austin.

▼▼▼

HANDY CONNECTIONS!

To those experimenters who delight in devising and trying out new circuits and who know the bother of having to change

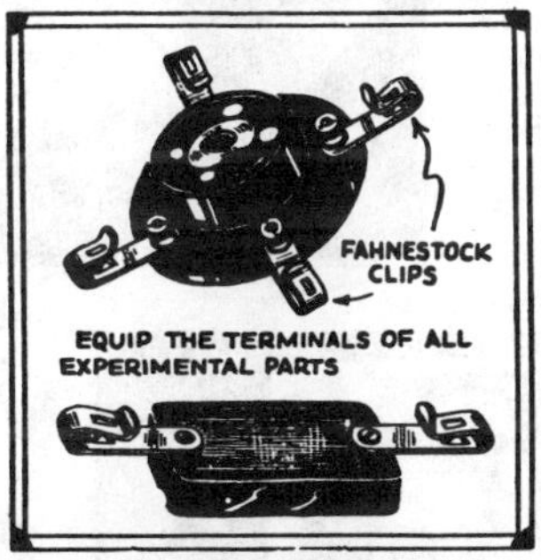

connections from one point to another by screwing and unscrewing thumb-nuts (in the life of an experimenter about every ten seconds) try the following kink: Equip the terminals of all your experimental parts (sockets, coils, rheostats, etc.) with Fahnestock clips. The parts thus equipped may be mounted bread-board fashion and minor circuits or complete circuit changes may be made quickly and without trouble. —M. C. Alexander, Jr.

▼▼▼

SMALL SOLDERING IRON

It is frequently difficult to lay your hands on a really small soldering iron suitable for soldering joints on fine mag-

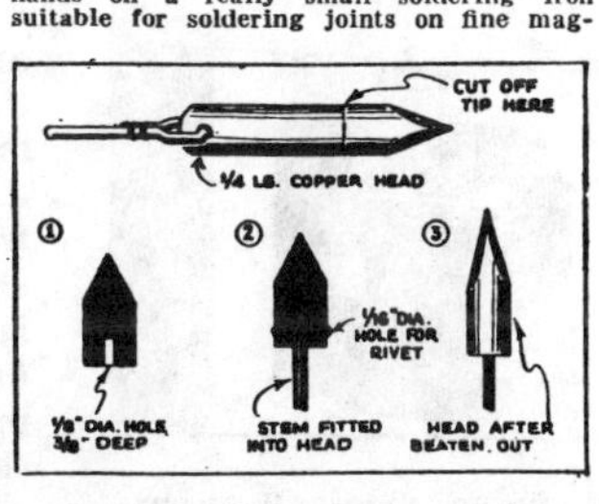

net wire. I made a one-ounce soldering copper as follows:

I bought a ¼ lb. copper and cut it in two about ¼ way back from the tip, drilled a ⅛ inch hole lengthwise in the base of the tip just cut off, to a depth of ⅜". In this I fitted the ⅛" stem of my old iron, and then drilled a 1/16" hole crosswise through the copper and stem. In this 1/16" hole I used a wire nail as a rivet to secure the assembly. To finish the job I heated the copper just a bit hotter than is used for soldering, and beat out the point to a long sharp one. This is easy as the copper is very soft and does not require much more than a few firm taps to do the job. The completed tool only weighs two ounces and is very light.

I beat out a point on the remainder of the ¼ lb. copper in the same way, and now have two nice size coppers, 3 ounce and 1 ounce.—Cliff Dawson.

▼▼▼

5-METER TRANSMITTER

I believe this to be the simplest and most efficient 5-meter transmitter. In the past six weeks over 200 contacts were made

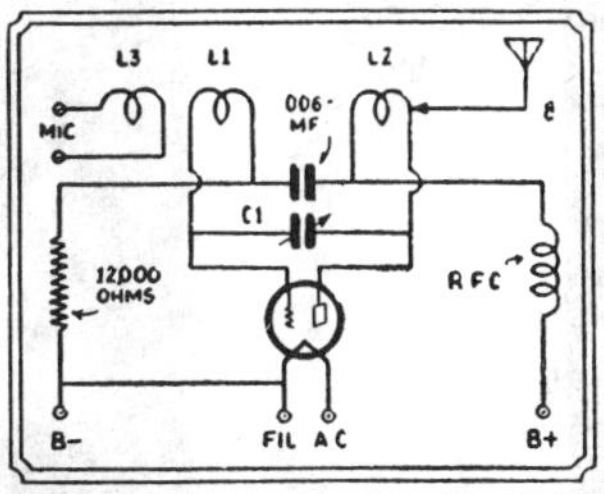

with 28 different stations. The "DX" was about 15 miles and an "R9" signal was reported from that station. Because position and not power is everything on this band, and the transmitter being located on the west slope of a hill, practically all contacts were with "Western" stations. The quality of modulation was good, although loop was used and the percentage of course was low. An indoor 8-ft. aerial was used and over 6 miles was covered with no antenna at all. The circuit is a variation of the split Colpitts and is as follows:

A '45 with 400 volts of "B" supply was found to work the best. L1 and L2 are one turn, 1¼" diameter of No. 16 wire. L3 is 1 turn of No. 32 wire wound around L1. C1 may be a 5 to 11 plate midget or any condenser of similar capacity.

The aerial is 8 feet long and may be connected to the plate coil directly. A neon bulb will show oscillation by touching it to either end of the antenna or a single turn of wire attached to a flash-light bulb will light within one-half inch of the plate coil. Any RFC (radio frequency choke) will suffice.—David Townsend.

▼▼▼

A HUSKY PHONE JACK

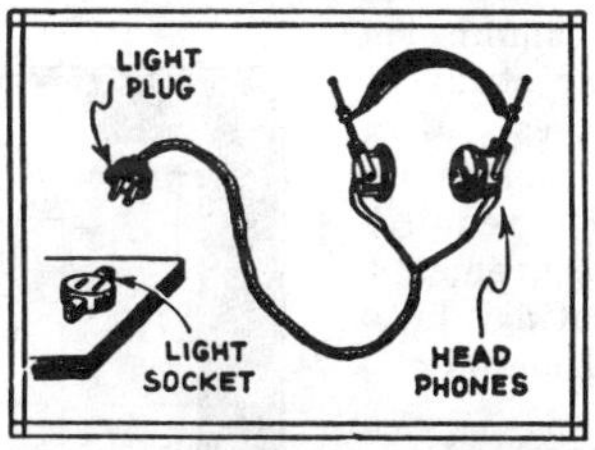

Use an electric light plug for a phone jack and a light socket of the prong type. I find it just as good. Solder one end of the lead from the phones to one of the screws of the plug and do the same to the other lead. Using a double-throw, single-pole switch makes a good way to switch the aerial to the ground, when not using the aerial for the set.—Warren W. Smith, Jr.

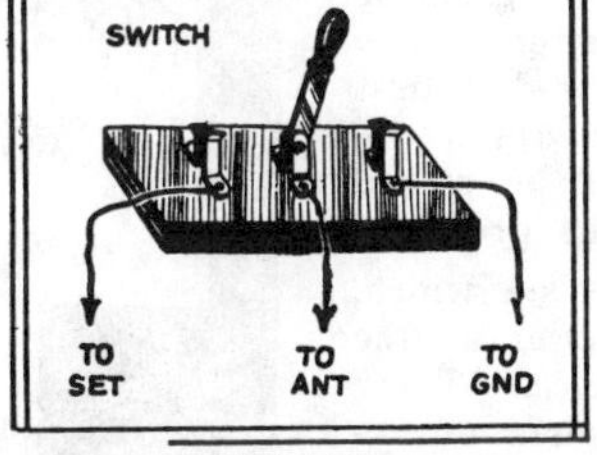

▼▼▼

ANTI-CAPACITY DEVICE

This is a method of curing those troublesome "body capacity" effects. It consists merely of "tuning" the earth lead to a point where the effects disappear, by means of a "variodenser," such as XL, of any capacity up to about .0005 mf. maximum; I use a .0003 mf. This will be found absolutely effective, and I have never known it to fail yet. If one moves

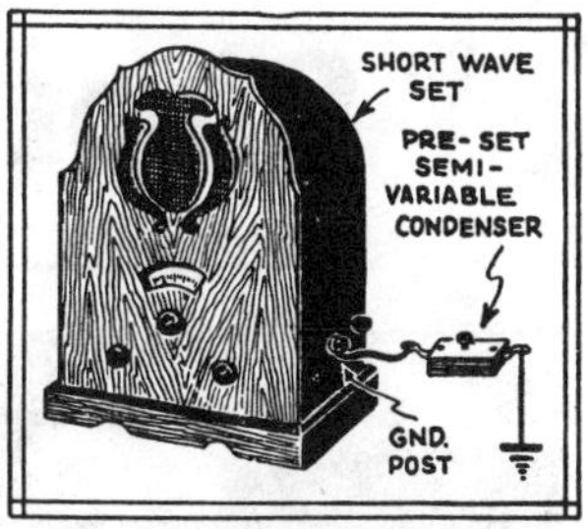

up from, say, the 20-meter to the 40-meter band, a slight readjustment may be necessary, but this is the work of a moment.—G. E. Gaunt.

▼▼▼

QUICK AERIAL CHANGE

I got tired of removing the antenna and ground leads from one set and placing them on another. Now I use baby phone jacks instead of the binding posts, and

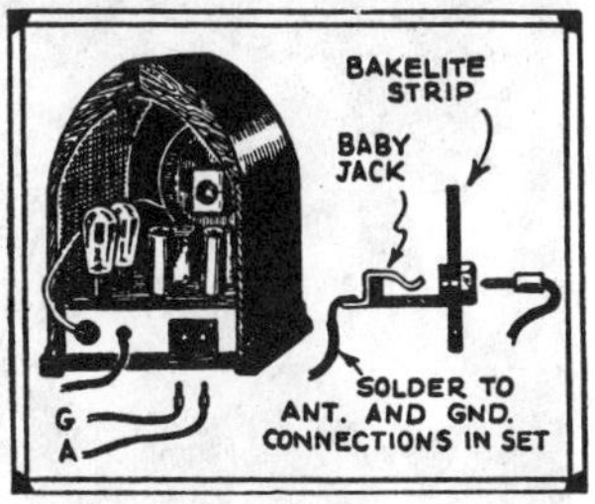

cord tips on the leads instead of the bare wire. The two baby jacks were put onto a small strip of bakelite as shown in sketch above and soldered to the antenna and ground connections in the set. Now it is a simple matter to yank these connections out of one set and plug them into the other.—J. T. Watkins.

▼▼▼

REDUCING "DEAD-END" LOSS

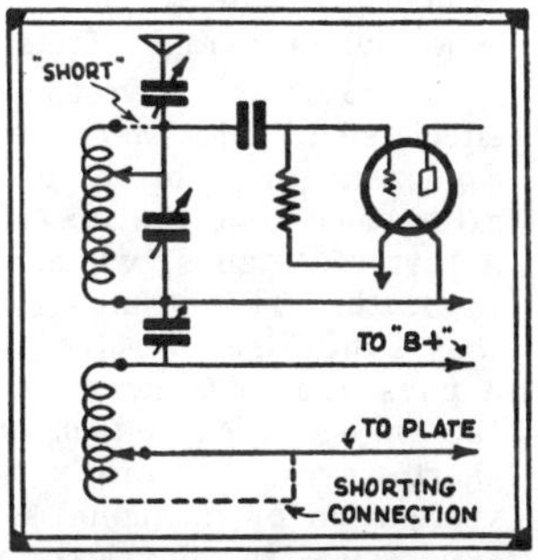

Referring to Mr. Haas' article, "Sliders Do The Trick," in the February-March issue of Short Wave Craft, I found the same trouble as with taps, the "grid" coil showed a loss and the "plate" coil went into oscillation too abruptly on account of the "dead-ends." Then I did for the sliders exactly what I had been doing with the taps, that is, "shorting" with a heavy piece of bus-bar from the pivot of the sliders to the "dead-end" of the coils. Presto! No loss to be noticed from the grid coil and the plate coil could be varied from a whisper to a maximum without oscillation.—W. H. Lord.

▼▼▼

WINDING TUBING

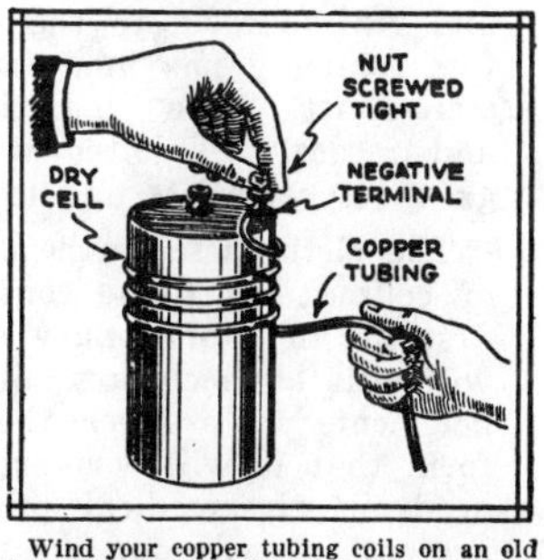

Wind your copper tubing coils on an old dry-cell. First flatten the end of copper tubing, drill a hole through it and slip over binding post. Wind the required turns while the other end is pulled tight in a vise or held by someone.—Floyd Gribben.

Short Waves in Medicine

ULTRA SHORT WAVES IN *MEDICINE*

By Dr. FRITZ NOACK
(*Berlin*)

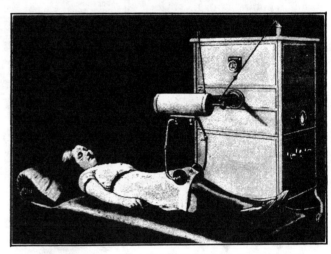

PROFESSOR ESAU, of the University of Jena, the well-known pioneer in the field of ultra short waves, was the one who nearly two years ago pointed out that very evidently the ultra short wave exercised beneficial medicinal effects.

For further investigation of this effect, small mice were put between the condenser plates of the sender; and Lo! after some time the mice fell victims to a sort of tetanus, which eventually led to death. The result of this fact was that Prof. Esau resolved to commission a physician to make a further study of the physiological effect of these ultra short waves. Naturally the serious investigation of medicinal effects requires experience in physical measurements, likewise an apparatus operating in a constant and easily controlled manner.

One of the "fair sex" receiving a treatment through the knee by ultra-short waves. Several treatments are given for the average ailment.

Siemens and Halske Interested

It is gratifying that a manufacturer which already possesses very great experience in constructing electro-medical apparatus and which also has the requisite experience in physical measurements, the prominent firms of Siemens and Halske took up the matter. Now this company has manufactured a first ultra-short wave *radiation apparatus* and recently exhibited it to a picked group of scientists and to the press.

The pictures show the new set which, of course, has to be considered as the first of an evolutionary series. There will still be necessary all sorts of experiments to produce the set in such form that it will correspond to all the needs of the medical profession. The present set will, for the time being, be used for the further study of the physiological effect of ultra short waves. It will be the problem of the doctors to

The ultra-short wave medical apparatus with its electrode arm adjusted to permit the waves passing through a solution for experiment.

investigate the fields of use and the conditions under which the set may be used.

Arrangement of U. S. W. Generator

In the large shielded box are the real ultra short wave generator—(actually resembling a radio transmitter) and the parts needed for its operation. The generator can work on two waves, four and eight meters. The wavelength is adjusted by inserting the proper tuning coils. The arm which projects out of the apparatus has inside it two lead wires which convey the oscillations outward; they are led to the two electrodes, which are placed at the ends of the two visible supports.

In the cross-beam at the end of the arm is a "tuning" device, which tunes the electrode oscillatory circuit exactly to the wavelength of the generator. The exact tuning can be read on a meter, which is above the arm on the box; this shows the direct plate current of the transmitter, which, with correct tuning of the electrode circuit, adjusts to a minimum; the value then indicated by the meter gives, after calibration of the a value for the electrode energy.

To be able to adjust the electrodes, as is necessary from one case to another, the electrode arms are set in ball-joints. To ascertain whether oscillations are actually present in the electrode oscillatory circuit, there is on one electrode arm a little *glow lamp* (Neon tube), which lights up in the presence of oscillations.

Electrodes Swing on Giant Arm

The arm projecting out of the apparatus can be swung in all directions, for the better use of the whole unit. To accomplish this, it is fixed to the box in a universal joint. The end cross-piece of the arm can furthermore be turned, so that the electrodes can assume any desired direction. The right side wall of the box contains the main current switch, as well as a resistor with which the filaments of the tubes are regulated. The apparatus is in fact constructed just like a vacuum tube radio transmitter. The electrodes may take various forms; in accordance with the most recent experiences, they are so shaped that they do not have to be put firmly against the body, but exercise their effect even in the presence of a layer of air between body and electrode. At the same time, the remarkable fact may be determined, that the thickness of the layer of air and the "depth effect" are dependent on each other.

Advantages of Ultra Short Waves

According to all experiences thus far, the medicinal ultra-short wave radiator is a sort of *diathermal* apparatus; therefore it serves to conduct to the part of the body, or the organ to be treated, electric wave energy, which is converted into heat and thereby exercises a healing effect. As compared with the usual diathermal sets (which operate on a wavelength of about 600 meters, using leaden electrodes, which have to be put right on the body), the *ultra-short wave* set offers a considerably better control of the direction in which the heat is to be applied, so that one can better reach any desired organ than was possible before. Furthermore, by changing the wavelength, one can also adjust for a definite deep-lying stratum. Unpleasant accompanying phenomena, which result with the usual diathermal sets, where the electrodes do not rest firmly on the body, do not occur in the case of the ultra short wave radiator. It is especially important to note that one can now, for the first time, perform a "therapy" or treatment directly on the head. The operation of the set is very simple.

Whether, besides the warming of the different strata in the body, still other biological effects occur or can be accomplished, with the new ultra short wave radiator, is, at the present time, still a matter of question.

Human Beings *as* Antennas

By DR. ERWIN SCHLIEPHAKE, M.D.
Of the Jena University Medical Clinic

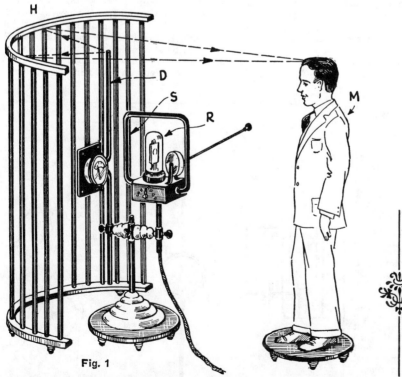

Fig. 1

Arrangement of the apparatus used by Dr. Schliephake in measuring and observing the effects of short waves on the human body, absorption, et cetera. H, parabolic reflector formed of wires; D, antenna, charged inductively by the coil S; vacuum tube at R, and beside it wave changing condenser; subject at M.

IF human beings remain in the vicinity of powerful radio transmitters, they are exposed to a very strong electromagnetic field. One must therefore ask what significance these radiations have for the internal organs, and how the body reacts to them. This question is justified by the fact that all other kinds of electromagnetic rays influence the human body more or less strongly.

The character, however, of the reaction is different, according to the wavelengths. Radium rays and X-rays, for example, because of the extraordinary shortness of the waves and the high effectiveness of energy under these conditions, exercise a very strong effect on the structure of atoms; so that serious injuries to the organism can arise (for instance, under certain conditions, cancer). Less serious are the effects of light rays which, however, in the ultra-violet part of the spectrum, can still occasion serious injuries; while the visible light which we perceive is of too low a frequency to injure healthy persons. There are, however, subjects, made sensitive by illness, who suffer serious affections of the skin from strong illumination. Above the optical spectrum, there follow the infra-red and heat waves, whose effect on atoms is relatively slight; in the case of still longer waves only molecular effects are to be expected. Here we are entering the range of the Hertzian waves, electric waves in the stricter sense. Of these, the shortest are about half a millimeter (.02-inch) long, and therefore just above the heat radiations.

From direct analogy, it is to be assumed that the shortest of these waves must be absorbed by human bodies, as in the case of heat rays, and

DO YOU KNOW —

❡ What effect short waves have on the hair and scalp in general?

❡ Whether the current in the exciting antenna increases or decreases when a person stands in its field? Why?

❡ If the person stoops will this affect the reading of the antenna current meter?

❡ Whether 'cold" or "heat" is produced inside the body by an overdose of the high frequency waves?

❡ The effect of short waves on the nervous system?

this absorption can be actually demonstrated.

Absorption of Radio by the Body

With specially-built transmitters, especially such as those which have been described by Kohl, electric waves only a few centimeters in length can be produced; these waves can be reflected by mirrors, exactly like light rays. If one puts the antenna in the focus of a concave mirror, then there is formed a parallel beam of radiation which can be concentrated again in another concave mirror. By means of a lens placed in the course of the radiation, the ray is made very sharp. If a person steps into this path, reception ceases. Since no reflection by the human body or only a very slight one is demonstrable, absorption of the radiation must have occurred; this effect is, moreover, demonstrable in the case of other organic substances or water. Unfortunately, not much can yet be said about the physiological effects of these radiations; the power of the transmitters of extremely short waves is still too low to produce perceptible changes.

It is otherwise with the wavelengths from three meters up, to the production of which powers of several kilowatts can be applied. Here the above described optical phenomena cannot be so well demonstrated, because the diffraction is much greater; but absorption by the human body can be very well shown.

When an antenna is inductively coupled to a 3-meter transmitter, the oscillations can be indicated by a detector even at a considerable distance away. If however, a person puts himself in the place of the antenna, the detector responds much more weakly, although the power consumed by the transmitter remains the same. Accordingly, a part of the power must have been used up in the body. The same phenomenon can be demonstrated as follows: a closed or open oscillation circuit is inductively coupled with the transmitter (see Fig. 1). The ammeter in the circuit shows a definite current strength. If a person places himself on an insulating stool beside it, the current in the oscillation circuit is reduced. This withdrawal of energy, however, depends on the length of the body; for if the subject stoops, or changes to a sitting position, the current in the other circuit increases (Fig. 2). Therefore, it appears as though, by the tuning of the subject, *to about half the wavelength, the power transmitted to him becomes much greater.*

Concentrating Power by Means of a Reflector

By means of a large concave mirror one can also collect the transmitted power to a focus. Such a mirror need consist only of parallel wires stretched between two wooden frames. Its height must be equal to the wavelength, the opening one and one-half times the wavelength. It is best to use elliptical reflectors, with the transmitting antenna at one focus; then the reflected radiation is at the other focus of the ellipse. With a "dipole" ("Hertzian" antenna) containing an ammeter, especially strong concentration of energy at this point can be demonstrated. Here a lessening of the current in the dipole is instantly shown if a person steps into the vicinity. Since the human body is to be regarded essentially as an electrolytic system, with regard to the electric wave, and I have tried to demonstrate the effect in the following manner in a model experiment.

A glass tube, of half the wave's length, was filled with an 0.5% sodium chloride

(salt) solution, to which gelatine was added to prevent convection. In this jelly the temperature was measured, at different places, by thermo-elements. It was shown that the heating was greatest in the middle and least at the ends; being half as strong at the quarter points of entire length as it was in the middle. Since the maximum strength of the current is at the middle, the greatest heating is therefore connected with this.

Physiological Effect of Short Waves

Especially noteworthy, also, was a feeling of vibration, which was particularly evident if the hand was raised in front. We could establish this sensation only at our transmitter, which is operated with 50-cycle alternating current. There must be, therefore, a direct influence on the nervous system. Here too we have, therefore, another proof that the ultra-short radio waves exercised an effect on the nerves.

These vibrations are also felt if the hand is placed in a condenser's field. Many persons who remain close to the transmitter also experience remarkable sensations on the head, near the roots of the hair; these are like a peculiar prickling, the hair likewise standing up a little. In many subjects we could also observe slight increases in bodily temperature, which however did not exceed 0.5 of a degree, Centigrade (0.9 degree Fahrenheit). Since the body contains extremely fine regulators, by which the temperature is always kept constant, and since also the amounts of energy which can be conveyed into the body, even by powerful radio transmitters, give (when translated into heat units) only a relatively low number of calories, this increase in heat is probably not attributable to the received energy alone. As is to be shown later, nervous effects may play an important part in this.

The effect on the nervous system is

Fig. 2.
One of the most interesting effects of placing a human being in the field of a short wave (high frequency) oscillator is that of absorption. The antenna "D" is excited by the power tube and loop circuit at the left. Do you think the antenna current is increased or decreased by bending the body?

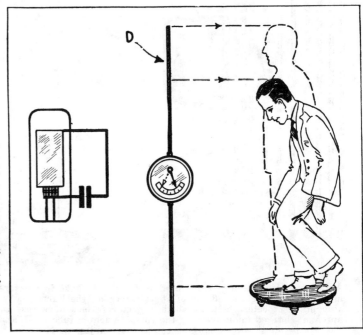

plainly felt by all persons who work a fairly long time with short waves. The sensations are different with the individuals; there are persons who are relatively insensitive, and others who very soon complain about the disturbances of their health. Usually there is first an increasing sleepiness; they are very tired by day, but at night they sleep badly. Several times a night they start out of their sleep, and they are tired and sleepy in the morning.

These phenomena increase more and more. Often there are also intensive headaches, particularly covering the back of the head. Many persons complain of digestive disturbances and pressure on the stomach. Most of them are furthermore easily excited and irritated, being inclined to complaining and to violence. This increased excitability of the nervous system can also be shown by electrical tests of the nerves.

Effects of the Electrostatic Field

It is much easier to study these changes by using the condenser field. Here it is not the electromagnetic wave which is used, but the electrostatic field, which always accompanies it. Here, however, the effect may be much more strongly concentrated. I have generally used plates four inches in diameter, between which the parts of the body in question were treated. The field between these plates suffices to heat 100 cc. of a 0.5% sodium chloride (salt) solution 5 degrees C. (9^0 F.) in one minute. If parts of the human body are introduced into this field, and the blood is then taken from some part of the body (for instance, from the earlobe) on investigation obvious changes are found.

The number of red corpuscles per cubic millimeter is very greatly increased; i. e., if they formerly amounted to 5 million, the number has risen to 6 million. The same is true of the haemoglobin (a constituent of the blood) and the white corpuscles, the number of which likewise increases. But this is not always true to the same extent, depending upon the strength of the radiation and the place treated. One frequently sees instead a lessening of these amounts, which would therefore correspond to a thinning of the blood. Likewise, the albumens of the blood undergo certain changes which I cannot discuss here.

In the case of these strong effects it is to be assumed that the tissues of the body also undergo changes; which, however, cannot be directly proved in a subject.

Results of Experiments on Animals

In experiments with animals, on the contrary, such changes are plainly recognizable. They occur particularly strong in projecting parts of the animals, such as in the ears and tips of the tails; since there the electric field is distorted. Very often, one sees, some time after the treatment, that the parts in question are dying and falling off. In the case of a rabbit whose leg had been too strongly exposed to the electrostatic field, I noticed a complete disintegration of the flesh in a ring-shaped region, so that only the bone remained; but then, after a while, that also fell off.

That the nervous sytem of the animals is also influenced is shown by the fact that many of them shudder on the switching on or off of the field, at a time when there can be no question of a strong heating effect.

Also very interesting are the disturbances of the internal heat regulation, which one can produce in animals. If the region of the neck and the back of the head of a rabbit are exposed to a limited capacitative field and the temperature is afterward measured, one can soon note a permanently increased temperature, which often lasts for some weeks. If a very powerful dose of "irradiation" is administered the opposite can occur: the bodily heat falls more and more, often below 35 degrees C. (95^0 F.), until these low temperatures are no longer compatible with life. At the same time, it is interesting to observe that almost all animals which have had such disturbances of their heat regulation after a few weeks developed inflammation of the lungs and pleurisy, afterward dying. It seems as though, by the disturbance of the heat regulation, the resistance of the animals to disease had suffered greatly; so that in this path throughout the central nervous system there was created a special susceptibility to colds. On investigating the spinal marrow of such animals microscopically, serious injuries to the nerve cells have been observed.

Dangers and Beneficial Possibilities

After these experiences, I have not dared to expose entire human beings to a condenser field and in this way produce artificial rises in temperature. The responsibility seemed too great. At the same time, effective heating of the body can be accomplished equally well in other ways; such as with the well-known Apostoli "condenser bed," which can be connected to any diathermal apparatus. With the method previously described, only serious dangers for the patients treated would have been conjured up, without the possibility of producing a fundamentally new effect.

On the other hand, the disturbances of the physical health, which we could observe in the field of free radiation of powerful transmitters, have never been serious. After a period of recovery of a few weeks, with no irradiation, all effects have been observed to vanish. For four years now, I have almost daily worked for several hours at a transmitter with 1½ kw. plate dissipation; and the effects, often very unpleasant, have always gone back to normal on stopping the work.

In these things the wavelength is also certainly of importance, and in fact we have the impression that the disturbances to health became stronger as the wave was shortened. Anyway, the unpleasant sensations appear much quicker with a three-meter wave than with longer ones.

From all these experiences, it is at any rate clear that treatment with electric waves can in no way be regarded as always harmless for the human body. It is plain that their incorrect use can cause serious injuries to health. Certainly such injuries are to be expected only when the frequencies are very high; that is, with ultra-short waves; even then there is nothing to fear except with fairly high transmitter power.

On the other hand, with proper use, the short electric waves seem to be a valuable means of treatment. According to our experiments to date, with bacteria cultures and infected animals, the germs of disease can be killed. There is the added point that certain defensive processes are stimulated in the body. I have also already repeatedly treated human beings; and, in about a hundred cases, I have been able to attain an extraordinarily quick cure of suppuration (pus formation).

Human Beings As Antennas
By Dr. Erwin Schliephake, M.D.

Ills Treated by Short Waves

● IN a dispatch to the N. Y. Times, a report from London states that cures for various ailments by the use of short radio waves were claimed by Dr. Erwin Schliephake, a German physician and scientist.

Writing in the *British Medical Journal* he described how he succeeded in treating deep-seated abscess in the human body by passing ultra-short wireless waves through the patient, who was not in immediate contact with any instrument. He found, he said, that various tissues exhibit different degrees of conductivity in the presence of these waves.

Dr. Schliephake declared he has used waves to treat pulmonary abscesses after pneumonia, in pleural empyema, pneumonic tuberculosis, in certain forms of peritonitis, in migraine and acute tonsilitis.

Dr. Willis R. Whitney, research director of the General Electric Company, revealed in April, 1930, that he had developed a radio type of apparatus for killing bacteria in the body.

Left—One end of the "artificial fever" short - wave apparatus.

The wooden box in which patient is placed is indicated by dotted lines.

A NEW tool has been made available to the medical profession for investigations concerning fevers and their use in the cure of certain diseases. At a joint meeting of the New England Physical Therapy Society and the American Physical Therapy Association in Boston, on April 18, the apparatus was shown and described by Charles M. Carpenter and Albert B. Page of the Research Laboratory of the General Electric Company.

The equipment, similar in principle to a short-wave radio transmitter, is featured by a tube which generates current oscillations at the rate of between 10,000,000 and 14,000,000 cycles per second (corresponding to those of 30- to 21-meter waves). This oscillating current is concentrated between two condenser plates, instead of being fed into an aerial, and the body to be heated is placed between the two plates.

Heat has been used throughout the history of medicine as a means of alleviating and curing diseases, and, more

How Short Waves Are Used To Produce Artificial Fever IN THE HUMAN BODY

recently, the causes and effects of fevers have been the subject of investigations and debates. Previously it was thought that fever temperatures were a sign of disease, just as pain is, and that the fever heat should be eliminated to make the patient more comfortable. Recent investigations, however, have indicated that, at least in the case of certain diseases, the fever is valuable in killing the germs of the disease; since many germs are unable to withstand the fever temperature of the human body.

The production of artificial fevers in the human body has been a difficult task, because man's temperature-regulating mechanism is so efficient. A fever results from a rise in temperature throughout the body, and local external heating is dissipated without raising the temperature of the whole body. Various methods of producing fever temperatures have been tried in the past; such as the use of hot-water baths and exposure of the body to artificially heated atmospheres. The injection of a protein re-

Hook-up of "artificial fever" apparatus.

Remarkable short-wave apparatus, developed in the Research Laboratory of the General Electric Co., induces heating effect in the body when it is brought under the influence of 21 to 30 meter waves. Frequency of waves used varies from 10,000,000 to 14,000,000 cycles per second. Brain action accelerated by short waves, research shows. Tomorrow we may use short wave to keep us warm, instead of furnaces.

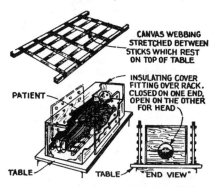

Wooden box and canvas webbing support for patient. Motor blowers circulate air in the box.

The opposite end of the "artificial fever" apparatus—the engineer is holding one of the 30 meter oscillator tubes. It is a 4-element screen-grid tube, with an output of 500 watts.

sults in a fever, and a high fever temperature for the treatment of paresis can be produced by the injection of malaria germs into the patient. The injection of a protein is hazardous because one is dealing with unknown factors and uncertain quantities, as Messrs. Carpenter and Page pointed out in their paper presented at the Boston meeting; the use of malaria or other germs often fails because of the immunity of the patient, and it is dangerous because a living virus has been introduced; while the hot-water bath and similar methods are time-consuming, difficult of application and not easily controlled. The new short radio-wave method, on the other hand, is at all times under control.

The Origin of the Idea

The development of the equipment for producing the artificial fever resulted from some experiments conducted by Dr. Willis R. Whitney, director of the Gen-

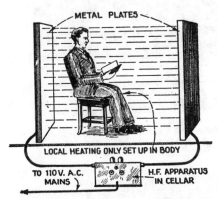

Tomorrow—instead of using furnaces we may heat "our body only" by sitting in a high frequency field.

eral Electric research laboratory, and Mr. Page, when studying the effects of high-frequency induction coils on fruit flies and mice. Shortly after these experiments, it was noticed, the blood temperature of research men working in close proximity to vacuum-tube oscilla-

"Short - wave" heating tomorrow! Room temperature is only 20 degrees above zero—but the body heat is 72 degrees.

tors delivering six or eight kilowatts of 5- to 6-meter waves were slowly raised. It was known that various ways of producing fever heat have been applied to human beings for therapeutic purposes; so it seemed worth while to study experimentally the electric fever, since it seemed to carry with it no danger and no discomfort. In addition, when the current is off, the fever quickly subsides—in other words, it is controllable.

Various forms of electrical diathermy have been extensively used for years; but they are methods of direct application of electrodes to the body, and have certain limitations which are not present in this new apparatus. With the new equipment, which is essentially more

costly, it is necessary only to place the body, or that portion to be treated, in the space between two insulated plates, and the body temperature is raised at a rate and to an amount dependent only on the controlling or generating apparatus.

"If there is merit in artificial fevers," says Dr. Whitney, "it seems worth while to study carefully the electrically-induced fever. If there are infections whose temperature tolerance is less than that of the host of the infection, it may be possible to destroy the infection. It is also customary to bake out, or heat by various means, stiff joints. As the radio method produces the heat within the tissues themselves, because of the electrical resistance of the body fluids, it seems probable that this method of applying heat should be studied in member and joint diseases."

Description of the Apparatus

The apparatus shown at the Boston meeting is enclosed within a case about 3 feet high, 3 feet wide and 6 feet long, mounted on small wheels so as to be portable. It is like a short-wave radio transmitter, except that the energy is concentrated between two condenser plates instead of being directed from an aerial. The heater consists of a vacuum-tube oscillator, with a full-wave rectifier which supplies the high voltage needed. The oscillator comprises two 500-watt vacuum tubes operating at a frequency of from 10,000,000 to 14,000,000 cycles; their output is concentrated between two plates mounted vertically on top of the cabinet. The rectifier, which changes the low-voltage A.C. house supply to direct current for use in the vacuum tubes, has an oil-immersed transformer with a 7,000-volt secondary which feeds two half-wave, hot-cathode, mercury-vapor tubes. In conjunction with a filter system, this unit furnishes the 3,000-volt direct-current supply for the oscillator. An auto-transformer is connected in the primary circuit of the high voltage transformer to provide plate-voltage regulation.

The condenser plates are of aluminum, 28 by 18 by ⅛-inch, covered by hard rubber plates 30 by 20 by ¼-inch, to prevent arcing, should the patient or attendant come in contact with the plates. In this field of undamped waves between the plates there is a rapid alternation of 3,000 volts drop of potential.

The patient is suspended on interlaced cotton tapes stretched across a wooden frame, the under surface of which is covered with composition boards, forming an air chamber beneath the body. A cover of the same material is fitted over the frame, so that the head of the patient projects through an opening at one end; thus there is formed a fairly tight air chamber around the body as it lies on the tapes. The patient rests on his back and the plates are placed at each side of the box; so that the waves oscillate through the body from one side to the other. The plates' separation can be varied, but as a rule has been kept at 30 inches. Two small hair dryers are placed in openings at the foot, one above and one below, to circu-late hot air around the body. These decrease the heat loss, and equalize the humidity, throughout the enclosed atmosphere. By applying the plates in this manner and by enclosing the body, it is heated rapidly without causing great discomfort to the patient. When the desired temperature is reached it may be maintained by decreasing the voltage, by increasing the plate distance, or by employing only the hot-air blowers.

The 10,000,000-Cycle Tube

The tube used for the production of the 30-meter waves is a four-element screen-grid tube, designated as the G.E. "Type PR-861 Pliotron." Especially adapted for use at the higher frequencies it has a nominal output rating of 500 watts.

The filament, grid and plate are supported on separate stems, with the leads brought out at separate seals; thus insuring high insulation and low electrostatic capacities between electrodes. The filament is of thoriated tungsten in the shape of a double helix, supported from a center rod, and requires no tension springs. The grid and plate are cylindrical; the plate has six wings for dissipation of heat.

The fourth electrode, the screen-grid, consists of a close mesh or winding placed between the control grid and plate, and extends the full height of the tube. It is supported by suitable means on the filament and control-grid stems. It has two leads; one of which is brought to the blade of the base on the filament arm, and the second through a separate seal to a base near the grid end of the tube.

For Experimental Uses

In connection with the exhibition of the equipment and the announcement of the work with artificial fevers, Dr. Whitney issued a statement emphasizing the fact that the equipment is being used at present for experimental purposes only:

"Our policy concerning this new method of high-frequency therapy has been to sell no apparatus but to study it ourselves and to assist research by others. We have built a number of outfits, and have lent most of them to competent research groups. The expense has been considerable, and we could hardly justify increasing the number of these loans.

"Therefore, if the medical profession, in view of the experimental results already announced, feels that such researches should be multiplied, while we are still unwilling to sell such outfits generally until their utility is more completely proved, we are now willing to sell apparatus to accredited medical institutions equipped for research work."

Medical Aspects
of
Ultra Short Waves
By C. H. West
U. S. Public Health Service

Mr. West here describes some of the effects and also some of the dangers of subjecting the human body to powerful ultra short wave fields

Photos courtesy J. G. Francis. U. S. Public Health Service. Member R. P. S.

Above—A rabbit about to be exposed to a powerful ultra short wave field; the sides as well as the top and bottom linings of the cabinet are of metal.

At left—The appearance of the ultra short-wave apparatus built and tested by Mr. West in his experiments.

● IT SEEMS that the general public is once again taking to radio and the art of building as a pastime. The amateurs are busy developing the five meter band, while the engineers are devoting a goodly portion of their time

to dabbling in the mysteries that the ultra short waves have created.

On the other hand, medical-scientific personages are making use of their medical knowledge, plus a large portion of electronics, in efforts to make the world a better place to live in with less aches and pains to bother us.

Many engineers employed in large electrical laboratories have made valuable use of their knowledge pertaining to the ultra short waves, and have constructed oscillators of high power for the purpose of producing fever in the human body at will. It is a known fact that fever is nature's doctor to a certain degree. If it can be produced in cases where there is no cause for a high fever, the patient's chance for recovery and cure is far better than combatting disease by injection of malaria germs, etc., which may produce reactions worse than the disease itself

Unknown Qualities

However, the production of fever by an oscillator is a simple procedure, but the application of these high wave speeds to the human body is an entirely different matter. Secondly, the output or radiation of an ultra high frequency oscillator has never been completely identified.

It seems that for every known element discovered there are many others within—undiscovered. An illustration of this fact was brought to light very recently, in which the writer witnessed a demonstration of photo-electric cell work.

A double-cell apparatus manufactured by the Weston Electrical Instrument Corporation produced readings in foot-candles from the output of a

10 k. w. X-ray tube in which the cells were completely covered by ⅛ inch lead sheet.

The roentgen rays do not penetrate lead, and as the tube was quite a distance from the cells, the reaction in the sensitive meter was not due to heat waves, but simply a current output which registered through the lead upon the faces of the cells. The X-ray is noted for producing very bad burns if not handled properly, and the deflection from the *target* is called and identified as the *roentgen ray* after the name of the discoverer. How does science know whether the burns are caused from this additional element as registered in *foot candles*, for want of a better name?

U. S. W. "Fever" Apparatus

Recently, the writer constructed and put into practice an *ultra short wave oscillator* for the production of fever or temperature. This apparatus consisted of two UX-852 tubes arranged in push-pull, but with the "back-to-back" arrangement of *Mesny-Valuri*. It is a persistent oscillator and is capable of running long hours on wavelengths from 2 to 10 meters.

An auxiliary cabinet was constructed with two leads to couple across the plate inductance. This cabinet has one adjustable side; the top and bottom being one lead, and the two sides the other and forming a large condenser.

Rabbits were used in experiments and their temperature reached 41 degrees (Centigrade) from the usual 38 degrees (normal). However, after a few weeks severe burns were noticed. Since that time research has been carried on by other factions in an effort to ascertain the direct cause of the burns, which do not seem to be due to collections of moisture, or from coming in contact with the condenser plates.

This brings to mind that experiments by the writer with the very short waves disclose the fact that some sort of a photo-electric ray could be produced and would highly expose small dental films. It is believed this elementary action was first discovered by Mr. John Reinartz, the noted radio amateur, who perceived a bluish glow within an auxiliary indicator at a certain frequency.

Before any actual application to humans may be applied, it is our duty to ascertain accurately whether other elements are prevailing. To all indications there are many, and one of them is severe burns, which may be caused by an unidentified element.

Constructional Data and Summary

As will be noted in the photographs, the apparatus is entirely enclosed and surrounded by glass sides to better view the "works" in case things tend to go the "hay-wire" route. The lower portion of the cabinet contains the necessary plate and filament transformers, which operate direct from the 110 volt, 60 cycle line. The plate power is raw A. C. at a potential of 2600 volts, the filament transformer being the customary 10 volt affair. Inductances are removable; the ones shown are at present used for 30,000 kc. work. The circuit is shown in Fig. 1, which is the conventional push-pull method.

However, the auxiliary cabinet is for purpose of holding an animal within the field and without strapping him and causing discomfort. The condenser plates are of sheet metal and insulated, the schematic diagram appearing in Fig. 2.

When a rabbit is placed in this pen and the power applied, he fails to react immediately; but as he "warms" up to the situation the veins in the ears show dilation, and in 20 minutes his temperature starts to rise and keeps going. It is only at a much later date that burns are noticed.

From the foregoing it would hardly seem advisable to subject a human to any lengthy treatment, at least, for the present; but the value of high frequency in various forms have proved essential in many cases. Secondly, the natural heavy perspiration that is produced is essential in one degree in opening the pores

(some of which have probably never been opened before) and allowing the natural poisons of the body to escape more quickly than could have been accomplished with potent medicines.

The majority of high frequency apparatus produces this result. In the case of pulsating currents, which produce a series of reactions or "jumps" within the muscular system, this often is very beneficial in activating those muscles which could not have been manipulated manually without a great degree of pain.

Probably the first reaction noticed with reference to the ultra high frequencies while early experimenters were conducting transmission tests with high power. It was noticed that the body temperature tended to rise at various times, due entirely to their proximity to a powerful oscillator. Science has been looking for a long time for some method to raise the temperature in the human body *quickly*. Ultra high frequency oscillators have solved that problem but have brought many other matters to light. Secondly, to produce the desired results the patient must be enclosed between two or more condenser plates of large proportions. The oscillator must be of high power, capable of delivering 500 watts output and the patient must be wrapped in a suitable covering to prevent his contact with any portion of the charged plates.

Various research workers who have volunteered their services as "trial patients" have received a goodly portion of burns as compensation, and are quite satisfied that ultra high frequency currents are "hot stuff!" Where heat is concerned, one could get equivalent results by setting on a hot steam radiator. The writer knows this to be true, and has experienced many a painful burn which he would have been willing to trade for a shock from a quarter kw. closed-core transformer.*

It could hardly seem probable that burns would prevail where the body is not in contact with any metallic object; but an examination under darkness of a large condenser tapped from the oscillator, will disclose a bluish aurora between the two elements, which is similar to that noted in Tesla coil experiments. Within this bluish discharge there can be many components not yet identified, which have a tendency to attack tender portions of the body and produce an internal burn, which does not come to the surface for identification until some days have passed.

It is open to discussion whether one-half of the electrical apparatus in present use actually cures the patient of pains, or whether it is the psychology of the matter in which the patient is surrounded by many cabinets of mysteries, of which he knows absolutely nothing. What cures one man does not seem to cure another with the identical complaint.

---

*Yes—but it is generally understood we believe, that the high frequency currents or field causes heat to be developed inside the body; even in organs such as the liver, etc., without heating up the whole body, which is not the case with direct application of heat as when seated on a hot radiator, etc.—Editor.

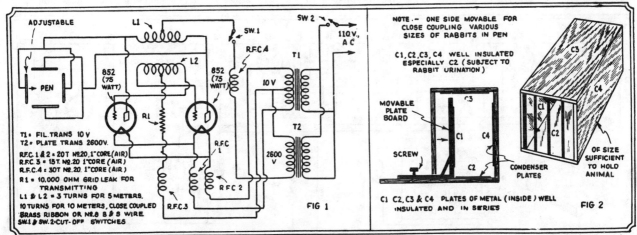

The drawings above provide details of the circuit used by Mr. West in building his ultra short-wave physiological apparatus, together with details of the apparatus itself.

SHORT WAVE CONVERTERS

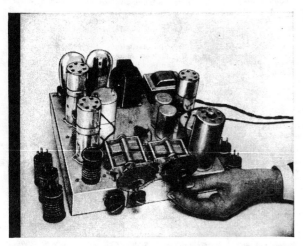

Above—General View of Mr. Cisin's DeLuxe Short-Wave Converter

Bottom View of the Find-All Short-Wave Converter; Can Be Used With Any "BC" Set

Find-All deLuxe S-W Converter

By H. G. Cisin, M. E.

● UP-TO-DATE radio receivers include short-wave reception as a matter of course. They are known as "dual wave" sets. They bring in the standard broadcast stations between 200 and 550 meters, and also foreign stations operating on short waves, radio amateurs, aircraft and police calls.

Hundreds of thousands of very excellent receivers, however, do not include the new "dual wave" feature. These can be modernized quite readily by means of the Find-All DeLuxe Short Wave Converter.

Converts B. C. Receivers to Super-het.

The Find-All Converter is used to

Here is a short-wave converter that should satisfy every "short-wave fan" who has ever thought of building one. It is a "superhet" converter provided with separate oscillator and it uses plug-in coils to change the wave-bands. This converter enables you to receive short waves on any broadcast receiver and makes an S-W superheterodyne of your present "BC" receiver.

change any broadcast receiver into an excellent short wave superheterodyne. The circuit consists of an R.F. stage

employing a 58 variable mu pentode (5), a screen grid 24 oscillator (18), a variable mu 58 detector (15) and an intermediate stage using a third 58 pentode (23). The converter has its own power supply, employing an 80 full wave rectifier, with a suitable filter system.

Two sets of Alden short-wave coils are used with this converter. The coils are of the plug-in type. There are four to a set, permitting coverage of the short wave band from 20 to 200 meters. The coils are precision wound on Makelot color-coded coil forms. Coil (3) serves as the antenna coup-

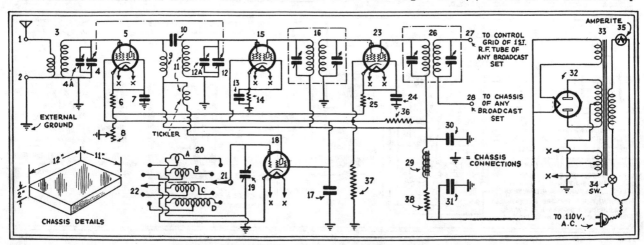

Schematic diagram of the Find-All DeLuxe Short-Wave Converter

ler. The secondary of coil (11) is used as a tuned impedance between the R.F. stage and the detector. The primary of (11) inductively couples the plate of the oscillator (18) to the grid of the detector (modulator) (15). In this way, both the signal voltage and the oscillator voltage are impressed on the grid of the detector. The resultant current, of the predetermined beat frequency, flows through the intermediate frequency transformer (16) from the plate of (15).

The secondary of the antenna coupler and the impedance (secondary of coil 11), are each tuned by a section of the dual .00015 mf. Cardwell "Midway" variable condenser. A single .00015 mf. condenser of the same type is used to tune the oscillator grid coil (20).

Separate Osc. Coil for Each Band.

A separate oscillator coil is provided for each of the four short wave bands covered by the Alden plug-in coil sets. The oscillator coils, however, are fastened in place permanently and the change-over from one coil to another is accomplished by means of a two-gang, four position selector switch of compact design. By using a six position switch and adding two additional oscillator coils, it is possible to cover the entire broadcast band with this converter, since two more Alden coils are available for use at (3) and (11) which permit operation of the converter up to 540 meters, using the same tuning condensers.

Volume is controlled in the Find-All Converter by an Electrad potentiometer (8) in the cathode return circuit of tubes (5) and (23). Hammarlund I.F. transformers are used at (16) and (26). These peak at 465 kc. Each transformer has a tuned primary and a tuned secondary. Tuning is accomplished by means of small adjustable mica condensers. The I.F. transformers are carefully shielded and the tuning condensers are mounted inside the shield, but in such a manner that they can be adjusted from the top of the shield. Since the success of a good "super" ultimately depends upon the I.F. transformers, only the best obtainable should be used. Resistor (38) in series with the audio choke (39) limits the plate voltage to the required value of 250 volts. An amperite, in series in the primary circuit of the power transformer, prevents fluctuations of the line voltage from affecting the operation of the converter.

Oscillator Coil Data.

Holes are drilled in the front chassis wall for selector switch (21, 22) and combined Electrad volume control-power switch and these are mounted. The chassis is now turned upside down and the four special oscillator coils are mounted. All four coils are wound on fibre forms, 1⅛" in dia. and 1⅞" high and all are wound with No. 28 single silk covered wire. Coil (20A) consists of 5 turns spaced ⅛" apart. Coil (20B) consists of 11 turns spaced 1/16" apart. Coil (20C) consists of 19 turns spaced 1/16" apart. Coil (20D) consists of 37 turns spaced appx. 1/32" apart. A tap is taken out at the center turn of each coil. Coils may be obtained ready wound if desired.

The R.F. choke (9) should be fastened on the underside of the chassis as shown in the illustration. Condenser (10) is fastened directly to the bottom of the chassis. For making the connections to the caps of the four tubes, use armored braidite. Wire the four oscillator coils to the selector switch first. Then wire filament circuits, grid circuits, plate circuits, cathodes, negative returns and by-pass condensers. When wiring up a compara-

tively unfamiliar socket, such as the six prong socket, a sketch showing respective socket terminals should be worked from. Such sketches are available from tube charts. When wiring the Alden coil sockets, refer directly to the coils, noting that the lower ends of primary and secondary connect to the filament (thicker) prongs. Primary and secondary coils of the I.F. transformers are identical.

Adjusting Converter.

The rectifier tube filament is wired in next, then the power supply circuit, including resistor (38), choke (29) and filter condensers (30, 31). Finally, the amperite and switch are connected in series with the primary winding of the power transformer. The tubes should be put in place and the converter should be connected to the 110-volt source. It is desirable to check voltages first. Plate voltages should be 250. If these are found to be too high, regulate resistance (38) by means of the slider. Screen grid voltages should be about 100 volts. Of course, the best way to peak the I.F. transformers is with an oscillator. If this is not available, connect ground wire to post (2). Connect

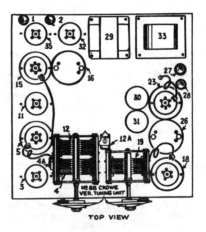

TOP VIEW

post (27) to the control grid of the 1st R.F. tube of the broadcast receiver. Put two similarly colored short wave coils in the converter. Connect post (28) to the chassis or ground of the broadcast set. Turn on converter and set, tuning the latter to a position where WEAF (or station of approximately similar wavelength) usually comes in. Turn set volume all the way up and have volume control (8) of converter similarly adjusted. Connect the antenna wire to the cap of tube (23) and adjust I.F. transformer (26) for loudest volume. Put screen grid clip back on tube (23) and connect antenna wire to cap of tube (15). Adjust I.F. transformer (16) for loudest signal.

Leave the broadcast receiver as it is and connect the antenna to post (1) of the converter. The latter is now ready to operate by tuning the variable condensers (4, 12) and (19). Equalizer condensers (4A) and (12A) should be adjusted for loudest volume. Crowe No. 88 tuning units help to bring in "hard-to-get" short wave stations.

LIST OF PARTS REQUIRED FOR FIND-ALL S-W CONVERTER

1—Cardwell Two-Gang "Midway" Variable Condenser, .00015 mf. each section, "Featherweight" type 405-B Double (4, 12)

1—Cardwell "Midway" Variable Condenser (single), .00015 mf., "Featherweight" type 405-B (19)

2—Sets Alden Short-Wave Plug-In Coils—4 coils per set covering bands from 20 to 200 meters (3, 11)

1—Set 4 Special Oscillator Coils (See Winding Directions) (20-A,B,C,D)

1—Electrad tapered Volume Control, 15,000 ohms, type RI-201-P (8) with Switch (34) Clarostat

1—Electrad 5000 ohm Truvolt Wire-Wound Resistor, type B-50 (38)

2—Hammarlund, 465 kc. Intermediate Frequency Transformers, complete with I.F. Coils, tuning condensers and shields, type TR-465 (16, 26)

1—Aerovox .00015 mfd., Mica Coupling Condenser, type 1460 (10) Polymet

4—Aerovox .01 mfd., 200 volt Cartridge By-Pass Condensers, type 281 (7, 13, 17, 24) Polymet

2—Aerovox 4 mf., 500 volt Dry Electrolytic Condensers, type G-5-4 (in TD cans) (30, 31) Polymet

2—Electrad 500-ohm Flexible Resistors, type 2G-500 (6, 25) Polymet

1—I. R. C. (Durham) 15,000 ohms, 1 watt Metallized Resistor, type F-1 (14) Lynch

1—I. R. C. (Durham) 20,000 ohms, 1 watt Metallized Resistor, type F-1 (36) Lynch

1—I. R. C. (Durham) 50,000 ohms, 1 watt Metallized Resistor, type F-1 (37) Lynch

1—Amperite Regulating Line Voltage Control, type 5A-5 (35)

1—Trutest Flush-Mounting Power Transformer, type 4C-1490 (33)

2—Hammarlund Equalizing Condensers, 2 to 35 mmf. (4A, 12A)

1—Hammarlund Isolantite R. F. Choke, type CH-8 (9)

2—Hammarlund Four-prong Isolantite Sockets, type S-4 (3, 11)

3—Hammarlund Triple-Grid Tube Shields, type TS-50 (5, 15, 23)

1—Hammarlund Screen Grid Tube Shield, type TS-35 (18)

1—Yaxley Two-Gang, Four position Selector Switch, type F-6514 (21, 22) Best

4—Eby "Ace" Binding Posts (1, 2, 27, 28) Cinch

1—Trutest R-196 Audio Choke, type 4A242 (29)

3—Na-ald Six-prong Wafer-type Sockets (5, 15, 23)

1—Na-ald Five-prong Wafer-type Socket (18)

2—Na-ald Four-prong Wafer-type Sockets (32, 35)

2—Crowe Short Wave Single Speed Tuning Units, No. 88—Ratio 14 to 1 in 180 degrees

1—Roll Corwico Braidite Hook-up Wire. stranded (Cornish Wire Co.)

1—Aluminum Chassis, 14 to 16 gauge, 12"x 11"x2" high—Blan, the Radio Man

3—Variable Mu R. F. Pentodes, type 58 (5, 15, 23)

1—Screen Grid 24-type Oscillator (18)

1—Full-Wave 80-type Rectifier (32)

Note: Numbers in parentheses refer to corresponding numbers marking parts on diagrams.

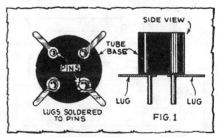

Improved "tube base" coil forms can easily be made as shown.

An A. C. Operated Short-Wave CONVERTER

By R. B. KINGSBURY

To hear S.W. stations on your broadcast receiver, you need a "tip-top", well-designed, S.W. converter. Here it is—a "prize winner"—110 volt, A. C. operated —has its own plate supply—also R. F. amplifier and coupling tube.

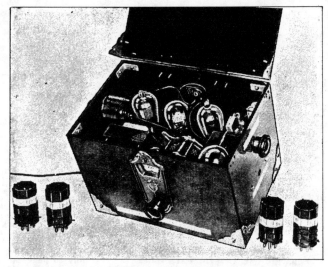

Five tubes, including the rectifier, provide this S.W. converter with a real "wallop." It enables you to hear S.W. stations on your "broadcast receiver."

● THE average "fan," after dabbling in the mystic realm of short waves for a considerable time, cannot tolerate sets and converters which produce only mediocre results. The writer, having graduated into this class, set about to build a converter which could be termed really efficient and yet be moderate in cost.

Single dial tuning, compactness and a tuned intermediate stage, permitting the transfer of the signal to the broadcast set with very little loss in transmission, are but a few of the features incorporated in the converter illustrated. The word simplicity has therefore been superseded by the more desirable one—*efficiency!*

Despite its seemingly complicated design, the average set constructor should experience no trouble at all in assembling the converter here shown.

Exclusive of the cabinet, the total cost should not exceed $10 or $12 and this moderate figure can be brought down even lower by substituting parts which are usually found in your "odd-parts" box.

Tested on Many Receivers

This set has been tried out with a dozen different broadcast receivers and one has yet to be found with which it will not "percolate" (meaning *work*).

Five tubes, including the rectifier, are used in this layout and as it supplies its own power, the problem of low filament voltage and incorrect "B" supply is entirely eliminated.

A switch is used to "cut out" the R. F. stage below 30 meters, as it has been found that this stage is only a *losser*, when tuning in stations covered by the 20 meter coil. The energy gathered by the antenna is transferred to the grid of the modulator tube. No direct connection is made, however, as the exact amount must be determined by experiment. To determine this take two pieces of enamelled wire about No. 18 or 20 gauge Solder one to the switch in the antenna circuit and the other to the grid circuit of the modulator. As this stage tunes very broadly, this condenser can also be used as an auxiliary *vol-*

ume control. This condenser may be eliminated, however, if the builder desires to use two dials for tuning.

The method of modulation is that the oscillation is introduced into the modulator tube by way of the screen. Grid bias detection is used.

The plate circuit of the modulator contains a circuit tuned to the intermediate frequency. This insures a high impedance load on the modulator at the intermediate frequency.

Intermediate Frequency

It is best to use an intermediate frequency at which the broadcast receiver is the most sensitive. If it is equally sensitive throughout the entire tuning range, then set the frequency either below 550 kc. or above 1500 kc. Therefore, no specific instructions are given as to the number of turns on T. This is left up to the constructor to determine, as he will know at which frequency the broadcast set he will use in conjunction with the converter is the most sensitive. The ratio of this transformer is 1 to 1 and the primary is tuned by a midget or balancing condenser.

This converter has a volume control of its own, as the signal intensity range will be very large and two controls are desirable. The control is the potentiometer P, which is connected between the screen return and ground, with the cathode of the modulator connected to the slider. When the slider is moved to the left, that is, toward the ground, the bias on the tube is lowered and at the same time the screen voltage is increased. The best place is a short distance to the right of the ground end, at which point the maximum sensitivity will be found.

Power Transformer Details

The filament and plate voltage are supplied by a power transformer T3 and a '80 rectifier. This transformer has one 2.5 volt winding which supplies the four tubes in the converter and one 5 volt winding for the rectifier filament. The high voltage winding gives approximately 325 volts after the choke

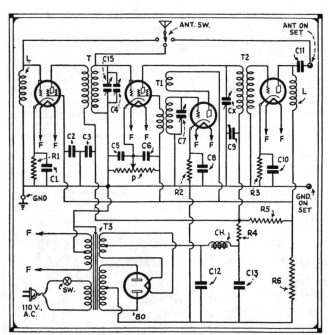

Wiring diagram for constructing Mr. Kingsbury's S.W. converter.

and filter. This voltage is cut down to 165 volts by R4, which has a value of 14,500 ohms and is connected in the plate circuits of the two screen-grid tubes. R5 is another section of the voltage divider and its value is 3,400 ohms. The output tube's plate, oscillator plate and screen voltages are taken from this point. R6 is a bleeder resistance of 17,900 ohms.

The matter of terminal connections on the modulator and oscillator coils is optional; however, the following instructions are given but they need not be followed.

Modulator coil T: G on the coil socket to P on the R. F. tube socket; F minus on the coil form to "B" plus 165; P on coil socket to G on modulator; F plus on the coil socket to ground. The coil terminals on the form should be made to the prongs in the corresponding manner.

The oscillator coil T1 should be wired as follows: G on coil socket to G on the oscillator socket; P on the coil socket to P on the oscillator socket; K on the coil socket to ground; HP on coil socket to "B" plus for screen voltage; HK on coil socket to grid on modulator tube socket. The terminals of the oscillator coil should be connected in the corresponding manner, two terminals being connected to HP.

To insure oscillation G and P should be far apart. The pick-up winding should be on the side of the tuned winding opposite to that of the tickler. Using Pilot coil forms, the number of turns is as follows:

20 Meter Band

Modulator coil:
   Primary 1 turn
   Secondary 4 turns

Oscillator coil:
   Primary or pick-up 1 turn
   Secondary 5 turns
   Tickler 3 turns

40 Meter Band

Modulator coil:
   Primary 4 turns
   Secondary 8 turns

Oscillator coil:
   Pick-up 3 turns
   Secondary 9 turns
   Tickler 7 turns

Coils for other bands may be wound at will. The size of the wire is not so important; anything between No. 20 and No. 26 will be satisfactory.

List of Parts for Converter

T—Set of modulator coils as described
T1—Set of oscillator coils as described
T2—1 to 1 ratio R. F. coil for midget condenser
T3—Power transformer
Ch—30 henry choke
L—2-85 millihenry chokes
C1, C2, C3, C5, C6, C8, C9, C10—Eight .1 mf. by-pass condensers.
C4, C7—Two .00015 mf. tandem tuning condensers
C15—midget condenser across C4, approximately .00005 mf.
C11—.00025 mf. fixed condenser
C12, C13—Two 4 mf. electrolytic filter condensers
CX—Small tuning condenser, such as a trimmer.
R1—300 ohm bias resistor
R2, R3—Two 1000 ohm bias resistors
R4, R5—Voltage divider tapped at 3400 ohms. Total value 17,900 ohms
R6—One 17,900 ohm bleeder resistor
P—25,000 ohm potentiometer with AC switch
6 UY sockets
1 UX socket
1 Dial (vernier type)
2 '24A tubes
2 '27 tubes
1 '80 tube
4 binding posts
1 25 foot roll "pushback" hook-up wire

A Handy Short Wave Converter
WITHOUT PLUG-IN COILS

By C. H. W. NASON

THE idea of a continuous tuning unit to cover the entire short-wave spectrum is by no means new. Since the high frequencies first came to the attention of the radio enthusiast any number of developments have been made which purported to accomplish this end—just as there have

There is no reason for making any changes in the broadcast set, or making any connections thereto other than to connect the output of the converter to the antenna post. If this is left con-

TWO SETS ON ONE AERIAL

A BROADCAST receiver and a short-wave receiver can be operated simultaneously on the same aerial, without interference. The lead-in wire is connected directly to the aerial post on the broadcast set; but is bridged to the short-wave

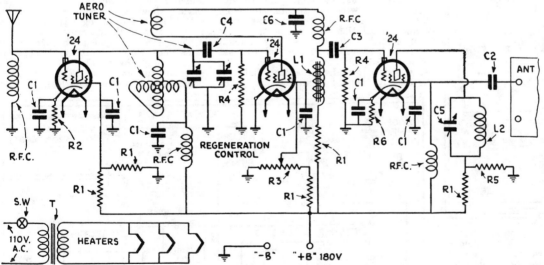

1—Aero tuner
RFC—Hammarlund 85 M. H. R. F. choke
T—2½ volt heater transformer
L1 — Amertran 200 hy. choke
L2—100 turns No. 36 d. s. c. wound on 1¼" tube
R1—50,000 ohms
R2—500 ohms
R3—10,000 ohm potentiometer
R4—5 megohms
R5—25,000 ohms
R6—1500 ohms
C1—.25 mf.
C2—.002 mf.
C3—.05 mf.
C4—.0001 mf.
C5—.0001 mf. trimmer
C6—.0005 mf.

been many devices for converting the broadcast receiver into a short-wave job. The device which the writer considers to fill the gap is a combination of two recent developments in the radio field. The Aero tuner which covers the short-wave spectrum up to 100 meters, and has a single additional coil to fill the gap between that point and the broadcast band, is employed in combination with the dynatron oscillator.

nected it will have but slight effect on the operation of the receiver at frequencies in the broadcast band. Three '24 tubes are required in the converter. The practice of drawing filament supply from an A.C. receiver for use in a short-wave converter is to be deplored, because of the danger of overloading the power transformer and either causing a burn-out or lowering the efficiency of the receiver.

outfit through a small condenser, of any capacity between .0001- and .00001-mf. In the absence of a condenser, the wire from the short-wave receiver may simply be wrapped around the lead-in, for a distance of about ten inches.

If the aerial is connected directly to both sets, the volume of the broadcast receiver suffers noticeably, although the short-wave signals do not seem to be affected very much. This is probably due to the relative impedances.

The Blanchard S-W Converter

This converter enables you to hear the "police" and other short wave phone signals on your broadcast receiver. It is at once simple and cheap to build.

Above—Front view of the Blanchard short wave converter, the "out-put" of which is connected to the "aerial" and "ground" post of the broadcast receiver. "Hi" and "Lo" switch at the right gives two-band control.

Above—Bottom view of the Blanchard short wave converter. Center photo shows rear view of converter.

THE converter here illustrated was built by Mr. Thomas A. Blanchard, of Reading, Pa., and it has a number of novel features that will appeal to the short wave "fan". By means

tested by one of the editors in conjunction with a screen-grid T.R.F. broadcast receiver and it worked very smoothly; in this test, the aluminum chassis was connected by a wire to the "ground post" of the broadcast receiver. The regular aerial lead-in wire was disconnected from its post on the "BC" receiver and was connected to the antenna post on the "converter." The tuning was smooth and the broadcast receiver dial was set to about 230 meters, but the best point at which to set the broadcast dial will be quickly found by simultaneously ad-

justing the converter and the broadcast dials until a station is heard.

The specially wound coil consists of a bakelite tube, 1¾ inches in diameter, on which are wound the coils L1 and L11 comprising 10 turns of No. 28 enameled wire (tapped at the 6th and 8th turns as shown in the diagram), and 8 turns of No. 28 wire in the coil L111, which is wound on the same tube and placed about ¼ inch from the end of coil 1, instead of being adjacent to L11 as indicated in the diagram.

The plate supply can be taken direct from the 110 volt circuit, passing the current through the rectifier tube. As to the detector, the usual .00025 mf. grid condenser and 2 to 4 megohm leak are used.

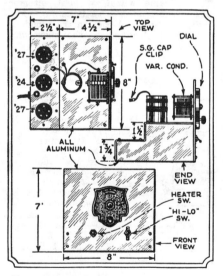

Top, side and front profiles of the short wave converter, with dimensions of the aluminum frame. The aerial and output posts are shown between the tube sockets and the ground connection is made from the chassis to the "ground" post on the broadcast receiver. Wiring diagram of the converter is shown at the right. The coil L111 should be placed above or beside L1, and ¼ inch from it, all on the same tube.

of a switch as shown in the diagram, Mr. Blanchard has arranged the circuit so that two different short wave "bands" may be tuned in without the use of "plug-in" coils. This converter was

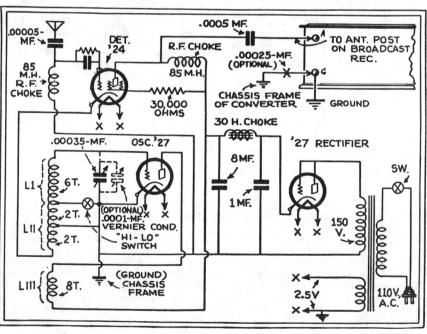

A Tried and Proven
S-W Adapter

By ROD PERRY

Radio Operator, S.S. "S. M. Spalding"

● PRACTICALLY all marine radio operators, like myself, carry their own "short wave" receivers; these vary from simple one-tube affairs to multi-tube supers. I have tried many types and find the adapter described very satisfactory and meeting the requirements of the sea-going operator.

Several factors had to be taken under consideration; the adapter must be compact so as to take up a minimum amount of space and be easily packed in a traveling bag. It must be sturdily built to withstand the conditions aboard ship; it must be enclosed to protect the set from the damp air and salt spray, and it must cover a wave band at least from 15 to 50 meters.

A 5" x 9" x 6" aluminum box was chosen for the cabinet; this is of standard size and easily obtained. Plenty of room is available to allow sufficient spacing between the parts; if the cabinet is too small there will be a tendency for feed-back to take place between the various circuits.

Four controls are on the face of the set; a 30 ohm rheostat, a .00016 mf. tuning condenser, a .00025 mf. plate condenser and a midget condenser in series with the antenna. The variable condensers are of great importance and should be of some reliable make and of the S. L. F. type.

The tube socket projects through a hole cut in the top of the cabinet and a socket to receive the plug-in coils projects through the back in a similar man-

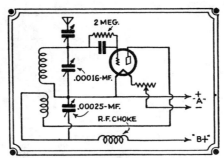

Circuit of the successful short wave adapter here described by Mr. Perry.

ner. The tube socket is fastened to the front panel with two brass angles, leaving the cover free to allow easy access to the set for inspection. The coil socket is bolted to the rear of the cabinet.

The circuit is of the conventional regenerative type found in most one tube adapters. Some trouble was experienced in controlling regeneration; the set had a tendency to go into oscillation with a "plop", and made it very difficult to tune in broadcasting stations. This was overcome by trying different grid-leaks and by varying the number of turns on the plate coil. It was found that a two megohm grid-leak worked best. If the adapter still persists in going into oscillation in an abrupt manner, place an external variable resistor of 50,000 ohms in series with the 45 volt detector lead of the long wave receiver. Even if the

adapter seems to work satisfactory without the variable resistor, it is advisable to use one to compensate for any drop in battery voltage.

Any "plug-in" coils can be used that will match the plate and tuning condensers, but a "one-tap coil" will cover the required band of 15 to 50 meters. For the grid coil wind 8 turns of number 18 enameled wire on a 2½ inch form and take off a tap at 3½ turns. The plate coil has 4 turns of number 22 cotton covered wire, spaced about ⅛ of an inch from the grid coil.

A plug fits into the detector socket of the long wave receiver to obtain the filament and plate supply and to couple the adapter to the receiver's audio amplifier. Flexible leads are attached to the plate, filament positive and filament negative prongs of a tube base that is used for the plug. A great many battery receivers have the old style UV sockets, and the tube base comes flush with the top of the socket, making it difficult to turn the base in the socket. It is advisable to make a cap for the plug in the following manner: Remove the brass fittings from an ordinary hard rubber electrical appliance plug, file the sides until it fits snugly into the tube base and glue it in place; this makes a neat finished looking job.

This adapter can be used with any battery receiver by plugging into the detector socket. No detailed list of parts is given as the constructor will probably wish to use parts already on hand,

---

A Short Wave Adapter Table

THE very attractive piece of furniture in the photo at the left is a cleverly designed short wave converter, which can be used in conjunction with any modern midget type broadcast receiver, and thus permit the reception of short waves. By combining a midget broadcast receiver with the Audiola short wave table illustrated you will increase your wavelength reception range from 20 to 550 meters. As the diagram shows, a detector and oscil-

lator are used, which in conjunction with the broadcast receiver forms a superheterodyne. A switch is provided for changing the antenna from the converter to the broadcast receiver. Coil data for similar converters have been given previously.

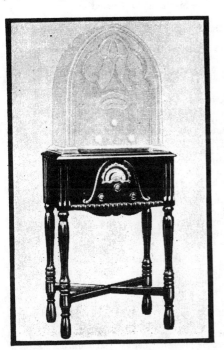

Left—The Audiola short wave converter which converts any midget broadcast receiver for short wave reception.

Right—Wiring diagram of the Audiola converter.

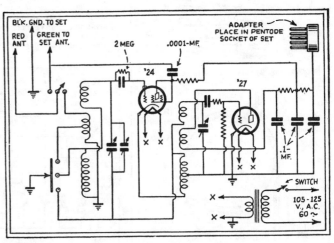

AN ULTRA
Short-Wave Converter

Details for building an ultra short-wave converter to be used with any broadcast receiver, for tuning in waves in the 5 to 10 meter region. It uses but one tube which acts as an oscillator and detector or "frequency changer," forming a superheterodyne when connected to your B. C. set.

The ultra short-wave converter which uses but one tube; it permits you to tune in waves of from 5 to 10 meters length with your broadcast receiver.

● IN some recent experiments conducted by the B.B.C. (British Broadcasting Corp.) in which programs were broadcast on a wavelength of about seven meters, a number of sets were designed to pick up the signals. Preliminary tests made it obvious that receiver technique on seven meters is very different from that of the normal broadcast bands. The length of the aerial, for example, may be no more than the battery leads and much less than the usual ground wire. It is generally better, therefore, to replace the usual ground connection by a counterpoise and couple the aerial to the grid magnetically, so that there is no direct connection. Again, as the tuning inductance for this wavelength consists of about 10 inches of wire, great care must be taken with the layout of parts to avoid long leads.

This receiver is an adapter, consisting simply of an oscillating tube which acts as a frequency changer. By this means, any existing broadcast set can be converted into an ultra short-wave superheterodyne receiver. The circuit of the adapter is shown in the diagram. The tuning circuit is very simple. An aerial coil L1 is coupled to the grid coil L2, which is tuned by a variable condenser C1, having .0001 mf. capacity. This condenser is made up of two parts, a main section and a vernier section. The main portion is adjusted by a small knob in the rear of the set and the vernier only is operated by the dial on the front panel. This makes the tuning sufficiently stable for ordinary purposes, and still allows the set to cover a reasonably wide wave band.

Regeneration is provided by coil L3 and the plate bypass condenser C2. A plate resistor of 25,000 ohms is used with an output coupling condenser of .001 mf. Four ultra-short-wave chokes are inserted in the battery supply leads.

Tuning Coils

The coils are wound with No. 8 B & S gauge bare copper wire. The aerial, grid and feedback windings each consist of about 1¾ turns. They are made by winding the wire as tightly as possible around a wooden form 1¼ inches in diameter, spacing the turns 1/16 inch. When the wire is released

from the form, the turns will spring and the diameter of the coil increase to about 1⅜ in. which is the required size. The direction of the winding is important and the sketches of the coils should be referred to in order to ensure the correct result.

The R.F. chokes are of the single-layer type, made by winding 24 turns of number 18 B & S gauge insulated copper wire, with adjacent turns touching, on a form one inch in diameter and 1¾ inches long. They may be supported by a machine screw inserted from beneath the baseboard into a threaded hole in the form.

(While the tube recommended for this set cannot, of course, be obtained in the U. S., it can be substituted by any triode with similar characteristics. The original type has an impedance of 4,000 ohms.)

Methods of Operation

To tune in the ultra-short-wave transmission, the first step is to adjust the broadcast receiver to maximum sensitivity with the tuning dial set to a point where *no interference* is produced from the broadcast band. The vernier condenser should then be set to the middle of its scale and the main section of the tuning condenser should be set approximately to tune in the desired transmission. Final adjustments can then be made quite satisfactorily with the vernier condenser. It will be found that there are two tuning points at which signals come in with equal clarity; this is the usual result with this type of adapter. There may also be found two other points at which signals are received, but at which they are much weaker and of poorer quality than the real tuning points.—*World-Radio, London, England.*

A Simple Indicating Scale

Many small control devices on a set are not important or critical enough to deserve a regular vernier indicating dial, but do require a scale of some kind on the front panel. Engraving on aluminum or bakelite is expensive. A simple and useful scale can be made of white bristol board (smooth, heavy cardboard), attached to the panel with rubber cement.

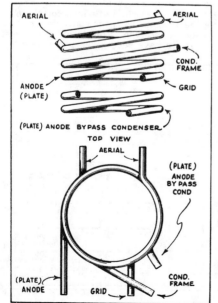

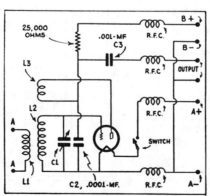

Diagram at left shows details of the ultra short-wave converter coils, while wiring diagram showing how to build this extremely simple converter appears below.

A Simple Short-Wave Adapter

By CHAS. SCOTT, JR.

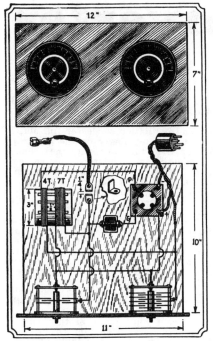

Front and plan views of the easily built short-wave adapter here described by Mr. Scott.

● This efficient short-wave adapter is exceedingly easy to construct, with very little or practically no expense to the builder.

The rig is used in conjunction with any type of tube receiver operated on battery current. Simply remove the tube from the receiver's detector socket, and replace it with the adapter plug-in base. Put the tube in the socket on the adapter. Then, put the battery clip on the antenna lead and you are all set for short-wave reception.

Most amateur radio enthusiasts will, no doubt, have all of the necessary parts for this easily constructed adapter. One can find enough parts in an old dismantled battery receiver to eliminate the necessity of making any expenditures for new equipment.

The list of parts required is as follows:

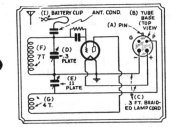

Wiring diagram for the short-wave adapter, the plug of which fits into the detector tube socket of the "B.C." receiver.

1 Panel, 7x12
1 Baseboard, 10x11
1 Var. cond., 3 plate
1 Var. cond., 11 plate
1 Vernier dial
1 Ordinary dial
1 Tuning coil
1 Grid cond., .00025 mf.
1 Grid leak, 7 megohm
1 Base from burned out tube
8 Ft. single lamp cord
1 Ant. cond.
1 Tube socket

The circuit diagram and the drawings show all the necessary details. The tuning coil consists of seven turns and the tickler coil of four turns of No. 18 bell wire wound on a bakelite or cardboard tube, 3 inches in diameter and 2 inches long. The turns are raised from the surface of the tube by match sticks placed at ¾ inch intervals.

The tuning condenser is made from one of the "One-Buck" low loss condensers cut down to three plates. Take the condenser apart and reassemble, using two stator plates and one rotor plate. The tuning is well spread out over a range of 33 to 45 meters. One meter will occupy a half-inch sector on

a 4-inch dial, making it easy to find stations. The 11-plate condenser controls regeneration and has little or no effect on the tuning so that the receiver can be calibrated directly in meters.

The "A" battery supply is taken from the receiver through the plug-in tube base which also carried the output of the short-wave set into the receiver's amplifier. The rheostat on the receiver controls the short-wave detector voltage.

The regular antenna is used. Size doesn't matter much on these waves and the antenna can be left connected to the receiver while the short waver is in use.

The little antenna condenser consists of two brass or copper angles mounted as shown. A ground can be added to the positive filament lead as shown in dotted lines, but in most cases it will make very little difference in tuning or signal strength and can be left off. There is already a high capacity ground through the filament batteries and wiring in most receiver installations and the addition of a straight ground connection will merely shift the tuning a degree or so on the dial.

The coil data given above is suitable for tuning in waves in the region between 33 to 45 meters. To the readers of this magazine, it will, of course, at once be apparent that any wave band desired may be tuned in by means of this simple adapter, by making use of the well-known "plug-in" coils. Data for winding these coils for the various bands and suitable for a certain specified capacity tuning condenser have been published in practically every issue. It is advisable to employ a small capacity midget condenser, of 25 to 50 mmf. capacity, in series with the antenna in place of the fixed condenser. This antenna variable midget will be found a great help when the set fails to oscillate or when the dead spots occur.

Making a Switch to Change Bands

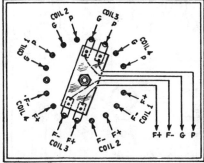

How the band coil selector switch is wired.

● What to do with plug-in coils when building a cabinet set is a question. The photograph and diagram show a satisfactory method of group mounting the coils.

The coils are not connected together, but are selected as originally intended. The four wires (F+, F—, G, P) from

switch rotor are connected in the circuit in place of the plug-in socket shown in the circuit diagram you are using.

The radio "junk box" should furnish the necessary material. The dimensions of the three pieces of bakelite I used are, coil base 3¼"x4½", switch panel 3¼"x3½", switch rotor 1"x2", all cut from an old set panel. Switch contacts, rotor bushing, rod and knob from a vario-coupler. The switch blades are from a tube socket. Four wafer sockets for coils which in my case are wound on tube bases. Two pieces of ½" angle brass each 3¼" long for holding base and panel together and to prevent coil prongs from striking set base. Six small bolts and nuts, a switch stop and some flexible rubber covered wire obtained from old lamp cord will complete the parts list.

By compact mounting, well soldered connections and wires crossing at right-angles for minimum inductance when possible, the length of wire in the circuit hook-up is increased very little over a single socket.

Photo, above, shows how the author built his band change switch, with the four coils grouped behind the switch.

A similar switching idea is easily worked out for five and six prong coils.
—S. M. Cook, Jr.

A Novel S-W Converter

By R. M. LEGATE

Here is a novel short-wave converter circuit which Mr. Legate ran across in some of his experiments with a superhet broadcast receiver. He has heard stations "all over the globe" at loud-speaker volume with his "stunt" circuit, when not a peep could be heard on a commercial 11-tube short-wave set which he used for comparison.

Being an unemployed electrical engineer and knowing nothing whatever about radio, I am sending you a few details on my receiver. I have tried all kinds of circuits and combinations and obtained excellent results with some of them, but the one that has absolutely astounded me is the one I give herewith.

There is nothing new in either the broadcast or the short wave circuits; the one is a circuit I took from SHORT WAVE CRAFT'S *Question Box* of a few months ago and the other is a model 801 Westinghouse superhet *broadcast* receiver, but experimenting in my ignorance I hit on the attached combination. The sensitivity and volume that this receiver gives is really wonderful and is the envy of all the local "hams."

The circuit explains itself; the change-over switch shown in the assembly sketch controls two "ganged" toggle switches marked 1-2-3 and 4 on the diagram. A D.P.D.T. switch transfers the converter output to the grid cap of the 1st detector tube in the superhet, and at the same time transfers the antenna from the B.C. (broadcast) to the short-wave converter, the other switch turns on the filament of the converter tubes and at the same time cuts off the filament of the B.C. oscillator.

Please don't ask me how or why—I don't know. I am simply passing on the results of my "discoveries." I have compared this receiver side by side to two well-known manufacturers 11-tube sets and in some cases, when I haven't been able to pick up the wave on the manufactured set, *it has come in loud and clear on my set!*

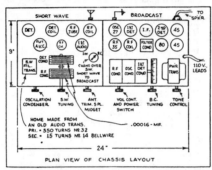

Top view of chassis layout used by Mr. Legate for his combination "short" and "broadcast" wave receiver. The short-wave receiver employs an R.F. stage, a regenerative detector, and also an audio coupling stage.

Today for example, GBS-GSA-and W8XK came in with such terrific volume that with my volume-control *full off* and oscillation condenser full out, I had to put my finger on the grid cap of the I. F. tube to reduce the volume sufficiently to understand the speech, and this isn't a tall story! This condition is no doubt due to the fact that with the way I am using the B. C. circuit I have only the one tube (I.F.) controlled.

I have tried feeding the converter output into the B.C. set antenna, but this requires that the B.C. dial be set as near 550 k.c. as possible while in my arrangement the B.C. dial does not affect the short wave tuning and the noise ratio is away down.

I wish you would publish a suitable and simple A.V.C. circuit for use with this hook-up and I would be glad to hear from you or any of the "hams" who might try this stunt circuit and let me know what they think of it. I am enclosing a partial log.

Stations Logged

Only stations received with good loud speaker volume are logged here and all these stations have been logged during the past 6 months.

	Phone	
PLE	15.93 m.	W8XK
W3XAL		W2XE
GSG		W3XAL
W2XAD		VE9GW
FYA	19.68	W9XF
W8XK	19.72 m.	W8XAL
DJB		W3XAU
GSF		GSA
XDA		W4XF
		All the GB stations

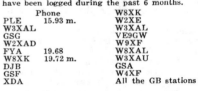

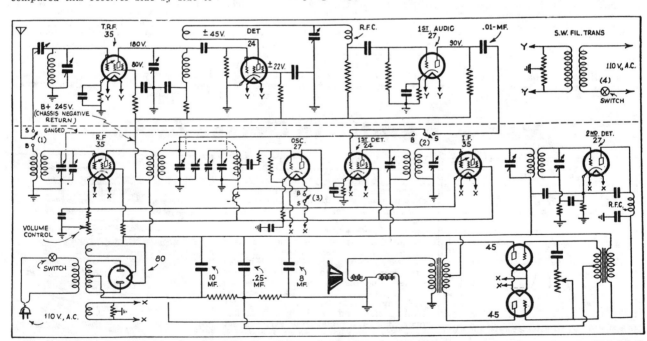

Complete wiring diagram of Mr. Legate's ingenious circuit on which he received short-wave stations from "all over the globe" with tremendous volume. The three tubes shown at the top of the circuit comprise the short-wave converter as constructed by the author, while the remaining tubes and associated apparatus indicated below the dotted line comprise the regular set-up of the standard commercial "broadcast" receiver he used. By means of the two single-pole double-throw switches indicated at 1 and 2, the operator is enabled to quickly switch from "broadcast" to "short-wave" reception, the output of included special converter being fed into the grid of the first detector of the broadcast set. By using the set-up of tubes as here shown, a very smooth and gradual amplification of the incoming signal is realized and when it finally passes through the two 45 A.F. amplifier output tubes into the dynamic speaker, the signal has some wallop!

SHORT-WAVE AERIALS

Some Things You Don't Know About S-W Aerials

By DON C. WALLACE

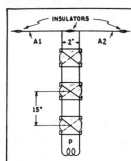

Here is shown the typical "compromise" short wave antenna system suitable for covering the entire band of from 20-200 meters. The flat-top portion, A1 and A2 must be cut to exact length. Each of the wires, A1 and A2, is 33'-6" long, No. 14 enameled copper. The 3 insulators are of glass. Note the "feed line" coming down the center. The feed line is also of No. 14 enameled copper wire. It is transposed about every 15 inches with a TRANSPOSITION BLOCK, as shown in the illustration to the right. The lead-in, or "feeders" as they are called in short wave practice, are to be 66 feet long (each wire). These feeders are spaced 2" apart and held in place by the transposition blocks. In the antenna illustration "P" is a coupling coil which couples the antenna to the receiver. The ground wire is removed from the receiver.

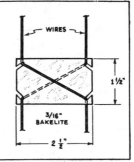

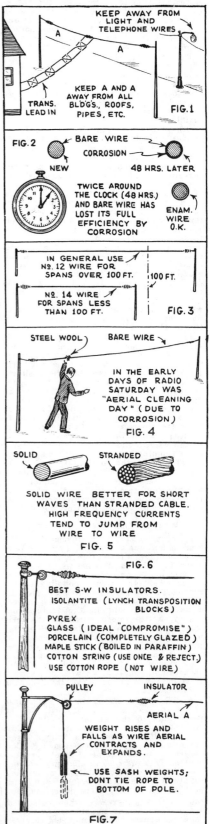

FIG. 1

FIG. 2 — BARE WIRE CORROSION — NEW — 48 HRS. LATER — TWICE AROUND THE CLOCK (48 HRS.) AND BARE WIRE HAS LOST ITS FULL EFFICIENCY BY CORROSION — ENAM. WIRE O.K.

FIG. 3 — IN GENERAL USE No. 12 WIRE FOR SPANS OVER 100 FT. — No. 14 WIRE FOR SPANS LESS THAN 100 FT. — 100 FT.

FIG. 4 — STEEL WOOL — BARE WIRE — IN THE EARLY DAYS OF RADIO SATURDAY WAS "AERIAL CLEANING DAY" (DUE TO CORROSION)

FIG. 5 — SOLID — STRANDED — SOLID WIRE BETTER FOR SHORT WAVES THAN STRANDED CABLE. HIGH FREQUENCY CURRENTS TEND TO JUMP FROM WIRE TO WIRE

FIG. 6 — BEST S-W INSULATORS. ISOLANTITE (LYNCH TRANSPOSITION BLOCKS) PYREX GLASS (IDEAL "COMPROMISE") PORCELAIN (COMPLETELY GLAZED) MAPLE STICK (BOILED IN PARAFFIN) COTTON STRING (USE ONCE & REJECT) USE COTTON ROPE (NOT WIRE)

FIG. 7 — PULLEY — INSULATOR — AERIAL A — WEIGHT RISES AND FALLS AS WIRE AERIAL CONTRACTS AND EXPANDS. — USE SASH WEIGHTS; DONT TIE ROPE TO BOTTOM OF POLE.

● FEW people realize what a pronounced improvement in reception is had from the use of a properly designed short-wave antenna system. It must be correctly laid out, correctly built and correctly installed in the proper place.

The best location for an antenna is on or over vacant property. A "back lot" antenna is superior to one that is stretched across the housetops. The unusually large network of house-wiring, all of which is directly, inductively or capacitatively coupled with all of the electrical devices in the city, picks up noises which are inherent in the wiring system but which are not picked up a few yards distant.

Too many treatises on antenna systems deal with the subject in a vague, general manner. Actual dimensions are left to guesswork. This article gives **exact dimensions,** their importance being such that the success of short-wave reception depends upon them to a greater extent than the average experimenter is aware of. A surprisingly large number of new stations

Several surprising facts concerning short-wave aerials are given in the accompanying article by Don C. Wallace, one of the best known short-wave experimenters in the country. If you want the best S-W aerial, it should be constructed with very heavy copper wire, such as No. 6, 8, or 10, with No. 12 for the feeder system. Furthermore, bare copper wire starts to corrode on the surface within forty-eight hours after erection and its efficiency is therefore impaired at the very start. Enameled wire or the new Chromoxide is ideal for the purpose. Reception noises can be mostly eliminated by using a transposition feeder system from the antenna to the receiver. Also, solid wire is preferable to stranded cable for S-W aerials.

will be heard if the proper short-wave antenna system is used.

The dimensions and placement of the antenna are more important than the kind of wire used. The ideal antenna wire is that of the largest size, consistent with the ability to erect and permanently suspend it in the proper place. Conditions too often do not permit the use of large wire, neither will the pocket book afford it. A compromise must be made. Radio, in all its branches, is a compromise . . . between convenience, cost, time, ease of construction and operation, availability of material, knowledge of the subject, inherent inhibitions against things "new" or those that differ from the traditional. This article deals with the successful and practical compromise of antenna systems that are within the reach of all.

The Size of Wire to Use

In order named are the practical sizes of antenna wire which are best suited for short-wave reception:

1. No. 6, No. 8 or No. 10 solid copper enameled wire for the flat top portion and No. 12 enameled wire for the feeder system.
2. No. 12 solid copper enameled wire for both the flat top and feeder system.
3. No. 14 solid copper enameled wire for both the flat top and feeder system.

Wire smaller in size than No. 14 is not strong, mechanically. It will not permit of "full stretching" when pulled taut. As a last resort No. 16 enameled wire could be used with perfectly satisfactory results. In general it is suggested that No. 12 wire be used for spans of more than 100 feet and No. 14 for spans of less than 100 feet. Enameled wire is the more practical to use. Radio frequency currents have a tendency to travel on the surface of the wire. Bright new copper wire would be best if it could be made to retain its shiny finish. The R.F. (Radio Frequency) currents travel with minimum loss on a bright surface, the antenna system radiates with greatest ease, and maximum efficiency is the result. However, corrosion on the surface of the wire will increase the resistance to these minute R.F. currents and within 48 hours the corrosion will be so far advanced as to lower the efficiency of the antenna.

In the early days of radio, Saturday was antenna cleaning day. Several of the more enthusiastic would lower the antenna and polish the wires with steel wool. Steel wool was not pleasant to handle and a pair of old leather gloves was donned to prevent the fine particles of steel wool from entering the flesh of the hands.

Theoretically, the corrosion of copper wire, if sufficiently corroded, is just as good an insulator as an enameled coating. But too often the corrosion is unevenly distributed and, therefore, of uncertain effectiveness. Consequently, **enameled covered wire is ideal for a short wave aerial.**

Other coverings may be used, such as rubber, weatherproofing, paraffin cloth, cotton or silk, or any other covering of a good insulating quality.

The span of copper is all-important, the covering of the wire of secondary importance. The covering for portable aerial of station W6ZZA is a double layer of silk cloth woven over a large number of strands of carefully cut-to-size loop wire. Both the flat top portion and the feeder system use this kind of wire. One of the feeder wires is green silk covered loop wire, the other feeder is brown, making it easy to prevent the feeders from becoming entangled when the portable aerial is erected on a hotel roof after dark. This flexible loop wire is not as good as enameled wire but it permits of speedy installation and enables the operator to wind the antenna around the lid of a cigar box when it comes time to check out of the hotel.

It is repeated that solid copper wire is specified for short-wave aerials. Stranded wire offers more surface, lower resistance to the R.F. currents on the broadcast band. **But it is not as good as solid wire for short-wave reception.** This is because the higher frequencies (short waves) alternate so many times per second that certain losses are introduced when uneven-surfaced wire is used. The high frequencies tend to jump from wire to wire (stranded wire is twisted) rather than to follow the twists of the wire. Solid copper wire eliminates this "jumping" tendency, thus making an easier path for the flow of currents. Therefore, solid copper wire is recommended.

These details may seem commonplace and "finicky" to some. But it must be remembered that improvements and corrections in radio design multiply rapidly.

A 2408% Increase in Efficiency

If we make a 2% improvement in the kind of antenna wire used, a 2% improvement in antenna insulation, a 2% improvement in antenna dimensions, a 2% improvement in antenna placement, a 2% improvement in antenna coupling to the receiver, a 2% reduction in noise pick-up, a 2% improvement in receiver coil design, a 2% improvement in the tuning condenser, a 2% improvement in the grid leak, a 2% improvement in the shielding, a 2% improvement in the placement of the receiver in its housing, a 2% improvement in the radio frequency choke coil, a 2% improvement in the tube and coil sockets and contacts, we will then have a total improvement of 2x2x2x2x2x2x2x2x2x2x 2x2x2x2=2048%.

A 2% improvement in six of these places, or 2x2x2x2x2=32%, will not be perceptible to the human ear. Individually, these 2% improvements will result in no audible increase in volume, individually they are of no consequence. Collectively, the sum total of 2048% is what counts. This increase in efficiency will enable you to hear more stations, from more countries, with more volume and with greater ease. It is evident, therefore, that these little 2% increases, when multiplied, are of far-reaching importance in the total effectiveness of the completed receiver.

Additional increases in efficiency are gained from the proper insulation of the antenna.

An antenna designed to deliver utmost performance at a certain frequency (wavelength) operates at peak efficiency only if tuned to its exact wavelength. At other wavelengths it does not deliver the same efficiency. Improper or poor insulation not only tends to distort the actual dimensions of the antenna but the antenna actually does not know where it terminates. Poor insulation is partly conductive. Thus the antenna has no definite terminating point. Like other things in radio, there is a difference of opinion as to the merit of various well-known insulating materials and the proper placement of the insulation. In practice we cannot resort to the last word in insulation because it is awkward, expensive and troublesome and the improvement which it offers over and above the accepted and commonplace method of insulation is not of sufficient importance to detract from the effectiveness of the properly designed all-around short-wave antenna system.

Insulating Materials

The best insulating materials for antenna are silk, linen, cotton, or woven strands of these materials. They should be free from coloring because the base of all coloring is of a conductive nature. When silk, linen or cotton become wet the impurities in the material, plus the natural impurities in the air, introduce conductiveness and a consequent lowering of the insulating qualities of the material. The quality of insulation can be preserved by boiling the material in vaseline. In time the sun will melt the vaseline and the useful life of our "perfect insulator" is from six months to one year.

Obviously, this perfect insulator is not practical and once more we resort to the time-worn radio compromise by using glass for antenna insulation.

Those who can afford to pay a little more for better insulators are advised to use PYREX. Good porcelain, finely grained, well baked and completely glazed, is the next best thing to use. Glass is the nearly perfect insulator and is an ideal compromise for short-wave antenna. Glass insulators can be procured from your parts supply house.

As a possible alternative a maple dowel stick can be used. It should be from 3/8" to 1/2" in diameter, one foot in length, boiled for an hour or two in paraffin. Like the vaseline-boiled linen insulator, these dowel sticks are at the mercy of the weather, dust and soot particles will accumulate on the dowel surface and the effectiveness of the insulator is then considerably reduced.

Portable W6ZZA uses cotton string for insulation. A ball of string is thrown over an elevator shaft or penthouse, hoisted to the top of a flag pole or attached to some other convenient support. Because the cotton string is used but once it is not affected by rain or moisture and a negligible amount of soot and dirt will accumulate on its surface. Cotton string makes a perfect short-wave antenna insulator, most convenient in its application, will retain its insulating qualities for an entire week. Given a quick jerk it will break easily and down comes the aerial. The aerial is then rolled over the lid of a cigar box and thrown into a suitcase when checking out of the hotel. But this cotton string insulation is intended for portable use only.

Glass, being our perfect compromise for a permanent antenna installation, can be had in the form of insulators 3" in length. The standard Pyrex Glass insulator is of that length. Longer glass insulators can be used.

Rope should be used for hoisting the antenna. Cotton rope is a better insulator than hemp. Do not use wire. It picks up noises from nearby wiring. The rope hoist is attached to the insulators on each end of the antenna. Real enthusiasts can boil this rope in vaseline, thereby weatherproofing the rope and preventing it from contracting and expanding with changes in humidity.

The hoist rope usually runs through a pulley, attached to a pole on the house or in a vacant lot. Do not fasten the rope to the base of the pole. Tie a window sash weight to the end of the rope, thus permitting the rope to contract several feet during a heavy rain or fog. The weights "go up the pole" as the rope contracts. The pole will not bend, the rope will not break, and there is less wear and tear than when the rope is attached to the base of the pole. Window sash weights can be obtained from any hardware dealer. They are good looking. The weights used at the Wallace station for holding a 612' antenna taut, are the 34-pound size. By using these weights the top of the pole is never subjected to a strain of more than 34 pounds.

Placement of the Antenna

An antenna to be most effective must be in the clear. The placement of the antenna is of utmost importance. It should be as high as possible, not too close to the houses or other large objects, as far removed from lighting circuits and telephone lines as possible. Too often such an ideal condition cannot be found for the erection of the average antenna.

Transposition Blocks

Transposition blocks for the antenna feeders can be made from various insulating material. Bakelite is cheapest, can be purchased in suitable block form, as shown in the illustration. Porcelain blocks are better than those made from Bakelite.

The feed lines are transposed by means of these blocks. Cancellation takes place throughout the length of the feed lines where insulation is not quite as important as in the antenna proper.

The ideal transposition blocks for short waves would be those of glazed porcelain.

To ascertain the correct dimensions of an antenna the use of a half wave is resorted to; the figure 1.56x the wavelength. Because of the size of the antenna wire used, capacity to earth and various other corrections, it is not possible to use the straight meter system and transpose it into feet and expect to find the wavelength of the antenna proper. The figure 1.56 is accepted as an average, being the result of a large number of tests made from antennas which have been carefully tuned by means of oscillators. Inasmuch as the amateur short-wave bands are in harmonic relation with each other, the antenna sizes can be selected with regard to their convenience. The two most widely used short-wave broadcast bands are not in harmonic relation to each other. It therefore becomes necessary to adopt the 26 or 49 meter band as a standard. However, we also want to hear all of the other stations.

The transposition blocks should be spaced from 15 inches to 36 inches apart. A space of 2 feet between blocks seems to be the accepted compromise. The exact size of the transposition blocks is not important. Any size, from 1" square to 8" square will suffice. The larger blocks must be spaced far apart, the smaller blocks close together. Large blocks offer added resistance to wind pressure. Small blocks are more suitable for general requirements.

Determine the proper size of the antenna by measuring the wires with a tape or yardstick. Stretch the enameled antenna and feeder wires. It is not necessary to cut the antenna wires where they meet the feeder wires. Reeve the antenna wires through the glass insulator in the center of the antenna and continue these for use as feeders. Fasten the aerial wires to the insulators with short pieces of wire, made into the form of a loop and soldered, thereby insuring a "definite ending," as explained previously in this article. (*Courtesy* "RADIO.")

Eliminating Man-Made Interference by
DOUBLET ANTENNAS

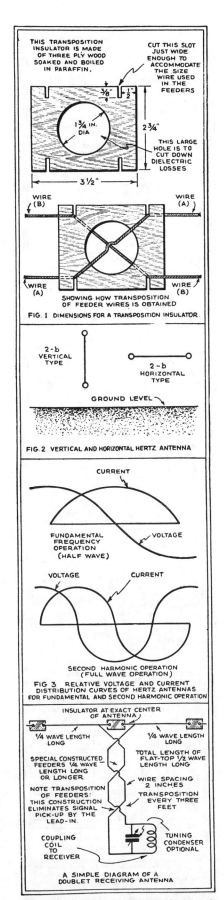

THIS TRANSPOSITION INSULATOR IS MADE OF THREE PLY WOOD SOAKED AND BOILED IN PARAFFIN.

CUT THIS SLOT JUST WIDE ENOUGH TO ACCOMMODATE THE SIZE WIRE USED IN THE FEEDERS

3/8"   1 1/2"

1 3/4 IN. DIA

2 3/4"

THIS LARGE HOLE IS TO CUT DOWN DIELECTRIC LOSSES

3 1/2"

WIRE (B)   WIRE (A)

WIRE (A)   WIRE (B)

SHOWING HOW TRANSPOSITION OF FEEDER WIRES IS OBTAINED

FIG. 1 DIMENSIONS FOR A TRANSPOSITION INSULATOR.

2-b VERTICAL TYPE

2-b HORIZONTAL TYPE

GROUND LEVEL

FIG. 2 VERTICAL AND HORIZONTAL HERTZ ANTENNA

CURRENT

VOLTAGE

FUNDAMENTAL FREQUENCY OPERATION (HALF WAVE)

VOLTAGE   CURRENT

SECOND HARMONIC OPERATION (FULL WAVE OPERATION)

FIG 3 RELATIVE VOLTAGE AND CURRENT DISTRIBUTION CURVES OF HERTZ ANTENNAS FOR FUNDAMENTAL AND SECOND HARMONIC OPERATION

INSULATOR AT EXACT CENTER OF ANTENNA

1/4 WAVE LENGTH LONG   1/4 WAVE LENGTH LONG

SPECIAL CONSTRUCTED FEEDERS 1/4 WAVE LENGTH LONG OR LONGER

TOTAL LENGTH OF FLAT-TOP 1/2 WAVE LENGTH LONG

NOTE TRANSPOSITION OF FEEDERS: THIS CONSTRUCTION ELIMINATES SIGNAL PICK-UP BY THE LEAD-IN.

WIRE SPACING 2 INCHES

TRANSPOSITION EVERY THREE FEET

COUPLING COIL TO RECEIVER

TUNING CONDENSER OPTIONAL

A SIMPLE DIAGRAM OF A DOUBLET RECEIVING ANTENNA

Illustrations above show the design of transposition insulators, together with diagrams showing action and also construction of complete doublet antennas.

THERE is much general knowledge on the design and construction of good high frequency receivers. It is now possible to buy several makes of excellent short wave sets at reasonable prices, all A.C. operated and complete and ready to work—except for the necessity of attaching an antenna. For those experimenters who care to build their own sets many good circuits of practical receivers are fully described in every issue of every good radio magazine. It is easy to find complete constructional details on any type of short wave receiver one might care to construct—whether it ranges from the portable "junk box" variety to the most advanced superheterodyne.

Yet, with all of this knowledge of receiver design, together with the abundance of good receivers obtainable, the average short-wave enthusiast considers that a good receiving antenna for high frequencies is just "any old kind" of a single wire thrown up in a haphazard manner. It seems strange that there is so much general knowledge of good receivers and so little general knowledge of how to design high frequency receiving antennas. They go hand in hand, yet on the average it is safe to say that the majority of short wave listeners are using antenna systems of the same general design that have been used for the past several years. Why are we content to put up a single wire affair for use with the best receiver that we can buy —and then blame the receiver and the manufacturer for all of the static, background noises, and all other sources of interference coming from the loudspeaker?

When Good Receivers 'Flunk"!

The best receiver in the world will give only mediocre results when attached to a poor antenna system, which has been erected with no particular thought in mind of adapting the antenna design to surrounding conditions or of minimizing background noise-level.

Have you ever tried listening to your high frequency receiver without the aerial attached? If you have, you will, no doubt, have noticed that with the disappearance of the signals you have also lost all of the crackling and popping sounds that are so commonly attributed to static and man-made interference. Old Man Static (true atmospheric interference) is blamed a great deal more for the numerous odd noises that we hear and call static than he actually causes. True static is not objectionable on frequencies above 5,000 kilocycles at any season of the year—except when reception is attempted during especially

stormy weather. Most of the crackling and frying sounds we hear are not true atmospherics generated by Nature, but are, instead, originated by many man-made sources of interference. Street car lines, transmission wires, electric refrigerators, washing machines, irons, flashing signs and at least several thousand other electrical devices are grinding away merrily every hour of the day and night, creating undesirable radio interference without the least regard for the patience of the short wave listener, the rules of the Federal Radio Commission, or a consideration of anybody else. Generally, the interference area of such interfering sources is limited to the immediate neighborhood in which they are located. Yet, for any given locality a few of these devices, creating their limited interference areas and picked up by the antenna and its lead-in attached to a short wave set, are the cause of most of the crackling and sizzling sounds we attribute so often to true static.

Country vs. City Reception

If you have listened to short waves in the country or any other location free from man-made electrical sources of interference, you will be amazed at the remarkably clear reception that is possible even during the summertime. Signals that you cannot even begin to hear in your city flat or apartment, will boom in on the loud speaker. Yet the same receiver in the city location will be unable to receive these signals because of the higher background noise level. If we will face the facts we shall find that we are blaming Mother Nature too much for our troubles and doing too little work ourselves to eliminate these undesirable noises. Until we take the necessary steps to minimize this objectionable interference, we are not getting the full benefits of really good short wave reception.

Enter—the "Doublet Antenna"

Radio engineers, for the past several years, have been using a type of antenna called the "doublet antenna" that is suited for high frequency reception. It is an antenna that reduces background noise levels to a minimum and yet does not weaken or diminish the signal from the station that is being received. Its design is simple and the doublet is extremely easy to construct. The doublet antenna can be built within the confines of the average city yard or apartment roof. It seems hard to believe that an antenna so admirably suited to short wave reception has to this day not been more universally adopted. The purpose

Transposition Lead-in Systems

Mr. Everett L. Dillard, a well-known radio engineer, here explains the theory and construction of an ideal doublet antenna system for short-wave as well as broadcast reception, together with data on "transposition feeder".

Short wave listeners are just beginning to realize that a lot of the noise and interference caused by electrical devices operating in the vicinity of their antenna lead-ins, may be eliminated by the use of carefully designed doublet antennas with transposition lead-ins.

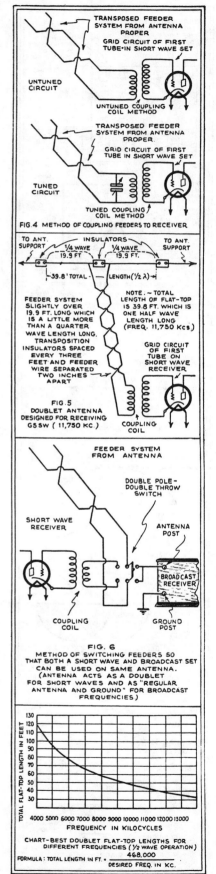

FIG. 4 METHOD OF COUPLING FEEDERS TO RECEIVER

FIG. 5 DOUBLET ANTENNA DESIGNED FOR RECEIVING G5SW (11,750 KC.)

FIG. 6 METHOD OF SWITCHING FEEDERS SO THAT BOTH A SHORT WAVE AND BROADCAST SET CAN BE USED ON SAME ANTENNA. (ANTENNA ACTS AS A DOUBLET FOR SHORT WAVES AND AS "REGULAR ANTENNA AND GROUND" FOR BROADCAST FREQUENCIES)

CHART—BEST DOUBLET FLAT-TOP LENGTHS FOR DIFFERENT FREQUENCIES (½ WAVE OPERATION)

$$\text{FORMULA : TOTAL LENGTH IN FT.} = \frac{468,000}{\text{DESIRED FREQ. IN KC.}}$$

Fig. 4, above, shows how feeders are coupled to receiving set; doublet design for 11,750 kc. signal and connection of doublet for S.W., B.C. reception.

of this article is to promote a more general use and understanding of the doublet for receiving purposes among short wave fans. Due to its superiority in high frequency work it is entitled to become the standard antenna design for short wave reception.

The doublet antenna is the simplest form of high frequency antenna to build that will improve reception over that of the ordinary single wire aerial. The secret of its success in high frequency work is the fact that a great deal of our background noise is eliminated with no sacrifice in the strength of the original signal. With our background level reduced it is possible to hear stations that it was previously impossible to hear. The doublet is truly the antenna to be used in the city or industrial districts, being simple to erect and doing its work effectively.

Mr. "Lead-in," the Villain

Most of us know that for best results we must place our antenna as high in the air as practical and as free from all obstructions as possible. This is to give us the best pick-up of radio signals. Yet, it is not the part of the antenna high in the air that receives most of our man-made static—instead, it is the *lead-in* portion running close to the house or garage, building or wiring that adds this type of interference to our signals, which is picked up by the antenna high in the air. Motors and most of the other sources of man-made interference do not radiate their interfering waves over any considerable distance, but the lead-in, running close to these sources of trouble, cannot help but pick up these interfering sounds because of its proximity to them, even though they are radiated over limited areas. It is safe to say, that, if the antenna itself were placed a considerable distance away from these sources of interference, our reception would remain unspoiled by them; i.e., if the interference picked up by the lead-in could by some means be eliminated from reception.

This is exactly what the doublet antenna does. Using the doublet it is possible to place the antenna high in the air where it, itself, will only pick up the desired signals. Then by means of a special type of lead-in construction there is absolutely no signal pick-up by the lead-in part of the antenna system —the lead-in merely furnishing a path for the radio signal received in the antenna proper to reach the receiving set. It plays no part other than this and does not act as a part of the antenna

proper in the sense of picking up any signals.

New Antenna Eliminates Interference

Thus, by placing our antenna away from man-made interference, even though our specially constructed lead-in must pass close to these sources of interference to reach the receiving set, we will pick up only the signal energizing the antenna. What an ideal aerial for crowded apartment hotels and factory districts where it is possible to get an antenna high enough for good reception, yet where the signal pick-up of the lead-in, by itself, of man-made static is too great to allow comfortable reception! With the doublet it is possible to use a short wave receiver in the next room to the family refrigerator, electric fan or curling iron without excessive interference from them. The only possible means whereby this interference might get into the receiver, other than from the antenna lead-in, would be through the power supply lines, and generally this can be effectively eliminated by adequate filtering.

Transposed Feeders the Secret

The doublet antenna, which we will explain in this article, is known as the *"current-fed"* doublet antenna. In reality it is nothing more than a half-wave, current-fed Hertz antenna, using *transposed feeders*. The transposed feeders are the secret of our successful elimination of those undesired signals that would ordinarily be picked up by the usual type of lead-in. Transposed feeders are feeders that are transposed at equal distances. Thus, each succeeding reversal along the entire feeder length of the feeder position of the wires cancels out any voltage induced in the preceding feeder section. Our fields in each section are 180 degrees out of phase with each other, a condition which results in cancellation of any induced signal voltages in any one feeder section. By a feeder section we mean the length of the feeder between any two consecutive transposing insulators. To assure fields exactly 180 degrees out of phase, all feeder sections must be of equal length. This is important.

Feeders transposed every three feet with the wires separated two inches will work nicely where the feeders are more or less in the open and not too close to the sources of interference. When it becomes necessary to run the feeder within a few feet of potential sources of interference pick-up, it becomes necessary to transpose more often and to decrease the distance between the two wires constituting the feeder. This affords more effective feeder cancellation of such interfering signals. Transposition every twelve inches with a wire spacing of one inch is correct under these conditions.

Transposition Insulators Needed

The construction of the transposing insulators is quite simple. Insulators made of ordinary three-ply wood boiled in paraffin are excellent and will withstand many months of our-door use without too much deterioration or breakdown in dielectric resistance. When boiling and soaking these wooden insulators in hot liquid paraffin, care must be taken to thoroughly soak the paraffin into all air pores of the wood. In this connection it will be noted that when first placing the wood into the melted paraffin, there will arise from the surface of the wood hundreds of small air bubbles. Only after these air bubbles have been completely boiled out are the insulators suitable for out-door use.

The drawing shown in Fig. 1 is fully explanatory and shows the construction of a transposition insulator, which will keep the wires separated two inches. The large hole in the center is to eliminate as much of the dielectric loss as possible. Where conditions demand closer spacing of feeder wires and shorter distances between transposition insulators, it is only necessary to reduce the dimensions of the insulator shown in Fig. 1 to the proper proportions.

We have explained in what manner the doublet antenna for receiving is different from the so-called Hertz antenna. In reality it is nothing more than the Hertz, except for the specially constructed lead-in. Before continuing it will be necessary to take up the facts pertaining to the Hertz antenna proper, which will be of assistance to us in designing an efficient doublet.

In the Hertz antenna the ground and its capacity to ground have little to do with its performance. The true Hertz antenna is suspended sufficiently high above the ground so that its capacity to ground is extremely small. The straight wire constituting the Hertz antenna is an open oscillatory circuit, and the inductance, capacity and resistance always necessary in an oscillatory circuit are distributed along the open wire. Having fixed values of inductance and capacity, our Hertz naturally will have a resonant frequency.

There are two types of Hertz antennas. They are both shown in Fig. 2; 2-a is the vertical type and 2-b is the horizontal type.

Operating at its fundamental frequency the true Hertz is a half wave antenna. Under these conditions the voltage and current curves are as shown in Fig. 3-a. Its lowest resistance, of course, is at the fundamental frequency. The Hertz will also respond to frequencies harmonically related to the fundamental frequency. The case of operation at its second harmonic is graphically shown in Fig. 3-b. The point of maximum current is no longer in the center as when operated on the fundamental frequency, but instead there are two current anti-nodes, each a quarter wavelength away from the two ends of the antenna. The exact center is now a point of no current, but instead, is a point of maximum voltage.

The doublet we are discussing is a current-fed doublet on its fundamental frequency and immediate frequencies adjacent to it, and will be considered as such in this discussion, except as brought out later.

Fundamentals

Here is a thought that, while entirely obvious, is not clear to many short wave listeners. Whether the same Hertz antenna is energized by a transmitter or by the extremely minute voltages of the radio signal in space, its relative voltage and current distribution is the same for a given exciting frequency. If we attach our feeders to the center of our antenna

---

* Insulators of this type are now available on the market.

with an insulator separating each side at the exact center and each half-section a quarter wavelength long, we are feeding at a point of maximum current and our feeder system is then known as a current-fed one. To be exact, it is current-fed at the fundamental frequency, but as our received signals approach the second harmonic frequency, our antenna assumes the voltage and current distribution as shown in Fig. 3-b, and our feeders are then at a point of highest voltage with each half-section of the antenna acting as a half-wave fundamental antenna to this second harmonic frequency signal. In this condition the feeders are voltage fed. The antenna has not been changed, yet, with the higher frequency exciting voltage our voltage and current distribution curves are different.

The doublet that we are designing is built to operate primarily as a current-fed system and accordingly we will confine most of our discussion to this mode of operation.

Feeders in receiving antennas do not have to be cut to the exactness required in transmitting circuits. Thus, we can say that for best results the length of each wire in the feeder should be at least a quarter-wave long. This insures the best distribution of voltage and current on the antenna for maximum efficiency at received signals on the fundamental frequency of the antenna. While feeders at least one-quarter wavelength long are specified, their length after this minimum has been obtained is not of serious consequence. Ideal conditions exist only at the fundamental frequency, but good reception will occur over all of the high frequency band. We do not want to become too exacting by saying that for a particular frequency such and such feeder lengths are an absolute necessity. Even at other frequencies where our voltage and current distributions are far from ideal, the doublet will give better results and eliminate more background noises than any single wire antenna ever devised.

The secret of the success does not lie in antenna and feeder lengths cut exactly to within fractions of an inch, but in the cancellation effect of the transposed feeders.

Fig. 4 shows two methods of coupling the feeder system to the receiver. The first method is an untuned coil arrangement, which is entirely satisfactory. The second method, though, if an extra tuning operation is permissable, will give superior results. It consists merely of a tuned circuit. This added control is not of a great deal of bother and, after once being set for a given band of frequencies, it does not need to be retuned, except for extreme changes in frequency. Where feeder lengths are much longer than the minimum of a quarter wavelength specified above, its use will allow tuning the feeder to such a correct frequency that the ideal voltage and current distribution for best operation can be more nearly met. This, of course, means better reception—and every time the signal level is increased with the background remaining at a definite level, reception is just that much better.

We have been more or less delving into general theory and before closing this article a few practical working figures will be given.

The proper length of a pure Hertz antenna operating at its fundamental frequency (half-wave) is easily found by the following formula:

$$\text{The Length in Feet} = \frac{468,000}{\text{Frequency in Kilocycles}}$$

This gives the *total length* of the Hertz portion. We have stated that our feeder should be somewhat more than one-quarter wave long, i.e., for the ideal condition of current-feed. Since the above formula gives us the proper length of a half wave antenna, the feeder must be at least half the length of that given in the formula in order to be at least a quarter wavelength long.

Design of a 11-750 kc. Doublet

Let us illustrate by designing a doublet for, say, best operation at 11,750 kilocycles, which is the frequency of G5SW. We assume that

this is the station we would like to hear most. Using the formula just given, we are able to compute the length of a Hertz antenna resonant to 11,750 kilocycles. The solution is this:

$$\text{The Length in Feet} = \frac{468,000}{11,750} = 39.82 \text{ feet.}$$

Then 39.82 feet is the *total length* necessary for the Hertz part of our antenna system. Since formulas are at the best only very good approximations, we will not cut our antenna to the hundredth of an inch as figured above, but will, instead, cut it to 39.8 feet, which is fully satisfactory for our purposes. We must break our antenna with an insulator at its exact center, making each section one quarter wavelength long. The figure of 39.8 feet is the total length of the Hertz part, which is, of course, one-half wavelength long at the frequency of 11,750 kilocycles. Since our feeders must be at least one quarter wave long, our feeders must then be equal in length to one-half of 39.8 feet, which is 19.9 feet. Thus, each wire in our feeder must be cut at least 19.9 feet long. They may be longer but the value of 19.9 feet should be the minimum length. The antenna shown in Fig. 5 is then best suited for picking up G5SW on his 11,750 kilocycle frequency.

The antenna of Fig. 5 would also give good reception on all stations whose frequencies are between 7,000 and 15,000 kilocycles. Reception equal to that on the average antenna could easily be had over the entire short wave spectrum now generally used, even though the voltage current distribution curves would be extremely complex and far from ideal.

Another Design Problem

Let us consider another location and assume that again the station we want to hear most is G5SW. The conditions this time are considerably different. The short wave set is located in an apartment building on the third floor from the roof of the building. We have poles twenty feet high to which the antenna can be attached; on the floor below the roof are the elevator motors; on the floor below this several electric ice machines and electric fans. We find that the only path that our lead-in can take to reach our window is within a few feet of these sources of potential interference. Here is a condition where the antenna can be placed high enough for excellent reception, yet there is going to be an almost certain amount of man-made interference picked up by the lead-in, due to its closeness to the electric motors. This is where the doublet antenna really proves its worth. Let us design our antenna to the surroundings.

Since we want an antenna again resonant to 11,750 kilocycles, the length of the flat top remains the same as above, i.e., 39.8 feet. We again place an insulator in the exact center, making each half of the Hertz 19.9 feet or one quarter wave long. So far our case is similar to the previous instance. However, we find that the shortest length of feeder we can use is around eighty-three feet. This is much longer than a quarter wave long, in fact several times longer. We will not let this worry us, however, and cut our feeders the necessary length to stretch down to the set, three stories below. Being considerably over a quarter of a wavelength long, we can couple to our receiver with the tuned coil coupling circuit. By tuning this coil we can then tune the feeder for best operation.

We know there will be little man-made interference picked up by that portion of the lead-in stretching from the antenna to the roof of the building, so for this length we will transpose our feeders only every three feet, and leave the feeder wires separated about two inches. We know, though, that the elevator motors will create interference, especially on starting. To successfully eliminate this we must transpose our feeders more often and keep the wires closer together on that part of the feeder running from the roof past the motors and ice-boxes, right down through the window and up to the set.

SUPER-REGENERATIVE
Experimental Receivers

An Improved Super-Regenerator

By J. A. GRATER

● AFTER reading Mr. B. F. Locke's article and circuit on a super-regenerative receiver, published in the July 1932 SHORT WAVE CRAFT, I decided to send in a circuit that I have experimented on for the past six months, but on second thought I have revised and combined Mr. Locke's circuit with the one I developed and find it by far superior to either circuit.

We will begin by adding an untuned R.F., (radio frequency) inductively coupled, as a booster circuit. This is really essential as it sends a strong signal to be tuned before reaching the detector. Both R.F.'s being inductively coupled means two high-gain R.F.'s that really work. Care should be taken to see that the screen-grid voltages on both R.F. tubes are constant and proper. Excessive voltage on the screen-grids will cause the R.F. tubes to oscillate.

A 35 tube is used as a detector as it proves out to be superior as a regenerative detector, over the more commonly used 24.

The 35's are also used in the R.F. as they are a standard high-gain tube. This tube seems best for all-around A.C. short-wave reception, although everyone has his own pet idea.

A 27 tube is used as an oscillator and proves very satisfactory. A switch in the plate circuit to cut out the oscillator while tuning is preferable over filament control, as the tube is ready to work immediately.

Another fine point is the 50,000 ohm variable resistor in the screen-grid circuit of the detector tube; this acts as a vernier on regeneration.

Motor-boating and fringe-howl are caused mostly by battery coupling. The 25,000 ohm resistor and the 2 mf. by-pass condenser eliminate this completely. This double resistance, by-pass, type of coupling in the detector plate seems to work out as the most efficient.

By using the volume control in the grid circuit of the first audio, you have your R.F.'s and detector working at a maximum, and only work your audio system on an average, below the noise level. This also gives good quality when using phones with the volume turned down.

Changing the 1st audio to a 35 space-charge, resistance coupled to a 47, you get rid of most of the 47's disadvantages and retain the wonderful amplifying quality of this tube. To get a positive bias on the grid of the 1st audio, a small flashlight battery works very good; if you wish you can take it off the voltage divider.

For the volume hound who likes this "ten-room apartment" stuff, he might try 47's in push-pull. Some may prefer a 45 to a 47 in the output.

Now we come to the most important factor in A.C. operation, on short waves the *power supply*. The power transformer should be of the electro-static shielded type. A little money spent on this item is well repaid by the results obtained. Tunable hums are eliminated by the by-passing in the power pack. The filament supply has a 20 ohm C. T. (center top) resistor with the C. T. grounded. One side of the filament is grounded by a .01 mf. condenser. This throws the oscillations set up by the filament circuit outside positive lead and the .1 mf. buffer condensers on the A.C. line help to keep out line noises. The R.F. choke in the positive lead and the .1 mf. buffer condensers in the high-voltage winding keep out the R.F. current sent out by the 80 tube. It must be kept in mind that this R.F. choke must be heavy enough to pass the required milli-ampere drain of the tubes and also the bleeder draw. An 85 mill (millihenry) choke will do, but it leaves only about 10 mills for the bleeder draw, which is rather small. A 90 mill choke is much better. The power pack cable should be shielded and the shield grounded to the set and also to the power pack chassis.

Both power pack and set are completely shielded.

An insulated condenser coupling should be used to eliminate dial scratching coming through the phones, due to a magnetic field set up by the condenser and metal dial.

The suggested arrangement of parts and the shielding arrangement are self-explanatory.

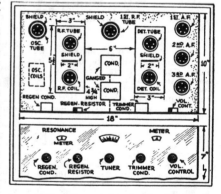

Suggestions for panel and subpanel layout of the "Improved Super-Regenerator" here described by Mr. Grater.

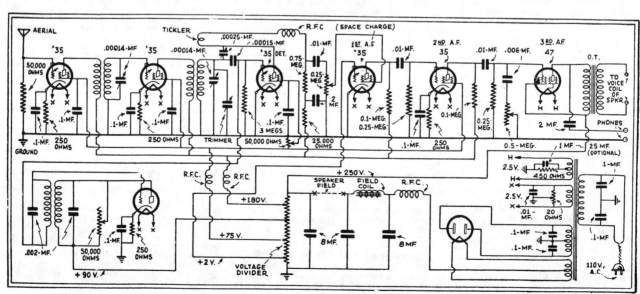

Schematic diagram showing how to wire Mr. Grater's Improved Super-Regenerative Short-Wave Receiver

The "EASY-BUILD" S-W Super-Regenerator

By C.E. DENTON

Above—looking down on the versatile short-wave receiver here described, which can be converted from regenerative to super-regenerative at the "flip of a switch."

● With the tendency of operators of S.W. transmitters to operate on high frequencies, it becomes necessary for the S.W. fan to have a receiver that will give better results at these frequencies.

While super-regeneration as a principle has been understood for a number of years and has been covered in many texts with all of its variations, it is now receiving the acknowledgment of the S.W. fraternity as the ideal set for use at all wavelengths below 20 meters. Readers interested in the study of super-regenerative receivers and circuits are advised to read SHORT WAVE CRAFT for December, 1932. The author at that time covered the various forms and types of circuits at length.

The receiver described in the present article is novel in several respects. First, it can be used as a straight regenerative receiver, and by the flip of the double throw, single pole toggle switch can be made to "super." When a signal is tuned in simply test for the circuit that will give the greatest output to the amplifier or phones. This is really a radio set with two types of circuits, both of which can be tested on the same signal at practically the same time. The maximum sensitivity will be obtained when used superregeneratively when tuning in "C.W." signals.

Band-Spread Tuning

The simplified method of band spread tuning as described by the author in several of the S.W. receivers built lately is used. Letters from and talks with set builders indicate that this system is simple and has the advantage of not necessitating special coils. The size of the tuning condenser should be smaller for the real high frequencies and is satisfactory for use above 50 meters. In fact, the size of this condenser as specified is a compromise but it really works quite well.

Additional Amplification

The output of the detector is trans-

> This "dual role" short-wave receiver, which can be changed by the flip of a switch from "regenerative" to "super-regenerative," is particularly efficient for the reception of CW or code signals when operating on the super-regenerator principle, particularly on the lower wavelengths, or those below twenty meters. Phone stations may be tuned in by means of super-regeneration and the change-over switch operated to change the circuit to the ordinary "regenerative" type. In other words, at the lower wavelengths and on code or CW signals, the super-regenerative circuit shows the most marked gain in efficiency. The cost of this set is very nominal and any wave band can be tuned in by using suitable plug-in coils.

former coupled to the audio stages that should follow, although a transformer with suitable characteristics can be used to couple the output of the detector to a pair of phones. Use an audio frequency transformer with a very high primary impedance. This is very necessary. A high impedance load in the plate circuit will give greater signal output to the phones or the audio amplifier input.

Panel Layout

The tank tuning condenser is mounted on the left hand side of the front panel and the double throw,

Wiring diagram for the dual-role short-wave receiver here described by Mr. Denton, its constructor.

Parts List

Two Eby 4 prong sockets (10, 15)
One set Na-ald (or Octocoils) for S.W. Bands. mount in socket 10.
One Hammarlund Equalizing condenser, 100 mmf. (9)
Eight Fahenstock clips, (1, 2, 3, 4, 5, 6, 7, 8,)
One Hammarlund MC-140-M condenser, 140 mmf. (11)
One Hammarlund MC-35-S condenser, 35 mmf. (12)
One Illini .000125 mf., mica condenser (13) ; (Polymet)
One International Resistance 1 watt, 3 megohms, (14) ;(Lynch)
One Aerovox .006 mf. mica condenser, 16; (Polymet)
One Flechtheim By-pass condenser, 1 mf., 250 volts D.C. (21) ; (Polymet)
One Flechtheim Tubular condenser .0015 mf., 1000 volts (20) ; (Polymet)
One Silver Marshall Type 240 audio frequency transformer (19)
One National tuning dial, midget type B
One Acratest S.P.D.T. toggle switch Cat. No. 4104 (17)
One Acratest toggle switch Cat. No. 4010 (23)
One Frost volume control, type 6158 Acratest, 100,000 ohms (22) ; (Clarostat)
One wooden baseboard 7x10 inches.
One aluminum panel 7x10 inches, Blan-the-Radio-Man
One Eveready-Raytheon type 32 screen-grid tube, (R.C.A.)

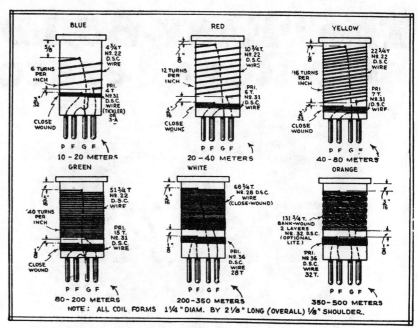

Details for Making Coils to Be Used in Super-Regenerator. (Alden type)

single pole switch that throws the circuit over from straight regenerative to super-regenerative action is mounted directly underneath. Tuning is done by means of the dial mounted on the panel in the center.

The regeneration control resistor and the filament switch are mounted on the right hand side of the set within easy reach of the operator.

Chassis Layout

Every other part not mounted on the front panel is fastened down to the wooden baseboard by means of wood screws. The exact location of each part can be seen by reference to the photographs. The small 100 mmf. antenna series condenser should be mounted off the baseboard by means of a small brass collar. The only piece of equipment not held down to the chassis by means of wood screws is the 3 meg. grid leak and that is held in place by the soldered pig-tail leads on the resistor. The photographs should be studied by the constructor so that all of the parts can be placed in the same relative position. No difficulty should be found in laying out the set and mounting the parts.

Wiring

There is but little to be said in regards to the wiring of the set. Use a good hot iron, and make sure that all connections are firmly and properly made so that there will be no sacrifice of signal due to poor connections. It is a good idea to wipe all connections with alcohol directly after soldering.

Operation

Anyone familiar with the operation and construction of a simple regenerative receiver will have no difficulty in

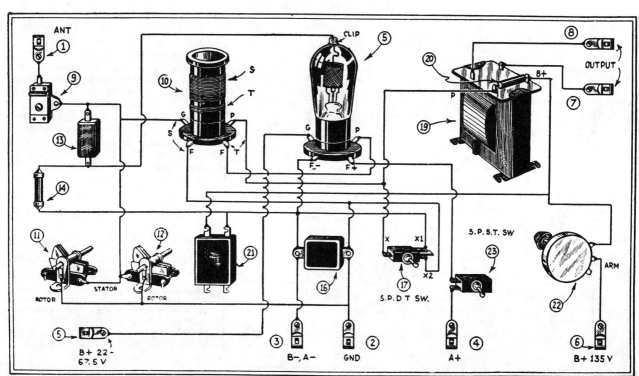

Physical wiring diagram showing how to connect the various component parts of the dual role receiver

tuning in signals with this set. Try the circuit as a straight regenerative receiver and then as a "super." Note that certain stations will come in with greater volume on the super circuit and that other stations will give better signals on the straight regenerative hook-up. In general the super-regenerative circuit will give better results on the very high frequencies and the straight regenerative circuit will be more satisfactory on the lower frequency bands. C. W. signals and super-regeneration go well together and very high values of amplification can be built up. This great "build-up" of signal will result in the distortion of the original pitch of the signal, but this will not be objectionable.

For 2-volt battery operation use the 32 type tube, for 6 volt operation use the 36 type tube and the 24 or 57 for operation on A.C. If a satisfactory filament transformer with a secondary voltage of 6 volts is available, then the 36 type tube can be used if desired. *Be sure that the tubes are in good condition. Poor tubes will ruin any short wave receiver.*

When using tubes of the 32 class do not place more than two volts across the filament or the life of the tube will be materially decreased. Keep the voltage at exactly TWO volts for maximum life. The plate voltage should be 135 and the screen voltage should be varied until the most

sensitive and smoothest operating point is found. This voltage will vary with different types of tubes and tests should be conducted when tuning to a weak station to determine the proper operating screen voltage.

Set builders who have looked at the many super circuits that have appeared in SHORT WAVE CRAFT in the past, and have hesitated to build them because they were doubtful as to the results, should try this one and then they will go after the more complicated and smoother operating jobs.

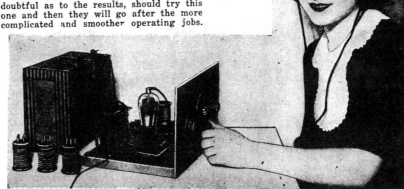

If the signal does not sound so "hot" on the "regen"—flip a switch and then listen to it on the "super-regen" circuit.

Super-Regenerator Rolls 'Em in
By BEN. F. LOCKE

After all is said and done, the super-regenerative receiver is one of those illusive and not so well-known circuits; but with it Mr. Locke has established some very fine reception records, bringing in European and other "DX" stations.

I WONDER how many ever give the Super-Regenerative Short Wave Circuit a thought? Well suppose that we try one and see how it comes out in the "DX" line. I give herewith a diagram for a receiver that I have designed and built myself and I say that it is there for the "DX" stuff. I bring in all the following stations on the loud speaker; G5SW, CJRX, W2XAF, W2XAD, W8XK, KWT, W6XN, W9XF, 7LO, YN, PCJ, PCL, JHBB. W3XAL, W6XAX, RFM and many amateur stations and trans-Atlantic stations. On the headphones I have brought in 3LO, EH9OC and EH9XD as my most distant "DX" stations.

The action of this receiver is very simple and it is easier than the ordinary short wave receiver to tune—that is, to me. There is no "body capacity" or "side-swiping" on the station that you are listening to.

To tune this receiver proceed as follows: Turn on the switch and leave the "super" rheostat turned off; then you tune in just like you would the ordinary receiver, till you hear the "whistle" of a station. Make the necessary adjustments till the whistle is at its loudest and then turn on the "super" rheostat and the "whistle" will entirely disappear; your station will now come in with plenty of volume. Once you get the H.C. Coils "set," they are to be left in that position.

I hope that you will publish the diagram of this wonderful receiver and I

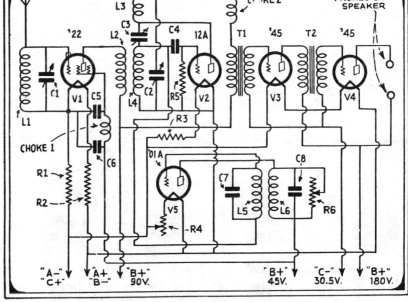

Hook-up of Mr. Locke's super-regenerative short wave receiver.

would like to hear about the results others obtain with this set.

The plug-in coils are wound as follows: Use 1¾" diameter tubing.

List of Parts for Mr. Locke's Super-Regenerative Receiver

2 Variable cons. .00014mf., C1, C2.
1 Variable con. .00025mf., C3.
1 Fixed con. .00025mf., C4.
2 Fixed cons. .006mf., C5, C6.
2 Fixed cons. .002mf., C7. C8.
2 15-ohm fixed resistors, R1, R2.
1 Amperite ¼ ampere, R3.
1 3 meg. grid leak with base clips, R5.
1 25-ohm variable rheostat, R4.

1 0-50,000-ohm variable resistor. R6.
2 A.F. transformers, 3½-1 ratio, T1, T2.
1 H.C. Coil, 1250 turns, L5.
1 H.C. Coil, 1500 turns, L6.
1 H.C. Coil mounting for two coils. Not shown.
2 R.F. Chokes, S.M. No. 277. Choke 1 : Choke 2.
5 UX tube sockets, V1, V2, V3, V4, V5.
3 Vernier dials. National Type B (0-100-0). Not shown.
1 Panel, 7x20x¼.
1 Baseboard, 7x18x¼.
3 Panel brackets.
1 Set of hardware, etc.
1 Phone or speaker plug.
1 Single Circuit jack.
1 Filament switch.

Tubes used are as follows :
1 UX 222 Cunningham as R.F.
1 UX 112-A Cunningham as Det.
2 245 Cunningham as A.F.A.
1 UX 201-A Cunningham as super regenerative circuit (or any other tube that you see fit). '

Wave Length Range	L1	L2	L3	L4
15 to 40m	3 turns #18 wire	9 turns #18 wire	4 turns #28 wire	Same as L1
30 to 90m	8 turns #18 wire	9 turns #18 wire	10 turns #28 wire	Same as L1
80 to 250m	24 turns #22 wire	9 turns #18 wire	15 turns #28 wire	Same as L1
240 to 550m	80 turns #28 wire	9 turns #18 wire	15 turns #28 wire	Same as L1

Front view of Mr. Tanner's 1-Tube Super-Regenerative Receiver.

The 3-in-1 Super-Regenerator

BY R. WILLIAM TANNER

● ONE of the first arrangements of the super-regenerative circuit demonstrated consisted of only one tube feeding directly into a magnetic speaker. The demonstration model operated in the 200-550 meter broadcast band and gave some excellent results.

In the broadcast band it is necessary to employ a very low variation frequency (in the audio range) but at frequencies from 3,000 kc. and up, a higher variation frequency can be used which can be heard faintly only occasionally. Properly designed and constructed, a one tube super-regenerator can operate a magnetic speaker on loud or medium signals, provided the plate voltage is sufficiently high.

In Fig. 1 is depicted such a circuit. It will be noted that a three-electrode tube is employed. While a screen grid or pentode tube would undoubtedly function better, no output transformer matching the plate impedance of these tubes with a speaker is ordinarily available. The tube 11 should pref-

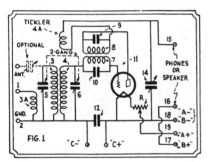

Schematic hook-up of the 1-Tube Super-Regenerator arranged for D.C. (battery) operation.

erably be a Western Electric VT2 or a '10 operated at a plate voltage of 300 to 350 volts; however a '12A will provide satisfactory reception with a plate voltage of 180.

The super-regenerator is a very broad tuning circuit and requires at least two tuned circuits between antenna and grid. This is easily accomplished by means of a simple band-pass filter. This circuit uses such a filter and is tuned by a two-gang condenser. The tuning capacities should be within the range of .000125 to .00016 mf. condenser. A broadcast unit may be cut down by removing plates, 7 plates in each section generally being of approximately the correct number. The antenna coupling condenser C may be a midget .000025 mf. or a compensating condenser of the same capacity. This is set once, generally at a very low value, and then left alone.

If the detector were of the grid-leak type, it would be necessary to employ a grid-leak not lower than 1 megohm, in which case it sometimes happens that an audio frequency is generated within the tube in addition to the variation frequency, resulting in interference.

This circuit employs grid bias rectification by means of a "C" battery connected as shown and by-passed with a .1 mf. condenser. The value of the "C" bias will depend upon the tube and plate voltage. With a '12A the value will be between the limits of 3 and 6 volts. Correct adjustment is where regeneration is smoothest.

The coils 3, 4 and 4A are wound on S-M type 130P midget forms, (details are given in table accompanying this article), and are mounted horizontally as shown in illustrations. The spacing between 3 and 4 when mounted should not be less than 1 inch.

The variation frequency oscillator coils 7 and 8 are both alike and peaked at approximately 20 kc. by means of .002 mf. condensers 9 and 10. The inductance value of the coils is 30 mh. and they are lattice-wound. These may be purchased at all jobbers and are generally used for R. F. chokes. The dimensions are quite small, ¼" thick and 1" in diameter. Holes are provided through the center so that mounting on a ⅜" wooden rod is possible. The rod should be about 2¼" long. The method of mounting these coils is shown in Fig. 2. One coil is made solid by applying a little coil dope, collodion or shellac to the rod before placing the coil on it. The other coil is loose so that the output of the oscillator can be adjusted by varying the coupling between the coils.

A suggested layout of parts is shown

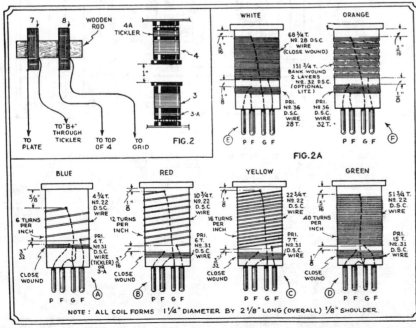

Coil winding data for the plug-in coils, covering both the short wave and broadcast bands; two sets of each are required.

Mr. Tanner, well-known to all SHORT WAVE CRAFT readers, here contributes one of the most interesting and timely articles of his career. "What can you do with one tube"—seems to be the motto of the day. This set was actually built and tested as the photos herewith testify.

One Tube Super-Regenerator

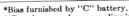

Type of Tube	Plate Voltage	Voltage of Grid Bias	Value of Resistor 13 in Ohms
01A	135V	—13.5V*	....
210	350V	—35.0V*	....
112A	180V	—20.0V*	....
22**	135V	—13.5V*	....
24**	250V	— 9V	5,000
27	250V	—30V	15,000
30	180V	—20V*	....
32**	180V	— 6.75V*	....
36**	180	—6V	5,000
37	180	—20V	15,000
56	250	—20V	15,000
57**	250	—10V	5,000

*Bias furnished by "C" battery.
**Requires very large coupling impedance.

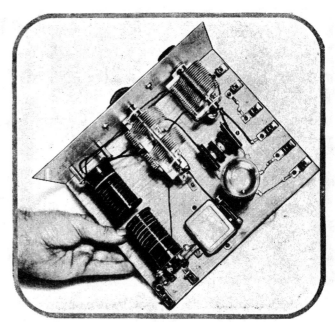

Rear view of the Tanner "1-tube" Super-Regenerative Receiver.

in photograph. This will depend considerably upon the type of parts used, as all makes do not have the same size or shape. It is merely necessary for the builder to be sure the leads carrying R.F. currents are as short and direct as possible. No shielding of any kind is necessary or even desirable. The variation frequency coils should not be located too close to the signal frequency coils. The parts not shown should be mounted wherever convenient. The two variation frequency tuning condensers 9 and 10 are located close to their respective coils.

The adjustments are simple: First loosen the coupling between 7 and 8 and tune in a signal, with any set of coils in circuit, exactly as with any regenerative tuner. Then adjust the 5 condenser trimmers to the best value with 14 set at a low value. If regeneration is erratic try different values of "C" bias. At this time it would be well to mention that a '12A tube with 180 volts on the plate will generally oscillate very strongly and the tickler coils may need a turn or two removed. The turns should be such that the tube just oscillates with the regeneration condenser set at nearly maximum when the tuning condenser plates are entirely in mesh.

After all adjustments mentioned have been made tune in a signal and then tighten coupling between 7 and 8 until the tube generates the variation frequency. The signals will then jump up to a value many times louder than with straight regeneration.

Not all stations can be brought in on a speaker, but those of sufficient strength will be.

Special Notes on Construction

The construction of this set is very simple and can be done with a screw-driver and a pair of cutting pliers if the panel is bought ready drilled.

After the front panel is drilled the base-board can be attached to it. Use one inch long wood screws for this.

Parts List for Both One Tube Super-Regenerator

One—Hammarlund Midget Condenser. Type MC250M. 250 mmf. (14)

One—Hammarlund Midget Condenser. Type MCD-140. 140 mmf. (5, 6)
One—Fletchtheim By-pass Condenser. Type GB-100. 1 mf, 200 volts. (12)
Two—Flechtheim Tubular Condensers. Type AZ-10. .002 mf. 1000 volts. (9, 10)
Two—Sets Alden Mfg. Co. Short Wave Coils. 15 to 200 meters.
Eight—Fahnstock Clips, (1, 2, 15, 16, 17, 18, 19, 20)
Two—Eby Chassis type sockets. (3, 4) Four prong for the plug-in coils.
One—Eby Socket, type depends on the choice of tube used for reception. (11)
Two—Blan Special choke windings. (7, 8)
One—International Resistor 10,000 ohms, 1 watt. (13)
One—Hammarlund Equalizer Condenser. Type EC-80. 25 to 80 mmf. (21)
(Optional method of antenna coupling.)
One—Blan 6 inch by 10 inch aluminum panel.
One—Wooden base-board 8x10x¾ inches.

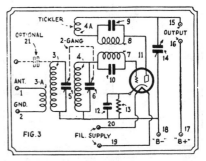

A.C. schematic diagram for the 1-tube super-regenerator.

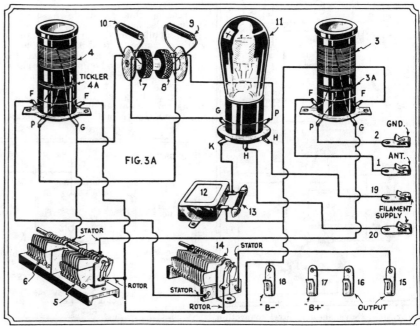

Picture diagram for building the A. C. operated 1-tube super-regenerative set.

A Balanced-Detector Super-Regenerative Receiver

By CLIFFORD E. DENTON

Tuning of the balanced-detector super-regenerator is made easy by the use of a National velvet vernier dial.

● ONE of the major difficulties encountered in operating receivers using super - regeneration is the action of the detector or super - regenerative tube when the circuit amplification is too high. Under this condition the tube breaks into self-oscillation and the super-regenerative action does not take place.

An interesting circuit was tried by the author which has the qualification of eliminating this tendency of self-oscillation in the Super-regenerative tube circuit. Here the tubes are used in a peculiar circuit. Note that the grid circuits are in push-pull and that the plate circuits are connected in parallel. A point in this circuit's favor is that of lowered plate impedance. The two plate circuits are in parallel so the impedance is half that of a single tube under the same operating conditions. This gives a more satisfactory match for phones and is more convenient when audio amplifiers are to be used to increase the output. It is hard to match tubes with transformers or a pair of phone so that the most satisfactory transfer of energy will take place under the normal operation of tubes as bias detectors.

Breaking this circuit down so as to study it, let us look at Fig. 1. A radio signal applied in the ordinary manner will not cause a feed-back of energy into coil L2 from coil L because the grids are equal and opposite in phase and the plate currents will be constant. Placing a center-tapped coil in the cathode leads, as shown in Fig. 2, will act as a means of supplying the quenching frequency from the local oscillator.

Tubes V1 and V2 should have the same approximate value of mutual conductance, as the action of the circuit is better when they are matched. This point is mentioned for the best results, though the circuit will work with tubes with mutual conductance variations of 25 to 30%. The closer the two tubes are matched the better will be the regenerative action.

The voltage output of the quenching frequency generator is supplied to the cathode circuits of the two detector tubes in opposite phase, as indicated in Fig. 2. This voltage developed by the local oscillator during one half of a cycle will increase the effective voltage on the grid of one detector tube and decrease the voltage on the grid of the other tube. With unequal voltages applied to the two grids there will be a change in the mutual conductances of both tubes as far as the signal frequencies are concerned, so that some feed-back will take place between L and L1 of Fig. 2. When the opposite half of the quenching frequency cycle appears across L3 it will cause a voltage in reverse phase, as shown in the dotted lines. Thus the two tubes are made unequal as far as their mutual conductances are concerned and some energy is fed-back, as stated above. This energy will be in reverse phase to that which flowed on the first half of the quenching frequency cycle.

As long as the quenching frequency generator is functioning there will be a regenerative and *degenerative* feed-back action from the coil L to coil L1. Now the value of the two feed-backs are equal, so all of the energy that may be present in the feed-back coil during the period of regeneration will be nullified during the *degenerative* period. This degenerative action will throw the circuit out of oscillation even though the values of the two feed-backs are equal, due to the resistance of the circuit, which aids the quenching action

Diagrams, above, show a novel connection of balanced-detector tubes; Fig. 3 shows hook-up of apparatus to Acme 30 K.C. transformer.

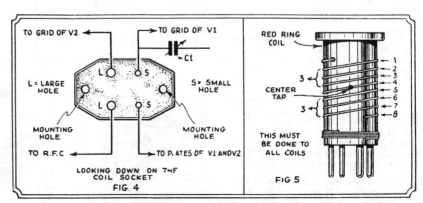

Diagrams, above, show the proper coil socket connections at Fig. 4; Fig. 5 above, how the center tap is soldered to the plug-in coils.

of the *degenerative* portion of the cycle and tends to prevent the regenerative portion from building up to values which would cause self-oscillation in the circuit.

Construction Details

The construction of the set is shown in the photographs and should offer no trouble to the builder. A wooden baseboard is used with an aluminum front panel so that hand-capacity effects are reduced to a minimum.

Parts mounted on the front panel include the phone terminal block, oscillator frequency selector switch and the oscillator power control. Of course the tuning dial is placed in the center of the panel with a rod of sufficient length to reach the tuning condenser coupling unit. The rest of the parts are mounted on the chassis with wood screws or are held in place by the wiring. The photographs should furnish the builder with all information necessary to duplicate the original set.

A winding must be placed on the core of the Acme transformer. This consists of 120 turns of No. 28 silk covered wire. Loosen the two top machine screws holding the core in place and slide the L shaped section out of the frame. Random-wind 60 turns of wire and then leave a 6 inch long loop. This loop will serve as the center tap connection of the oscillator pick-up winding. Then continue with the additional 60 turns, which will give the required number of turns.

Wiring

Run all wires directly to their points of connection. Do not try for looks, but for results! *Make all connections clean ones.* Poorly soldered connections make receivers noisy especially if they are high-gain short-wave receivers.

When making the special connections to the Acme transformer follow the detail wiring diagram of Fig. 3. If the diagram is followed correctly there will

Rear view of Mr. Denton's latest short-wave receiver—the balanced-detector super-regenerator; the third tube has been left out of the socket at the right for clarity.

be no difficulty in getting the tube to oscillate. Figure 4 shows the proper coil socket connections. *Remember that the tube tends to go into oscillation only when the local oscillator is working.*

Operation

The set is placed into operation by connecting a power supply unit that will give the required voltages. The author used the power supply unit described on page 82 of the June, 1932, issue of SHORT WAVE CRAFT. This unit

was designed for use with the "STAND-BY" receiver and was found to be very satisfactory for use with this receiver.

Place three type 56 tubes in the sockets after making sure that all connections are properly made. Plug in the tips of a pair of phones and connect the antenna and ground. Select one of the plug-in coils and locate the center of the winding, as shown in Fig. 5. Note that there should be three turns on each side of the tap and that each section of three turns has an equal additional fraction of a turn. The way to take this tap is quite simple. Scrape away the insulation with a sharp knife or a

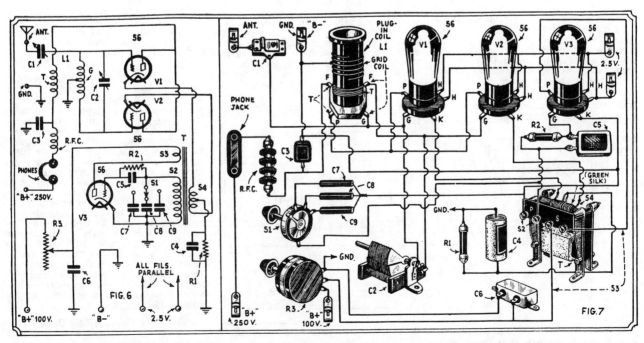

Schematic and picture diagrams are presented above, so that the uninitiated as well as the experienced short-wave fan can build this latest creation of Mr. Denton's—a super-regenerator receiver with balanced detector.

razor blade; then solder a piece of wire about five inches long to the wire which has been bared. This is the center tap of the coil and is connected to the ground and "B"—return.

After a signal has been tuned in, adjust the antenna series condenser C1 for the best results. Try various values of capacity at C7, C8, C9. Each band will have a value that will be best suited for maximum amplification. Lots of time must be spent in adjusting and trying out the circuit because the best results will only be had after considerable work has been done. This does not mean that the set is hard to get working, but that maximum results will be the product of the builder's efforts.

Further information regarding super-regeneration appears in SHORT WAVE CRAFT, December, 1932, issue.

If the builder does not have the new type 56 tubes he will find that the older type 27 will work very well. This point is mentioned because the new tubes are not yet available all over the country.

Use a good pair of phones and do not worry about the plate current of the tubes flowing through the windings of the phones, because the detector tubes are biased to plate current cut-off and the total plate current is less than one milliampere.

Conclusion

It is, so it seems, the desire of every *short-wave* "bug" to build a super-regenerative re-

ceiver that will work. The letters received by the author during the past year show that about one out of three get satisfactory results. After looking at several of the sets, one really wonders why they ever work at all! Use good tubes! Solder carefully! Connect leads as shown! If possible use the parts recommended by the author.

Parts List

1—Alden S.W. Coil Kit (L1).
1—Hammarlund Equalizing Condenser 100 mmf. (C1).
1—Hammarlund MC-140M Midget Tuning Condenser (C2), Capacity-140 mmf.
1—Hammarlund S4 Isolantite Socket (used for L1).
1—Flechtheim Midget Condenser .00025 mf. (C10).
1—Flechtheim .25 mf. bypass condenser. Type GB-25 (C6).

3—Flechtheim Midget Tubular Condensers. Type AZ. One .0001 mf. (C7); One. 0002 mf. (C8); One .00025 mf. (C9).
1—National, R.F. Choke Type 100 (RFC).
1—Acratest Mica Condenser .001 mf. (C3).
1—Acratest Electrolytic Condenser 25 mf. 25 volts (C4).
1—Carter 4 position switch (S1).
1—Frost potentiometer 50,000 ohms (R3).
3—Pilot (or Alden) sockets (V1, V2, V3).
1—Acme 30 kc. transformer (T).
1—International Resistor Co. 25,000 ohm, 1 watt resistor. (R1).
1—International Resistor Co. 1 meg. 1 watt resistor. (R2).
1—Blan coupling for the tuning condenser.
1—Blan Aluminum Front Panel.
1—National Tuning Dial.
1—Eby twin tip-jack terminal unit.
1—Base-board 8" by 10" by ¾" wood.
6—Fahnstock Clips for connections.
3—Eveready-Raytheon type 56 tubes.

Data on Alden Plug-in Coils

	Number of Turns		
(1)	4¾	6 Pitch No. 22 D.S.C.	Primary 4 turns No. 31 D.S.C.
(2)	10¾	12 Pitch No. 22 D.S.C.	Primary 6 turns No. 31 D.S.C.
(3)	22¾	16 Pitch No. 22 D.S.C.	Primary 7 turns No. 31 D.S.C.
(4)	51¾	40 Pitch No. 22 D.S.C.	Primary 15 turns No. 31 D.S.C.
(5)	68¾	Close wound No. 28 D.S.C.	Primary 28 turns No. 36 D.S.C.
(6)	131¾	Bank wound, 2 layers, No. 32 (Optional Litz)	Primary 32 turns No. 36 D.S.C

WAVE BANDS:
　(1) Blue—10 to 20; (2) Red—20 to 40; (3) Yellow—40 to 80; (4) Green—80 to 200; (5) White —200 to 350; (6) Orange—350 to 550.

D.S.C.—double silk covered. Pitch—turns per inch.

A Super-Regenerator With PENTODES

By R. WILLIAM TANNER

● THE super-regenerator, while not a new circuit, has been given little attention by either the amateur or short-wave broadcast listener, due no doubt to lack of information on the subject. In the last few months a number of articles have appeared in this magazine and other publications describing this remarkable circuit. However, these articles are entirely lacking in data which would help the builder in the elimination of "kinks" and the super-regenerator has a number of such

"kinks." In the first place, there is the broad tuning feature which can be improved ONLY by the use of a sufficient number of tuned circuits between the antenna and detector grid. Second, the super-regenerative "hiss" or "mush," which can be suppressed by means described later. Third, distortion of voice or music, which can be cured by an adjustment of the variation frequency oscillation. Fourth, unstable operation due to insufficient tickler turns or coupling, the cure of

which is readily apparent. Fifth, when an R.F. stage is employed the detector generally fails to oscillate over a portion of the dial. This is due to absorption by the plate winding on the detector R.F. transformer, the cure being very loose coupling between R.F. plate coil and detector grid coil.

Before describing this new version of the super-regenerator, the writer wishes to settle the question always asked by the novice: "Can this set bring in Europe, South America, etc.?" Any short-wave receiver having one to twenty tubes can receive from any point on the globe, *providing* the time, season, weather, location, etc., are right (also providing the receiver is not located in the "skip" area). The *human element* must be given some consideration, as considerable patience and careful tuning are most important.

R.F. Pentode Ahead of Detector

Referring to the circuit in Fig. 1, it will be noted that one of the new '58 R.F. pentodes is employed in the tuned R.F. stage, resulting in a fair degree of gain even down to 15 meters. The super-regenerative effect is obtained through the use of an oscillator tuned to a comparatively low frequency and a regenerative screen-grid detector designed to oscillate more strongly than in the usual regenerative set. The output of the detector is generally sufficient to operate the '47 pentode power audio amplifier to full capacity, except on weak signals. Experiments which employed a '35 R.F. amplifier in place of a '58 proved that the overall sensitivity was equal to nearly all short-wave superhets with two I.F. stages. Volume was not quite as loud when compared to a superhet having two audio stages.

Selectivity Improved by Band-Pass Filter

Selectivity with a super-regenerator is, *in the usual form of circuit, notoriously poor!* With the tuned circuits,

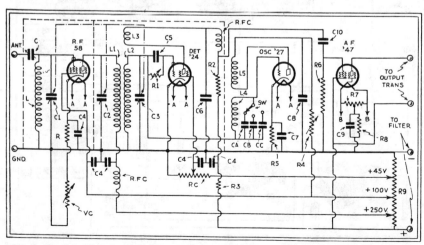

Fig. 1—Complete diagram of Mr. Tanner's super-regenerative receiving circuit which employs a screen grid detector with pentodes in the R.F. and A.F. amplifier stages, which gives some "wallop" to the signal.

R.F. grid and detector grid, tuning is still too broad. However, by adding a band-pass filter either between antenna and R.F. grid or between R.F. plate and detector grid, very satisfactory selectivity can be obtained. In the original model this filter was placed between R.F. plate and detector grid to help in the reduction of regenerative feedback and to prevent unstable detector oscillations.

In order to permit best super-regenerative action over the range of 15 to 100 meters, the variation frequency oscillator is provided with three fixed condensers of different values, controlled by means of a three-point tap switch. With a grid coil of 20 mh. inductance, the frequencies are 25, 35 and 50 kc. at capacities of .002, .001 and .0005 mf. respectively. If the 100 to 200 meter band is desired it will be necessary to add capacity in parallel with the .002 mf. section until the frequency is approximately 100,000 to 15,000 cycles. The output of the oscillator is controlled by the variable resistor R4

Coil Construction

Due to the use of three tuned circuits and to the fact that no form of tapped coils can equal plug-in coils in efficiency, the plug-in type was employed. These are of different construction than the usual coils. Figure 2 shows how the various coils are arranged on strips which are provided with General Radio coil plugs. A pair of strips having G.R. jacks are used as mounting bases. The layout of the strips is given in Fig. 3. Drill the base and coil strips exactly alike, in fact, it would be well to drill them all at the same time. The stock for the coil strips should be 3/16" thick and 1/4" for the base if hard rubber is used. If bakelite is used for the base, the thickness may be 3/16" since it is much stronger than hard rubber.

The three-gang tuning condenser was an old broadcast unit with all but three rotor and four stator plates removed from each section. The resulting capacity is uncertain but it is in the vicinity of .0001 mf. With this capacity condenser the coil values given in the accompanying table were employed. It must be remembered that variable condensers vary in capacity even though the number of plates is the same. The coil table may require modification to cover the exact range.

It will be noted that the tickler specifications are somewhat greater than usual. This is for the purpose of providing a stronger oscillation in the detector.

The method of coupling between L1 and L2 is somewhat different than is generally employed but the results are the same. In this case, the two coil sections do not require wide spacing.

The older types of inductances used in the variation frequency oscillator, such as honeycomb coils and iron core transformers, were eliminated in favor of a much more compact type consisting of a 20 mh. grid coil and one of 8 mh. for the plate. These were both of the lattice wound types having diameters of 1" and 3/4" for the 20 and 8 mh. coils respectively. Each coil has a center hole of 3/8" which allows them to be mounted on a 3/8" wooden or hard rubber rod, making it possible to slide one coil along the rod for the purpose of adjusting the oscillator output. Both coils MUST be wound in the same direction and placed on the mounting rod correctly. These coils may be purchased at nearly any radio store in the larger cities. The mounting arrangement of this unit is shown in Fig. 4. The oscillator coils, switch and fixed tuning condensers may all be mounted on a small bakelite base if desired.

R.F. Chokes

The R.F. chokes are easy to construct and are sometimes better than manufactured ones. The choke in the R.F. "B" positive lead may consist of 1,500 turns of No. 36 enamel or silk covered wire wound in three slots, 500 turns per slot. The choke in the detector plate circuit must have a higher inductance, since no R.F. currents can be allowed to get into the A.F. amplifier. This choke should be wound with 5,000 turns No. 36 wire in 5 slots, 1,000 turns per slot. Dimensions of the wooden forms are given in Fig. 5.

Shielding

The dotted lines in Fig. 1 shows how much of this circuit requires shielding. Box shields are necessary; Alcoa 5" by 6" by 9" aluminum.

The R.F. bias resistor R, bypass condensers C4 and R.F. choke are located within the R.F. shield. The detector grid condenser C5, leak R1.

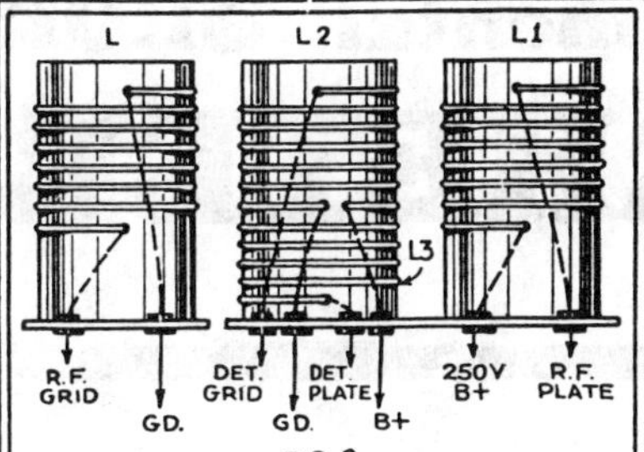

FIG. 2

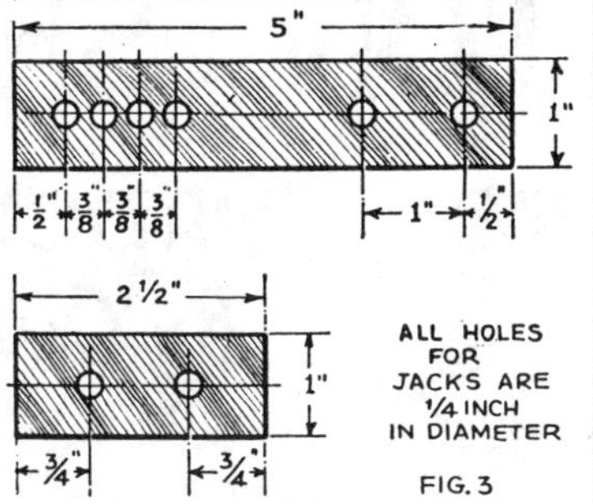

FIG. 3

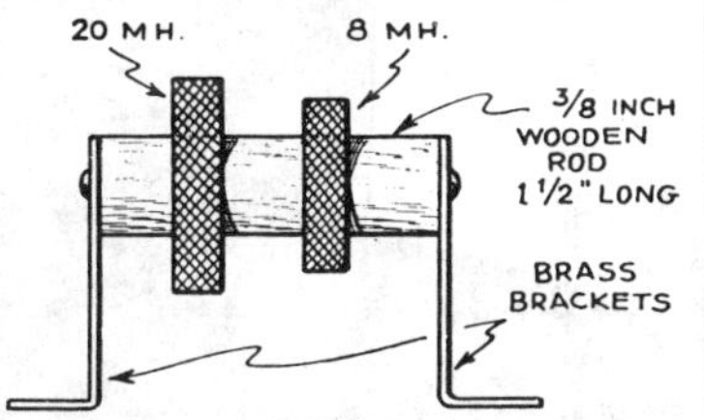

FIG. 4

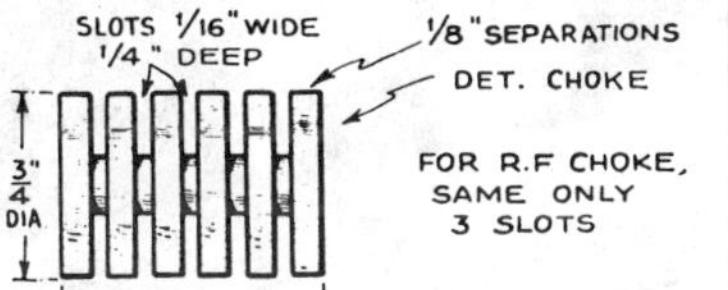

FIG. 5

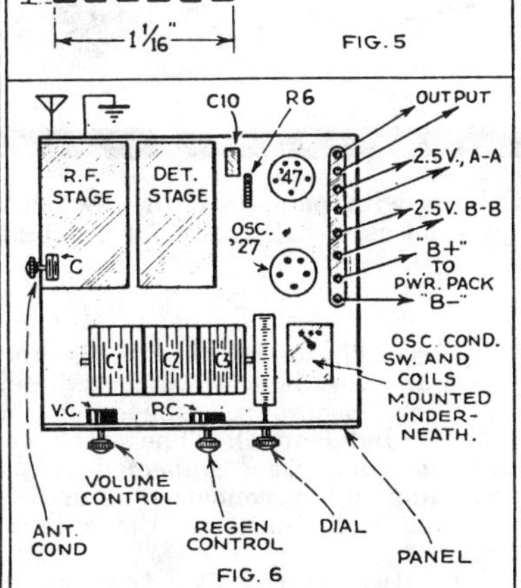

FIG. 6

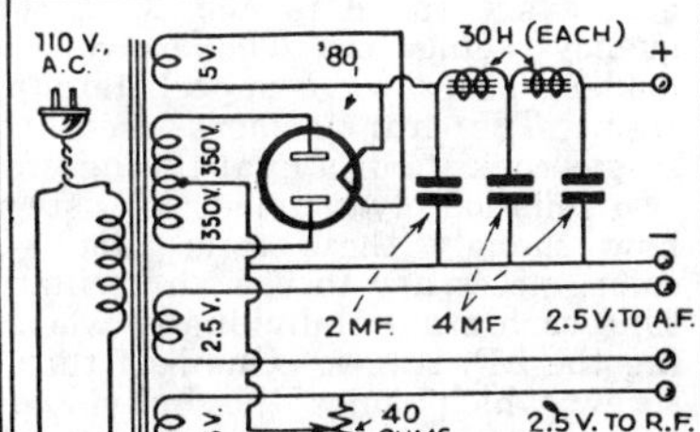

FIG. 7

Details of the various coils, arrangement of the apparatus and the hook-up of plate supply unit.

RF choke, bypass condensers C4 and plate condenser C6 are placed within the detector shield. Other resistors, condensers, etc., are located underneath the sub-base. If the oscillator variable plate resistor R4 is used, it may be mounted on the panel at the right of the tuning dial.

Adjustment and Operation

While the adjustments are not difficult, they should be made with care in order to realize maximum efficiency. First, plug in the 80-meter coils and set the antenna coupling condenser C at approximately halfway in. Separate the two coils in the oscillator as far as possible, or, if R4 is used, set at maximum resistance. Tune in a loud signal by means of the tuning dial and regeneration control. Adjust trimmers on the gang condenser to best point. Now turn volume control down so that signal is just above audibility. Then bring oscillator coils close together with switch set on the .002 mf. condenser. The regeneration control RC should now be turned considerably beyond the *normal*, point of oscillation although no oscillations should be present until RC is still further increased. The signal should now be extremely loud.

If a loud rushing noise is heard, first adjust the variable grid leak R1. This should help matters considerably, but if the noise still persists, try adjusting the coupling between oscillator plate and grid coils or the oscillator plate resistor R4. Sometimes both adjustments are necessary. When all adjustments are made correctly, music and speech should be of fine quality. C.W. code signals can be picked up by increasing the regeneration control.

If the coils have been made so that L, L1 and L2 are exactly alike, no additional trimmer condensers will be necessary. If not, it would be well to shunt .000025 mf. midgets across all three sections of the tuning condenser. These would be adjusted only once for each band shift.

Power Supply

The power pack may be any type giving an output of 300-350 volts and having two 2.5 volt filament windings. For the benefit of those who desire to build this unit as well as the super-regenerator, details will be given. The circuit is shown in Fig. 7. The filter chokes should be capable of carrying 85 ma. The 2 mf. input filter condenser should have a rating of at least 600 volts. The other two condensers may be 450 to 600 volts. The leads from the filament windings to set *must* be heavy, No. 14 to No. 8 wire, and should not be longer than 18 inches.

Like any other rather complicated receiver, the builder will probably not make it work as soon as it is hooked up. Use your head and radio knowledge if some kink develops.

Data Applying to Figure 1

C—.000025 mf. antenna coupling condenser
C1, C2, C3—.0001 mf. three gang tuning condenser
C4—.1 af. bypass condensers
C5—.00001 mf. detector grid condenser
C6—.0005 mf. detector plate bypass condenser
C7—.25 mf. bypass condenser
C8—.25 mf. bypass condenser
C9—1 mf. bias bypass condenser
C10—.1 mf. audio coupling condenser
R—270 ohm R.F. bias resistor
VC—50,000 ohm volume control
R1—Variable grid-leak such as Pilot or Clarostat
RC—50,000 ohm regeneration control
R2—100,000 ohm detector plate resistor
R3—25,000 ohm detector plate resistor
R4—50,000 ohm resistor variable
R5—1,000 ohm oscillator bias resistor
R6—500,000 ohm pentode grid resistor
R7—20-60 ohm center tapped filament resistor
R8—400 ohm pentode bias resistor
R9—25,000 ohm 75 watt Electrad Truvolt resistor with 3 taps
RFC—R.F. chokes
SW—Three-point tap switch
CA—.002 mf.
CB—.001 mf.
CC—.0005 mf.

Coil Table

Band	L	L1	L2	L3
20	4	4	4	4
40	9	9	9	7
80	20	20	20	11

Use No. 26 enamel wire on L, L1 and L2 and No. 30 on L3. All coils close wound.

Experimental SHORT-WAVE
SUPERHETERODYNES

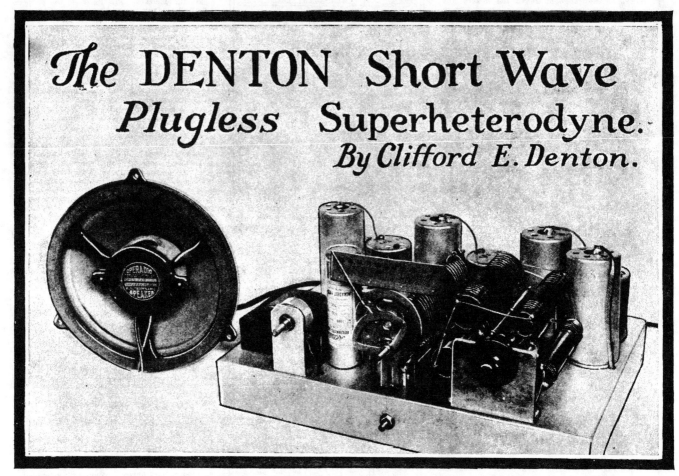

The DENTON Short Wave
Plugless Superheterodyne.
By Clifford E. Denton.

Here is the complete super-heterodyne, with band-changing switch at right, together with midget dynamic speaker—all ready for a cabinet.

IT seems that most everyone today wants a *short wave superhet* so let's see what can be done. Many readers have written to the author in regard to the various versions of short wave adaptors using the *Best* switch and coil assembly, and most of the letters close with "when will a complete receiver circuit come along?" Well, here it is:

The 7-Tube Circuit

The receiver has 7 tubes, including the rectifier, with a total of eight tuned circuits. An examination of the circuit diagram shows that the antenna is connected to the grid coil of the detector tube through a small semi-variable condenser. This condenser is quite critical in adjustment and should be changed with each change of antenna. If it is too large the incoming local stations will have strong harmonic *repeat points* which are not desired. Of course, if the condenser is too small, there will be insufficient signal input with a loss of sensitivity.

The oscillator is tuned in the plate circuit and the detector and oscillator tuning condensers are ganged together for single dial control. The small compensating condenser connected across the antenna tuning condenser is used to give critical adjustment of the antenna circuit.

The two intermediate frequency (I.F.) stages are tuned to 465 K.C., this frequency being that generally used in modern short wave superheterodyne design. The transformers used in these stages consists of a dual tuning unit with two coils loosely coupled; it is surprising how sharply these coils tune. It becomes necessary to use an oscillator and tune each stage individually when aligning the I.F. stages. One half turn of the screw which tunes the individual units is enough to throw the signal out completely. In the original model illustrated in this article, the author adjusted the voltages on the intermediate frequency tubes, so that with the volume control about eighty per cent on, the I.F. sec-

tion will oscillate. Thus at any ordinary signal level the I.F. amplifier is stable, but by turning the volume control full on, C.W. (code) signals can be heard and *voice carriers* tuned in easily.

The second detector is an ER224, with plate-bend rectification; the only novel thing to be noted here is the *resistance-capacity filter* used in the plate circuit, instead of the more common choke and condenser unit. The output of the detector is *resistance-capacity coupled* to the pentode, with condenser 31 used to buck out *hum* in the detector circuit.

Many times readers ask for loudspeakers to use with a given set, so a speaker of the *midget type* has been specified and used. The field coil of this small dynamic speaker is used as the *filter choke* in the "B" negative lead of the power transformer. The 800 ohms resistance of the field winding drops the voltage output of the power unit to 260 volts, which is the voltage applied to all plate circuits except the oscillator. The screens and oscillator have a potential

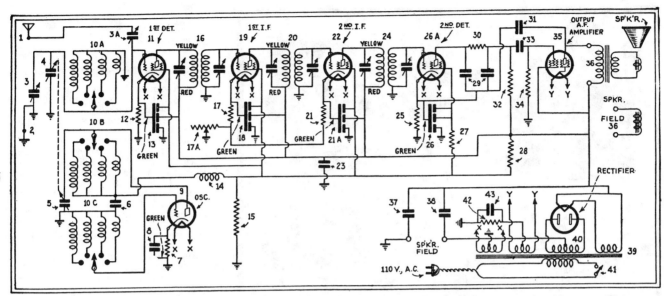

Here is the schematic diagram of the plugless S.W. super-het, in which two I.F. stages are used, with an oscillator, first and second detectors, and pentode power output stage.

of 55 volts, which is supplied by the resistor 28. It is necessary to use at least one insulated, dry electrolytic condenser at 38 in the circuit diagram, due to the fact that while the chassis is at the most negative value, as far as the receiver is concerned, the center-tap of the power transformer is actually more negative. due to the voltage drop in the loudspeaker field choke.

Mechanical Layout

Simplicity is the keynote in the design of this receiver. This manifests itself in analysis of the cost of the parts as well as the attractiveness of appearance.

The tuning dial with the master tuning condensers are located front and center, while the power transformer, electrolytic condensers and intermediate frequency amplifier *volume control* are grouped to the left.

The 12 coils and 3 switches for changing to the various short wave bands, which by the way are 10-20, 20-40, 40-80, 80-200 meters, is placed on the right of the tuning condensers.

The oscillator and first detector tube are placed right back of the tuning coils with the first I.F. transformer mounted in back of the first detector socket.

In the back row, looking from right to left, we find the first I.F. transformer, first I.F. tube, second I.F. transformer, second I.F. tube, third I.F. transformer, and second detector.

Directly in front of the second detector is mounted the pentode output tube and the remaining socket is used for the ER280 rectifier. All of the small by-pass condensers are mounted near the tubes which they serve to by-pass, thus insuring short leads.

The under-part of the chassis conceals the various resistors and the wiring. The resistors used are of the pigtail type and were held in place by using bus-bar, where conditions permitted, and the common push-back wire where insulation was imperative.

As the wiring can be plainly seen in the photographs no special mention need be made on this subject, except—*make all connections good ones.*

Operation

Due to the sharp tuning intermediate transformers, it is necessary to use an oscillator of some sort to tune the I.F. amplifier before signals will be heard. In testing this receiver it was found that the intermediate frequency transformer will tune from 800 K.C. to 425 K.C. So that if no oscillator is within reach, tune the intermediates to some local station which is received on the higher wavelengths of a *broadcast receiver.*

The way to do this is to connect an aerial to the control-grid terminal of the second I.F. tube and with an insulated screwdriver tune in the signal from the

nearby broadcast station as loudly as possible on the third I.F. transformer. Then move the antenna over to the first intermediate frequency tube and tune the signal in on the second I.F. transformer. Place the antenna on the detector tube and tune the first I.F. transformer. The receiver is ready for operation.

Connect the antenna and ground to the binding posts, making sure that the loud-speaker cable is plugged into the chassis receptacle and after permitting the tubes to heat up, move the master tuning dial *slowly* until a signal is heard. Clear up the signal by adjusting the com-

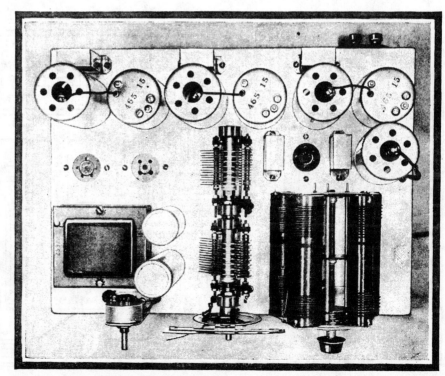

An airplane view of the Denton S.W. super-het, showing the ganged tuning condensers and wave band coils, etc.

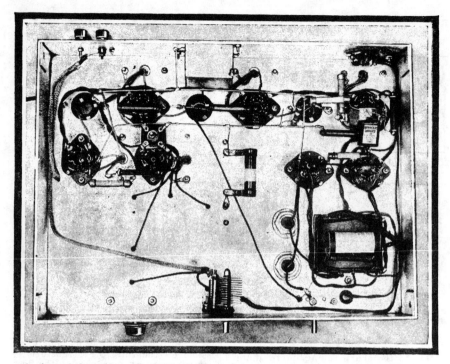

A view of the under-side of Mr. Denton's "plugless" short wave super-heterodyne.

pensating condenser and vary the volume control for desired gain.

Conclusion

It is desirable to note the dial settings on a chart of some kind, thus increasing the speed with which the operator can familiarize himself with the receiver.

The volume of sound emitted by the midget speaker, even on distant stations, is extremely satisfactory, and considering the low cost and simplicity of the total receiver, it will prove a pleasant surprise for the builder.

The only point in the operation of this receiver that deserves special precaution is the proper adjustment of the antenna coupling condenser 3A. Each aerial and every different location needs a variation of this capacity for maximum results. Experimentation gives the correct answer.

Parts List for Simple Super-het

1 Eby Antenna, Ground Post (1, 2)
1 Best S.W.C. 1 kit (10A, 10B, 10C) coil and switch assembly
2 National tuning condensers (see text), (4, 5)
1 Pilot 80 mmf. trimming condenser (3)
8 Eby wafer sockets marked for tubes (9, 11, 19, 22, 26A, 35, 36, 40)
5 Blan .1 mf. by-pass condensers, 3 in each can (6, 8, 13, 18, 21A, 26)
1 Electrad volume control and fil. switch R1-202-P (17A, 41)
1 International or Lynch resistor, 2500 ohms, 1 watt (7)
1 International or Lynch resistor, 10000 ohms, 1 watt (15)
1 International or Lynch resistor, .1 meg., 1 watt (32)
1 International or Lynch resistor, .5 meg., 1 watt (34)
2 International or Lynch resistor, 500 ohms, 1 watt (17, 21)
2 International or Lynch resistor, 25000 ohms, 1 watt (12, 25)
3 International or Lynch resistor, 10000 ohms, 1 watt (30, 15, 27)
1 International or Lynch resistor, 416 ohms, 1 watt, (42)
1 International or Lynch resistor, 50000 ohms, 1 watt (28)
3 Acratest 465 K.C. I.F. transformers (16, 20, 34)
1 Blan RF. choke (14)
1 Flechtheim Filter Condenser 2 mf. 450 volts (23)
1 Sprague midget condenser .04 mf. (33)
2 Aerovox .000125 mica condensers (29)
1 Aerovox .001 mica condensers (31)
4 Hammarlund tube shields
1 Acratest power transformer 2532 (39)

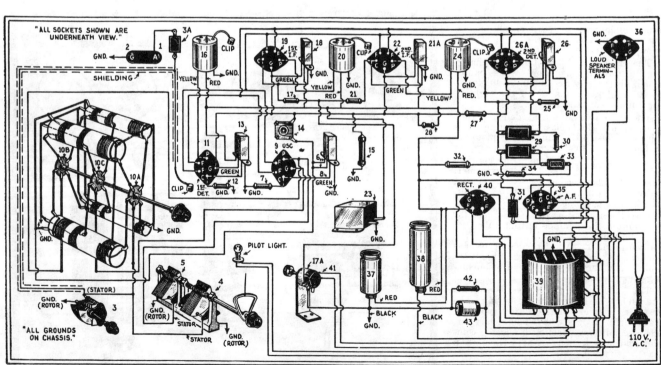

"Picture" diagram that even the novice can follow in building the Denton Short Wave Super-heterodyne. If you use a power transformer (39) other than the one specified, it's a simple matter to follow the connections as given by the manufacturer of the particular power transformer you purchase.

2 Acratest Electrolytic condensers (1-8 mf. No. 5308 and 1-4 mf. No. 5304)
1 Blan chassis and volume control m.t.g. bracket, completely drilled and folded
1 Crowe "full vision" dial and light holder
1 G.E. power cord and plug
1 Acratest 25 volts, 25 mf. No. 6646 (43)
1 Eveready Raytheon 280 tube
1 Eveready Raytheon 247 tube
2 Eveready Raytheon 224 tubes
2 Eveready Raytheon 251 tubes
1 Eveready Raytheon 227 tube
1 *Operadio* loud-speaker with output transformer for '47 pentode, with 1800 ohm field
1 Blan 4 wire plug and beads for speaker cable.

In the event that the builder wishes to use an ordinary loud-speaker, he may procure a pentode output transformer having a secondary of low impedance to match the average speaker.

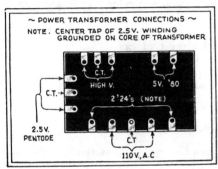

Connections of power transformer for S. W. super-het.

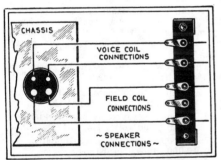

The four connections to the dynamic speaker are taken off through a socket mounted on the side of the chassis as shown above.

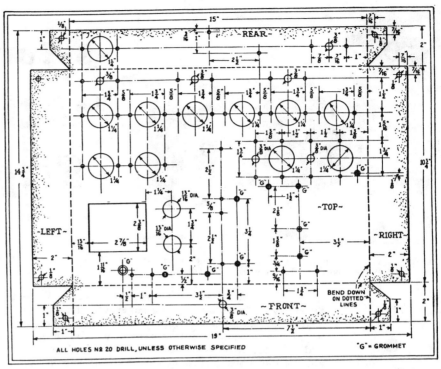

Above—Subpanel drilling layout for the "Plugless" Short Wave Superheterodyne here described. Most radio shops sell subpanels drilled and undrilled at a nominal price.

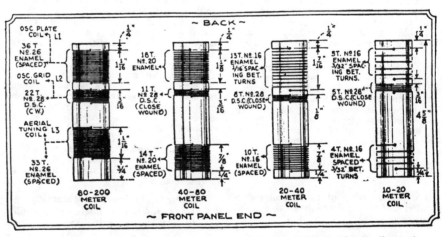

Converter Coil Data—2 coils at left 1¼" O.D.; 2 coils at right ⅞" O.D. (Outside diameter.)

New Ultra Short Wave Apparatus

THE first, specially designed, ultra short wave apparatus has recently shown itself on the American market and is revealed in the accompanying illustrations. This new U.S.W. apparatus has been designed by the well-known National engineers. In the photo at the left we see a

New National tube shield designed for use with the new 56, 57, and 58 tubes.

new tube shield, designed for use with the new series 56, 57 and 58 tubes. Below, photo shows at the left a new isolantite socket which will reduce losses in ultra short wave circuits to a minimum. This socket is available in 4, 5 and

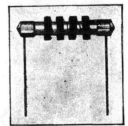

New ultra S-W R.F. choke wound on isolantite core.

Below — New ultra S-W socket; midget R 39 coil forms; midget 270° S. F.L. condenser.

6 prong styles and also for special 6 prong National coils. At center the new midget R-39 coil forms. The midget 270-degree SFL tuning condenser shown is insulated with isolantite; capacity 18 mmf.

H. W. Secor is here shown trying out the 8-Tube Portable "All-Wave" Superhet designed by Mr. Denton.

8-Tube Portable "All Wave" Super-Het

By CLIFFORD E. DENTON

Mr. Denton described the general line-up and circuit of his latest 2-volt "portable" 8-tube superhet. This super is a real job and will appeal to hundreds of people who want a good portable receiver using high-gain tubes, and a set, moreover, which can be carried easily. Diagram and data are here given for building this super also with the new 6-volt automobile tubes. This set works a loud speaker on either 2-volt or 6-volt tubes. Regeneration is used to provide maximum "pick-up" range.

MORE detailed information covering the connections of the detector and oscillator systems would not be amiss at this point, as this is generally a stumbling block for the average constructor.

In the original circuit the only coupling between the oscillator and first detector was the inductive effect between the two coils located two inches between centers. The reason underlying the use of the inductive coupling was that it offered the easy way out of a hard problem. It is a simple matter to couple circuits when the tubes are of the cathode type. But with the two-volt type there are no cathodes.

A detail drawing of the circuit incorporated in the receiver covers this point clearly.

Do not wind the pick-up coils on the oscillator coil forms and subtract these turns from the total number of turns wound on the secondary of the first detector coil. These additional turns are for the purpose of coupling only, and actually add but little to the inductance of the detector secondary.

If the coupling is too great between the oscillator and the detector, then a small aluminum shield should be placed between the coils, so that the greatest coupling takes place through the coupling coil winding on the oscillator.

This small shield should be placed in the center of the small coil mounting platform. The physical details are given in the sketch.

Changes for Six-Volt Operation

There are many builders who would like to use the six-volt tubes in place of the two-volt type, especially for service in the automobile or the motor boat.

A complete circuit is given with the necessary changes for operation with these cathode type tubes.

It will be necessary to change over the four-prong sockets to the five-prong type. There will be eight fives instead of six four-prong sockets in the parts list.

The following changes in the parts list will also have to be made.

25 becomes 15,000-ohm, 2-watt resistor (International Resistance Co.).

27 becomes Acratest by-pass condenser, .02 mf., 200 volts, No. 2817.

26 becomes Frost potentiometer, 0-10,000 ohms, No. 6188.

18 becomes 500-ohm, .5-watt resistor (International Resistance Co.).

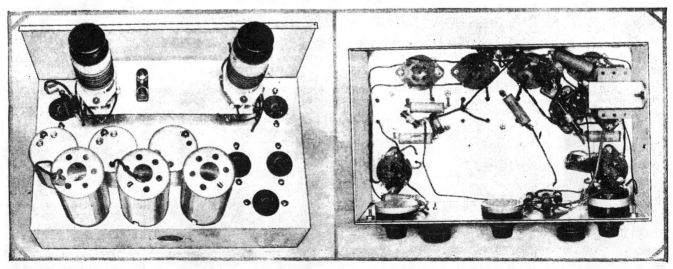

Rear view of the new Denton Portable Superhet.   A view of the superhet "looking up from under."

22 becomes 500-ohm, .5-watt resistor (International Resistance Co.).

39A is a 2,500-ohm, 1-watt resistor (International Resistance Co.).

40A is a 700-ohm, 1-watt resistor (International Resistance Co.).

The speaker should be wound with a winding suitable for operation with the 38 type power pentodes connected in push-pull. If such a speaker cannot be obtained, then an output transformer should be used to match the output of the tubes to the speaker to be used. It is important to use the proper output matching transformer, if the maximum efficiency is to be obtained from the set. Proper matching will permit the builder to obtain the same quality of reproduction which is to be had with the parts as specified.

It is amazing that 1.2 watt connected to a good speaker in a proper manner should give such satisfactory reproduction. For example, signals from a sta-

If the reader cannot build the chassis, due to lack of equipment, the author will be glad to advise where the completed chassis can be obtained.

Mount the tuning condensers on the front panel. After the condensers are mounted the small coil platform should be bolted down on the condensers. Mount the coil sockets; that is all the work that should be done on the front panel at this time.

The first detector socket should be mounted on the right-hand front of the chassis. Then, running along the back, place the first, second intermediate frequency amplifier tubes, and the second detector. The oscillator tube is mounted on the left and near the front of the chassis, this socket being backed up to the rear by the sockets of the output pentodes and the first audio tube. The first audio tube is mounted in line with the second detector.

A six-prong wafer socket is mounted

der one of the wafer socket mounting bolts and solder the small end of the lug to one of the socket filament connectors. The A-plus connection is then run and it is the one wire common to all sockets.

Wire all grid leads. Run the grid leads in the shortest possible manner. Be sure and mount the intermediate frequency transformers so that the grid and plate leads furnished with these parts are run in the shortest possible manner to their respective socket terminals.

Wire in all of the grid returns. Included under this section there are the by-pass condensers and the isolating resistors. As these parts are self-mounted, care must be used in soldering. At this time complete all grid and grid return connections. This holds true for both the audio and the radio sections of the receiver.

Wire in all plate leads. This is the

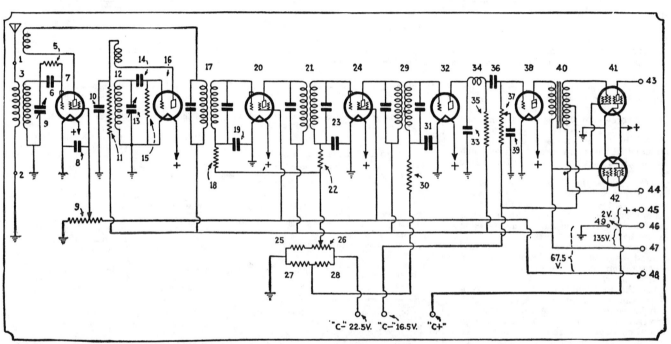

Wiring diagram of Mr. Denton's 8-tube portable superhet which utilizes 2-volt tubes.

tion over four thousand miles away come in loud enough to be heard all over the house. One would believe that the signals were from some broadcast receiver tuned to a station located within four or five hundred miles.

Short-wave reception is erratic at best, compared with reception on the broadcast band, but the "kick" of good loud speaker reception from stations more than 5,000 miles away is worth while. Reception from stations on the conventional broadcast band are tuned in a manner that would give due credit to any good broadcast receiver. For complete broadcast band coverage one must use two sets of coils, as the size of the tuning condensers will not provide the correct capacity ratio to cover the entire band.

Construction

The first thing to do is to lay out the front panel and the chassis. This should be done in accordance with the drawings accompanying this article. Use *half-hard* sheet aluminum about 1/16 inch thick.

on the back of the chassis, so that connections can be made to the battery-speaker case through the six-wire cable. The "C" battery is carried in the space under the receiver chassis, thus insuring short leads and reducing the number of wires running in the cable.

Mount the push-pull transformer underneath the chassis as shown in the photographs. The rest of the small parts, such as the resistors and the filtering or isolating condensers, are held in place by the wiring.

Wiring Details

The proper wiring of any radio set is half the battle on the road to a real radio receiver. Use a clean iron and heat the joint being soldered well. *See that all surfaces are clean!*

A receiver which is to be used as a portable needs particular care when it comes to wiring. Every connection should be good mechanically before it is soldered.

Wire all filaments. This is easy in the two-volt model. Use soldering lugs un-

last thing to be done under the chassis and nothing more need be said on this point.

Solder in three pieces of stranded, insulated wire to be used as "C" battery connections. The "C" battery lies in the open space to the rear of the sensitivity control. Have these leads about six inches long for convenience sake. There is plenty of space to place a standard 22-volt "C" battery in this space.

This completes the wiring of the chassis and the front panel can be mounted in place. This panel is held in position by the three variable resistors and the two lower holding screws of the National dial.

Testing

Of course, the first thing to do is tune the intermediate frequency amplifier to 465 kc. The method by which this is done will depend upon the equipment on hand which can be used for this purpose.

DO NOT USE A METAL SCREWDRIVER FOR TUNING OR ADJUSTING I.F. TRANSFORMERS WHEN THE BATTERIES ARE CONNECTED!

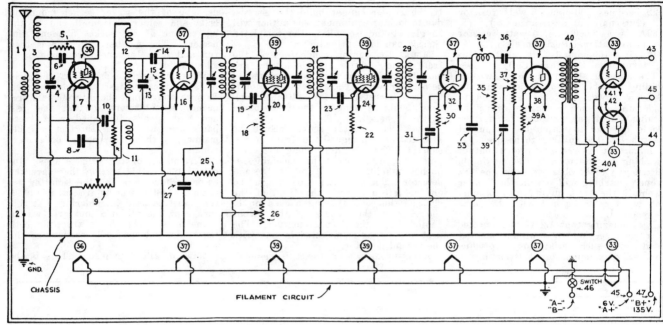

Six-volt automobile tube wiring diagram for Mr. Denton's 8-Tube Portable "All-Wave" Superheterodyne receiver.

The above statement is very important! The use of the metal screwdriver must be avoided; when one is adjusting the set screws on the tops of the shielded cans and the screwdriver slips, bang goes all of the tubes, unless a small fuse is included in the plate circuit. Fuses such as the Little fuse should be use for safety any way.

After the set has been adjusted, check the battery voltages, and if the filament voltage is too high on the tubes, place a small resistor or rheostat in series with the filament line and carry this resistor in the battery-speaker case.

Connect the set to an antenna and ground, tune in a signal and adjust the regeneration control below the point of *spilling over*.

Set the tone control for the required pitch and see if the action of the sensitivity control is smooth over the entire range. If this control is not smooth in action, then vary the "C" bias voltage at the battery to some lower value, although experience proves that the best action will be obtained from the 22.5-volt connection.

The wavelength ranges of the five sets of coils and the number of turns required are as follows:

14 TO 30 METERS

Detector

Primary...... 4 turns No. 30 D.S.C.
Secondary.... 7 turns No. 20 Enam.
Regeneration.. 5 turns No. 30 D.S.C.

Oscillator

Plate........ 4 turns No. 30 D.S.C.
Secondary.... 6 turns No. 20 Enam.
Coupling..... 1 turn No. 20 Enam.

Detector secondary is space wound 6 turns to the inch with a winding length of 1⅛ inches scant.

Oscillator secondary is space wound six turns to the inch with a winding length of 1 inch.

28 TO 60 METERS—Detector

Primary...... 4 turns No. 30 D.S.C.
Secondary.... 14 turns No. 20 Enam.
Regeneration.. 6 turns No. 30 D.S.C.

Oscillator

Plate........ 4 turns No. 30 D.S.C.
Secondary.... 12 turns No. 20 Enam.
Coupling..... 2 turns No. 20 Enam.

Detector secondary is space wound 12 turns to the inch with a winding length of 1⅛ inches scant.

The oscillator secondary is space wound 12 turns to the inch with a winding length of 1 inch.

55 TO 125 METERS—Detector

Primary...... 5 turns No. 30 D.S.C.
Secondary.... 33 turns No. 20 Enam.
Regeneration.. 16 turns No. 30 D.S.C.

Oscillator

Plate........ 5 turns No. 30 D.S.C.
Secondary.... 24 turns No. 20 Enam.
Coupling..... 3 turns No. 20 Enam.

The detector secondary is space wound 24 turns to the inch with a winding length of 1⅜ inches.

The oscillator secondary is space wound 24 turns to the inch; length of 1 inch.

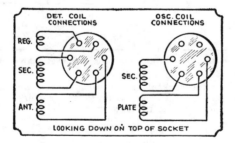

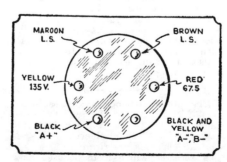

Diagram above shows the connections of detector and oscillator coils to 6-pin forms, and lower diagram shows the line-up of battery cable plug terminals.

120 TO 300 METERS—Detector

Primary...... 6 turns No. 30 D.S.C.
Secondary.... 78 turns No. 30 D.S.C.
Regeneration.. 30 turns No. 30 D.S.C.

Oscillator

Plate........ 14 turns No. 30 D.S.C.
Secondary.... 44 turns No. 28 D.S.C.
Coupling..... 5 turns No. 30 D.S.C.

The detector secondary is space wound 56 turns to the inch with a winding length of 1⅜ inches.

The oscillator secondary is space wound 56 turns to the inch; length of over ¾ inch.

240 TO 550 METERS—Detector

Primary...... 10 turns No. 30 D.S.C.
Secondary....114 turns No. 30 D.S.C.
Regeneration.. 48 turns No. 30 D.S.C.

Oscillator

Plate........ 30 turns No. 30 D.S.C.
Secondary.... 70 turns No. 28 D.S.C.
Coupling..... 8 turns No. 30 D.S.C.

Parts List

1, 2—Alden antenna-ground assembly.
3, 12—Hammarlund S6 isolantite sockets.
4, 13—National 150-mmf. tuning condensers ST-150.
2 National type "B" dials VB-D.
5—International resistor, 1 meg., 1 watt.
6—Illini mica condenser, .000125-mf.
7, 16, 20, 24, 32, 35—4-prong Eby sockets.
8, 10, 19, 23, 31, 39—Acratest by-pass condensers No. 2817, capacity .02-mf.
9—Frost 250,000-ohm potentiometer No. 6189.
11—Micamold resistor, 30,000 ohms, 1 watt.
15—International resistor, .1 meg., .5 watt.
17, 21, 29—Acratest 465 kc. transformers.
18, 22, 30—Acratest resistors, 50,000 ohms, .5 watt.
25—Acratest resistor, 15,000 ohms, .5 watt.
26—Frost potentiometer, 100,000 ohms, No. 6188.
27—Acratest resistor, 30,000 ohms, .5 watt.
28—Acratest resistor, 60,000 ohms, .5 watt.
33—Micamold condenser, .001-mf.
34—Acratest I.F. choke No. 2871.
35—Acratest resistor, 150,000 ohms, 1 watt.
36—Acratest coupling condenser, .075-mf.
37—Frost potentiometer, 500,000 ohms.
40—Acratest push-pull input transformer No. 5834.
41, 42—Eby 5-prong sockets.
43, 44, 45, 46, 47, 48—Pins of Eby 6-prong socket.
49—Power switch mounted on (37).
Special drilled and folded panel and chassis, Blan, the Radio Man.
3 Hammarlund tube shields, type TS.

PARTS LIST FOR BATTERY AND SPEAKER CASE

1 "Best" 9-inch magnetic speaker, with special winding for push-pull pentodes, 7,000 ohms per side.
3 45-volt "B" batteries.
2 No. 6 dry cells.
1 Eby 6-prong socket.
Connecting wire, etc.
1 small case, 8.5 x 12.5 x 10.5 inches.
2 Alden 106 plugs.
3-foot 6-wire cable.

The coils are tuned by individual tuning condensers so that the maximum resonance peaks can be easily tuned in. Individual control of the oscillator and detector tuning dials is very handy in tuning in weak or fading signals. Dial calibration can be accurately accomplished when two dials are used, doing away with the error in logging caused by the use of a large antenna compensating condenser.

Little need be said in regards to the wiring, as the various leads are run in a straightforward manner. There is one little kink which should be watched closely. It will be seen that there is a network of resistors shunted across the "C" battery and whether the set

is in operation or not the "C" battery will discharge through the network unless the circuit is opened when the receiver is not in operation. The drain through these resistors is low enough so that it equals the drain of the receiver on the "B" batteries. Thus when a set of "B" batteries is worn down to the point of replacement it is time to replace the "C" battery. This has the effect of properly proportioning the voltage applied to the grid of the second detector to its plate voltage.

If only 135 volts of "B" is used there is plenty of room to carry all of the coils in the detector and oscillator sockets, as well as the battery-speaker cable.

If it is desired—and the writer does it—coil up about 75 feet of enamel covered wire with a couple of insulators for a semi-portable antenna, with a 20-foot length of wire with a clip to ground to some convenient water pipe. All of these things can be carried without any trouble in the battery compartment. Thus the set is complete for use anywhere.

One may gather the impression from the above that a long antenna is necessary for this set. Short-wave stations in Venezuela were received on an antenna 30 feet long. This was at loud speaker volume and one did not have to put his head in the speaker to hear it.

It will be found that as the padding condenser is rotated, the background noise will sharply increase at two points. At these points the oscillator is aligned with the detector, the lower capacity setting of the padding condenser being the correct adjustment, since the oscillator is designed to work on the high frequency side of the detector. In other words, while there are two points where the oscillator and detector may be aligned (when the oscillator is tuned either above or below the detector by the amount of the intermediate frequency), the correct setting of the oscillator is equal to the detector frequency plus the intermediate frequency.

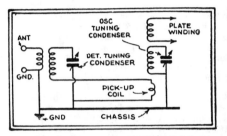

Simplified diagram showing "pick-up coil" linking detector and oscillator.

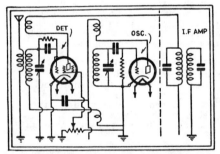

Circuit of detector and oscillator for Denton super-het showing coupling coil.

COIL NUMBER	1		2		3		4	
	DET.	OSC.	DET.	OSC.	DET.	OSC.	DET.	OSC.
TOTAL TURNS	3⅛	2¾	2⅜	2⅛	1⅞	1⅝	1¼	1⅛
FRACTIONAL TURN BETWEEN TOP AND BOTTOM OF COIL	⅔	⅔	⅜	⅜	⅓	⅓	¼	¼
RANGE MC	40 TO 46		46 TO 53½		53½ TO 62½		62½ TO 75	

TUNING CONDENSER, CAP. (DET. AND OSC.) = 18 MMF EACH. (TYPE, NEW 270° STRAIGHT FREQUENCY LINE

I.F. COIL DATA :-
FORM 1¼" DIA 2½" LONG
PRI. = 50 TURNS NO. 32 D.S.
SEC. = 100 TURNS NO. 28 ENAM
(SAME DIRECTION)
64 TURNS PER INCH
TUNED WITH A 4-70 MMF COMPRESSION TYPE I.F. TUNING CONDENSER

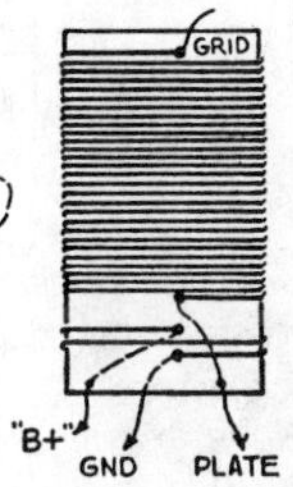

I.F. TRAP CCT TUNED TO 1,550 KC.
COIL L2 = 100T 10-38 LITZ WIRE ON A ⅝" FORM.
CAP C2 = 4-70 MMF

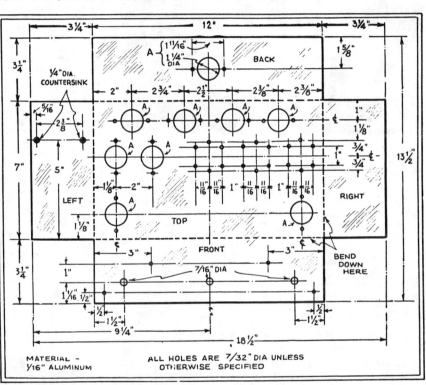

Subpanel layout for the 8-tube super.

MATERIAL - 1/16" ALUMINUM
ALL HOLES ARE 7/32" DIA UNLESS OTHERWISE SPECIFIED

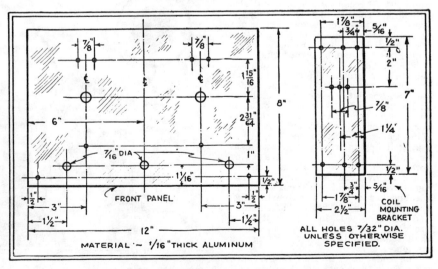

MATERIAL - 1/16" THICK ALUMINUM
ALL HOLES 7/32" DIA. UNLESS OTHERWISE SPECIFIED.

Layout of aluminum front panel for the Denton 8-Tube Super.

This 3-Tube "Super"
; Has "It"

Very "professional looking" indeed is this easy-to-build, 3-tube Superhet which brings in the "foreign" and other DX short-wave stations with marvelous ease—and on a loud speaker!

Loud-speaker reception of "foreign" short-wave stations is a regular performance for this 3-tube Superhet, has an extra tube which acts as a C. W. oscillator for code reception; one of the new pentagrid-converter tubes acts as first detector and oscillator for "phone" reception.

● THE new type tubes now on the market have paved the way for many changes in our short-wave receivers. Among these is the much discussed *Pentagrid Converter*, the type 2A7 or 6A7, which works very nicely on the "High Frequency" bands.

The receiver presented in this article employs one of these tubes as the high frequency *first detector and local oscillator*, and this tube is partly responsible for the minimum number of tubes used to make up a very simple and efficient short-wave superheterodyne receiver, which can be built by the average short wave "Fan" at very nominal cost. The receiver described here, provides all that any one could want for general short-wave reception, including various amateur activities. Although no provision is made for band-spread, and would be necessary should one wish to use it for amateur work.

Separate Beat Oscillator for "Code"

This receiver is really a three tube set, so far as ordinary reception is concerned; the fourth tube is provided to allow "CW" reception on code and also provide an easy method of locating the various short-wave broadcast (phone) stations; after the modulated signal has been located the beat oscillator is no longer needed and is turned off.

The line-up of tubes is as follows: the 2A7, as stated before, is used as the frequency converter, a type 58 for the intermediate frequency amplifier, and a type 2A5 as the second detector tube. The type 2A5 used as the second detector gives sufficient audio amplification to operate a speaker, either magnetic or dynamic, at regular speaker volume. That is, any of the major foreign short-wave stations can be heard all over the house and one does not have to stand with one's ear in the speaker either. The fourth tube, the type 57, is the *beat oscillator* tube and plays no part in the reception of broadcast (phone) reception, other than to aid in tuning or locating the station.

The coils used in this receiver are very easy to construct; they are all close-wound and the two sets, that is,

the first detector and oscillator coils are identical in number of turns. The coils used in the set shown are wound on small isolantite forms. Complete coil data is given in the appended "Coil table."

In designing this set, size and simplicity were among the main considerations. The front panel is made of 1/16th inch aluminum and is 12 inches long by 7 inches high. The base or chassis is made of the same stock and is 8 inches wide by 12 inches in length and 1 inch deep. All bypass condensers and resistors are mounted in this space under the set. The layout of parts in this receiver cannot be changed very much without the necessary addition to the size of the chassis.

A drum type dial is used in order that the two tuning condensers can be mounted on either side, allowing short leads to the two coils and the frequency-converter tube. The 2A7 converter tube is mounted directly behind the drum dial. If the 2A7 were put in any other position, the length of the connecting leads to one set of coils and condenser would be much too long. The layout used in a set using separate tubes for the oscillator and first detector cannot be used with the new tube with any great success, if simplicity of wiring and short leads are taken into consideration. Many other layouts were tried with very little success—in fact it made a very awkward looking job.

Description of the Circuit

In describing the circuit we will start with the high frequency end of the receiver first. Many different circuits were tried in the high frequency unit and that shown in the diagram proved to be the most stable in oscillation. The circuit as can be seen in the diagram, permits the current from the plate and the anode-grid to return through

This top view of George Shuart's latest set—the 3-tube Superhet, shows the excellent layout of the parts.

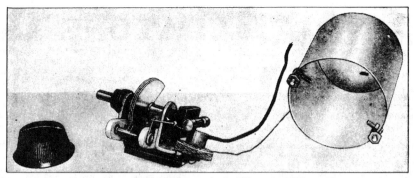

By GEORGE W. SHUART
(W2AMN-W2CBC)

The parts used, together with shield can, in building the CW oscillator for "code" reception.

the oscillator plate coil. In this circuit it is surprising how uniform the output of the oscillator circuit is; with only the few tickler turns shown in the coil table, there is no change in oscillator output over the entire tuning range of the oscillator grid coil, and it shows no tendencies of going out of oscillation as the capacity of the grid circuit is either increased or decreased. No increase in stability could be noticed when the oscillator grid circuit was changed to a high capacity with lower inductance. Even with the grid condenser turned to minimum capacity, there are no signs of instability. The values of resistances shown in the diagram of the converter circuit are those that work best with 250 volts applied to the plate. If a lower voltage is used, it is suggested that a change in the values be made, if the full gain of the 2A7 is to be had. This is mentioned because there is very little pickup in the first detector and the "over-all" gain in the converter circuit is very small.

When using *single-control* in a superheterodyne some sort of provision is necessary to get the two circuits to *track*. With the coil data given and the use of a 1000 MMF. condenser in series with the oscillator grid-tuning condenser, the two circuits were made to *track* very evenly, the 35 mmf. variable condenser used in the detector circuit is for compensating for small changes which may occur in the antenna circuit. The tuning condensers used in this receiver are of the 270 degree type. These condensers aid in tuning considerably because of the added 90 degrees in the tuning range giving less cramped tuning than the

regular 180 type, a very worthwhile feature indeed.

Air-tuned I. F. Transformers Used

Referring to the diagram it can be seen that the intermediate frequency amplifier stage is of conventional style using a pentode tube. The I.F. transformers are the new style having *air-dielectric* tuning condensers. These transformers represent a decided improvement in that they can be adjusted and will hold their calibration indefinitely. If the builder wishes to build his own IF. transformers this can be done quite easily. The coils for the primary and secondary should have an inductance of from 1 to 1.3 millihenries and should be tuned with a 100 mmf. midget variable condensers. The above values are for frequencies between 550 and 465 kc. The regular universal-wound inductances are used and should not be coupled too close or the selectivity of the transformer will be destroyed; about one and one quarter inches is a good degree of coupling. The volume is controlled by a 20,000 ohm variable resistor inserted in series with the 300 ohm cathode bias resistor of the type 58 IF. amplifier tube. The .01 cathode bypass condenser is large enough to render the volume control quiet in operation. The set is equipped with a voltage

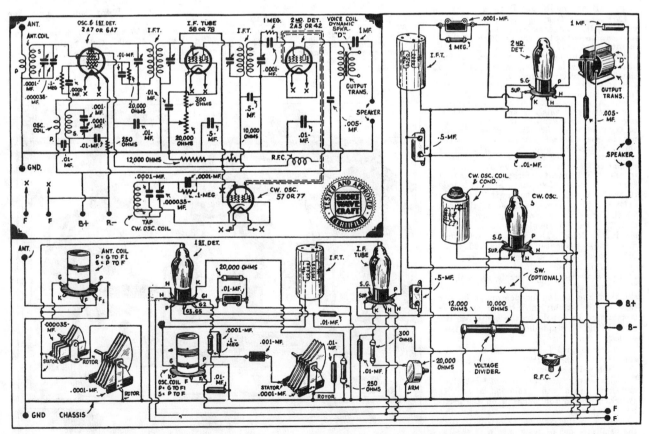

Schematic and also picture wiring diagrams are given above for the 3-tube superheterodyne short-wave receiver, the fourth tube being used as an extra oscillator for CW or "code" reception.

divider of 22,000 ohms, tapped at 12,000 ohms for the various screen voltages. The screen voltages are taken off the divider at a point 12,000 ohms from the high potential side of the resistor.

Grid-leak Detection

Grid-leak detection is used in the 2A5 second detector tube because this method gave by far the greater audio output, a one megohm leak seems to be about right and can be of the one-half watt variety.

Modulation of the second detector for CW reception is obtained by hooking the detector screen-grid in parallel with the oscillator plate and fed through an R.F. choke from the 100 volt tap on the voltage divider. In order to get full audio output from the 2A5 detector, it is necessary to bypass the screen with a .01 mf. condenser. To prevent the output of the oscillator from getting into the radio frequency stage, the connection from the plate of the oscillator to the screen of the detector is run through flexible shielding; this shield should be grounded to the base of the receiver at several points to hold it firm. Should this shield be left loose and rub against the chassis, it would cause considerable scratching noise in the speaker.

The power from the output tube is fed to the speaker by an output transformer or a choke and condenser arrangement, according to the type of speaker used. The output transformer is an affair used to work on a pair of pentodes and has a 12 ohm secondary for use with a dynamic speaker, the center tap on the primary is unused; the plate is connected to one side and the B plus is connected to the other. If it is desired to use a magnetic speaker this can be done by connecting one side of the speaker through a 1 mf. condenser to the plate of the tube, and the other speaker lead to the chassis; this keeps the high plate current of the tube out of the speaker or phones, should one wish to use them.

Beat Oscillator Details

Last but not least, is the beat oscillator, used mostly for code reception and this unit can be left out of the set if the builder is interested only in short wave broadcast (phone) reception. The *tuning unit* of the beat oscillator is a home-made affair and can be constructed very easily. The entire unit is inclosed in a shield-can measuring 2½ inches in diameter and 3 inches high. A general idea of the construction and assembly of this can be seen by glancing at the photograph. It is not advised to try and wind the coil because it could not be made small enough to fit in the shield with all the other parts. The easiest way is to "buy, beg or borrow" a *universal-wound* coil of the same value as that used in the "IF." transformers; unwind about twenty or twenty-five turns and tie a loop in the wire and start to rewind, or, wind back the turns removed, keeping the loop which forms the cathode tap "out in the clear." The tap should be about five or six inches long. This coil, the 100 mmf. fixed tank condenser and the 100 mmf. grid condenser and grid-leak are all mounted on a Hammarlund 35 mmf. midget tuning condenser. The whole assembly is then mounted in the shield can, with the shaft of the tuning condenser projecting through the shield; in fact the condenser is mounted in a hole in the center of the can, being a single-hole mounting condenser.

Bring the grid lead out the top of the can on the rim, so that the operation of the knob will not interfere with or disturb it. This lead by the way should be shielded and the shield can grounded also. No switch is shown for shutting off the oscillator, but can be added if desired. The operation in the receiver shown is to turn the oscillator out of resonance with the IF. frequency when the oscillator is not desired, and back again when needed.

Hints On Operating Set

Operation and adjustment of this little set is very simple and requires very little experience. If the following explanation is followed carefully no difficulty should be had in getting the set to "perk" right off.

Check all wiring before any voltages are applied, to make certain no error has been made in conections and that all have been soldered firmly. Then apply filament and "B" voltages and connect a pair of phones to the out-put terminals; all adjustments should be made using phones. Set the condenser across the plate coil of the first IF. transformer at about half capacity. Then start up some "noise-producing" electrical instrument such as a buzzer or vacuum cleaner, near the set and proceed to adjust the other IF. tuning condensers for maximum volume. When this stage has been reached, tune the *beat oscillator* condenser until a slight hissing sound is heard in the phones, this indicates that the oscillator is in tune with the IF. frequency; then proceed to locate some kind of short-wave station. Of course the *noise-generator* has to be shut off at this point. When a station is once located then adjust the IF. *trimmers* until full efficiency or loudest signal has been reached. During this adjustment however, the condenser in the plate circuit of the first IF. transformer should not be touched unless the frequency of the IF. unit is to be changed. This is mentioned because the adjustment of this primary tuning condenser has a decided effect on the high frequency oscillator frequency, and will most certainly result in maladjustment. Therefore do all adjusting with the other three IF. condensers unless the intermediate *frequency* is to be changed.

Coil Winding Table

Make two of the following

COIL	Tickler or antenna	GRID
No. 1—	4 turns No. 34 wire	5 turns No. 26 wire
No. 2—	5 turns No. 34 wire	10 turns No. 26 wire
No. 3—	8 turns No. 34 wire	24 turns No. 26 wire
No. 4—	10 turns No. 34 wire	45 turns No. 26 wire

All coils close-wound, Diameter of form 1¼ inch. The above coils cover all of the popular S. W.-broadcast and Amateur bands.

Any standard commercial SW coils will work if designed for 100 mmf. condensers. Otherwise change tuning condensers to match coils that are designed to work with 140 mmf. condensers.

Spacing between grid coils and tickler or antenna coil is ¼ inch.

Parts List—Shuart 3-Tube Superhet

1—8x12x1 Inch Chassis 1/16 in. Blan.
1—7x12 Inch Panel 1/16 in. Blan.
1—drum dial—National.
1—100 mmf. Variable Condenser, Clockwise, National—270°
1—100 mmf. Variable Condenser, Counter Clockwise, National—270°
2—35 mmf. Variable Condensers, Hammarlund.
8—5 Prong coil forms, small Hammarlund.
2—National Isolantite sockets (5 prong).
1—National Isolantite sockets (7 prong).
2—National "Airtuned" IF. Transformers.
3—Tube shields, Hammarlund.
3—6 prong tube sockets, wafer, Eby.
2—.5 MF. Bypass condensers.
1—.5 MF. Bypass condensers.
7—.01 MF. Bypass condensers.
3—.0001 MF. Mica grid cond.
1—22,000 ohm voltage divider, tapped at 12,000.
2—100,000 resistors—1 watt, Lynch (International).
1—20,000 resistors—1 watt, Lynch, (International).
1—300 ohm resistors— 1 watt, Lynch (International).
1—250 ohm resistors—1 watt, Lynch (International).
1—1 meg. resistor—1 watt, Lynch (International).
1—20,000 Volume control, Acratest.
1—Antenna-Ground binding post strip, Eby.
1—Speaker binding post strip, Eby.
1—4 wire cable.
For "Beat Oscillator tuning unit, see text."
1—2A7 or 6A7 tube, Gold Seal.
1—2A5 or 42 tube, Gold Seal.
1—58 or 78 tube, Gold Seal.
1—57 or 77 tube, Gold Seal.

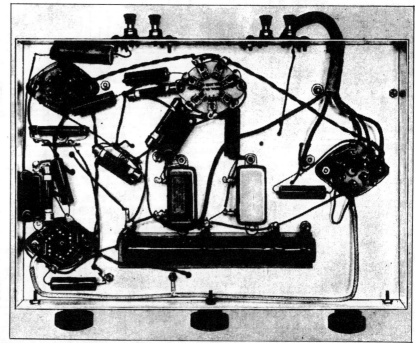

Bottom View of Mr. Shuart's 3-Tube Superhet.

SHORT WAVE PHYSICS

The directional antenna used by Dr. Karl Jansky in detecting galactic radio waves is mounted on wheels and rotated by a synchronous motor, so that it makes one complete rotation every twenty minutes.

Short-Wave Signals from • • Interstellar Space

● MYSTERIOUS radio waves which appear to come from the direction of the center of the Milky Way have been discovered by Dr. Karl G. Jansky of the Bell Telephone Laboratories and were described by him in a paper delivered before the *International Scientific Radio Union* in Washington, D. C., on April 27, 1933. They are short waves (14.6 meters) at a frequency of about twenty million cycles per second, and

Dr. Karl Jansky at one end of the antenna used for receiving galactic radio waves.

were discovered in the course of radio studies carried on as a regular part of telephone research. The intensity of these waves is very low, so that delicate apparatus is required for their detection.

An investigation of their nature and source has been carried on for some time, and a preliminary report was published in December of last year.* Unlike most forms of radio disturbances, these newly found waves do not appear to be due to any terrestrial phenomena, but rather to come from some point far off in space—probably

*Proceedings of the Institute of Radio Engineers.

far beyond our solar system. By a series of investigations carried on over a considerable period, the direction from which these waves arrive has been determined. Measurements of the horizontal component of the waves were taken on several days of each month for an entire year, and by an analysis of these readings at the end of the year, their direction of arrival was disclosed.

Directions such as northeast or southwest have no application, of course, except to things on the earth. Objects in space surrounding us are located by their right ascension, measured in hours to the east of the vernal equinox—the position in the sky in which the sun appears at the beginning of spring—and by their declination in degrees above or below the Equator. The coordinates determined for the

newly discovered radio waves are a right ascension of 18 hours and a declination of about 20°. The right ascension has been determined quite accurately but there is still some uncertainty about the declination.

The position indicated by these coordinates is very near to the point where the plane in which the earth revolves around the sun, crosses the center of the milky way, and also to that point toward which the solar system is moving with respect to the other stars. Further verification of this direction is required, but the discovery, like that of cosmic rays, raises many cosmological questions of extreme interest.

Took 40,000 Light-Years to Reach Earth

Electrical energy in the form of radio waves, which scientists believe come from a point so remote in space that it requires between 30,000 and 40,000 light-years for the waves to reach the earth, was heard by radio listeners throughout the United States in a recent broadcast. It was the first such experiment ever carried out. The sound, generated by the waves arriving at a supersensitive receiving set operated by Dr. Karl G. Jansky, research engineer of the Bell Telephone Laboratories' experimental station at Holmdel, N. J., sounded like steam escaping

The automatic recorder which made a continuous ink record on a moving paper strip, like that shown just opposite, of the galactic short-wave signals received by Dr. Jansky's special revolving antenna and ultra-sensitive S.W. receiver.

American broadcast listeners were recently entertained by short-wave signals originating far out in interstellar space. The signals, which have also been recorded graphically on a paper chart, were picked up on an ultra-sensitive short-wave receiver on a wavelength of approximately 14.6 meters, or a frequency 20,550 kilocycles at Holmdel, N. J. A special antenna rotated by motors was used to pick up the signals from space which seemed to emanate from the region of the constellation of Sagittarius (the Archer).

from a radiator. Wires carried the sound from the New Jersey receiving station to the WJZ coast-to-coast network.

Dr. Jansky, speaking of his work carried on secretly for more than a year, said an immense amount of electrical power would be necessary to transmit waves over such distances. Some of the stars, however, have been found to radiate as much as 500 sextillion horsepower, he added.

Signals Emanate from Region of Sagittarius

Dr. Jansky was introduced by O. H. Caldwell, former Federal Radio Commissioner, who explained how the research engineer, using an antenna rotated by motors, determined the point in the sky from which the waves apparently arrive through space. The rotation of the earth on its axis causes the waves to strike the earth at different angles, depending upon the time of day and the season of the year. By carefully checking the gathered data it was discovered that the waves were arriving from the region of the constellation of Sagittarius (The Archer).

Mr. Caldwell, in introducing, Dr. Jansky, said:

"These radio impulses from the stars were discovered by Karl G. Jansky of the Bell Telephone Laboratories while he was studying the faint static hiss that can be heard on a sensitive radio set when its amplification is turned up so as to get the faintest possible signal. At Holmdel, N. J., where the Bell Laboratories have a 400-acre tract in the woods, Mr. Jansky has a tremendously sensitive receiving set, with a long antenna system mounted on wheels so it can be turned in any direction.

"Using this elaborate, sensitive equipment to listen to the faint static hiss that is always present in such a sensitive receiver even on the best days or nights, Mr. Jansky noticed that the hiss was always a little stronger coming from one direction that from other directions, and also that this directional maximum was continually rotating around the horizon, approximately once every day.

Accurate Records Pile Up Evidence

"At first he thought naturally that this maximum of his had something to do with the sun's position and with the earth's daily rotation. But when he began to keep accurate records of the shifting of position of this stronger hiss, which is recorded by automatic measuring instruments, Mr. Jansky noticed that each day its position was just a *little bit ahead* of the position at the same hour the day before. That is, in a week there would be a difference of half an hour in the position of maximum hiss. In a month a difference of two hours. So apparently this strongest hiss was not following the sun's position at all, but was following something which gained on the sun about 4 minutes a day, or two hours a month, or a whole rotation of the heavens in a year. Mr. Jansky said nothing in public but continued to keep his records carefully over a whole year, and at the end of that time, the maximum hiss was back again, once more coming from exactly the same direction as it did on the same date 12 months before.

"Evidently then, the radio waves or hiss Mr. Jansky was picking up was coming from some definite spot in the sky of stars, entirely independent of the sun's changing position among the stars. The instruments were detecting a stream of radio coming from some fixed point in the universe of stars, outside of the earth, the sun or the solar system—*radio impulses from the stars themselves!*"

Simplified map of the Southern sky for July 21 (10 P. M.) showing point from which interstellar short-wave signals may be expected.

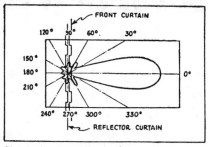

Chart above shows direction of maximum reception with respect to plane of antenna array used by Dr. Jansky.

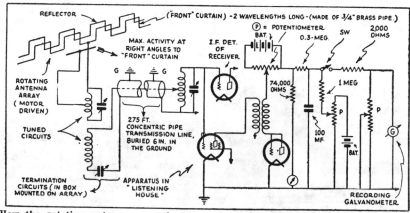

How the rotating antenna array is connected through tuned termination circuits, the short-wave signal currents then being led through a shielded cable buried in the ground to the recording amplifiers and the ink recorder itself.

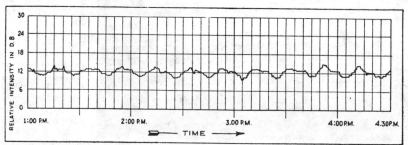

The mute evidence of the reception of short-wave signals from space is presented above, this chart showing just a small section of one of the long records made during Dr. Jansky's tests.

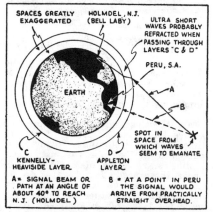

Angles at which ultra short-waves reach the earth from point in space.

How ARE SHORT WAVES PROPAGATED?*

By F. BODIGHEIMER

The author gives high credit to short wave amateurs who have contributed greatly to the data here presented on short wave phenomena. The question of whether short waves penetrate the Heaviside layer, thus making possible radio communication with other planets, is here considered.

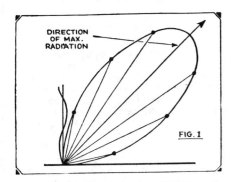

This diagram shows the direction of maximum radiation from a vertical antenna excited by harmonics.

BEFORE the extraordinary range of short waves was discovered by amateurs, it was held as *incontrovertible* that the electric waves followed the surface of the earth, and that the strength of the field decreased in proportion to the distance. It was assumed as simply natural without its causing any more surprise and attention, that for communication at a very great distance only long waves were serviceable, with the expenditure of correspondingly great energies. Operation was carried on with wavelengths of 2 to 3 kilometers (that is, with frequencies from 150,000 down to 100,000 cycles) and with energies of many hundred kilowatts.

The shorter the wave, the less suitable it seemed for distant communication. Waves of a few thousand meters were used in continental communication, but not in transoceanic. Waves of about 1,000 meters and less were intended for internal communication and for neighboring states. Finally came the waves of 600 and 300 meters for communication of ships with one another and with coast stations; that is, mostly for very short distances.

Waves Below 300 Meters Were Considered Useless

Waves of less than 300 meters were considered entirely useless, because they actually proved very unreliable in communication at short distances; for which at any rate, they appeared in question. It did not even cause thought that, during the war, weak German ship and field stations in Turkey were occasionally heard on the 300-meter wave by crystal receivers located in Germany. Likewise, the fact that the ships with their resounding transmitters disturbed or drowned out the first 300-meter radio

* The following is a section from the book "Radioamateurstation für kurze Wellen," by F. Bödigheimer. This should be of great interest to all short wave amateurs.

stations at night from "impossibly" great distances, received no consideration. The fact was established: waves of 300 meters and less are absorbed by the influence of the sun's rays in their course along the surface of the earth. That they were more serviceable at night and, under certain circumstances, audible at very great distances, was attributed to the absence of the solar radiation.

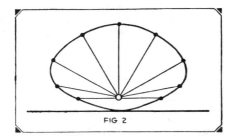

The radiation from a horizontal antenna is evenly distributed over almost 180 degrees, as shown.

Amateurs Pioneers in Short Wave Work

Now, against considerable resistance, these views have fundamentally changed. The pioneers of the new conception were the amateurs, who even today have at their disposal the greatest experience and in part stand preëminent in the clarifying of still doubtful problems. Below is a brief outline of the now fa-

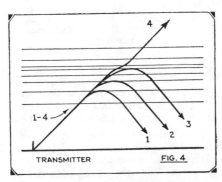

Radiation at various frequencies, 1 to 4, some of which are partly bent downward, some passing through the Heaviside layer with parallel deflection.

miliar laws for short waves, which touch on the new problems of propagation foremost in interest. The general laws here given rest on the personal investigations of the writer in the years 1926 and 1927; but, with reference to their general physical basis, on previously known facts or theories. The special data regarding the influence of the weather are based on independent researches performed by Dr. Karl Stoye and the writer, who have had occasional interchanges of ideas. These investigations are still going on.

(1) The maximum radiation from a vertical antenna, especially if it is stimulated by harmonics, projects obliquely upward at an angle. (See Fig. 1.)

(2) A horizontal antenna radiates evenly, over an angle of nearly 180 degrees (Fig. 2),

(3) At a height of 50 to 100 kilometers (30 to 60 miles) above the surface of the earth, there is, according to Heaviside's theory, a stratum of atmosphere which, because of the sunlight and the electron radiation of the sun, is distinguished by a very large number of free negative electrons per unit of space and, because of the slight atmospheric density, by a very great number of heavy ions or positive particles. In view of the great open stretch, there takes place, by *impact ionization*, a further increase in the number

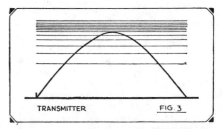

The space radiation is bent downward; more exactly it is refracted and totally reflected (at certain frequencies).

of free electrons. The electron density gradually increases in a vertical direction and again decreases. The dielectric constant of the Heaviside layer is smallest where, in consequence of very great electron density, the electrical conductivity of the layer is greatest. This gradual change in the dielectric constant effects a refraction similar to astronomical refraction (also analogous to the formation of the "Fata Morgana" and mirages) and finally total reflection of the electromagnetic radiation (see Fig. 3). The space radiation is thus bent downward.

Ultra Short Waves Pierce Heaviside Layer

(4) The refraction is, as in the case of light, dependent on the frequency. High

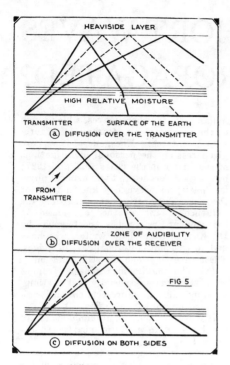

Assumed diffusion of energy by strata of high relative moisture; normal course of radiation in dotted lines. For the sake of simplicity, a straight course of radiation and reflection was drawn, instead of indicating refraction.

frequencies (short waves) are less strongly refracted than low frequencies (long waves). A pencil of electric waves of different frequency, increasing from I-IV (cf. white light) would behave as in Fig. 4. (This is similar to the production of rainbow colors in the refraction of white light.) The range is smaller in the case of long waves than in the case of short ones. Very high frequencies (ultra-short waves) are no longer refracted, *but pass, with a parallel deflection, through the Heaviside layer;* since, in consequence of the slight refraction, the limiting angle for total reflection is not reached. Rays striking the Heaviside layer perpendicularly pass through it unrefracted.

(5) The energy of ground radiation, whose proportion of the total radiation is great (especially with horizontal antennas) is quickly absorbed in consequence of the ion density being high near the ground, and because of other sources of loss. On the contrary, the space radiation moves along in the Heaviside layer almost without loss, because of the slight ionic density.

(6) The absorption in consequence of the greater ionic density near the ground is less, with high frequencies, than with the lower ones. The fact that the ground wave is nevertheless (as a rule) more quickly dissipated, with high frequencies, than with lower, is attributable to other sources of loss.

The Cause of "Dead Zones"

(7) Since the ground radiation is used up after a few miles, while the space

radiation descends again to the earth only after a greater distance, there results a *silent zone,* in which there is no reception or only weak signals are heard.

(8) **The height or make-up of the Heaviside layer, or perhaps both factors, changes with the time of day and of year and with the changing activity of the sun spots. Therefore these factors have a great influence on the propagation of the short waves.**

Best Frequency Varies With Seasons

With equal frequencies, the range is greater at night or in the winter than by day or in the summer; hence, for example, for these wavelengths:

20 meters by day in the summer: European communication,

by day in the winter: DX (distance) communication;

40 meters by night in the summer: still European communication,

by night in the winter: DX (distance) communication;

80 meters by day in the summer: almost useless,

by day in the winter: places very near at hand;

by night in the summer: European communication,

by night in the winter: also DX (distance) communication.

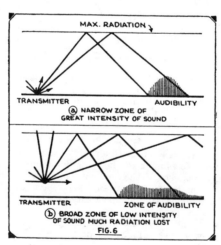

Vertical antenna (a) excited into harmonic oscillation and horizontal antenna (b) with its characteristic radiation.

(9) The shorter the wave (the higher the frequency), the better it is suited for communication by day and in the summer; but the less it is suited for communication at night and in the winter.

(10) Ultra-short waves are not deflected downward; with regard to their usefulness for communication, they behave almost like light waves. (*In so far as communication with other heavenly bodies might be considered, then ultra-short waves would be the most suitable.*) The limit between the ultra-short waves

and those still serviceable for "DX" (distance) is not sharp, but varies with the time of day and of year. It lies at about 10 meters, as calculation and practical experiments have shown. The present experiments with 10-meter waves therefore lead toward "DX" communication in summer and by day, which should be noted.

Condition of Atmosphere Affects Short Waves

(11) Considerable influence seems to be exerted according to investigations not yet completed, by the weather or, more correctly, the condition of the atmosphere at the edge of the *stratosphere.* In fact, there evidently is a considerable significance in the "moisture content" in the higher strata of air; shorter waves show themselves most sensitive to these influences. The influence of the weather is therefore stronger on 20-meter waves than on those of 40 or 80 meters length.

(12) Uniformly dry air over transmitter and receiver seems to be the best condition for good "DX" (distance) radiation (by day there is strong interference by increased absorption).

(13) Meteorological conditions, and probably also the Heaviside layer, are subject to marked changes (particularly the Heaviside layer) at twilight, and at times of disturbances in the earth's magnetism. The results are more or less rapid displacements of the zones and, therefore, changes in signal strength. This gives an explanation for "fading" which, according to the current explanation that it is caused by the difference in phase between *space* wave and *ground* wave, would be inexplicable in the case of short waves.

(14) At places in the middle of the zone of maximum sound intensity, the power of the transmitter received plays a small part. With favorable atmospheric conditions, one hears very slight energies (weak signals) with the sound intensity R9.

(15) The form of antenna, vertical or horizontal, is of distinct significance. From Fig. 6 it is evident that the horizontal antenna is more favorable for close communication (Europe); the vertical antenna, *excited on a harmonic,* is better for "DX" communication, though to be sure over a relatively narrow zone.

(16) From the viewpoint of short waves, it is also possible for us to look differently at long waves. Here too the *ground wave* is far from playing the part still assigned to it today. It does not reach far; with our chief German stations, in the autumn of 1930, not even 200 kilometers (125 miles).

Reception improvement in the local zone is a question of the *antenna,* likewise a question also of *frequency!* This effect should be studied carefully by those who are seeking the salvation of long-wave radio by utilizing tremendous transmitting powers.—*Funk Bastler.*

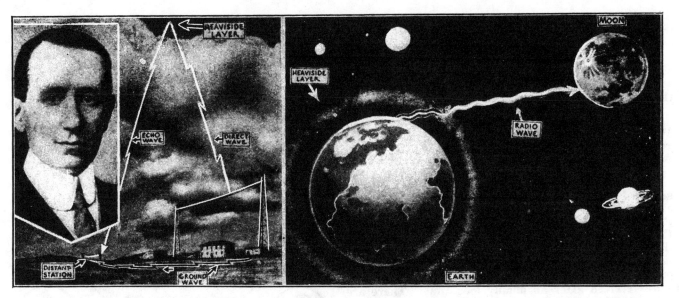

Guglielmo Marconi recently stated that he believes directed "beam" waves can and do penetrate the Heaviside layer, reaching out into space to distances as great as 48,000,000 miles. Picture at left shows "earth-bound" ground and echo waves, the latter reflected by the Heaviside layer.

Can We RADIO the Planets?

A New Use For Amateur Radio

IF someone had made the statement, forty years ago, that you could send a message clear around the world without the use of wires, even the greatest electrical experts of the day would have strongly doubted the sanity of the man making such an assertion.

We are getting rapidly over the wonderment created by radio and are becoming accustomed to the impossible.

Fig. 2—A "celestial" vacuum tube, in which the sun gives off the electronic stream in the direction of the arrows. The various planets may be taken respectively as plates and grids in a multi-element tube; while the atmospheres of the planets are analogous to the gas adhering to the plate and grid, as shown in Fig. 1.

In the February, 1927, issue of RADIO NEWS (of which I then was the publisher) appeared an illustrated article written by myself and entitled "Can We Radio the Planets?" which is reproduced at the end of the present article.

The 1927 article is chiefly interesting because it came in for a good deal of ridicule at that time; while today Marconi, in conjunction with many other scientists, has reached the conclusion that it will not be long now before it will indeed be possible to send radio signals to the heavenly bodies.

As I said in my former article, again I do not wish to have the meaning misconstrued when I speak of signaling the planets. There is no idea in my mind, at the present time, of communicating with imaginary inhabitants of the Moon, Mars or Venus. The purpose of sending signals is purely for scientific research which will, in due time, give us a much better understanding of radio waves. It is admitted today that we know practically nothing of what happens to the radio wave after it leaves the transmitting aerial and until it impinges on the receiving aerial.

The thoughts which I advance here, I believe to be new so far as radio amateurs are concerned, and I also believe that in due time something will come of them.

No doubt, the radio amateurs will remember that I am the one who in 1910 fought their battles in Washington; and an editorial which I published in the February, 1912, issue of MODERN ELECTRICS became the basis of the present radio law, whereby the radio amateurs

THE author of this article presents some new views on how radio amateurs, with modest equipment, can now actually experiment by sending signals to the Moon, which, reflecting the radio waves, should make it possible to have other amateurs receive them at another point on our planet.

Our front cover depicts this idea graphically. While revolutionary, the idea is by no means new, and recent scientific investigations tend to show that it may already have been accomplished.

Scientists have believed until recently, that the so-called Heaviside layer would serve as a barrier to any radio waves which might try to leave the earth. Recent research and experiments, however, have shown that certain short-wavelengths apparently penetrate the Heaviside layer and are reflected by some heavenly body, or possibly by streams or bands of electrified particles in space, for the simple reason that the signals were received after an unduly long time. There are other theories as to where the delayed signals may have spent their time, but the most plausible is, that they actually penetrated the Heaviside layer, sped outward into space where they struck some heavenly body, for example, and were reflected on a long journey back to earth, where a receiving instrument recorded them.

By HUGO GERNSBACK

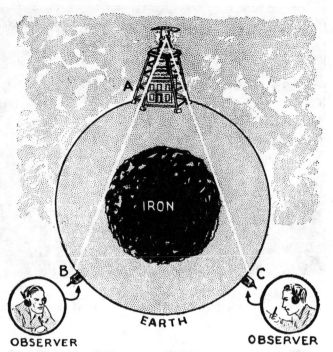

Fig. 4 shows a radio beam-transmitter "A" at some point on the globe, transmitting a beam to "B" or "C". As the angle of the beam is varied, the respective observers will get the signal either strongly or not at all. By this system it will be possible to explore the interior of the earth, to ascertain the size of the earth's iron core.

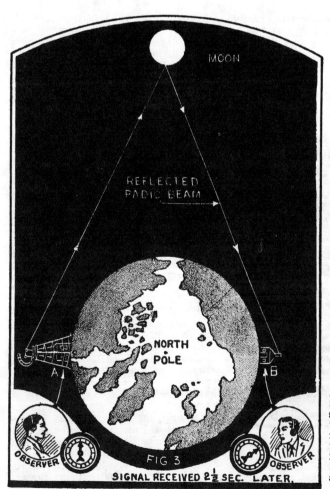

were allotted the territory below 200 meters. At that time, the authorities in Washington thought this was a good joke on the amateurs; because no one knew what to do with wavelengths below 200 meters and they were thought then to be of little use.

It may well be that history will repeat itself and interplanetarian signalling will open up avenues totally unsuspected today.

Fig. 3 shows how, by means of a powerful beam transmitter, located at some point "A" on our globe, we can send a beam of radio waves to the moon, which, being more or less metallic, will reflect the beam at the same angle. An observer, located at "B" on the opposite side of the earth, will receive the signal back from the moon, a distance of 238,000 miles, in two and a half seconds after it leaves the transmitter "A".

I need not tell the transmitting amateur that with very little power it is possible today to literally radio around the world. Five- and seven-and-a-half-watt short-wave transmitters have been heard half-way around the world, repeatedly Of course, we still have the old "Heaviside layer" bugaboo to contend with; it is even doubted by some authorities that it is possible for any sort of short wave to penetrate the Heaviside layer and leave our planet. I emphatically disagree with these authorities for the following simple reasons: radio waves are electromagnetic waves, like light. Little proof is required that light waves from regions of other worlds do pass through, not only the Heaviside layer, but the atmospheric layer as well. It is admitted that light waves in their passage through these layers are refracted; and this refraction is the counterpart of the "reflection" of radio waves in the Heaviside layer.

When, however, radio waves become sufficiently short, I do not believe that the Heaviside layer will stop them, any more than it stops the light or heat waves. Indeed, we have already joined, in the electromagnetic spectrum, radio waves and heat waves; so that no longer is it possible to distinguish between them, since they merge at some indefinable point.

No one knows, at the present time, what kind of short radio waves will penetrate the Heaviside layer; they may be waves of 5 meters or of 5 centimeters or 5 millimeters. No one knows, because no one has as yet tried it; and here is where the radio transmitting amateur comes in. The problem is not as difficult as we might believe. We know from experience that even minute power makes it possible to communicate over great distances on earth by means of short waves. If the right wave is finally selected, I sincerely believe that, with power from 100 watts upwards, it may be possible to

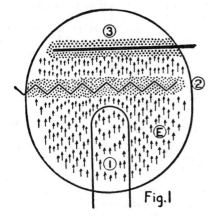

Fig. 1 shows the interior of a vacuum tube where (1) is the filament, E is the stream of electrons going in the direction of the arrows, (2) is the grid, (3) is the plate. The small dots surrounding the grid and plate indicate a layer of gas always adhering to metals, a miniature Heaviside layer, which the electrons must pierce.

send a wave to the Moon, and have it reflected back to earth without undue difficulties.

If this belief is correct, it will then be possible for a group of amateurs to do the following: the system used in Fig. 3 can easily be used. The transmitting amateur, let us say at some point in North America, by means of a direct aerial (as shown also on the front cover of this magazine) directs the radio wave at the Moon when it is at the proper elevation. The Moon, as well as most of our planets, is partly metallic in substance. The Moon, therefore, will serve as a gigantic reflector and, if the wave leaves the earth and penetrates the Heaviside layer, and is enabled to travel the distance of some 240,000 miles, the wave will then be reflected; just as the sun's light rays are reflected from the Moon, making it possible for us to see the Moon.

If, now, another amateur stationed on the other side of the globe, let us say Australia or South America, has also a directive aerial pointed to the Moon overhead, he should be able to record the signal, if this theory is correct.

If the transmitting and receiving amateurs are possessed of chronometers and signals are sent out, let us say, at the beginning of every minute on a pre-selected day, the receiving amateur will note the incoming signals, which should be received by him in about 2½ seconds. The reason is that radio waves, like light waves, travel at the rate of 186,000 miles a second. A little calculation will show, therefore, that for them to go out to the

Moon, some 238,854 miles, and return, will take a trifle over 2½ seconds. If a great number of observations are made in this manner, it is possible to come to an exact result; and we will be enabled to learn a lot more about radio than we know today.

I would advise amateurs making these tests that everything is not as simple as it might seem from this rough outline. One thing the amateur should realize is that he should have the assistance of a competent astronomer, who will guide him in minor problems that come up during these experiments. One of the problems, for instance, is that neither the earth nor the moon is at rest, with regard to the other, and that they move constantly. All this must be taken into consideration when tests are made. Another point to remember is that, while ultra-short waves may pierce the Heaviside layer, yet they will be affected by it by some sort of refraction, just as light waves are refracted through the atmosphere. Unless these factors are compensated, serious errors are liable to creep up; and, not only that, signals may be missed entirely, because they may never reach the Moon, which is the target, and therefore they cannot be reflected or received.

Yet, the cost of these experiments is not prohibitive, even to the modest amateur; and the fame that will come to him and his co-workers, when sufficient proof is received that signals have actually been sent to the Moon and back, will compensate them for all the time, money and effort that have been expended on the experiments.

Just what benefits we will derive from turning the trick, no one can even remotely perceive today. It is certain that, sooner or later, man will travel, not only to the Moon, but to other planets, by means of space-fliers; and before that happens, science should certainly be in a position to say whether it will be possible or not to communicate with such fliers, once they have left the confines of the earth.

Can We Radio the Planets?
By HUGO GERNSBACK
Member of American Physical Society

WHEN Jan Lippershey built his first telescope in 1608, he came in for severe condemnation, because it was argued that such an instrument of the devil could never do any good. When our own Percival Lowell first propounded his theory of the Martian canals and Mars as the abode of life, he, too, was greatly ridiculed as a visionary; and even today orthodox astronomers do not share his views. When the first telescope was built, the then intelligentsia could not see any good in it, except as an instrument of the devil; so when I ask the question, "Can we radio the planets?" I know that I shall be subjected to not a little ridicule.

The telescope and spectrum analysis have opened the heavens to us to a tremendous extent, and enriched our scientific knowledge immeasurably. Spec-

trum analysis has shown us that stars, millions of light years removed from us, are constituted of identically the same matter as that found in our own earth; making it, therefore, reasonably certain that the entire universe is composed of practically identical matter, with little possibility of exception.

As you will see further on, when I propound the question, "Can we radio the planets?" I do not necessarily imply that in doing so we can send intelligence to Mars or to Venus, or to the Moon, with the expectation of getting an answer—although the latter may not be as impossible in a hundred years as it is now. I am simply trying to show that, by making a start, the greater the art of radio and our knowledge thereof will become.

The largest telescopes have been made possible through the generosity of our wealthy people, and it is, therefore, not impossible to hope that what has been done in building telescopes can be duplicated in building super-power stations for radio for research purposes. I might say, right here, that the benefits derived from such a super-power radio station will no doubt be vastly greater than from building a telescope, and for the following reasons:

The telescope is useless when it comes to exploration of our own earth. It is built to explore the heavens. A super-power radio plant can be used, not only to explore the heavens, if I may call planetary space such, but also for tremen-

TABLE OF VIBRATIONS		
WHOSE EFFECTS ARE RECOGNIZED AND STUDIED		
	Number of Vibrations per Second	
1st Octave	2	
2nd "	4	
3rd "	8	
4th "	16	
5th "	32	
6th "	64	
7th "	128	
8th "	256	SOUND
9th "	512	
0th "	1,024	
15th "	32,768	
20th "	1,047,576	UNKNOWN
25th "	33,554,432	
30th "	1,073,741,824	ELECTRICITY
35th "	34,359,738,368	
40th "	1,099,511,627,776	UNKNOWN
45th "	35,184,372,088,832	
46th "	70,368,744,177,644	
47th "	140,737,468,355,328	HEAT
48th "	281,474,976,710,656	
49th "	562,949,953,421,312	LIGHT
50th "	1,125,899,906,842,624	CHEMICAL RAYS
51st "	2,251,799,813,685,248	UNKNOWN
57th "	144,115,118,075,855,872	
58th "	288,230,376,151,711,744	
59th "	576,460,752,303,423,488	X-RAYS
60th "	1,152,921,504,606,846,976	
61st "	2,305,843,009,213,693,952	
62nd "	4,611,686,018,427,389,904	UNKNOWN

In this table, beginning at the 25th octave and ending with the 45th, we have what may be termed the radio band of vibrations. It is thought that, as we approach wavelengths or frequencies near those of heat, it will be readily possible to pierce the Heaviside layer.

dously important radio research work between points on our own planet.

Penetrating Short Waves

I am fully aware of the criticism that will at once be raised, that it is not possible for us to send a radio beam beyond the confines of our own atmosphere, due to the so-called Heaviside reflecting layer effect, which is supposed to exist a hundred or so miles above the surface of the earth. According to the researches of the eminent scientist, Oliver Heaviside, the upper layers of our atmosphere are supposed to be so conductive electrically, due to the ionizing effect existing at such heights, that the radio waves are reflected; and it would thus seem impossible that we could project a radio beam outside of the confines of the earth.

This may be perfectly true when it comes to the usual radio waves, such as have been used in the past varying from some 15 meters up to 25,000 meters; although I maintain, along with many other physicists, that the Heaviside effect has never been proven conclusively. I am equally certain that at lower wavelengths, say from two meters downwards, entirely different conditions appear, for the following reasons:

We know that radio waves are nothing but an electromagnetic activity, the same as light waves or heat waves. It is believed that, the lower down we go in the wavelength scale (that is, the higher the frequency), the easier it becomes to penetrate the Heaviside layer, if we grant its existence at all. Light comes to us, from the sun and the planets, through the Heaviside layer; so we know that the Heaviside layer cannot stop light waves. To be sure, the frequency of light waves is enormously higher than that of even the shortest radio waves; but it still seems reasonable that for waves of the length of two meters, or even less, the Heaviside layer should not cause us undue worry.

Incidentally, interplanetary conditions are about the same as we find in our present vacuum tubes. Fig. 1 shows a vacuum tube, in which (1) is the filament; (2) the grid; (3) the plate. Electrons are given out by the filament (1), and shoot in the direction of (3); but, surrounding the grid and the plate, there is a miniature Heaviside layer, composed of a slight amount of gas, which surrounds all metallic and other matter, and which the electrons must first pierce before they can reach the grid or the plate.

In the Interplanetary Vacuum

Given a reasonably strong bombardment of electrons, this internal tube "Heaviside layer" can be broken down as is well known. Conditions on earth seem to be similar. If we employ the right radio wave, with sufficient power behind it, it should be possible to pierce the supposed Heaviside layer and shoot the waves out into free space. In this we would be assisted by the force of the solar radiation itself. This is made plain

in Fig. 2, which shows that our own planetary system is nothing but a vacuum-tube arrangement on a large scale. We have the sun in the center, with the planets outside, which in this case become the plate and grid of our celestial vacuum tube.

It will be noted that the solar radiation is in the direction of the arrows. It would seem, therefore, that a beam of the correct radio wave sent, let us say, from the earth to Mars, when in "conjunction," would stand a better chance of being transmitted than vice versa. For that reason, it would seem that a supposed signal emanating from Mars earthward would find it necessary to work against the stream of solar emanations, and encounter more resistance than if the case were reversed and the signal were sent from earth to Mars.

Marconi, in his recent researches, has shown that it is possible to conserve a great deal of energy by using his so-called "beam system." The beam system is nothing but a reflector arrangement whereby practically all of the energy is sent in one direction to the exclusion of other directions.

A Super-Reflector

Suppose we should now erect a tremendous power plant, which would use, let us say, 100,000 kilowatts, radiating the power on a wavelength of 2 meters or less, using the beam system of reflection. What would happen if such a tremendous amount of energy were let loose into the ether, we do not know today.

The question of sending intelligible messages to Mars or Venus need not be dealt with here at all, although it opens up interesting speculations. To the contrary, this discussion confines itself to practical scientific research, as will be apparent from a study of Fig. 3. It is known that radio waves can be reflected, just as light can be reflected, by means of a mirror. Hertz was the first to point this out, and Marconi is making use of the system by reflecting his beam, using metallic-screen reflectors to do so.

Scientists today are pretty well agreed upon the fact that the interior of the earth is composed largely of iron; practically every meteor that falls from the heavens is composed of iron; and practically every star investigated shows a large proportion of iron in its makeup. The conclusion, therefore, may be drawn that the moon, for instance, must therefore be largely composed of iron. It would therefore make an excellent reflecting medium.

Suppose now, that we proceed to erect our 100,000-kilowatt radio beam-transmitter at a point somewhere, indicated at A, on our globe. It will now be possible to direct a beam towards some point on the moon, where the angle of incidence will be suitable. The radio beam reflected, therefore, would come back to earth somewhere, shown at point B. It would be a simple matter for an astronomer to calculate the exact angle at which the beam should be sent, and it should be possible for an observer at B to detect

the reflected beams, if reflected they are. This could be easily proven as follows:

The distance between the earth and the moon is, on the average, 238,854 miles.

Radio waves travel at the rate of, roughly, 186,000 miles each second. If both observers, at A and B, were using chronometers, and if a signal were sent from A at a certain time, the signal going out to the moon and reflected from it would be found to return to the earth in a little more than two and a half seconds. This would afford, therefore, a complete proof of the theory.

The same method might be used, perhaps, with other heavenly bodies, such as Mars or Venus; and would be of tremendous assistance to science in general, if found practical. What immediate benefits would be derived, in dollars and cents, I am in no position to state; although I believe many valuable discoveries, incidental to the effects produced, would no doubt be made sooner or later.

It has often been proposed to use the well-known Goddard rocket to explore the heavens, and have this rocket equipped with radio instruments which could send back a signal, thereby proving or disproving the Heaviside theory. It is possible, by means existing today, to construct such rockets. Indeed, the Society for the Exploration of the Universe, which is now being founded in Vienna, Austria, proposes to build such a rocket. Dr. Franz Hoeff, the noted Viennese scientist, and chief promoter of the plan, states that the first experiment, which is to be followed by others, can best be carried out with the rocket containing several kilograms of explosive flashlight as the only load. The shock accompanying the landing on the moon would bring the flashlight to explosion and, with the help of the modern telescopes, the explosion would be noticeable from the observatories of the earth.

It would seem quite simple to incorporate a radio set in such a rocket, at no excessive cost. Of course the radio set would no doubt be smashed to atoms when the rocket landed on the moon; but this need not worry us. The experiment is supposed to be made only to prove or disprove the Heaviside theory; and the signals, providing the apparatus functioned properly, could be sent back to the earth for a distance of some 238,000 miles, until the rocket actually struck the surface of the moon, when the signals would cease.

Many other important facts, would surely be discovered through experimentation of this kind on a large scale;

And last, but not least, the subject of communication between the planets can then be undertaken in earnest. If we find out, by experiment, that we can reflect a radio beam from the surface of the moon, we can be reasonably certain that, given sufficient power, the same beam system can be used to send signals to either Mars or Venus; as these two planets hold forth the greatest hope of being the abode of some sort of life.

SERVICE SECTION

ATWATER KENT MFG. CO.

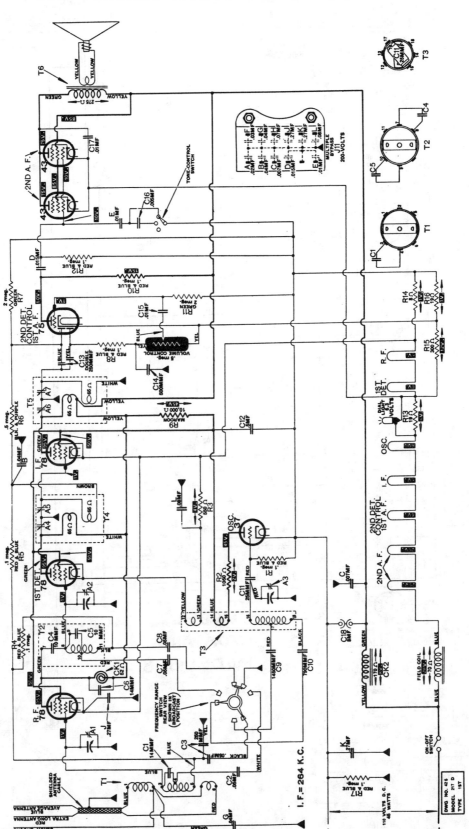

DIAGRAM OF MODELS 217D, 427D AND 667D

ATWATER KENT MFG. CO.

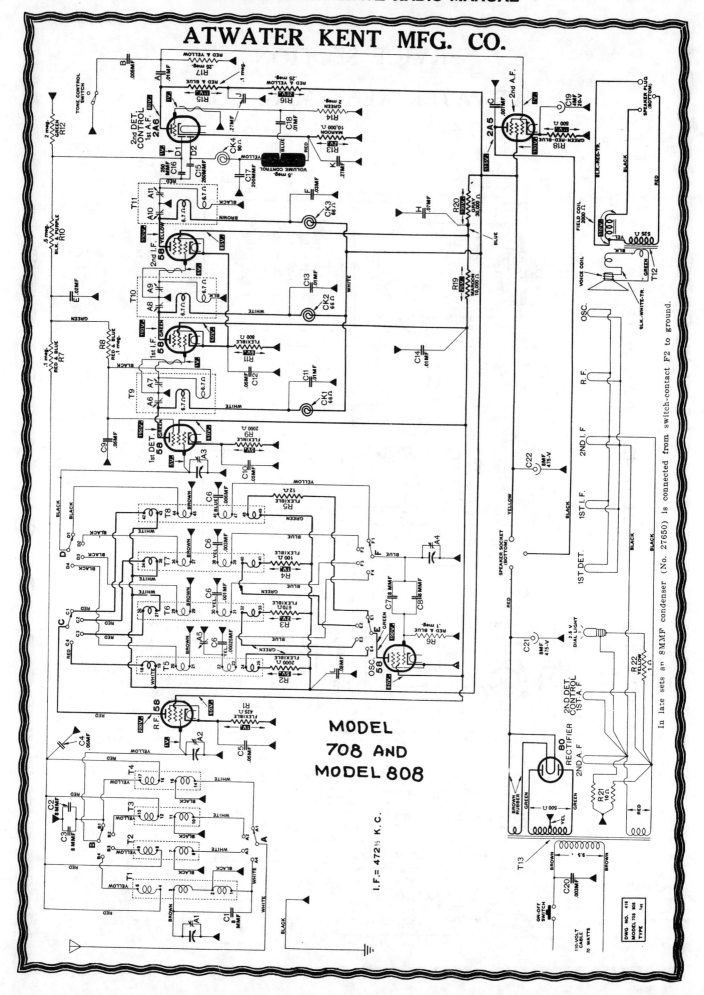

MODEL
708 AND
MODEL 808

I.F. = 472½ K.C.

In late sets an 8MMF condenser (No. 27650) is connected from switch-contact F2 to ground.

ATWATER KENT MFG. CO.

DIAGRAM OF MODEL 711

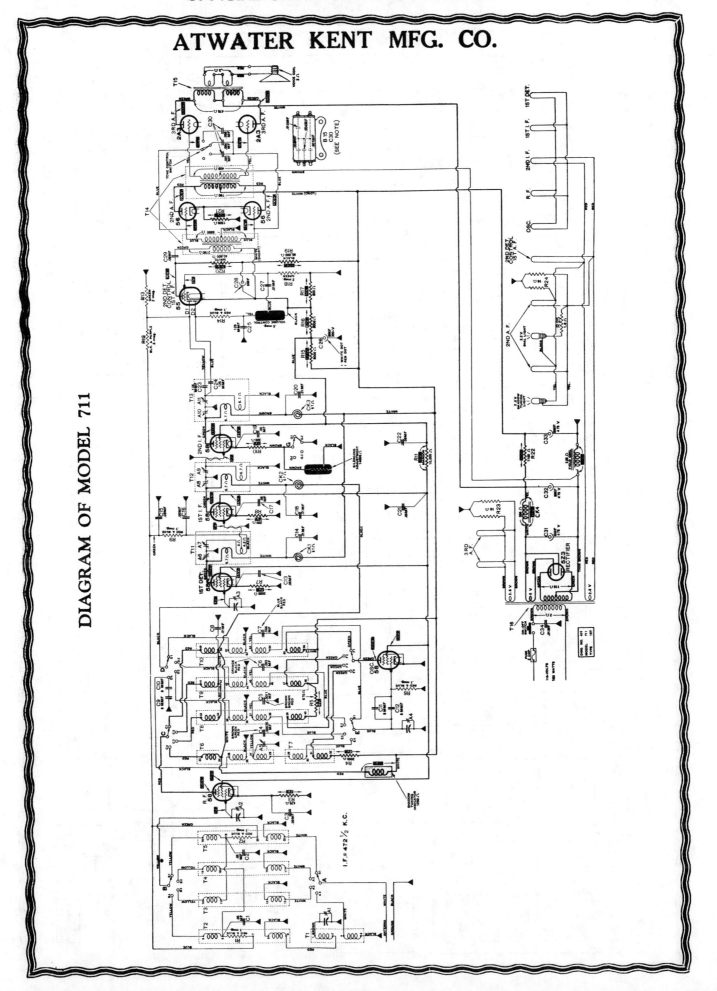

ATWATER KENT MFG. CO.

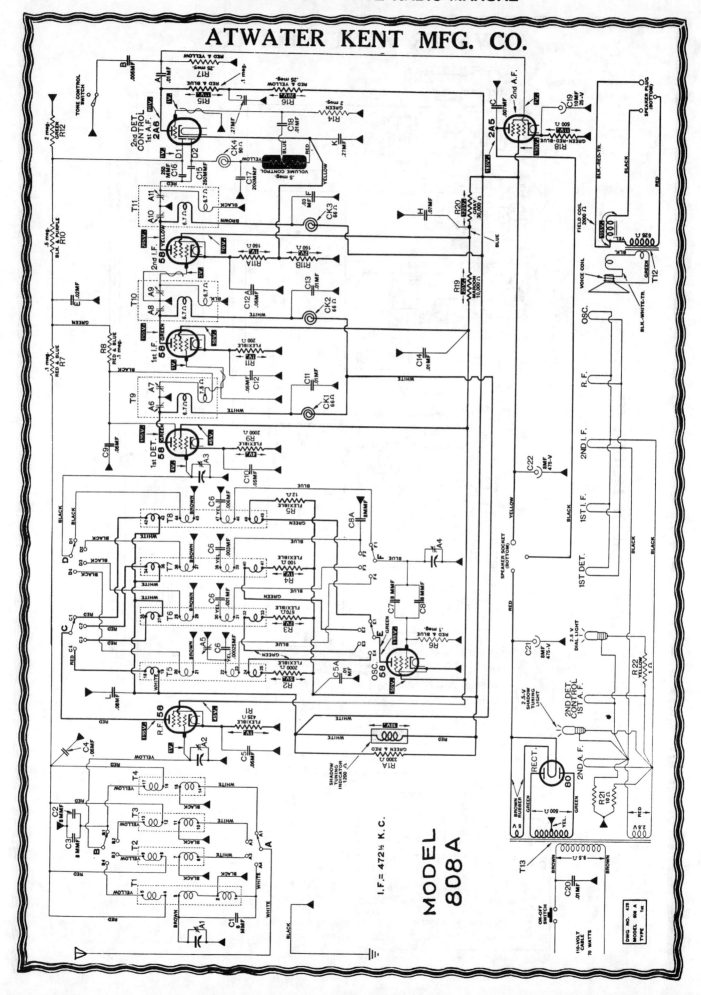

MODEL 808 A

I.F. = 472½ K.C.

BAIRD TELEVISION COMPANY

UNIVERSAL SHORT-WAVE SET

First line up the main sections of the variable gang condensers and tighten the set screws on the shaft.

Then let out the adjusting screws on the trimmer condensers C5 and C6 of the tuning condenser sections C2 and C3 respectively, and adjust trimmer condenser C4 to about its midposition (half-meshed).

Next tune in a weak station by adjusting the dial which controls the main variable condensers and the regeneration condenser C7. The regeneration con-denser C7 should be adjusted until the characteristic regeneration whistle is heard. Then it should be adjusted just below the spill over or whistling point, and the station tuned in by adjusting the main tuning condensers, until the station comes in as strong as possible.

The final adjustment can then be made with trimmer condensers C4, C5 and C6.

The adjustment of trimmer condensers C5 and C6 should be tried on different stations at different points of the dial.

Parts Required for Baird Universal Short Wave Receiver

Quan.	Part and Type	No. on Diagram
1	Chassis, all mounted, with 8 sockets riveted	S1, S2, S3, S4, S5, S6, S7, S8
	3 coil shields	CS1, CS2, CS3
	3 tube shields	TS1, TS2, TS3
	2 condensers	C10, C13
	and 1 3-post binding strip	Gnd., Short Ant., Long Ant.
3	Coil Sockets	S9, S10, S11
2	Pig-tail Resistors	R1, R2
3	.0001 Mfd. Condensers	C8, C11, C17
3	Screen Grid Clips	
2	.02Mfd. Moulded Condensers	C9, C12
2	Jacks	J1, J2
1	Block Condenser	C18, C19, C20, C21, C22, C23
1	Block Condenser	C24, C25, C26, C27
3	R. F. Chokes	CH3, CH4, CH5
1	3-Gang Baird Variable Condenser	C1, C2, C3, C5, C6
3	Electrolytic Condensers	C14, C15, C16
1	Baird Power Transformer	T
1	Baird Power Choke	CH1, CH2
1	Baird Gang Resistor	R5, R6, R7, R8, R9, R10, R11, R12, R13, R14
1	3-Condenser Strip	C28, C29, C30
4	Knobs	
1	Toggle Switch—2 pole	SW1
1	Speaker Terminal	Speaker
1	Combination Potentiometer and Switch	P, SW2
1	No. 9 Baird Midget Condenser	C4
1	No. 15 Baird Midget Condenser	C7
1	Buffer Condenser	C31, C32
1	Voltage Divider	VD
1	Baird Dial and Escutcheon	
1	Baird Front Panel	
2	Grid Resistors	R3, R4
	40' Wire	
	3-Pole Switch	SW3
1	Dial Bracket and Lamp	DL
	AC Cord and Plug	P1
	Hardware Assembly	
15	Octocoils 15-520 meters Wavelength	
1	Cabinet (optional—not furnished with kit)	

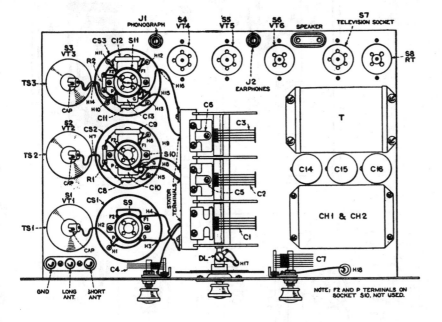

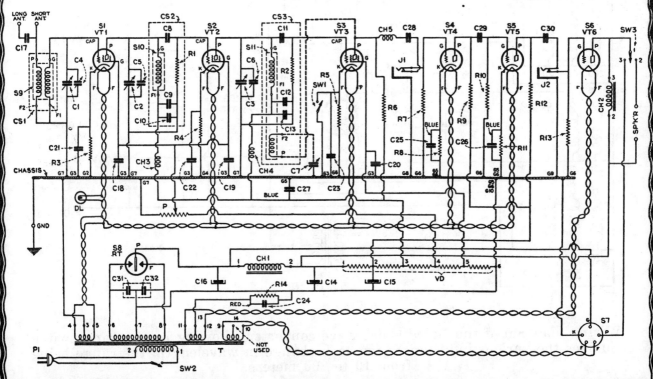

BAIRD TELEVISION COMPANY
BAIRD POWER CONVERTER

THE Baird *power converter* is the result of four years' experience in the design and construction of short wave converters and receivers. A short wave converter, to be satisfactory, must fulfill the following conditions: First, ease of tuning; second, sufficient sensitivity to work with any broadcast receiver; third, have no whistles or tuning noises when looking for stations.

The S-W converter is designed to operate when the broadcast receiver is tuned in to 1000 kilocycles. In some locations, this setting on the broadcast receiver may bring in a nearby local station, and the converter will operate efficiently on any dial setting on the broadcast receiver from 950 to 1050 kilocycles.

The 1,000-kc., tuned impedance in the "shield can" shown in the diagram, may comprise about 300 turns of No. 30 S.S.C. or enameled magnet wire, wound scramble fashion, on a small form. Note that the two main tuning condensers are ganged, as indicated by the dotted line. All the switches changing the coils for the different wave bands are also ganged, and a single knob mounted on the front of the converter.

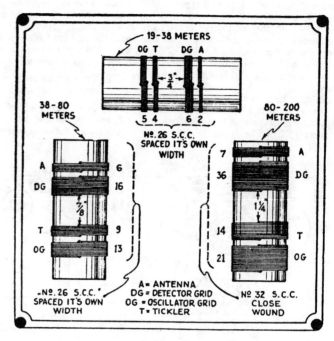

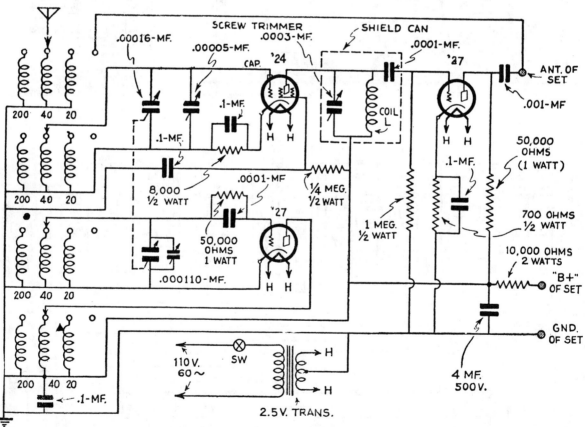

Wiring diagram of the Baird short wave converter, showing how a gang switch changes the various inductances for each change in wavelength, the range being from 19 to 200 meters.

BEST MFG. CO.

The BEST Short-Wave Converter

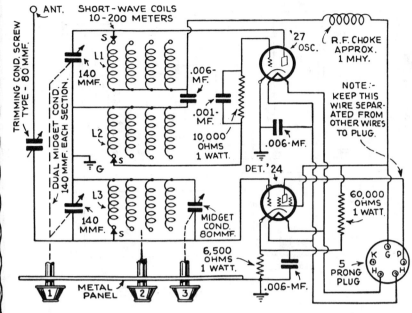

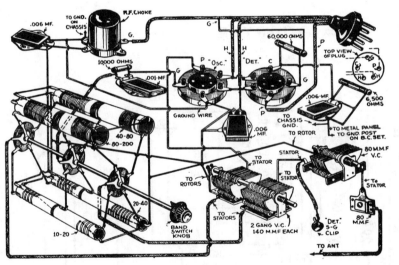

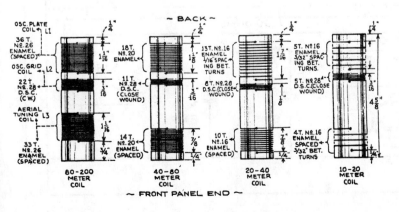

This converter employs only two tubes, a '27 for the oscillator and a '24 for the detector. It will be seen that the plate and heater supply for the two tubes in the converter is obtained from the power-supply unit in the broadcast set, through the medium of the five-prong plug. Very little drop in voltage was registered when using this converter with the average broadcast receiver; but, if the converter must be used with a cheaply-built receiver, which has no surplus power in the heater and plate-supply unit, a separate source of "B" supply can be used for the converter—or even "B" batteries, for that matter.

One of the accompanying drawings gives the winding data for the four coils used to cover the various short-wave bands from 10 to 200 meters. Bakelite tubing (1¼ and ⅞ inches diameter and 4⅝ inches long) is used to wind the coils on, in the manufactured unit. If you can obtain no bakelite, the coils may be wound in the manner illustrated, on wooden rods previously boiled in paraffin; or on cardboard tubes of the diameter specified, soaking the tubes in hot paraffin before using. If you do not have the exact size of wire, there will be but slight change in wavelength, if you use wire one or two sizes different from that called for. It goes without saying, that all joints in wiring up the converter should be thoroughly soldered with a non-corrosive flux.

Looking at the tuning arrangement, we note that a Hammarlund two-gang midget condenser, each unit of 140 mmf., serves to tune in the station; one unit tunes the aerial inductance, and the other the plate coil of the oscillator. The grid coil of the oscillator is not tuned. A second 80-mmf. midget condenser, having a separate knob of its own at the right of the panel, acts as a vernier to finish up the fine tuning.

The R.F. choke built and tried out in the original converter has about 1 millihenry inductance, comprising 100 turns No. 29 D.S.C. magnet wire, wound on a bakelite tube or rod ⅞ inch in diameter.

List of Parts

1—"Best" short wave coil and switch assembly, including calibrated escutcheon plate.

1—Two-gang Hammarlund midget condenser; each unit 140 mmf.

1—Hammarlund midget condenser, 80 mmf.

1—Hammarlund equalizer condenser, screw type, 80 mmf.

1—Shield box, 6 by 7 by 10 inches (Blan).

2—Tubes—one '27 and one '24.

2—UY-5 prong sockets.

1—Five prong plug.

3—.006 mf. bypass condensers—Flechtheim

1—.001 mf. bypass condenser—Flechtheim.

1—10,000 ohm, 1 watt resistor—Electrad or other make.

1—60,000 ohm, 1 watt resistor—Electrad or other make.

1—6,500 ohm, 1 watt resistor—Electrad or other make.

1—R.F. choke wound as specified, or 85 M.H. Hammarlund (as used in model here illustrated).

1—National Baby vernier dial.

CROSLEY RADIO CORP.

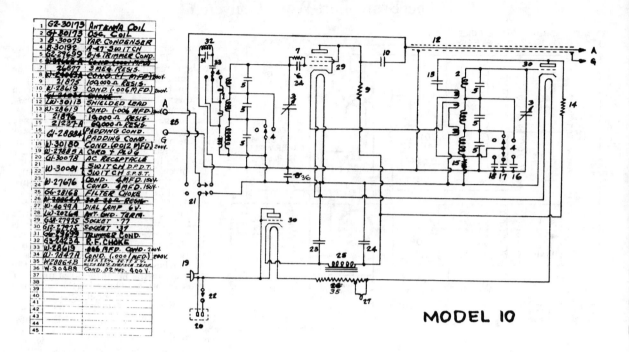

1	G2-30175	ANTENNA COIL
2	G1-30175	OSC. COIL
3	B-30079	VAR. CONDENSER
4	B-30192	A-47 SWITCH
5	G2-28699	B-14 TRIMMER COND.
6		COND. (.0001 MFD)
7	21677	3 MEG. RESIS.
8	W-29449A	COND. (.1 MFD) 200V.
9	21875	100,000 Ω RESIS.
10	W-28619	COND. (.006 MFD) 200V.
11	G1-31034	
12	W-30113	SHIELDED LEAD
13	W-28619	COND. (.006 MFD) 200V.
14	21896	10,000 Ω RESIS.
15	21237A	50,000 Ω RESIS.
16	G1-28884	PADDING COND.
17		PADDING COND.
18	W-30180	COND. (.0012 MFD) 200V.
19	W-27889	CORD & PLUG
20	G1-30078	AC RECEPTACLE
21	W-30081	SWITCH D.P.D.T.
22		SWITCH S.P.S.T.
23	W-27676	COND. 4 MFD. 150V.
24		COND. 4 MFD. 150V.
25	G6-28168	FILTER CHOKE
26	W-28064A	305-28 Ω RESIS.
27	W-4099A	DIAL LAMP 6 V.
28	W-20264	ANT. GND. TERM.
29	G38-27975	SOCKET '77
30	G12-27975	SOCKET '37
31	G6-29899	TRIMMER COND.
32	G3-14234	R.F. CHOKE
33	W-28619	.006 MFD. COND. 200V.
34	W-7847A	COND. (.0001 MFD) 200V.
35	W28864B	305A ESS., DC 73.5 V. WITH 505's SURFACE TEMP.
36	W-30488	COND. .02 MF. 400 V.
37		
38		
39		
40		
41		
42		
43		
44		
45		

MODEL 10

SHORT-WAVE ADAPTER

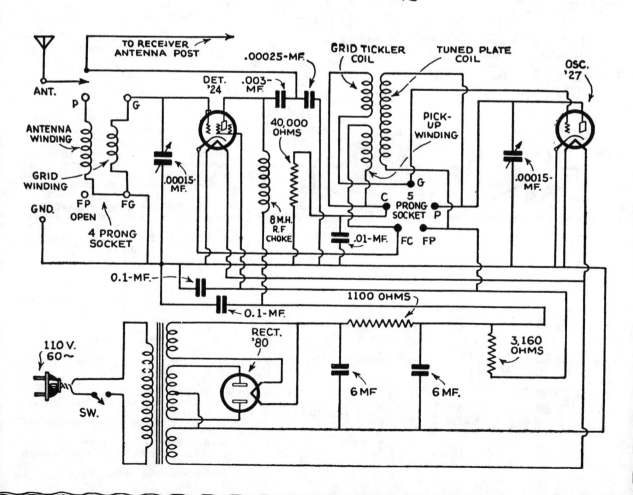

CROSLEY RADIO CORP.

SHORT-WAVE SUPERHETERODYNE
Model 136-1

Specifications

Model 136-1 is a ten tube superheterodyne for operation from A. C. electric circuits. Five sets of coils give the following frequency ranges: 550 to 1500 KC, 1500 to 3500 KC, 3500 to 6500 KC, 6500 to 12000 KC, and 12000 to 20000 KC. The intermediate frequency used is 456 KC.

Tubes And Voltage Limits

The following are the voltages measured with the receiver in operating condition but with no signal to the antenna circuit. Use a high resistance D. C. volt-

meter (1000 ohms per volt, or more) for all but filament voltages. In measuring filament or heater voltages use a low range A. C. meter. The voltage limits are + or — 10% of values given in the following table.

Line voltage—117.5 volts (235 for 220 volt receivers).
Plate voltage measured from plate contact to cathode contact.
Suppressor grid voltage measured from suppressor grid contact to cathode contact.
Bias voltage measured from cathode contact to chassis.

Tube	Position	Plate	Screen Grid	Voltages Supp. Grid	Bias	Fil.
-56	Oscillator	45			0	2.5
-58	1st Detector	275	100	0	10.0	2.5
-58	1st I. F. Amplifier	275	100	0	2.5	2.5
-58	2nd I. F. Amplifier	275	100	0	4.0	2.5
-56	Diode Detector	0			0	2.5
-56	Push Pull A. F. Amplifier	135		0	7.0	2.5
-56	Push Pull A. F. Amplifier	135		0	7.0	2.5
-42	Output	270	275		20.0	6.3
-42	Output	270	275		20.0	6.3
-80	Rectifier	370				4.8

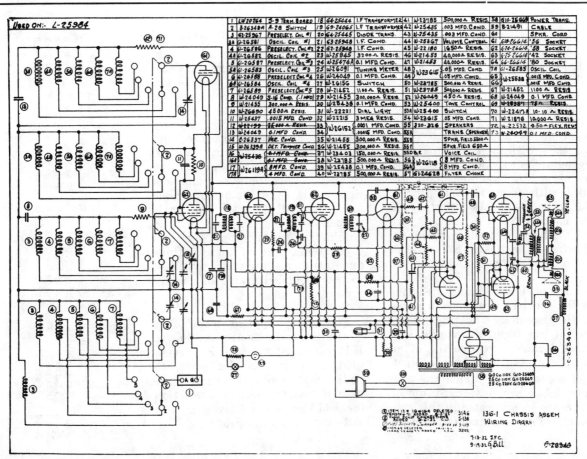

136-1 CHASSIS ASSEM
WIRING DIAGRM

DE FOREST RADIO CO.
4 TUBE S.W. SET

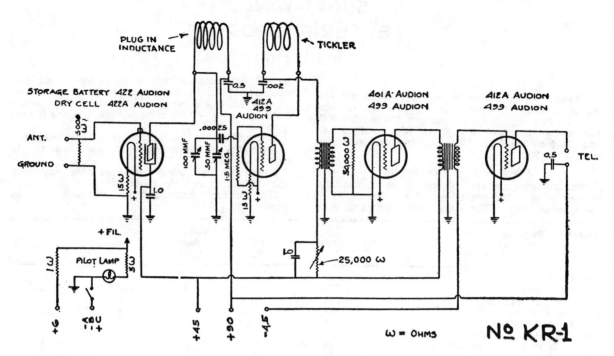

This diagram shows how the various parts of the new DeForest Short-Wave Receiver are connected.

The standard production model of the new receiver includes four plug-in coils with a tuning range of 14 to 195 meters. Of course the ambitious amateur may wind up other coils to cover higher or lower wavelength to suit his individual requirements and desires. These four coils, with their accompanying regenerative tickler coils, are wound on thin bakelite tubing provided with four pin-jack prongs. The three lowest wavelength coils are made of No. 18 enameled wire and the largest of silk covered No. 28 wire. The tickler coils are all of No. 28 silk-covered wire.

The unusually small dimensions of the receiver are made possible by the ingenious mounting assembly, in the form of bakelite sub-panel, located three-fourths of an inch below the top of the cabinet. The sockets from which the vacuum tubes are suspended upside down are incorporated in this sub-panel. The two audio frequency transformers are mounted on the two end plates of the aluminum case, and the two midget variable tuning condensers as well as the variable resistance regeneration control are attached to the front panel. The resistances and remaining $2\frac{1}{2}$ microfarads of bypass condensers occupy the remaining space at the rear of the cabinet. For further simplification, the leads of the battery cable are soldered directly to their proper positions in the wiring, eliminating all binding posts and connections. All of these battery and telephone leads are bypassed to the grounded frame of the receiver.

The voltage applied to the tubes of the new receiver are as follows: R.F. plate, 90 volts; R.F. screen-grid, 45 volts; detector plate, 45 volts; first audio, 45 volts plate with a filament minus connection for grid bias; second audio, 90 volt plate with $4\frac{1}{2}$ volts grid bias.

Coil	Primary	Tickler	Wavelength	Threads Per Inch	Tickler Spacing
No. 1	3 turns No. 18	3 turns No. 28	14 to 25 meters	6	1/8 inch
No. 2	7 turns No. 18	3 turns No. 28	23 to 49 meters	11	1/8 inch
No. 3	18 turns No. 18	4 turns No. 28	45 to 95 meters	11	1/4 inch
No. 4	26 turns No. 28	6 turns No. 28	93 to 105 meters	wound tight	1/4 inch

DETROLA RADIO CORP.
SHORT-WAVE CONVERTER

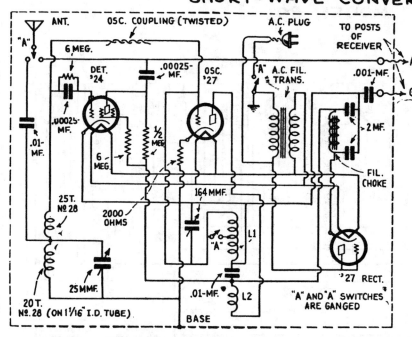

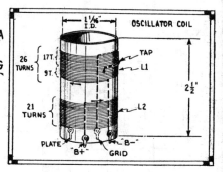

Winding details for the oscillator coil used in the Detrola Short Wave Converter. The coils are furnished already wound.

Left—Wiring diagram for the Detrola Short Wave Converter, which is supplied in "kit" form.

EMERSON RADIO & PHONOGRAPH CORP.

~SCHEMATIC CIRCUIT MODEL-S-7~

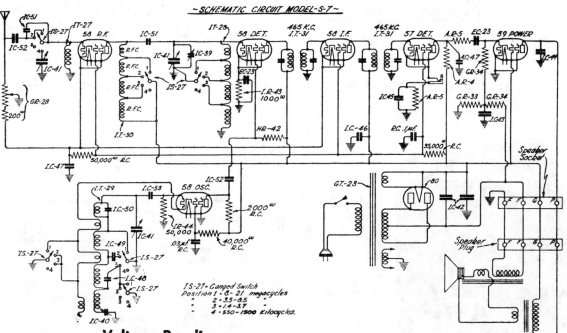

Voltage Readings:

Readings should be made using a D.C. voltmeter having a resistance of 1,000 ohms per volt. Volume control should be on full.

		Screen	Cathode
58 Oscillator	Ground to plate 230-245	Screen 120-140	Cathode
58 R.F. Amplifier	" " " 235-250	" 90-110	" 3-4
58 1st Detector	" " " 235-250	"	" 1-2
58 I.F. Amplifier	" " " 235-250	" 90-110	" 3-4
57 2nd Detector	" " " 100-125	" 90-110	" 4-6
59 Output tube	" " " 230-245	" 235-250	"

Line voltage, 115v.

The bias on the 59 and the screen voltage of the 1st detector cannot be read with the usual voltmeter.

DE WALD PIERCE-AIRO, INC.
MODEL 801

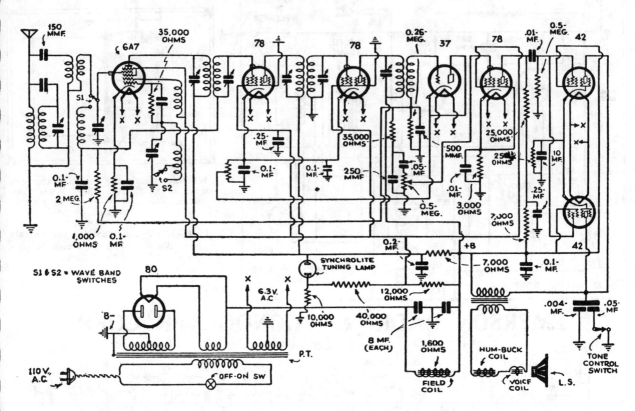

Wiring diagram of model 801 "BC" and "SW" receiver—an 8-tube superhet.

The DeWald midget broadcast and short-wave receiver tuning from 550 down to 60 meters.

● ONE of the best sounding midget radio receivers we have heard in some time is the model 801 DeWald, which covers the broadcast and short-wave band down to 60 meters. Another model tunes clear down to 15 meters and includes the broadcast band from 200 to 550 meters. This receiver is encased in a handsome walnut cabinet; the chassis is an 8-tube superheterodyne.

Among other features found in this very smooth working receiver we find 100 per cent automatic volume control, full-range tone, control, visual neon tube tuning, electron coupled circuit, a pre-selector antenna circuit, high sensitivity and diode detection.

The model 801 uses a well selected set of high gain tubes and by careful design of the various coupling circuits used to link the stages, a very high amplification is obtained, together with a very well balanced circuit, so that a surprisingly fine quality of reproduction is obtained. A dynamic speaker of the latest type and of the proper impedance to work with the particular tube used in this set provides an extra fine quality of sound reproduction. Liberal sized condensers and chokes are used so that a minimum of noise results. Moreover the set occupies only a small space and will find favor for many requirements, including small apartments, cabins, study dens, etc. The size of the cabinet housing the model 801, and also its close relative, the model 811—which tunes down to 15 meters, is 16¾" high by 14½" wide, by 8¾" deep.

Oscillator Coils

Tickler............37 turns   No. 38 enameled wire.
Grid................approximately 74 turns, No. 32 enam. tapped at 37 turns wound on ¾" tube with 1/64 space between the two windings all close wound.

Detector Coils

Short Wave
Antenna.......... 8 turns   No. 32 single cotton covered.
Grid...............30 turns   No. 32 enam.

Coils wound in same direction on a ⅞" form. Antenna coil is wound over the ground end of the grid coil, all coils close wound.

Detector Broadcast coil
Antenna coil...1.52 mh.   No. 38 enam.
Band pass......245 mh.   No. 40-7 strand litz.
Grid...............212 mh.   No. 40-8 strand litz

Above coils are universal wound and mounted on a ½" dowel stick, spacing between grid and band pass coil is ½ inch.

ELECTRICAL RESEARCH LABS.
(ERLA)

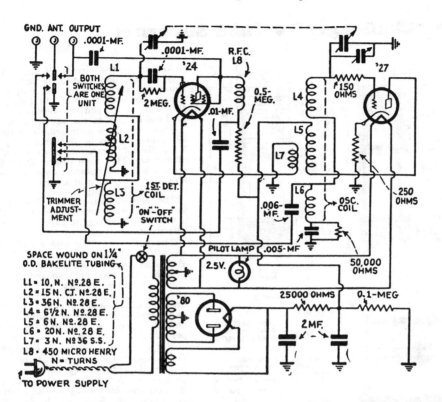

ERLA S.W. CONVERTER

It covers the short wave bands from 200 to 60 and from 60 to 15 meters; a wavelength chart is furnished. The condenser capacity joining L1 to ground is .00027 mf., and the capacity of the condenser joining L4 to ground is .00037 mf. with a trimmer across it having a capacity of 20 mmf. The vernier is an adjustable copper disc sliding within coil L1.

EMERSON RADIO & PHONOGRAPH CORP.
MODEL L-755

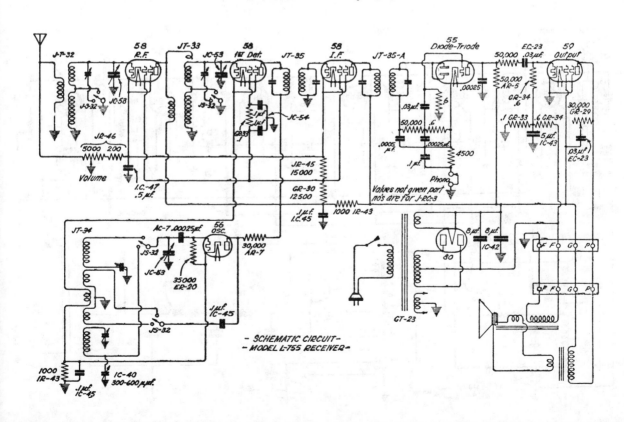

— SCHEMATIC CIRCUIT —
— MODEL L-755 RECEIVER —

EXPERIMENTERS RADIO SERVICE (ERDEL)

The "Challenger" 9-Tube Superhet

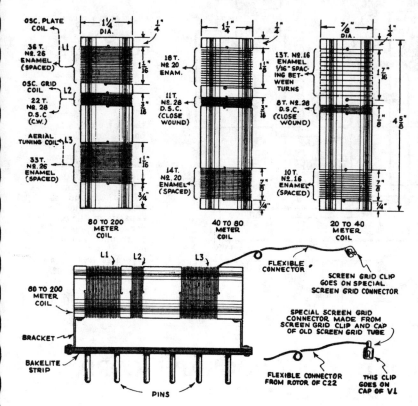

Analyzing the circuit of the "Challenger," we find that it is really unusually straightforward and simple for a nine-tube set. There is a first detector V1 using a 57 pentode tube and coupled inductively to the 56 oscillator V8.

There are three intermediate frequency stages employing 58 tubes (V2, V3, V4). The I.F. transformers are peaked at 465 kc. This high intermediate frequency has been selected as most desirable since it results in a close approach to "one spot" tuning.

The second detector employs a 24 tube (V5), with power detection. This is coupled resistively to the single audio output stage.

The full-wave 80 rectifier (V9) and its attendant filter circuit are of conventional design. Following standard practice, the 1000 ohm speaker field is made to serve as one of the filter chokes. Electrolytic filter condensers are used at C24 and C25 and a 25 mf. 25 volt cartridge-type electrolytic condenser is used at C26 to by-pass the pentode bias resistor.

A 1,000 ohm potentiometer at R6 in the cathode return circuit of the three intermediate frequency tubes provides an excellent sensitivity control. Volume is controlled by means of the Electrad potentiometer R14, which in effect varies the load resistance in the detector circuit—that is, the resistance coupling between the second detector and the output stage.

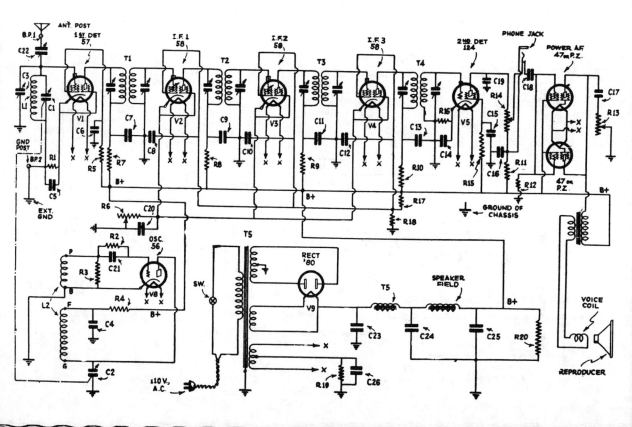

FADA RADIO & ELECTRIC CORP.
MODEL 66 (KY)

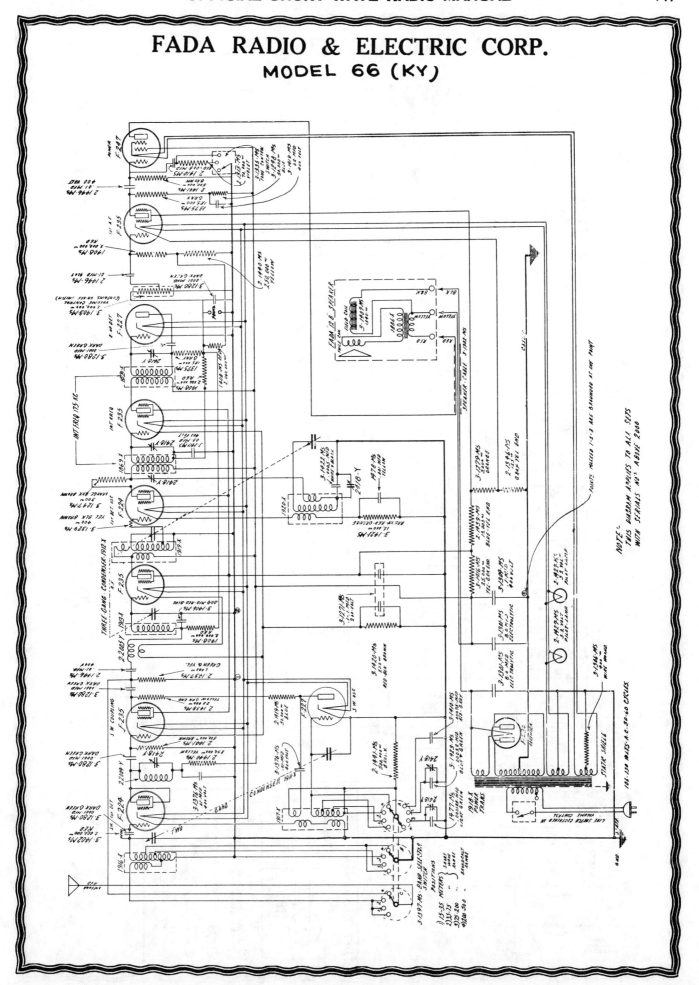

FADA RADIO & ELECTRIC CORP.

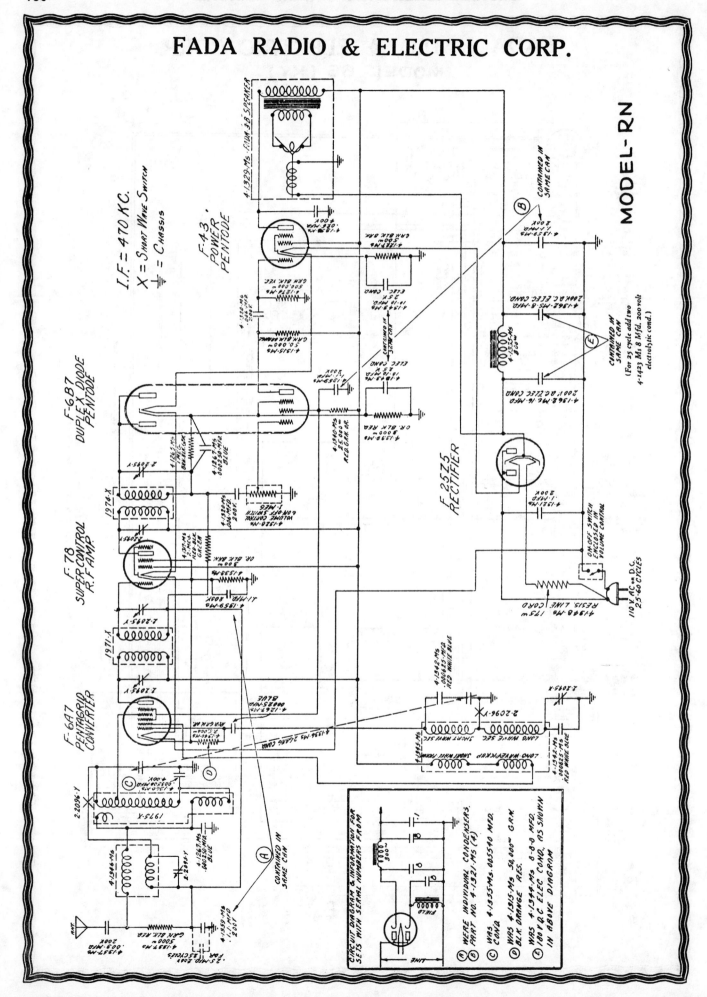

MODEL-RN

FEDERATED PURCHASER, INC.

FEDERATED "AIR-ROVER"

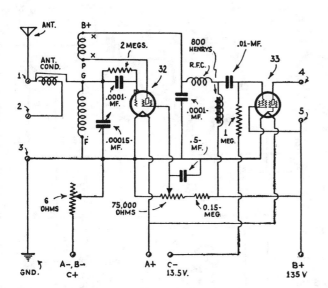

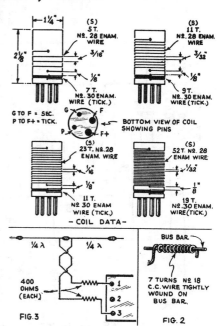

– COIL DATA –

FIG. 3   FIG. 2

Parts List for "Air-Rover"

Acratest Triple Binding Post. Aerial & Ground Connections
Acratest Twin Phone Tip Jack, Speaker or Phone Connections
1¼" Piece of Bare No. 14 Wire wound over with appx. 14 turns of No. 18 insulated push-back hook-up wire.
Set of Four Plug-in Short Wave Coils. These are accessories. Not furnished with kit.
    Coil A—200 to 80 meters
    Coil B—80 to 40 meters
    Coil C—40 to 20 meters
    Coil D—20 to 10 meters

A four-prong wafer type socket, for the short-wave plug-in coil, is riveted to the chassis
Acratest Short Wave R.F. Choke
High Impedance Acratest Audio Choke
2 meg., ½ watt Resistor
75,000 ohm Potentiometer
150,000 ohm, 1 watt Resistor
1 meg., 1 watt Acratest Resistor
6-ohm Acratest Rheostat
.00015 mf. Acratest Variable Tuning Condenser
.0001 mf. Acratest Mica Condenser
.00025 mf. Acratest Mica Condenser
.01 mf., 400 volt Acratest Cartridge Condenser
.5 mf., 200 volt Acratest Metal Case Condenser

Four-Prong wafer-type socket, marked for '32 Tube, riveted to chassis
Five-Prong wafer-type socket, marked for '33 Tube, riveted to chassis
Four-Conductor Battery Cable
Drilled Metal Chassis and Drilled Metal Front Panel, three sockets riveted to chassis
1-Screen grid clip
Three Knobs
Dial Escutcheon Plate
Hook-up Wire
Piece of Bare No. 14 Wire for Item 3
Spaghetti
Hardware Assortment

FEDERATED "ARGONAUT"

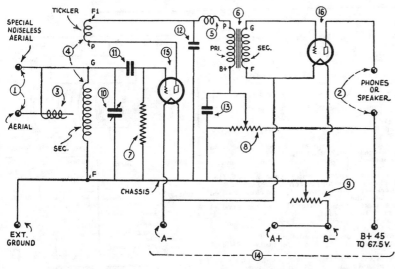

ARGONAUT PLUG-IN COIL DATA

Meters Wavelength	Grid coil turns	Tickler turns	Distance between 2 coils
200-80	52 T. No. 28 En. Wound 32 T. per Inch	19 T. No. 30 En. Close wound (CW)	⅛"
80-40	23 T. No. 28 En. Wound 16 T. per Inch	11 T. No. 30 En. C.W.	⅛"
40-20	11 T. No. 28 En. 3-32" between turns	9 T. No. 30 En. C. W.	⅛"
20-10	5 T. No. 28 En. 3-16" between turns	7 T. No. 30 En. C.W.	⅛"

Coil form—2⅛" long by 1¼" dia. 4-pin base.

Parts List "Argonaut" Two Tube Short Wave Receiver

1—Acratest Triple Binding Post. Aerial and Ground Connections.
2—Acratest Twin Phone Tip Jack, Speaker or Phone Connections.
3—1¼" Piece of Bare No. 14 Wire wound over with appx. 14 turns of No. 18 insulated push-back hook-up wire.
4—Set of Four Plug-in Short-Wave Coils.
    Coil A—200 to 80 meters
    Coil B— 80 to 40 meters
    Coil C— 40 to 20 meters
    Coil D— 20 to 10 meters
A Four-prong wafer-type socket for the short wave plug-in coil is riveted to the chassis.

5—Acratest Short Wave R.F. Choke, 4 mh. inductance.
6—Acratest 4 to 1 Audio Frequency Transformer.
7—5 Megohm, ½ watt Acratest Resistor.
8—100,000 ohm Acratest Potentiometer.
9—6 ohm Acratest Variable Resistor.
10—.00015 mf. Acratest Variable Tuning Condenser.
11—.0001 mf. Acratest Mica Condenser.
12—.00025 mf. Acratest Mica Condenser.
13—.1 mf. 200 volt Acratest Tubular Condenser.
14—Four-Conductor Battery Cable.
15—Four-Prong Wafer type Socket, marked for 30 Tube, riveted to chassis.
16—Four-Prong Wafer type Socket, marked

for 30 Tube, riveted to chassis.
17—Drilled Metal Chassis and Drilled Metal Front Panel. Three Four-Prong Sockets riveted to chassis.
Three Knobs
Special Acratest Short Wave Dial
Dial-Escutcheon Plate
Hook-up Wire
Piece of Bare No. 14 Wire for Item 3
Spaghetti
Hardware Assortment
1—Pair Headphones, or Extra Sensitive DX Phones, or Magnetic Speaker.
1—45-Volt "B" Battery.
2—No. 6, 1½ volt Dry Cells.
2—Triad or equivalent 30 type 2 Volt Tubes.

FEDERATED PURCHASER, INC.
"DISCOVERER"

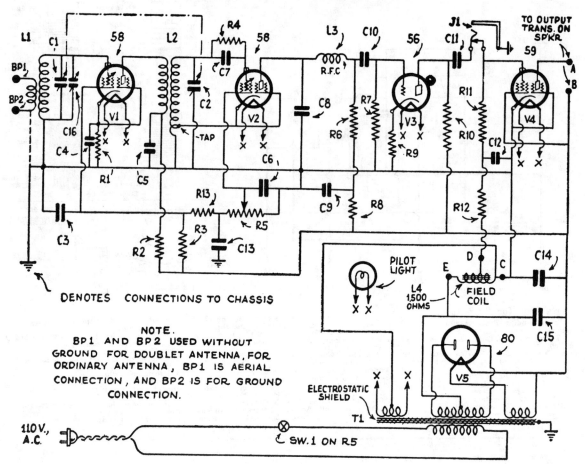

DENOTES CONNECTIONS TO CHASSIS

NOTE.
BP1 AND BP2 USED WITHOUT
GROUND FOR DOUBLET ANTENNA, FOR
ORDINARY ANTENNA, BP1 IS AERIAL
CONNECTION, AND BP2 IS FOR GROUND
CONNECTION.

In the ordinary regenerative receiver, the tube is pushed into oscillation by means of energy feed-back from the plate circuit to the grid circuit or by some other means. The fundamental action is that of a stable tube and the tube is actually forced into oscillation by means of any one of the commonly known methods. In this receiver, we use the opposite method, whereby the tube is constantly oscillating and has to be pulled out of oscillation for voice reception or music.

Therefore, keep the antenna trimmer condenser C16 constantly in step with the main tuning condenser C1, C2, so that exact resonance is obtained in both of these tuned circuits. Turn the regeneration control dial and after the whistle of the incoming carrier has been heard, carefully tune C1, C2, which are on a common shaft, and adjust C16 for absolute resonance. Then you will find at this time that you can advance C16 control with greater signal output.

Parts List Acratone "Discoverer" 5 Tube Set

1—Acratest Two-Gang variable condenser, .00014 mf. each section, (C1, C2)
5—.02 mf. 300 volt Acratest cartridge condensers, (C3, C4, C5, C10, C11)
1—.1 mf., 200 volts Acratest condenser, (C12)
1—.5 mf., 200 volts Acratest condenser, (C6)
2—.0001 mf. Micamold Mica Condensers, (C7, C8)
2—2 mf., 400 volts Acratest Electrolytic Condenser, (C9, C13)
1—Acratest Dual 8 mf. (ea. section) Electrolytic condenser, (C14, C15)
1—25 mmf. USL Midget variable condenser, (C16)
1—300 ohm ½ watt Acratest resistor, (R1)
2—2000 ohm, ½ watt Acratest resistors, (R2, R9)

1—25,000 ohm, ½ watt Acratest resistor, (R3)
1—3 meg., ½ watt Acratest resistor, (R4)
1—25,000 ohm potentiometer (R5) and switch (Sw1)
2—.25 meg., ½ watt Acratest resistors, (R6, R12)
2—.5 meg., ½ watt Acratest resistors, (R7, R11)
1—.1 meg., ½ watt Acratest resistor, (R8)
1—50,000 ohm, ½ watt Acratest resistor, (R10)
1—25,000 ohm, ½ watt Acratest resistor, (R13)
1—Full-Vision vernier tuning dial and escutcheon with pilot light V6
4—Acratest 6 prong wafer type isolantite sockets, (V1, V2, L1, L2)
1—Acratest 5-prong wafer type socket (for speaker connection) riveted to chassis.
1—Acratest 5-prong wafer type socket, marked for 56 tube (V3)
1—Acratest 7 prong wafer type socket, marked for 59 tube (V4)
1—Acratest 4 prong wafer type socket, marked for 80 tube (V5)
1—Acratest power transformer, (T1)
2—Acratest tube shields, for tubes V1 and V2
1—Acratest short-wave R.F. choke (L3)
1—Dual antenna-ground binding post, (BP1, BP2)
1—Drilled metal chassis, cadmium plated, 11½"x9⅜"x2" high
1—Metal shield plate 11⅜"x5" No. 20 gauge (Plate 'C')
2—Metal shield plates, 5"x5" No. 20 gauge (Plates "A" & "B")
1—Power supply cord and plug
2—Screen grid clips
4—Small knobs
1—Earphone Jack (J1)
1—Set 8 Special short-wave coils, covering 10 to 200 meters (L1, L2)
1—5-prong speaker plug
1—Dynamic Speaker, 8½" Dia. with 1800 ohm field tapped at 300 ohms; matched to output of 59 tube

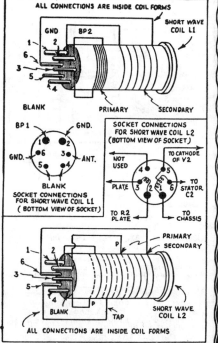

Details of Plug-in Coils and Sockets.

FEDERATED PURCHASER, INC.
VOYAGER No. 45A

The "B" supply can be any voltage from 180 volts to 250 volts. The maximum power can be obtained with 250 volts and the most sensitive condition for detecting in the 57 type tube will be obtained when the voltage is 180 volts, or a little higher.

The filament supply is A.C., of course, and should be capable of delivering 2½ volts at 3 amperes. Most every experimenter has a small 2½ volt transformer, which will safely carry the load of 3 amperes, so the filament supply should offer little difficulty to the constructor.

No batteries are necessary in the operation of this receiver, as the bias for the power tube is automatically taken up by a bias resistor under the chassis.

Under normal operating conditions a pair of phones will be uncomfortable as far as volume is concerned. In fact, most stations, especially the high-powered ones, can be received and tuned in directly on the loud speaker. When using the phones it is well to use a coupling transformer so that the direct current is kept out of the phone windings.

The circuit of the Voyager is very simple and essentially follows the older 2-tube All-Waver. The small antenna coupling condenser, which is mounted on the front panel, gives the maximum values of coupling between the antenna used and the tuned circuit, which consists of the plug-in coil and the .00015 mf. tuning condenser.

Grid-leak and grid-condenser detection is used and the grid return of the tuned circuit goes directly to the cathode, which is connected to the ground. The standard feed-back plate coil is in the conventional place, and the radio frequency energy is prevented from getting into the audio amplifier by means of the radio frequency choke and the .0001 mf. mica by-pass condenser. Regeneration is controlled by means of the 75,000 ohm potentiometer connected between the chassis and the 150,000 ohm resistor. This series resistor drops the maximum voltage of this circuit to the point that will permit smooth regeneration control and provide the maximum sensitivity.

The screen circuit is by-passed to the chassis and ground by means of the ½ mf. metal-cased by-pass condenser, connected between the middle arm of the potentiometer and the ground. Smooth control of regeneration is obtained by means of this method and has the added advantage of minimizing the detuning effects. The critical point and maintenance of smooth regeneration control can best

be obtained by variation of the grid-leak, which in most cases seems to be most satisfactory when a 2 meg. leak is used with this receiver. Variation of the capacity of the antenna series condenser will also effect the smoothness of this control and experience in operation will enable the set builder to obtain the maximum and smoothest results.

Coils are available which will permit this receiver to tune from 200 meters down to 15. Additional coils may be obtained to use with this tuning condenser which will permit tuning in any of the stations in the broadcast band from 200 to 550 meters. If the receiver does not oscillate reverse the terminals XX in the diagram shown in Fig. 1. In general, all these plug-in coils have their sockets and terminal connections made as shown in Fig. 1, i. e., BA goes to radio frequency choke, B the grid condenser and F to ground. Coils made by the Alden Mfg. Company must have the B connection to the plate of the detector tube and P connection of the coil socket to the R.F. choke in order that they can be made to oscillate properly. All the connections on these coils are for radio frequency amplification and unless this point is understood, the connections as indicated will not give the regenerative effect which is so desired. Care should be taken at this point to see that the proper connections are made to the feed-back coil so that the set will go into oscillation and give the best results.

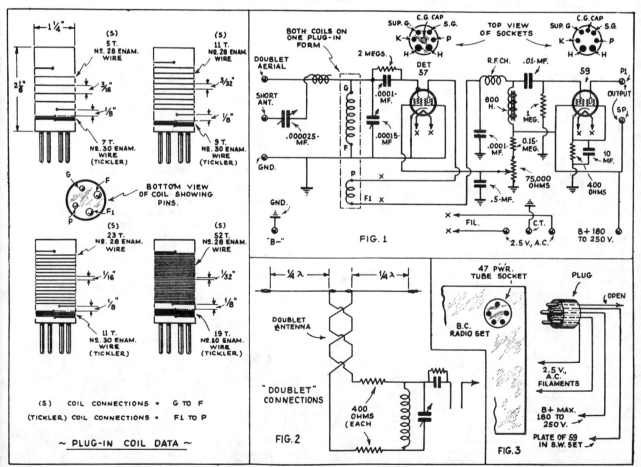

Above, to the left of the diagram, we have details of the plug-in coils used with the "Voyager" 2-tube receiver. The schematic wiring diagram is shown at the top of the drawing at right. One of the smaller drawings below shows connection of the doublet antenna with transposed lead-in and Fig. 3, lower right, shows how SW set can be plugged into output socket of your "broadcast" receiver.

GRIGSBY-GRUNOW CO.
SHORT-WAVE CONVERTER

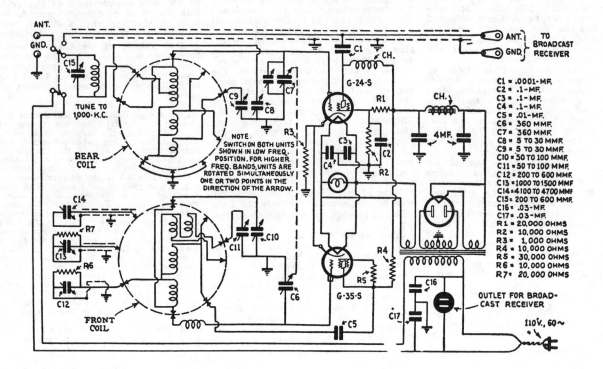

C1 = .0001-MF.
C2 = .1-MF.
C3 = .1-MF.
C4 = .1-MF.
C5 = .01-MF.
C6 = 360 MMF.
C7 = 360 MMF.
C8 = 5 TO 30 MMF.
C9 = 5 TO 30 MMF.
C10 = 50 TO 100 MMF.
C11 = 50 TO 100 MMF.
C12 = 200 TO 600 MMF.
C13 = 1000 TO 1500 MMF.
C14 = 4100 TO 4700 MMF.
C15 = 200 TO 600 MMF.
C16 = .03-MF.
C17 = .03-MF.
R1 = 20,000 OHMS
R2 = 10,000 OHMS
R3 = 1,000 OHMS
R4 = 10,000 OHMS
R5 = 30,000 OHMS
R6 = 10,000 OHMS
R7 = 20,000 OHMS

NOTE
SWITCH ON BOTH UNITS SHOWN IN LOW FREQ. POSITION. FOR HIGHER FREQ. BANDS, UNITS ARE ROTATED SIMULTANEOUSLY ONE OR TWO POINTS IN THE DIRECTION OF THE ARROW.

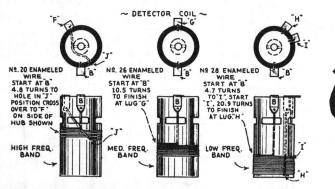

~ DETECTOR COIL ~

Nº 20 ENAMELED WIRE START AT "B" 4.8 TURNS TO HOLE IN "J" POSITION CROSS OVER TO "F" ON SIDE OF HUB SHOWN — HIGH FREQ. BAND

Nº 26 ENAMELED WIRE START AT "B" 10.5 TURNS TO FINISH AT LUG "G" — MED. FREQ. BAND

Nº 28 ENAMELED WIRE START AT "B" 4.7 TURNS TO "I", START "I", 20.9 TURNS TO FINISH AT LUG "H" — LOW FREQ. BAND

THE short wave converter here illustrated has been developed by the engineers behind the well-known Majestic line of broadcast receivers. When this converter is connected to a broadcast receiver, the tuning dial of which is set to 1,000 K.C. or 300 meters, then your broadcast receiver R.F. stages serve as the intermediate frequency stages of a super-heterodyne for short wave reception.

You may not realize it at first but if you couple such a converter as this to a ten-tube broadcast receiver, you have eleven "working" tubes in your short wave combination receiver, two of the tubes being rectifiers, of course. The value of such a combination, which results in a short wave super-heterodyne receiver, is just beginning to make itself known to broadcast listeners and short wave fans as well, and the tremendous amplifying power of the combination of S.W. converter and B.C. receiver brings

in far-distant stations on the loud speaker, which would be almost or quite impossible with any ordinary short wave receiver using but a few tubes.

The coil winding data and condenser as well as resistor values are given in the diagrams. With regard to the inductance shunted across the condenser C15, this combination may be a regular broadcast unit, which can be picked up in most any radio store. If C15 has a capacity of .00035 mf., then the coil may comprise about 81 turns of No. 30 enameled wire, close wound, on a ⅞-inch diameter tube. The R.F. choke ch may be about .5 millihenry. The iron core choke ch, connected between the two 4-mf. condensers in the plate filter, may be of the size usually employed in "B" eliminators, or about 32 henries. In the schematic diagram the rotation of the "rear" coil is counter-clockwise and clockwise for the front coil.

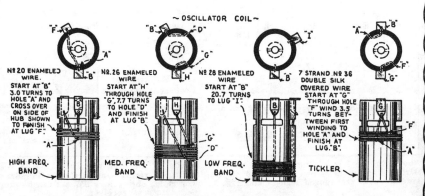

~ OSCILLATOR COIL ~

Nº 20 ENAMELED WIRE. START AT "B" 3.0 TURNS TO HOLE "A" AND CROSS OVER ON SIDE OF HUB SHOWN TO FINISH AT LUG "F" — HIGH FREQ. BAND

Nº 26 ENAMELED WIRE START AT "H" THROUGH HOLE "G", 7.7 TURNS TO HOLE "D" AND FINISH AT LUG "B" — MED. FREQ. BAND

Nº 28 ENAMELED WIRE START AT "B" 20.7 TURNS TO LUG "I" — LOW FREQ. BAND

7 STRAND Nº 36 DOUBLE SILK COVERED WIRE START AT "G" THROUGH HOLE "F" WIND 3.5 TURNS BETWEEN FIRST WINDING TO HOLE "A" AND FINISH AT LUG "B" — TICKLER

GROSS RADIO CO.
EAGLE BAND-SPREAD SET

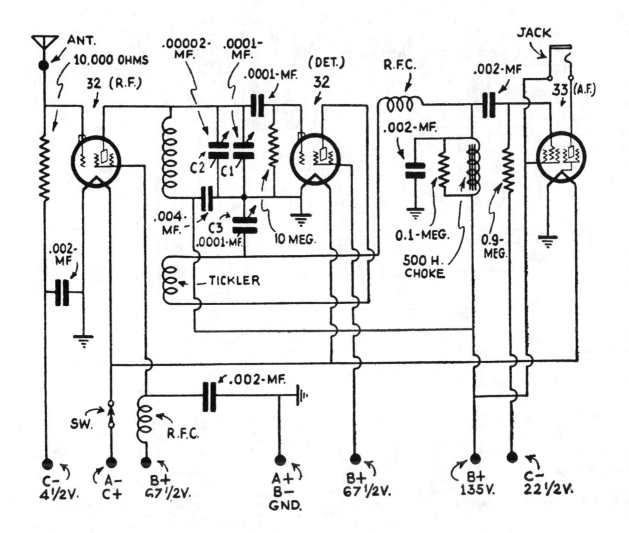

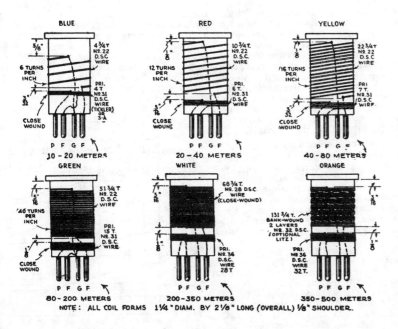

BLUE
4 3/4 T.
Nº 22
D.S.C.
WIRE
6 TURNS
PER INCH
PRI.
4 T.
Nº 31
D.S.C.
WIRE
(TICKLER)
OR
3 1/2 "
CLOSE
WOUND
P F G F
10 - 20 METERS

RED
10 3/4 T.
Nº 22
D.S.C.
WIRE
12 TURNS
PER INCH
PRI.
6 T.
Nº 31
D.S.C.
WIRE
CLOSE
WOUND
P F G F
20 - 40 METERS

YELLOW
22 3/4 T.
Nº 22
D.S.C.
WIRE
/16 TURNS
PER INCH
PRI.
7 T.
Nº 31
D.S.C.
WIRE
CLOSE
WOUND
P F G F
40 - 80 METERS

GREEN
51 3/4 T.
Nº 22
D.S.C.
WIRE
'40 TURNS
PER INCH
PRI.
15 T.
Nº 31
D.S.C.
WIRE
CLOSE
WOUND
P F G F
80 - 200 METERS

WHITE
68 3/4 T.
Nº 28 D.S.C.
WIRE
(CLOSE-WOUND)
PRI.
Nº 36
D.S.C.
WIRE
28 T.
P F G F
200 - 350 METERS

ORANGE
131 3/4 T.
BANK-WOUND
2 LAYERS
Nº 32 D.S.C.
WIRE
(OPTIONAL
LITZ)
PRI.
Nº 36
D.S.C.
WIRE
32 T.
P F G F
350 - 500 METERS

NOTE: ALL COIL FORMS 1 1/4" DIAM. BY 2 1/8" LONG (OVERALL) 1/8" SHOULDER.

Four plug-in coils give the "Eagle" a range of 17 to 200 meters, in these steps: 17 to 30, taking in the 20-meter amateur and 25 meter relay broadcasting channels; 30 to 63 meters, covering numerous commercial radiophone, aircraft and relay broadcasting stations; 62 to 110 meters, covering the 75 meter amateurs and many airport transmitters; and 100 to 200 meters, covering the extremely active "police" channels, amateurs and many experimental stations. These coils are of the two-winding type, and use four-prong forms.

The small condenser C2, controlled by a very smooth action vernier dial, is in parallel with the "tank" condenser C1. The idea is to spot the approximate location of any particular short-wave channel on C1 and then to do the actual tuning with C2. The band spreading action takes place at any position of C1, and is not merely present on limited portions of the tuning scale, as is the case with most "ham" band-spreaders.

Condenser control of regeneration, which is now returning to merited popularity, is used in preference to the resistance method because it is absolutely noiseless and positive. The detector slides into oscillation with that slow, gradual hushing sound characteristic of the perfectly functioning circuit.

HAMMARLUND MFG. CO.
"COMET" RECEIVER

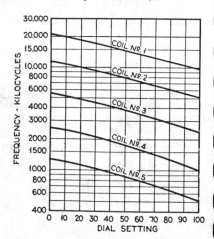

The interesting all-wave superheterodyne circuit developed by Hammarlund engineers. The wave bands from 14 to 550 meters are provided for by changing two plug-in coils at sockets "WL" and "OSC," insuring maximum freedom from dead-end losses. All circuits are fully by-passed and with the carefully designed I.F. stages, and A.F. amplifier, a signal of great volume and high quality is assured on all wave lengths.

For maximum efficiency throughout the entire wide range of frequencies which this receiver covers, special Isolantite form plug-in coils are used. These provide complete coverage of all the bands, including broadcast. There are five sets of coils, two coils to a set, having the following ranges: 14-30, 28-60, 56-125. 120-300, and 240-550 meters.

~ DATA ON COMET COILS ~

COIL No.	WAVE LENGTH RANGE METERS	PRIMARY No. TURNS	PRIMARY WIRE SIZE	SECONDARY No. TURNS	SECONDARY WIRE SIZE	LENGTH OF WINDING	T.P.I
1-OSC.	14-30	4	No 30 D.S.C.	6	No 20 D.S.C.	1.000"	6
1-W.L.	14-30	4	" "	7	" "	1.167"	6
2-OSC.	28-60	4	" "	12	" "	1.000"	12
2-WL.	28-60	4	" "	14	" "	1.167"	12
3-OSC.	56-125	5	" "	24	" "	1.000"	24
3-W.L.	56-125	5	" "	33	" "	1.375"	24
4-OSC.	120-300	10	" "	70	No 28 S.S.C.	.785"	56
4-W.L.	120-300	5	" "	78	10/41 LITZ	1.393"	56
5-OSC.	240-550	14	" "	70	No 28 S.S.C.	1.250"	56
5-W.L.	240-550	8	" "	114	TWO BANK 10/41 LITZ	1.062"	

ALL COILS WOUND ON ISOLANTITE FORMS
1½" DIAMETER. ALL PRIMARIES WOUND IN GROOVE
3/16" WIDE, 1/64" DEEP SPACED 1/16" FROM
SECONDARY

INTERMEDIATE TRANSFORMER COILS

TWO "UNIVERSAL" WINDINGS OF 10/41 LITZ WOUND
ON TREATED WOODEN CORE 9/16" DIAMETER.
EACH COIL HAS INDUCTANCE OF 1.2 MH. AND
THE TWO COILS ARE SPACED 31/32" APART

INTERMEDIATE (LONG WAVE) OSCILLATOR COILS

TWO "UNIVERSAL" WINDINGS OF No 32 D.S.C.
WIRE WOUND ON TREATED WOODEN CORE 9/16"
DIAMETER. EACH COIL HAS INDUCTANCE
OF .5 MH. AND THE TWO COILS ARE SPACED
¼" APART.

Although it is obvious that a lower intermediate frequency would afford even greater selectivity, by reason of a further increase in the percentage frequency difference, there is another considera-tion which makes a high intermediate frequency desirable. All superheterodyne receivers are subject to "image" inter-ference, which stated briefly, means an undesired signal whose frequency differ-ence from the desired signal is exactly equal to twice the intermediate frequency used in the receiver. It naturally follows that a high intermediate frequency les-sens interference from this source. A maximum spread between a desired sig-nal and its image interference is especi-ally important in *short wave* reception. On the other hand modern design neces-sitates the use of an intermediate fre-quency materially lower than that of any of the signals to be received. For these reasons 465 K.C. was chosen as the inter-mediate frequency for the "Comet".

Discussion of Circuit

The super-heterodyne has often been referred to as the "king" of radio re-ceivers, chiefly because its circuit simpli-fies the problem of obtaining *uniform* radio frequency amplification of almost any desired amount and at the same time a high order of selectivity which is also substantially uniform over a wide band of signal frequencies.

With the superheterodyne principle this difficulty disappears. By means of the local heterodyne oscillator, the 1,000 K.C. signal (which we shall assume to be the one desired) is changed to 465 K.C. At the same time, the undesired 990 K.C. signal is changed to 455 K.C., and both signals are impressed on the intermediate amplifier. The intermediate

amplifier then has the task of amplifying the 465 K.C. signal (for which it is tuned) and reducing (or rejecting alto-gether) the 455 K.C. interference. This is comparatively easy as the percentage difference here is 10/465 or over 2%. The effective selectivity in this case has been more than doubled. This effect increases rapidly with increasing signal frequen-cies. In the case of the 15,000 K.C. (20 meter) and the 14,990 K.C. signals the same process takes place. The 15,000 K.C. signal is changed to 465 K.C. and the 14,990 K.C. interference to 455 K.C. This also results in a percentage differ-ence of more than 2% (as was the case with the 1,000 K.C. and 990 K.C. sig-nals) which corresponds to a gain in selectivity of over 30 times as the origi-nal percentage difference between the two signals was only 1/15 of 1%.

HAMMARLUND MFG. CO.
"COMET PRO"

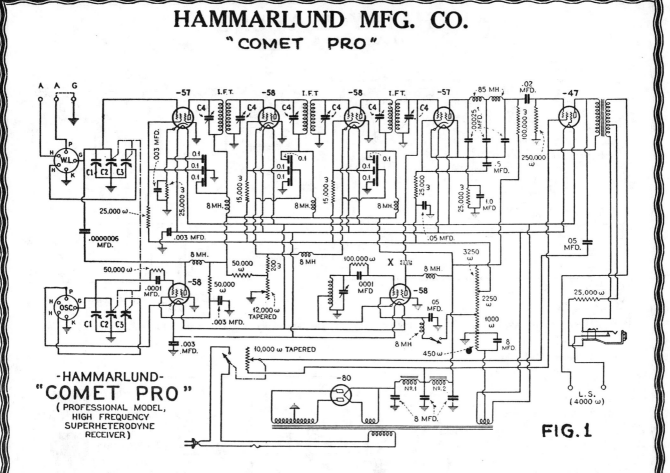

-HAMMARLUND-
"COMET PRO"
(PROFESSIONAL MODEL,
HIGH FREQUENCY
SUPERHETERODYNE
RECEIVER)

FIG. 1

An intermediate frequency of 465 kc was chosen as a compromise. It is below the broadcast band, and at the same time is high enough to provide a large spread between a desired signal and its "image" interference. By using Litz wound intermediate coils the selectivity and sensitivity are kept high.

"Band-Spread" Feature

The arrangement of the tuning condensers is interesting and unique. The fundamental circuit is shown in Fig. 1, and although designed primarily to give a band-spreading action on the four amateur bands of 20, 40, 80, and 160 meters, the same effect is obtainable throughout the entire range from 15-250 meters (20,000 to 1200 kc). Condensers C1, of 138 mmf. each,

constitute *tank* condensers and are individually controlled by separate vernier dials, one at left center and one at right center of the panel. By means of these two condensers, together with the appropriate set of coils, the receiver may be tuned to any frequency within its range. After this has been done, the main tuning dial, which controls condensers C2 and C3, will provide substantially true *single control* over a relatively narrow band of frequencies. If the main dial is set at 50 when the adjustment of the two tank condensers is made, approximately half of the spread band will be above and the other half below the mean frequency determined by the choice of coils and the setting of the two tank condensers. If the main dial is at zero when the tank condensers are

adjusted the entire spread band will be above that frequency. Conversely, setting the band with the main dial at 100 will throw the spread band on the lower frequency side. The dials on the two tank condensers are finely and accurately calibrated to facilitate precise logging. While calibration curves are furnished with each receiver, the operator should make an accurate calibration of his own receiver by means of standard frequency signals, certain stations known to be well controlled, etc.

This type of band spreading circuit necessarily results in a non-uniform band width at various frequencies, and this fact should be taken into consideration by the operator. At 20 megacycles the band is approximately 1500 kc wide and narrows to 300 kc wide at 10 mega-cycles (using the "AA" coils). With the "BB" coils the band width is 1000 kc at 10 mc. and 150 kc wide at 5 mc. The band spreading on these two ranges is accomplished by the 15 mmf. condensers C2 and C2, Fig. 1, on the main tuning dial. These condensers alone are inadequate for proper band width in the 5 mc. to 1.5 mc. range covered by the "CC" and "DD" coils. In this range, the 26 mmf. condensers E and F (Fig. 1) are connected into the circuit also. However, no switch is necessary, as this additional connection is automatically made when the "CC" and "DD" coils are inserted in their sockets. The fifth coil prong (which is not used in Coils "AA" and "BB") is used for this purpose in Coils "CC" and "DD." In this frequency range the band width varies from approximately 1200 kc. at 4.5 mc. to 225 kc. at 1.5 mc.

WINDING DATA ON COILS FOR NEW COMET "PRO"
W.L. Coils (to be wound on standard forms)

Coil No.	Wavelength Range	Primary Turns	Wire Size	Secondary Turns	Wire Size	T.P.I
AA W.L.	15-31	3	No. 30 DSC	7	No. 20 DSC	6
BB W.L.	28-61	3	" "	16	" "	12
CC W.L.	56-120	4	" "	29	" "	24
DD W.L.	115-250	5	" "	55	10/41 Silk Litz	56
EE W.L.	250-550	8	" "	136*	10/41-two bank, Silk Litz	

TPI equals Turns per Inch.
*The turns given are a guide only—the inductance should be 1½ or 2% greater than our present No. 5-W.L. coil.
OSC Coils (to be wound on new forms with holes for tap—these coils have no primaries).

Coil No.	Wavelength Range	Turns	Wire Size	T.P.I	
AA—OSC	15-31	7	No. 20-DSC	6	Tap at 1 2/3 turns from bottom
BB—OSC	28-61	14	" "	12	Tap at 2 2/3 turns from bottom
CC—OSC	56-120	23	" "	24	Tap at 4 2/3 turns from bottom
DD—OSC	115-250	39	28-SSC	56	Tap at 9 2/3 turns from bottom
EE—OSC	250-550	80	28-SSC	60	Tap at 16 2/3 turns from bottom

All taps to be soldered to the "P" terminal of coils CC-W.L., DD-W.L., CC-OSC and DD-OSC; coils also to have jumpers between the "G" terminal and the "H" terminal next to the "K" terminal.

Forms 1 and 7 sixteenths inch diameter

HATRY & YOUNG
HY-7B A.C. SET

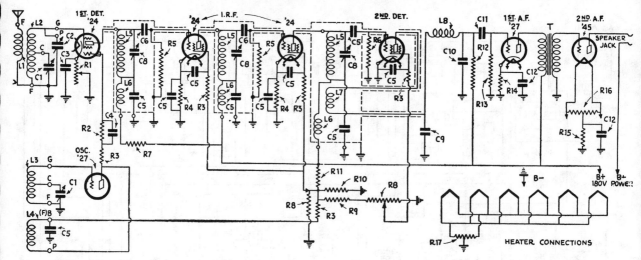

Complete wiring diagram of the Hatry type HY-7B, short-wave super-het, designed for A.C. operation.

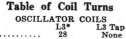

Table of Coil Turns

OSCILLATOR COILS

Coil	L3*	L3 Tap	L4
01	28	None	14
023	18	None	10
045	10	6	5

ANTENNA OR DETECTOR COILS

Coil	L1	L2*	L2 Tap	Lv
A1	5	35**	None	None
A2	3	19	None	None
A3	2	12	5	None
A4	2	9	7	6
A5	2	7	4	4

* L3 and L2 are wound with No. 22 D.S.C.; all other windings are No. 30 D.S.C.
Windings are placed not more than an eighth-inch apart on National coil-forms and are wound as close to the bottoms of these forms as is reasonable.

** For this coil L2 is 35 turns of 30 D.S.C. instead of 22 gauge.

L1—Antenna winding of A coils.
L2—Grid winding of A coils.
L3—Grid winding of O coils.
L4—Plate winding of O coils.
L5-C6-C8-C5-L6-R5 comprise an inter-mediate R.F. transformer and are all included in one can. L5-C8 must tune to 1,500 kc. or slightly higher. See original HY-7 article for suitable dimensions.
L6—R.F. choke.
L7—Tickler in Detector I.F.T. which also includes L3-C8-L6-C5-C7.
L8—R.F. choke such as Hammarlunds 85mh. or SPC.
C1—National 50 Mmfd. midget short-wave condenser, ST- or SE-50.
C2—Same as C1 but used as vernier and range extender.
C3—Sangamo .01.
C4—.00025 Mfd.
C5—.25 Mfd. non-inductive 200v., Sprague.
C6—.0005 Sangamo.
C7—.0002 Mfd. Sangamo.
C8—100 Mmfd. Hammarlund equalizer EC-80.
C9—.0001 Mfd. Sangamo.
C10—.0005 Mfd. Sangamo.
C11—.01 Sangamo.
C12—1 Mfd. Flechtheim, 250v.
R1—5000 ohms Clarostat potentiometer.
R2—.25 megohm Electrad metallic leak.

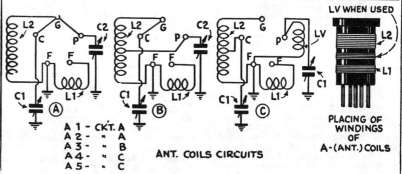

Details of oscillator and antenna coil circuits.

R3—2000 ohms Electrad flexible.
R4—400 ohms bias resistor.
R5—2 megohm Electrad metallic leaks.
R6—3 megohms Electrad metallic leak.
R7—.1 megohms Electrad metallic leak.
R8—50,000 ohms Electrad Royalty po-tentiometer.
R9—.15 Megohms Electrad metallic leak.
R10—.01 Electrad metallic leak.
R11—5000 ohms Electrad Truvolt type B50.
R12—.25 Megohm Electrad metallic leak.
R13—1 Megohm Electrad metallic leak.
R14—2700 ohms 2 watt Durham resistor.
R15—1500 ohms Electrad B15.
R16—20 ohms centertapped, Clarostat.
R17—10 ohms centertapped, Clarostat.

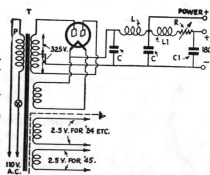

Wiring diagrams of the HY-7B plate, filament and heater supply.

HATRY & YOUNG
HY-7 SUPERHETERODYNE

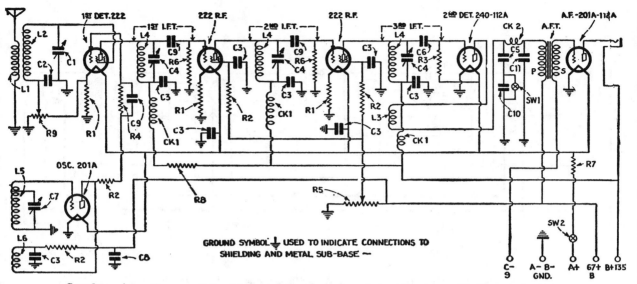

GROUND SYMBOL ⏚ USED TO INDICATE CONNECTIONS TO
SHIELDING AND METAL SUB-BASE ⏚

Complete wiring diagram of the 6-tube short-wave receiver designed by Mr. L. W. Hatry, and which embodies double detector principle with intermediate amplifier.

List of Constants

L1—Detector coil primary or antenna winding;

L2—Detector coil grid winding;

L3—Second-detector tickler. 8 turns of No. 30 D.S.C. on top of L4 at "B+" end;

L4—I.F.T. winding, 140 turns No. 30 D.S.C. on 1-inch (outside diameter) tubing. Three required;

L5—Oscillator-coil grid winding;

L6—Oscillator coil plate winding;

CK1—R.F. chokes, Hammarlund shielded type. In manufactured I.F. transformers for the "HY-7," CK1, L4 (and L3 in one), and C3 are within the shielding can along with C9 or C6. Thus live circuits are fully shielded;

CK2—Hammarlund shielded R.F. choke. One or three required;

A.F.T.—National A-100 audio transformer.

C1—50-mmf. Hammarlund or Pilot midget condenser for first-detector tuning; two required;

C2—.01-mf. Sangamo fixed condenser, by-pass for R9;

C3—0.25 mf. Sprague midget fixed condenser, six required;

C4—100-mmf. mica variable condensers (Hammarlund "EC80" equalizers), three required;

C5—.001-mf. Sangamo fixed condenser, two required if second detector is '40. (C5 becomes .0002-mf. if second detector is '12A);

C6—200-mmf. Sangamo fixed condenser. Grid condenser for detector;

C7—Same as C1, but tuning condenser for oscillator;

C8—1-mf. Tobe fixed condenser;

C9—500 mmf. Sangamo fixed condenser, three required unless second detector is '12A (see C10);

C10—500-mmf. for '40 second detector,

or 200-mmf. for '12A;

C11—.001 mf. Sangamo fixed condenser;

R1—15-ohm Yaxley filament resistor, three required;

R2—2000- or 3000-ohm Electrad flexible resistors. Used for R.F. choking or filtering effect, 4 required; if substituted in place of CK1s in home-made job, 7 would be required;

R3—Second-detector grid-leak, Electrad metallic type. 7 megs. for the '12A as second detector, or 4 megs. for the '40;

R4—100,000-ohm Electrad metallic fixed resistor, leak type. Reduces D.C. voltage placed on space-charge grid;

R5—50,000- or 100,000-ohm Electrad "Royalty" variable resistor, potentiometer type for volume control;

R6—2-megohm Electrad metallic leak, two required;

R7—Resistance to set filament voltage on tubes. Yaxley 4L or any 2-ohm rheostat or resistor adjustable to approximately 0.9 ohm;

R8—10,000-ohm Electrad metallic grid-leak resistor;

R9—50,000-ohm Electrad "Royalty" potentiometer. Biases first detector for best detection;

SW1—Battery switch, Yaxley "Type 10." Turns second detector in and out of oscillation;

SW2—Same as SW1 but turns battery current on and off. Two required.

Additional requirements in parts: two 5-prong tube sockets for coil sockets; two 4-prong sockets (Pilot) for first detector and oscillator tubes. If the standard "HY-7" aluminum chassis is purchased with the kit, the four additional sockets are riveted in place. A dial is needed for C7, and about 18 inches of shielded wire for wiring.

COIL-TABLE

Detector Coils

No.	L1	L1 tap	L2	L2 tap
A4-	2	4	33	8
A5-	2	5	18	11
A6-	2	no tap	50	13

Oscillator Coils

No.	L5	L6	Tap
01-	7	4	4
02-	18	10	no tap
03-	31	14	14

(Frequency ranges: 1700-3000 kc.—03-A6; 3000-4800 kc.—02-A4; 4700-6000 kc.—A5-03; 6000-7800—02-A3; 10,000-12,000 kc.—01-A2; 13,000-15,000 kc.—A1-01.)

INSULINE CORP. OF AMERICA
A.C. CONQUEROR

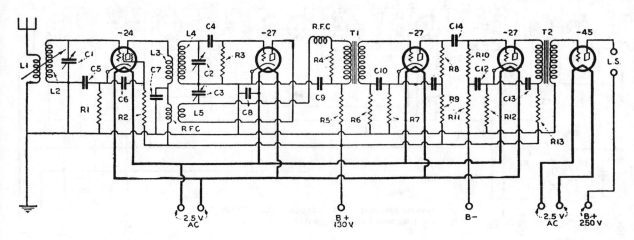

Diagram of connections for the A.C. "Conqueror" short wave receiver.

The circuit comprises a one-stage radio-frequency amplifier, regenerative detector, and three-stage audio-frequency amplifier. It is the typical short-wave type, with plug-in coils to cover the entire short-wave and broadcast bands. Five tubes are used: A screen-grid type '24, three type '27's and a type '45. In the power pack is a type '80 rectifier.

Accurately made, rigid, well-balanced coils are essential to properly cover the entire range and maintain selectivity and proper regeneration control. The coils are probably the most important part of the set. The form is of genuine hard rubber, 2" in diameter, rigidly held on metal end supports. The wire is wound in grooves in the hard rubber and cannot slip. The wire ends are connected to socket terminals on the coil base. Plug contacts are placed in the set. The absence of plug contacts on the coil makes it less vulnerable to damage when lying about not in use.

THE COILS
17-28 Meters
R.F. coil 6 turns.
Det. coil 4 turns on the secondary, 1 turn on the primary, 4 turns on the tickler.

27-45 Meters
R.F. coil 11 turns.
Det. coil 8 turns on the secondary, 3 turns on the primary, 6 turns on the tickler.

40-80 Meters
R.F. coil 19 turns.
Det. coil 18 turns on the secondary, 4 turns on the primary, 8 turns on the tickler.

75-150 Meters
R.F. coil 34 turns.
Det. coil 30 turns on the secondary, 10 turns on the primary, 15 turns on the tickler.

145-300 Meters
R.F. coil 54 turns.
Det. coil 54 turns on the secondary, 15 turns on the primary, 18 turns on the tickler.

295-600 Meters
R.F. coil 107 turns.
Det. coil 139 turns on the secondary, 30 turns on the primary, 50 turns on the tickler.
Antenna coil is 1-15/16" in diameter 10 turns.

The complete set of coils covers the range of from 14 to 600 meters. Two coils are used to cover the broadcast band. This is an advantage; it spreads the low wavelength broadcast stations over the entire dial and they are more easily separated and tuned in. As far as broadcast reception is concerned, this set ranks with the best. This type of coil is the same as that used in almost all ship and shore stations.

The two 2,000-ohm resistors are for obtaining the proper "C" biasing voltages. The r.f. chokes shown in the radio-frequency amplifier and detector plate leads are also important for efficient filtering.

Excessive a.c. hum in short-wave sets was traced to the detector. Probably the a.c. magnetic field surrounding the detector heater causes modulation of the plate current by affecting the electron stream. However, this has been completely wiped out by proper filtering. In the first place, the detector heater is maintained at a positive bias of 65 volts—by means of the bias resistor in the power pack, Fig. 2. This biasing voltage, together with the .001 mfd. by-pass condenser connected between the detector heater and cathode and placed directly at the detector socket, reduces a.c. hum to a point where it is imperceptible—even with headphones connected to the output of the three-stage audio amplifier.

The Power Supply
The power supply is clearly shown in the diagram of Fig. 2. It supplies the 180- and 250-volt plate leads and the two 2½-volt filament leads to the set by means of a cable connecting it to the set. It is built separately in a metal case as shown in the photographic illustration.

An objectionable feature with many sets is the fluctuation of voltage supplied by the power pack when operating the set. With a regenerative detector, the detector plate current varies considerably, depending upon the degree of regeneration. On power packs in which an unusually high ohmage resistor is used to cut down the voltage to the desired amount, the variation in load, caused by the variation in regeneration, produces a great voltage fluctuation, resulting in unstable and unreliable set operation and difficulty in tuning. In this power pack an unusually low ohmage bleeder resistance is used. This causes a rather large current drain from the power pack—large in proportion to the drain caused by the detector tube of the set. Therefore, any change in detector plate current has little effect on the supply voltage and steady operation is obtained.

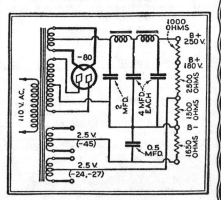

Circuit of "B" supply unit.

INSULINE CORP. OF AMERICA
2 TUBE SCOUT

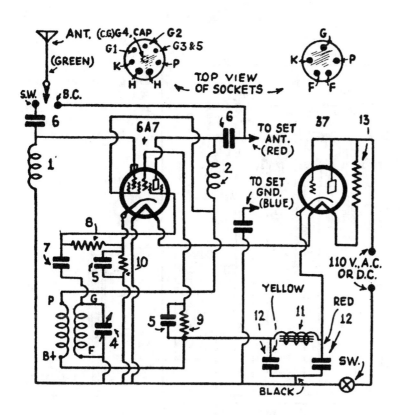

the ICA Short Wave Scout is here illustrated; one of its features is that it utilizes but two tubes and it can be instantly switched into circuit by means of a switch mounted on the front of the converter. The same switch, when thrown in the opposite direction, connects your broadcast receiver for regular 200 to 550 meter reception. This converter, which is mounted in a very neat and small walnut cabinet, may be connected to any broadcast receiver and complete instructions and wiring diagram come with each converter.

It is a very simple matter for the purchaser to quickly connect the converter to his broadcast receiver. The antenna wire is disconnected from the regular broadcast receiver and is connected to a post on the converter. A marked wire from the converter is connected to the antenna post of the "BC" receiver. The ground connection is left on the "BC" receiver and after that

Hook-up of S-W Converter

one has simply to plug the cord from the converter into a convenient 110 volt AC or DC lamp socket. The dial on the BC receiver is set to approximately 650 KC or 460 meters, or as close to this point as freedom from BC interference will permit; for tuning the short waves the center dial on the converter is used, the dial on the BC receiver being left set at 650 KC.

The range of this ICA *Scout* S-W Converter is from 200 meters down to 60 meters and by means of an additional plug-in coil supplied at slight extra cost, the short-wave range can be extended down to 20 meters.

ALAN RADIO CO.
"PRIZE WINNER"

Though primarily intended for headphone operation, this set will operate a magnetic or small dynamic speaker on many local signals. The circuit utilizes a sensitive type 78 grid-leak detector, a 43 power amplifier, and a 25Z5 rectifier, a combination of tubes which adapts itself readily to both A.C. and D.C. operation with practically identical performance.

Aside from the use of the 78 tube, the detector circuit is a conventional regenerative grid-leak arrangement. The grid-leak found most satisfactory was 5 megohms, and the grid condenser .0001 mf. Regeneration is smoothly controlled by means of a 50,000 ohm potentiometer of the carbon element type in the .78 screen lead, by-passed with a .5 mf. condenser. The tuning condenser has a maximum capacity of 165 mmf. or .000165 mf. and about 9 mmf. (.000009 mf.) minimum capacity. This, coupled to a 10:1 tuning dial, makes for fairly easy tuning.

In the antenna circuit, a small "trimmer" of 40 mmf. maximum has been included for the elimination of *dead spots*.

Coil Data For Use With 165 Mmf.

Coil Forms: 1¼ inch diameter, five prong type.

"G" terminal of grid winding at top of form.

"P" *terminal* of tickler winding at bottom of form.

All windings close wound except "A." Spacing on "A" equal to diameter of wire.

All windings wound in same direction. Grid terminal of *form* not used.

Wavelength (meters)	Grid		Tickler		Separation
	Type of wire	Turns	Type of wire	Turns	
'A'-14 to 35	No. 20 enam.	5	No. 28 SCC	3	5/32"
'B'-34 to 63	No. 20 enam.	11	No. 28 SCC	4	5/32"
'C'-62 to 112	No. 24 SCC	19	No. 28 SCC	5	1/8"
'D'-110 to 195	No. 26 SCC	48	No. 28 SCC	7	3/32"

List of Parts

COILS
1 set four "Prize Winner" coils to cover 14 to 200 meters or as described in table.
1 AUDIOFORMER (National, etc.)
1 15 h., 100 ohm filter choke
1 25 h., 350 ohm filter choke

CONDENSERS
1 .00004 mf. Ant. Trimmer (Hammarlund)
1 .01 mf. bypass, Aerovox
1 .02 mf. bypass Aerovox
1 Prizewinner 165 mmf. short-wave condenser
1 .0001 mf. mica bypass condenser (Elmenco)
1 .00035 mf. mica bypass condenser (Elmenco)
1 10 mf. 25 volt electrolytic bypass (Dubilier)
1 Filter block consisting of two 8 mf. and one 16 mf. 200 v. electrolytic condensers (Aerovox, Wego)

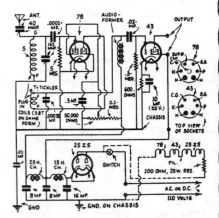

Wiring diagram of the new "Alan" Short-Wave Receiver—it employs plug-in coils to change the wave bands. The tubes are operated in series with a small resistance across the line.

RESISTORS
1 5 meg. resistor
1 600 ohm one watt resistor (Elmenco)
1 100,000 ohm ½ watt resistor (Elmenco)
1 50,000 ohm potentiometer (Centralab)

INTERNATIONAL RADIO CORP.
ALL-WAVE DUO RECEIVER

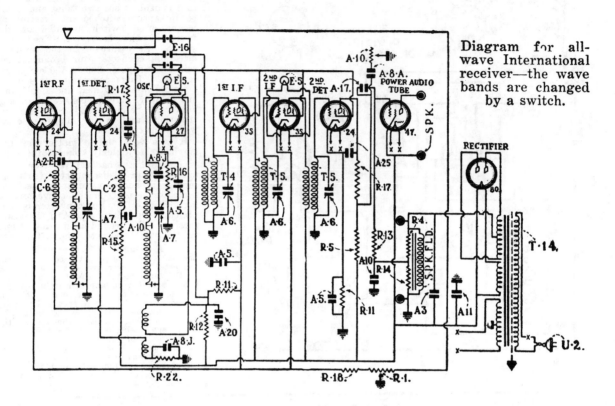

Diagram for all-wave International receiver—the wave bands are changed by a switch.

T HE International All-Wave Duo is a long and short wave receiver built in one complete chassis. There are two illuminated tuning dials, one for short waves, which will receive from 20 to 200 meters—the other for long waves, from 200 to 600 meters. One tone control which operates for both short and long wave reception; one switch for changing from long to short wave reception. There are no "plug-in" coils to change. Short wave coils are wound on a drum type selector which operates from a knob on the front of the panel.

There are three positions on the short wave band—No. 1, 20 to 75 meters; No. 2, 75 to 125 meters; No. 3, 125 to 200 meters.

The Duo is an eight-tube "super" het chassis; the tube equipment consists of: 2—235 Variable Mu; 3—224 Screen Grid; 1—227 Oscillator; 1—247 Pentode (output); 1—280 Rectifier.

Two-dial "dual" receiver, tunes from 20 to 600 meters; without "plug-in" coils.

The Duo short wave receiver *operates with oscillation under control at all times.* You can hear the carrier waves on the short wave stations, without the "whistling sound". In other words, the Duo tunes in short wave stations the same as a regular long wave distant station on any standard set; you only hear the voice or music.

The International All-Wave Duo can be obtained in a mantel type cabinet or chassis only for export. 110-220 A.C. voltage transformer which operates on 50 to 60 cycle current is furnished as standard equipment.

Simple Operation—The A.C. Switch is combined with the volume control, which controls the volume for both short and long wave bands. After set has been turned on, you can switch from short to long waves instantly, or back to short waves again, by turning the lower left hand knob.

COLIN B. KENNEDY CORP.
COMBINATION L.W. & S.W. SET

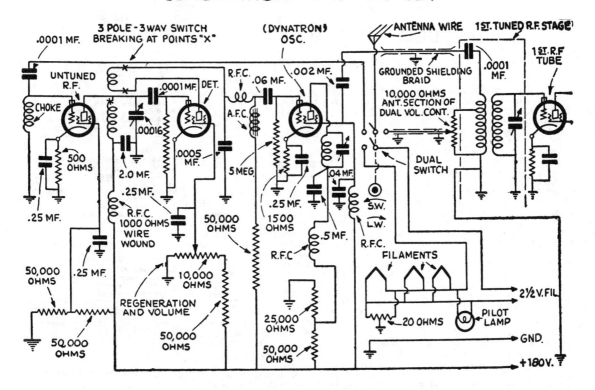

The receiver is not of the super-heterodyne type, but uses a dynatron oscillator, the output being taken from the screen of this tube through a condenser as shown. The frequency of the oscillator is fixed at approximately 1530 K.C., but can be varied.

For short wave reception it is only necessary to once set the dial of the long wave receiver at the output frequency of the short wave oscillator, and then tune the short wave receivers' single dial, as well as operate the regeneration control.

This long and short wave receiver thus employs a total of six screen grid tubes, four being tuned amplifier tubes and thus producing a tremendous overall amplification of the short wave signals. Including the audio stages, a total of 11 tubes are employed for short wave.

KENNEDY GLOBE TROTTER
CONVERTER

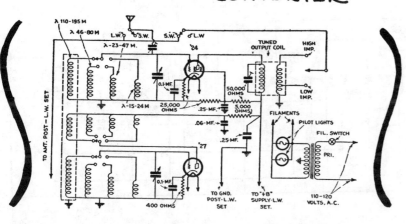

It will be noted that on the back of the short wave unit a wire has been brought out which may be connected to either one of the two binding posts near the end of the base. The purpose is to enable the user to adapt the unit to his particular type or make of broadcast receiver for obtaining the best results. After the short wave unit is in operation, either one of the two binding posts may be tried first on one and then on the other of these two posts, and permanently left where best results are obtained.

The lower right hand knob on the short-wave unit may be turned to five different positions. One of these positions is marked "long wave" and when the switch is turned to this position, the antenna is automatically connected directly to the long-wave receiver and the output of the short-wave unit is disconnected. The broadcast receiver may then be operated exactly as though the short-wave set or converter were not there at all.

When short-wave stations are desired, this switch is turned to the particular point, which includes the desired wave-length, and the filament switch is turned (lower left hand knob) to the "on" position, which will turn on the short-wave unit tubes and light up the dials.

The long wave receiver dial must then be turned to the end of the scale, or 1,500 kilocycles.

LEOTONE RADIO CORP.

4 TUBE DRY BATTERY SET

Improved selectivity is afforded by providing a well-designed tuned radio frequency stage. The plug-in coils used in the antenna circuit are of the same 4-pin type as those used for coupling the plate circuit of the R.F. tube with the grid of the detector tube. These plug-in coils may be of any standard make, data for winding which are given herewith, and designed for use with .00014 mf. tuning condensers. The two tuning condensers are ganged and a small variable trimming condenser of about 30 to 50 mmf. is connected across the second tuning condenser in the plate circuit of the R.F. tube.

To prevent the possibility of short-circuiting the B battery should the plates of the second tuning condenser accidentally touch, a small fixed condenser of .01 mf. is connected in series with the plate of the R.F. tube and the stator of the second tuning condenser as the diagram clearly shows. Due to the high value of this blocking condenser there is no appreciable reduction in the tuning capacity.

Any average size antenna system may be used with this short-wave receiver or approximately the same size antenna as that used for broadcast reception. Instead of using a series trimming condenser in the antenna circuit the trimmer in this case is connected across the second tuning condenser in the plate circuit of the R.F. tube, as indicated in the diagram.

Detector Circuit Features

In this particular receiver grid-leak detection is employed, which provides high sensitivity and also smooth operation of the regeneration characteristic, the detector operating approximately on the square law. Regeneration is controlled in this detector circuit by means of the 100,000 ohm rheostat (potentiometer) which is connected in series with the 22.5 volt B plus feed, as illustrated in the diagram.

A volume control is also provided in this interesting circuit. In series with a 15-ohm filament resistor in the positive wire feeding the R.F. tube, we find a 20 ohm variable resistance, which is used as a *volume control*. This 20 ohm variable resistor also serves the purpose of checking any oscillations which might be set up in the R.F. circuit.

As in all other short-wave receivers, the radio frequency choke, RFC, is an important item and it should have low distributed capacity with a maximum of inductance, the value of the R.F. choke in this case being about 30 mh. One may try R.F. chokes having inductances from anywhere from 20 to 100 mh. A by-pass condenser of .00025 mf. is connected from the tickler to the filament circuit; the plate load resistor for the detector is 0.25 megohm, which is coupled through a .01-mf. fixed condenser to the 1-megohm grid resistor of the first audio amplifier, a 30 type tube. The plate load resistor for this first audio tube has a value of 0.25 megohm. This couples through another 0.01-mf. condenser to the 0.25-megohm grid resistor of the second audio or output tube, also of the 30 type.

List of Parts for Leotone Receiver

COILS:
2 sets of short-wave plug-in coils for 0.00014 mf. capacity. Alden, (Bruno).
1—30 millihenry honeycomb R.F. choke coil.

CONDENSERS:
1—two-gang 0.00014 mf. tuning condenser.
5—0.01 mf. mica condensers.
1—0.5 mf. bypass condenser.
1—0.00025 mf., mica condenser.
1—0.0001 mf. mica condenser.

RESISTORS:
1—15 ohm fixed filament resistor.
1—20 ohm rheostat.
1—2.7 ohm fixed resistor (may be improvised from a 6 ohm rheostat set to apply 2 volts on filament, when the 20 ohm rheostat is set at zero resistance).
1—3 meg. pigtail resistor.
1—100,000 potentiometer, with switch attached.
3—0.25 meg. (250,000 ohm) pigtail resistors.
1—1.0 meg. pigtail resistor.

OTHER REQUIREMENTS:
3 UX and one UY sockets.
1 antenna-ground connector.
1 speaker connector.
1 six-lead outlead cable.
1 drum dial, scale, escutcheon.
1 shielded box with hinge cover, overall 9¾" wide x 8¾" high x 8¾".
1 chassis with shield compartments, to fit inside shield cover.
2—"C" batteries; 1—3 volt, 1—22½ volt (small "B" unit).

LEOTONE-ALDEN PLUG-IN COIL DATA			
Meters Wavelength	Grid coil turns	Tickler turns	Distance between 2 coils
200-80	52 T. No. 28 En. Wound 32 T. per inch	19 T. No. 30 En. Close wound (CW)	⅛"
80-40	23 T. No. 28 En. Wound 16 T. per inch	11 T. No. 30 En. C. W.	⅛"
40-20	11 T. No. 28 En. 3-32" between turns	9 T. No. 30 En. C. W.	⅛"
20-10	5 T. No. 28 En. 3-16" between turns	7 T. No. 30 En. C. W.	⅛"
Coil form—2¾" long by 1¼" dia. 4-pin base.			

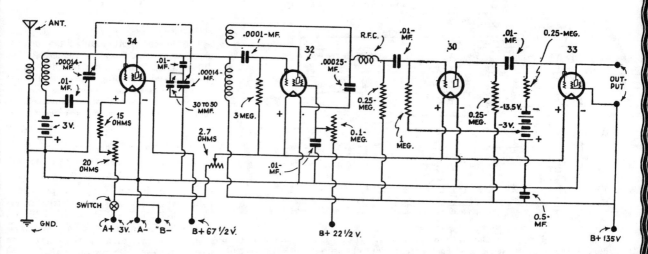

The interesting hook-up devised by the Leotone engineers, whereby maximum strength of signal is obtained, with a minimum of battery consumption. An A. C. model is also available.

EASTERN RESEARCH LABS.
"LEUTZ" YACHT RECEIVERS

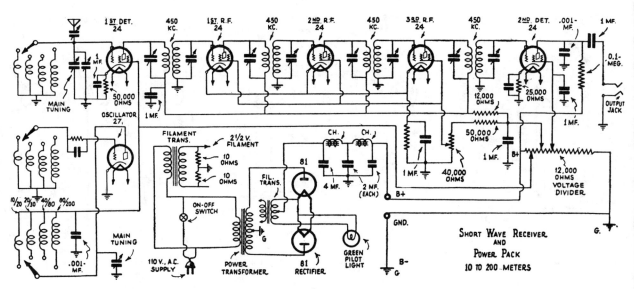

Wiring diagram of the short-wave receiving set installed on the yacht, "Aras"

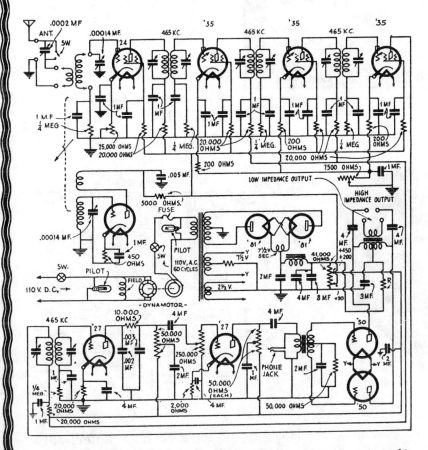

Wiring diagram for the "short-wave" superhet receiver installed on the yacht "Migrant."

● THE successful design of broadcast receiving equipment for yacht installations is a problem all by itself and is not generally understood. With the craft at dock and near broadcast transmitters, any average broadcast receiver works fairly well. With the yacht under way, it is another matter and each installation must be given individual consideration

Range 15 to 2000 Meters

The "ARAS" installation is divided into three receiver sections. First there is the main *broadcast* receiver covering 200 to 570 meters, together with a power amplifier. Second, there is a *short wave* tuner covering 15 to 200 meters, the output of which can be switched through the power amplifier of the above *broadcast* receiver. The third section is an *auxiliary receiver*, entirely separate from the above and having its own power pack. The auxiliary receiver in addition to tuning 200 to 570 meters also tunes from 550 to 2,000 meters, allowing the reception of foreign broadcast wavelengths when the vessel is in European waters.

The main broadcast receiver has three stages of high gain tuned radio frequency amplification using -24 tubes. One band pass filter stage precedes the first radio stage. The detector stage is also tuned using a -24 tube with plate rectification. This makes a total of five tuned circuits. An antenna series variable condenser is provided to adjust the antenna electrically to the optimum value for the different wavelengths received.

The plate circuit of the detector is resistance coupled. Following the detector is an initial stage of audio amplification. Ordinarily, it would seem that this stage of audio would not be necessary. It has a low ratio of amplification and becomes very useful when receiving relatively weak signals and it is also important in connection with the electric phonograph.

Yacht "Migrant" Installation

This short wave receiver covers 13.8 to 200 meters with six sets of coils. Other coils are provided to tune wavelengths up to 1,000 meters so that the apparatus can also be used for regular broadcast band reception as an auxiliary.

Tubes are arranged as follows:
-27 Oscillator
3 -35 Intermediate radio frequency stages at 465 K.C.
-27 Second Detector
-27 First Audio
2 -50 Power Push Pull tubes
Power Pack (in separate case) has
2 -81 rectifiers

EASTERN RESEARCH LABS.
MODEL "C" & MODEL "L" SETS

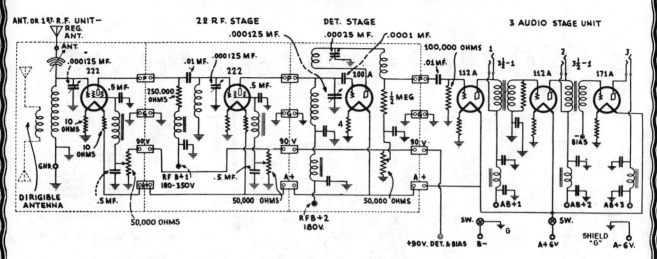

Complete wiring diagram of the Leutz short-wave receiver; three stages of audio amplification are used.

These two receivers, "Model C" and "Model L," are of unit construction, and resemble in appearance the famous Leutz "Transoceanic Silver Ghost," premier long-wave receiver. The short-wave receivers, however, are not quite as lengthy, having at most only four units; whereas the "Silver Ghost," when all the units are assembled, comprises six.

The unit form of construction allows extreme flexibility. The units comprising the detector and the audio stages may be combined to form the receiver; if greater volume is desired, one or both of the R. F. stages may be added.

LEUTZ
S.W. CONVERTER

Appearance of Leutz adapter that makes short-wave reception possible with your broadcast receiver. It has its own "B" and filament-heater supply.

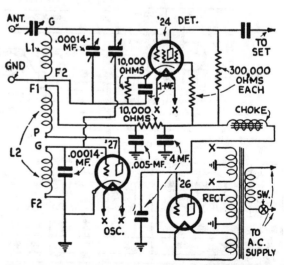

Hook-up of Leutz S.-W. adapter; only two wires connect to B.C. set.

LINCOLN RADIO CORP.
R-9 — 9 TO 200 METER SET

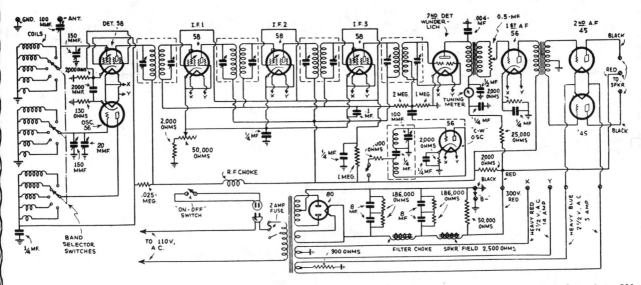

Diagram of the new Lincoln R-9 Superheterodyne, especially designed for reception in the short-wave spectrum from 9 to 200 meters. Switches select the proper coils for each band.

Photo above—business-like appearance of the newest Lincoln off-spring, the R-9 short-wave super-het, with 11 tubes.

● OWING to the increased demand in both commercial and broadcast listeners fields, many months of development work have been put on the new Lincoln R-9 receiver, designed to register wavelengths from 9 to 200 meters. While the field has been well covered with commercial types of *short-wave* receivers and combinations of *short-wave and broadcast*, yet the *strictly* short-wave receiver design has not had the attention that a few of the receivers ranging from 15-550 meters have had.

Silver-Contact Band Selector-Switch

The general plan of the R-9 employs the proved design of the DeLuxe SW-33, utilizing the silver-contact selector-switch, which independently selects the desired inductances for group frequency range. The grouping of the frequencies is as follows: Starting from the lower wave end—

1st	position	8.8 meters to	16.8
2nd	position	14.6 meters to	27.7
3rd	position	27.4 meters to	51.6
4th	position	48.2 meters to	99.
5th	position	86.2 meters to	216.

The circuit uses an intermediate frequency amplifier of three powerful stages, with tuned first detector stage. The coils are of Litz wire; eight tuned circuits are used. The last I.F. stage is of balanced-grid push-pull construction, feeding into the Wunderlich tube (2nd detector), the output of which feeds into the transformer coupled 56 first audio stage and through large transformer coupling into two 45 push-pull output tubes. Tubes used are 4—58; 3—56; 1—Wunderlich; 2—45; and 1—80.

Due to the remarkable action of the Wunderlich tube, perfect *automatic volume control* is had by controlling the complete I. F. amplifier. This feature can be eliminated and the set will work without A.V.C. by manipulating a switch on the front panel.

"C.W." Reception Provided for

A *beat oscillator* is employed for reception of "C.W." signals; this is also operated by a switch on the front panel. Full indication of signal is had with the meter mounted in the center of panel, allowing accurate tuning to the exact center of the carrier wave and also indicating unmodulated carriers which can be tuned with perfect accuracy. This valuable feature is of paramount importance, as many stations are "standing by" temporarily and would be entirely overlooked without this method of location.

Two dials are employed, the one to the left being the main *tuning dial* and the one to the right for *band-spreading*, which is effective on all frequencies. In commercial work, where the operator only works a specified band of frequencies, this bandspread dial is very desirable, allowing wide separation on the dial.

Sensitivity and Volume Controls

The regulation of *sensitivity* and *volume* is identical with the Lincoln DeLuxe SW-33, and it is one of the most satisfactory systems for the broadcast listener as well as the commercial operator. In order to get distance in the conventional type of receiver, one is required to advance the sensitivity control to a point where heavy noise and signal can be heard loud enough to disturb the whole neighborhood; while in the new Lincoln system, the sensitivity control can be advanced to a maximum, giving power to reach any distance and the volume control can be at minimum, with speaker volume only loud enough to be heard a few feet away from the speaker. This system also allows absolutely *silent tuning* by the use of the *signal indicator*.

The R-9 model is of the table mounting type, having a heavy metal removable cover and heavy metal front panel. The chassis is mounted on a wood base with moulding at bottom, and the whole unit is attractively finished.

A separate power pack is used, identical with the large DeLuxe all-wave model, together with a dynamic speaker. A head phone jack is incorporated.

LINCOLN RADIO CORP.

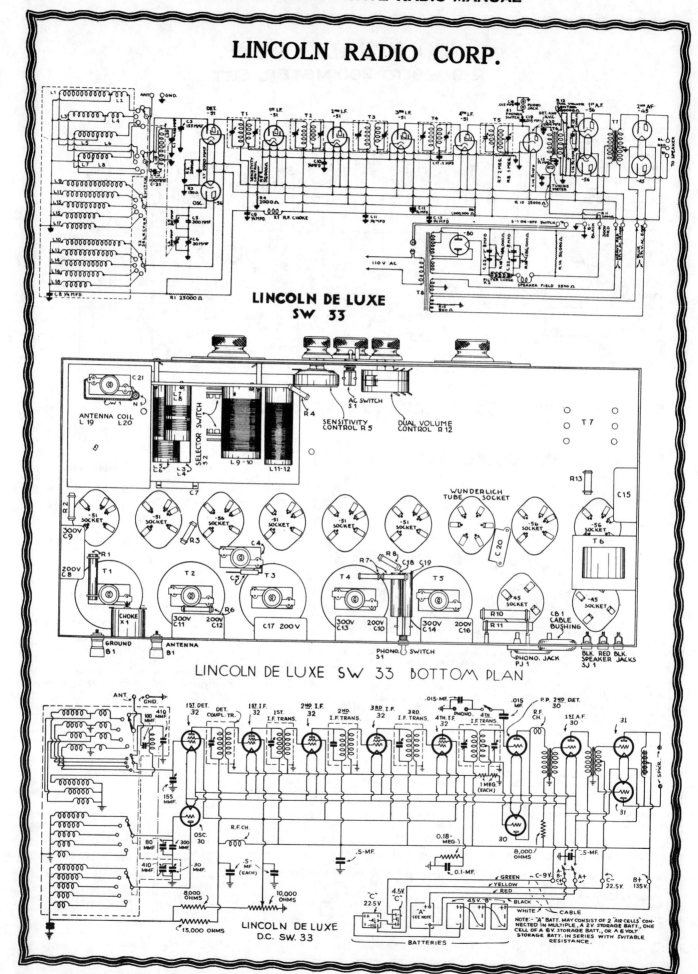

LINCOLN DE LUXE
SW 33

LINCOLN DE LUXE SW 33 BOTTOM PLAN

LINCOLN DE LUXE
D.C. SW. 33

LINCOLN RADIO CORP.
DELUXE 32 MODEL

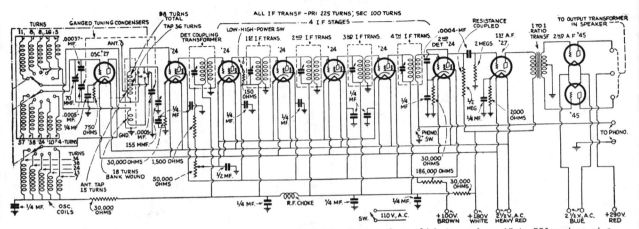

Complete wiring diagram of the new Lincoln De-Luxe "All-Wave" super-het, which tunes from 15 to 550 meters at a single "twist of the wrist". The various coils used for tuning in the different wavelengths and their connections to the "ganged" control switch, are shown at the extreme left of the diagram.

Tuning In Short Waves a Cinch!

Whether the purchaser of a Lincoln new All-Wave Super is interested particularly in short-wave reception, or only listens in occasionally on short waves, this cleverly designed and highly efficient receiver, will appeal irresistibly. Tuning in short waves is a "cinch" with the new Lincoln—a simple twist of the wrist on the band-selector switch puts immediately at your service the following wavelength ranges: 15-30; 30-50; 50-100; 100-200; and finally the broadcast range of 200-550 meters.

The length of antenna depends on local absorption. Ten feet of antenna in one locality will equal a 50-ft. antenna in another. As a general thing, buildings of steel construction require a longer antenna to compensate for loss of absorption in steel, state the Lincoln experts. Antennas from 15 to 100 feet will work satisfactorily.

Operation Hints

First you connect A.C. line to wall socket; turn right-hand control knob until switch is turned on, and see if all heaters are operating.

See that all adjusting knobs on top of I.F. (intermediate amplifier) cans point to marks on top of can. Turn volume control (right hand knob) about halfway on and tune in a station with main dial (center knob), adjusting trimmer (second knob from left) to loudest signal. Then reduce your volume control to a weak signal and carefully go over all of the adjustments on top of the transformers, peaking each one at the loudest point. Next, tune in a distant signal and go over the operation again. Once this adjustment is done, no further attention need be paid to it. Adjustments of the I.F. transformer should be made with the *low-high power switch in high power position.*

The total amplification of the receiver need not be used for the majority of long-distance reception, and for that reason there are two regulations for power; allowing the use of low amplification for the nearby more powerful stations. In fact, a range up to 2,000 miles in the broadcast band can be brought in with very good volume on low power, and only when extreme distance or short-wave reception is desired need the high power position be used.

How to Use Volume Control

Never advance the volume control past the point of stability. Tremendous amplification is available before oscillation will occur. The strength of the incoming signal is directly proportional to the amount of amplification required; and when a signal is so weak, due to atmospheric conditions, that same cannot be heard with a reasonable amount of amplification being used, nothing can be gained by advancing the volume control further.

Selecting Wave Bands By Switch

The range of the Lincoln De Luxe "32" equipment is from 15-550 meters. This range is divided into five groups to

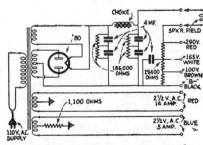

Diagram of Lincoln power supply.

be selected by switch at left side of control panel and the range is divided as follows:

15— 30 meters	30— 50 meters
50—100 meters	100—200 meters
200—550 meters.	

Each group will register from 0 to 100 on the dial. For instance, in the 15—30 meter band the 15 meters will register at low end of dial and the 30 meters at upper end of dial, and likewise throughout all of the group. The De Luxe "10" and "32" models tune very sharp and dial must be accurately set in resonance with the station as it is very easy to pass over a short-wave station, especially in the higher frequencies. Also, it is very essential that the left-hand antenna trimming knob should be kept in resonance to bring in the signals.

Now assume that you wish to tune in G5SW (Chelmsford, England); from your log book you note that this station uses a wavelength of 25.53 meters which falls in the 15-30 meter group. So merely turn the selector switch to the 15-30 meter position; turn your dial to about 65; bring the antenna trimmer to resonance; and tune slowly both sides of this position until G5SW is located. When W8XK is broadcasting on 25.25 meters, G5SW can be quickly located about one dial division above them.

The above procedure is followed for all short-wave reception. Stations in the 31-meter band are found on the lower portion of the dial when selector switch is set on the 30-50 meter position. The 49-meter band stations will then be found around 70 on the dial. The 85-meter band will be found near 60, with switch in the 50-100 meter position. One group of police calls will be found near 30 and another group near 70, with switch in the 100-200 meter position.

MIDWEST RADIO CORP.
SELF POWERED
SHORT-WAVE
CONVERTER

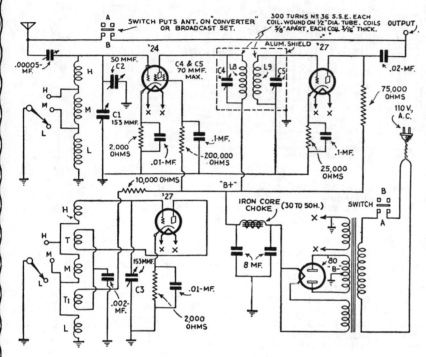

The Midwest S-W converter circuit is shown above; the "output" post connects to "aerial" post on broadcast set; chassis is grounded to "G" post on "B.C." set.

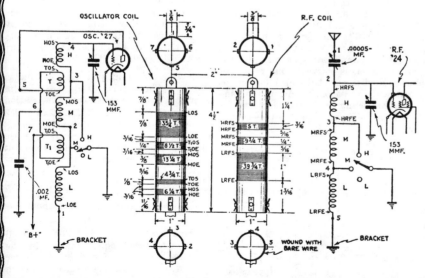

Data for the oscillator and antenna coils used in the Midwest S-W converter are given above; all coils wound with No. 31 D.S.C. (double silk covered cotton.)

Self-Contained "B" Supply Desirable

It is assumed that the converter should be of the self-sustained type and that it should contain all of the materials and of such quality as is necessary to produce the results expected under favorable conditions of course. It is therefore necessary that this converter should *contain its own power-pack* and should feed energy to the radio set and subtract nothing from it. For this reason both A and B supply units are included in the design.

Intermediate Frequency

Assuming that a modern broadcast set of the highest quality is to be used, the choice does not rest with the matter of selectivity, because these sets are rarely sensitive all over the band; the only other choice is one of selectivity. It is well known that sets are more selective at the low frequencies and that the modern broadcast set is better able to find a clear spot between stations at about 90 than at any other point on the dial. It is for this reason that the intermediate frequency was chosen at about 575 kilocycles.

In the event that the customer cannot find a clear spot at this point on account of local stations, he should be able to efficiently operate at some other frequency and provisions are made for this by having the "i.f." transformer on the converter of the top adjustable type. These adjustments are broad enough so that he can select frequencies up to 540, which is just outside of the broadcast band and might be an ideal point for him to operate provided he can meet it on his radio set.

This changing of the selected i.f. frequency requires a corresponding change in the relation between the oscillator circuit and the r.f. tuning, and the "trimmer" condenser must be large enough to compensate for this change.

Coil Requirements

We now have our coil requirements pretty well in mind. The r.f. coil should be tuned by the variable condenser for 200 meters to as low a point as is possible and desirable. Experience shows that the 80 meter "ham" band can easily be included in this first point on the switch. Correspondingly, the oscillator coil must cover this same band, plus 575 kilocycles, the intermediate frequency.

The second point on the switch should begin at this 80 meter point with a slight lap and again carried down as far as possible, and this is known to be about 35 meters. Correspondingly the third point should begin with a slight lap and go down to at least 15 meters.

MIDWEST RADIO CORP.
16 TUBE SET

Midwest Data

R1—.5 meg 0.5 W. (watt); CA—.01 mf. 200 V. (volts); R2—30,000 ohms 1 W.; R3—50,000 ohms 0.5 W. C1—.05 mf. 200 V.; R5—30,000 ohms 1 W. C3—.0005 mf. Mica; C4—.00036 mf. 3 gang; C5—.001 mf. (Mica) Pad 4 gang; C6—.05 mf. 400 V.; R6—700 ohms flex.; C8—.05 mf.; C9—.05 mf. 400 V. 200 V.; C10 20, mmf. Trim.; R7—200,000 ohms 0.5W; R8—30,000 ohms 1 W.; R9—30,000 ohms 1 W.; C12—.05 mf. 200 V.; R10—50,000 ohms .05W.; R11—50,000 ohms .05W.; C14—.05 mf. 400 V.; R13—30,000 ohms 1 W.; R14—50,000 ohms 0.5 W.; C15—.05 mf. 2000 V.; R15—275 ohms flex.; C16—.05 mf. 200 V.; R16—200,000 ohms 0.5 W.; R17—30,000 ohms 1 W.; C17—.01 mf. 200 V.; R18—50,000 ohms 0.5 W.; R19—200,000 ohms 0.5 W.; C18—.05 mf. 200 V.; R20—275 ohms flex.; C19—.01 mf. 200 V.; C 20—.1 mf. 200 V.; R21—200,000 ohms 0.5 W.; C21—.01 mf. 200 V.; R22—200,000 ohms 0.5 W.; C22—4 mf. spec.; C23—.05 mf. 200 V.; R23—200,000 ohms 0.5 W.; C24—.05 mf. 200 V.; C25—.05 mf. 200 V.; C26—.05 mf. 200 V.; R24—275 ohms flex.; R25—200,000 ohms 0.5 W.; C27—2 mf. 500 V. ELEC. R26—5 meg. 0.5 W.; R 27—2,500 ohms 5 W.; C28—20 mmf. Trim; R28—.5 meg. 0.5 W. C29—.05 mf. 200 V.; C30—.05 mf. 200 V.; R30—1,000 ohms 0.5 W.; C31—.001 mf. 400 V.; C33—.1 mf. 200 V.; R31—100,000 ohms 0.5 W.; R32—10,000 ohms 0.5 W.; C35—.05 mf. Spec.; C36—.05 mf. 200 V.; R33—200,000 ohms 0.5 W.; R34—.5 meg. Pot.—A.C. Switch Spec.; R35—600 ohms ± 5% flex.; C37—.05 mf. 200 V.; R36—12,000 ohms 0.5 W.; R37—5 meg. 0.5 W.; C38—.05 mf. 200 V.; C39—.05 mf. spec.; R38—12,000 ohms 0.5 W.; R39—100,000 ohms 0.5 W.; C40—8 mf. 450 V. Elec.; C41—.05 mf. 400 V.; R40—25,000 ohms 0.5 W.; R41—410 ohms flex. 0.6 W.; C42—4 mf. 450 V. Elec.; C43—1 mf. 400 V. Elec.; C44—8 mf. 450 V. Elec.

If a signal is tuned in and a negative voltage generated by the rectifier, this negative voltage is applied to the grid "Statomit" tube, completely blocking it, and the voltage across Rx falls to zero unblocking the audio tube and permitting it to amplify and pass the signal through to the loud speaker.

This action is accumulative and self-locking in such a way that when once started it carries itself through. This is accomplished in the following manner:

The voltage Et on the grid of the "Statomit" tube is composed of the drop across Ry, plus the automatic volume-control bias voltage. These voltages are in in series and equals about 80 volts. The voltage across Ry is produced by the total current composed of several constant currents plus the variable currents Is and Ia. When the set is blocked, Is is at a maximum and Ia is zero; therefore, only Is is effective in helping to produce the voltage across Ry. When the set is operative Is is zero and Ia is at a maximum; therefore maximum Ia is greater than maximum Is. The voltage across Ry will be greater when the set is operative than when the set is blocked. Now the voltage across Ry is aiding the voltage from the automatic volume control tube; therefore, when this automatic volume control bias voltage builds up to the point where it operates to unblocking the audio system, it is locked in this position by this aiding voltage.

This locking section is illustrated in curve shown in Fig. 2. Assume that the automatic volume control bias voltage is zero and the set is operative. Assume the automatic volume control bias to increase until it reaches the point marked L; its effect is to produce a drop in voltage across Rx, permitting passage of a small current to the audio tube, so that Is is decreasing and Ia is increasing. As soon as Ia increases greater than Is, the locking action takes place and instantly the curve shifts to point N.

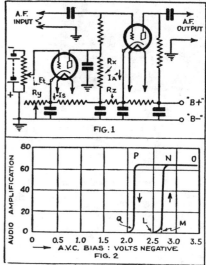

FIG. 1

FIG. 2

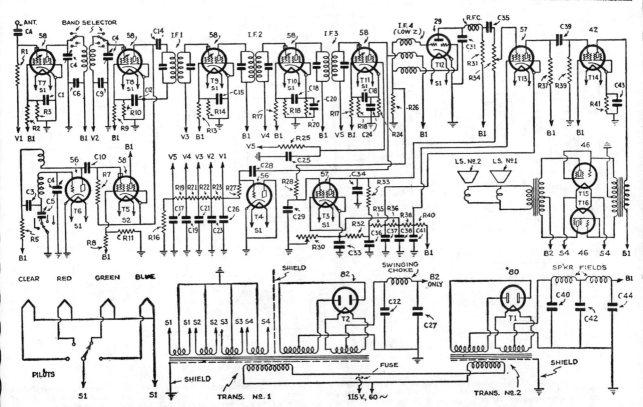

Wiring diagram of the newest Midwest creation—the 16-tube short and long wave receiver. A superheterodyne with special automatic volume control system, which enables the operator to adjust the sensitivity to high or low degree. With the sensitivity adjusted "low," reception with a minimum of static and other interference is obtained.

THE NATIONAL COMPANY
TYPE FB-7
SUPERHETERODYNE

Little need be said concerning this top view of the new receiver. Its compactness, symmetry of design and complete expressability are immediately apparent.

General Characteristics

This new receiver carries the designation "FB-7". This designation is particularly applicable to a receiver especially suited to the needs of the amateur communication enthusiasts. FB stands for *phone band* and in the vernacular of the "ham" it also means *fine business*, which is an expression commonly employed to indicate satisfactory results.

The FB-7 is essentially a short-wave

superheterodyne of the most advanced type, incorporating many of the features only to be found in the most expensive and elaborate receivers of the strictly commercial type. As may be seen from the accompanying illustration, the entire receiver is comparatively compact, while all of the component parts are completely accessible. The tuning scale is of the full vision type and is thoroughly illuminated. Tuning is accomplished by a single knob and there are no additional adjustments of any kind, other than the volume control. The tuning range of the receiver is from 15 to 200 meters or 20,000 kilocycles to 1,500 kilocycles. Five different sets of coils, with suitable overlap, are used to cover this range; they are of the regular National commercial type and plug directly into the front panel of the receiver. Provision is made for both telephone and loud speaker operation and the receiver may be operated from the regular National power supply unit or from batteries. "Hams" who desire to use this type of receiver for communication purposes sometimes find it desirable to operate from a small filament transformer and "B" batteries. This enables them to duplicate the performance of the receiver operated from the regular power supply, at slightly reduced cost.

To be more specific:

Determining upon the circuit which would most nearly meet all of the conditions required for the communication services, for which this type of receiver was designed, was the subject of a

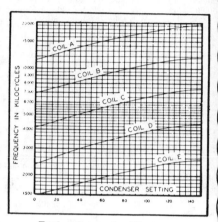

Tuning Curves for FB-7, with "general coverage" coils.

great deal of study. Another important subject was the determination of the particular types of tubes which would best function in a receiver from which so much was to be demanded. From antenna to loud speaker, we believe that the FB-7 is the satisfactory solution to a great many receiver problems. The following tubes have been selected because they seem to suit the requirements admirably. The first detector is the type 57; the high frequency oscillator and the beat oscillator are of the 24 type; the two intermediate frequency amplifier tubes are 58's; the second detector is the 56 and the output tube is the type 59 pentode.

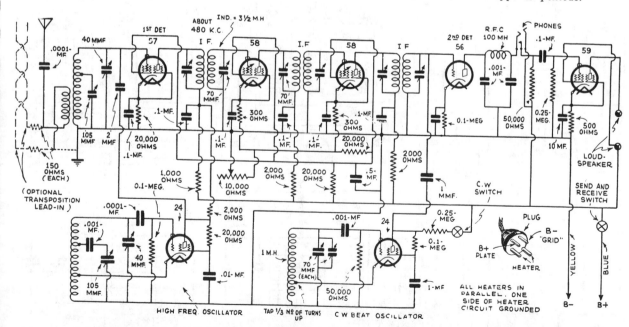

The complete circuit diagram of the new National FB-7 Short-Wave Superheterodyne Receiver. Seven of the latest tubes are employed and the receiver is ideal for use in connection with many services as a study of the circuit will disclose. All of the heaters are connected in parallel. It will be noticed that one side of the heater circuit is grounded to prevent radiation from the beat oscillator. Other systems, commonly employed, were found inadequate.

THE NATIONAL COMPANY
5 METER RECEIVER

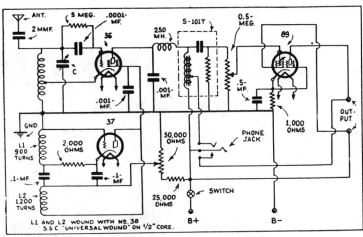

Here we have the interesting circuit diagram of the new National 5-meter super-regenerative receiver, the other popular bands being covered when desired by the use of suitable plug-in coils, including the 4 to 7½ meter band.

A Study of the Circuit

Referring to the circuit diagram, Fig. 1, a number of rather unusual features will be noted, including those discussed above. The *interruption-frequency* oscillator, employing the type 37 tube, is arranged in a split Hartley circuit with the grid at ground potential. Grounding the grid in this manner produced a maximum plate swing and as the plate is connected directly to the screen-grid of the detector it, in turn, produces the maximum interruption frequency coupling.

The .001 mf. condenser connected from the screen-grid of the detector to ground acts not only to complete the detector circuit but also to furnish the necessary tuned circuit capacity for the interruption frequency oscillator.

It may also be noted that the regeneration control is wired in such a way that the detector screen voltage and the oscillator plate voltage vary together. This gives a constant and efficient degree of superregenerative action, regardless of the operating point of the detector.

Progressing to the detector output circuit, it will be seen that impedance coupling is used. The choke coil is of special construction and has a total inductance of 700 henries. A tap is brought off at the correct point in the winding so that headphones, when plugged in the phone jack, correctly match the plate impedance of the 36 tube due to the auto-transformer action of the plate choke.

The audio volume control at the input of the 89 tube is very useful, especially if the operator desires to connect a pair of phones in the output circuit.

POWER SUPPLY—The heater circuit requires approximately 6 volts at .9 ampere. The voltage is not critical and may be between 5.5 and 6.5 volts. The supply may be either A.C. or D.C. except as noted under instructions for the *Low Frequency Coils.*

The plate supply voltage normally required is 180 volts and this may be obtained either from "B" batteries or from an A.C. operated power supply. The National type No. 5886 AB power unit fulfills these requirements and is supplied with a single receptacle for the 4-prong cable plug. As little as 135 volts of "B"-battery may be used with good results, provided the 25,000 ohm resistor, mounted near the center of the chassis (underneath) is changed to 10,000 ohms. Fair results may be obtained with 90 volts of "B"-battery, in which case this resistor should be "shorted" out entirely.

OUTPUT CIRCUITS—The output tip jacks for speaker operation are located at the back of the receiver on the righthand side. The speaker requirements are not at all critical and any conventional *magnetic* or *dynamic* type of unit will give good results. The output impedance of the receiver is approximately 7,000 ohms, requiring a speaker impedance of between 3000 and 15,000 ohms.

The phone jack for headphone operation is located in the left-hand side of the front panel and is connected to the plate circuit of the detector tube by means of a step-down auto-transformer.

CAUTION: At all times when the heaters are lighted and when "B" power is connected to the receiver, either the headphone jack must be plugged in or a loudspeaker connected to the output terminals. If this is not done, the 89 tube may be seriously damaged.

ANTENNA—The antenna binding post is located at the left hand side of the receiver, the lead being brought through the rubber bushings beside the post. A series antenna coupling condenser is located directly below the antenna post near the chassis. The success of any 5-meter work depends largely upon the receiving antenna and antenna coupling employed. In most cases it is advisable to experiment with several antenna arrangements, but as a general rule the antenna described herewith will be found efficient.

The antenna proper should be as high as possible and may be a single vertical wire approximately 8 feet in length. The lead-in consists of a single wire connected to the antenna 13½ inches from the

center and should be run at right angles to the antenna for a few feet before being brought down to the receiver. The length of the lead-in is not critical in any way but should be well insulated and sharp turns should be avoided. It should *not* be shielded!

When the receiver is put in operation with certain types of antennas, it may be found that the detector will not oscillate over certain portions of the range. This indicates too much antenna coupling and the coupling condenser plates should be spread apart slightly until the *dead-spot* just disappears.

When the more conventional type of *untuned* antenna is used, the coupling condenser plates should be moved closer together for best results.

As a general rule, a ground connection is not necessary but under certain conditions its use may be beneficial.

LOW FREQUENCY COILS—Coils are available for covering the 10, 20, 40, 80 and 160 meter bands. When using these coils the low frequency oscillator (37) should be removed from the socket.

The heater circuits must be supplied from a D.C. source, such as a storage battery, in order to eliminate A.C. hum.

If A.C. operation is desired on these bands, it will be necessary to change the tubes to the 2.5 volt type. A 24 may be substituted for the 36, a 27 for the 37, and a 2A5 for the 89—altogether this last substitution will require some rewiring of the output tube socket. The bias resistor required for the 2A5 tube is approximately 500 ohms and should replace the 1000 ohm resistor used for biasing the 89.

Due to the fact that as a general rule superregeneration cannot be used on the low frequency bands, the sensitivity of the receiver will be considerably less than on the 56-60 mc. band and it is, therefore, advisable to use headphones connected in the output circuit instead of the loud speaker.

ADDITIONAL HIGH FREQUENCY COILS—Additional coils are available for covering the frequency range between 40 and 75 megacycles (7½ to 4 meters).

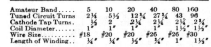

Amateur Band	5	10	20	40	80	160
Tuned Circuit Turns	2¼	5½	12¾	27¾	43	96
Cathode Tap Turns	½	2	2¾	2¾	2¾	2¾
Coil Diameter	1″	1″	1″	1″	1½″	1½″
Wire Size	#18	#20	#20	#26	#26	#30
Length of Winding	¼″	¼″	½″	¾″	1″	1½″

Capacity of main tuning condenser is 18 mmf. (or .000018 mf.)

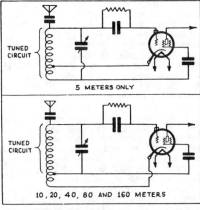

Note the difference in the tuned circuits for 5 meters and above.

THE NATIONAL COMPANY
FBX SINGLE SIGNAL
VARIATION OF FB7 RECEIVER

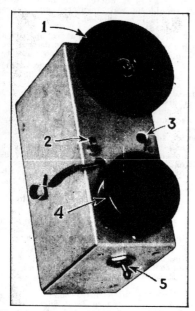

Close-up of "single signal" attachment.

only a few cycles, say 50, the signal will be completely detuned. The beat note resulting when the signal circuits are tuned to 10,504 kc. will now be so weak as to be negligible. In other words, any given signal may be tuned in at only one definite adjustment of the signal oscillator, and the audio response will depend solely upon the detuning of the beat oscillator from the I.F. In the above case, all signals will peak very sharply at 2000 cycles. While the receiver sounds to the ear similar to the older regenerative detectors with a sharply peaked audio amplifier, the principles involved are quite different, as witnessed by the fact that there is no "other side of zero beat."

The selector switch, referring to the diagram, is used to connect the crystal in series for true *single signal* reception, remove it from the circuit entirely, or connect it in parallel. The parallel connection is useful, particularly in phone reception, since the crystal will now reject a narrow group of frequencies and may, in consequence, be employed to eliminate heterodyne interference by adjusting the high frequency circuits so that the unwanted signal sets up an I.F. equal to that of the crystal.

The single signal receiver is said to represent the finest "C.W." (code) receiver

1—Selectivity control; 2—Phasing condenser; 3—I.F. "peaking"; 4—Crystal in plug-in mount; 5—"Series"-"Parallel"-"Off" switch.

yet developed and it has in addition numerous advantages for *phone* reception. Further details on the single signal receiver are to be found in Q.S.T. (Aug. and Sept. 1932). This type receiver is marked by its extensive selectivity and also by the fact that the annoying *double beat* characteristic of autodyne detectors is eliminated.

Some additional data on the FB7 receiver is here presented: The capacity of the main tuning condensers, both 1st detector and oscillator circuits, is 105 mmf. each. The trimmer condensers have a maximum capacity of approximately 40 mmf.

The beat oscillator coil consists of a winding of about 1 millihenry inductance, tapped ⅓ of the way from the grounded end. (The ⅓ referring to turns and not inductance.) The two standard 70 mmf. I.F. tuning condensers are connected in parallel to obtain a high-"C" circuit.

The inductance of the I.F. coils is approximately 3½ millihenries and the tuning condensers are 70 mmf. maximum. The intermediate frequency is about 480 kc. This data was kindly furnished by James Millen, of the National Company.

The use of a crystal filter connected in the I.F. amplifier in order to obtain an exceptionally high order of selectivity is desirable under certain circumstances. The idea is by no means new, having been incorporated in the Stenode receivers for several years; but its application to high frequency C.W. reception is comparatively recent.

Briefly, a properly designed and adjusted filter connected in series with the input of the I.F. amplifier, will pass only a very narrow band; the width being measured in cycles rather than kilocycles. The fundamental circuit, as shown in the accompanying diagram, is seen to be in the form of a capacity bridge, the function of C_N being to balance (or neutralize) the capacity of the crystal holder, and that of C_T being to tune and center tap the secondary circuit. In addition, the adjustment of C_T has a marked effect upon the width of the response characteristic, enabling the operator to vary it at will from a few cycles to several hundred.

It is evident, from the foregoing discussion of selectivity, that such an extremely narrow I.F. response characteristic will allow the complete separation of stations differing in frequency by only a small fraction of a kilocycle, provided the beat oscillator is correctly adjusted. To carry the discussion further, suppose the beat oscillator is tuned to 502 kc.; that is, 2 kc. from the I.F. (crystal); the 10,500 kc. signal will be tuned in as before, but now should the signal circuits be changed

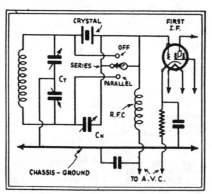

Connections of the quartz crystal filter used for "single signal" operation.

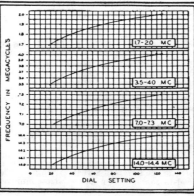

"Band-spread" coil tuning curves.

Coil Data: National FB-7
General Coverage Coils
DETECTOR

Secondary		Primary		Detector	
Turns	Size Wire No.	Turns	Size Wire No.	Form grooved per in.	Range KC
6 1/3	16EN	3	24EN	5 th'rds	19,500-11,400
11 5/6	18EN	3 1/2	24EN	8 th'rds	11,700-7000
21 5/6	18EN	5 1/6	34DS	14 th'rds	7300-4000
34 5/6	24EN	7 5/6	34DC	24 th'rds	4200-2400
58 5/6	28EN	8 5/6	32DS	40 th'rds	2500-1500

OSCILLATOR

A	B	C	Total No. of turns	Size Wire No.	Form Grooved per in.
2 1/6		4 1/6	6 1/6	16EN	5 th'rds
2 1/6	———	9	11 1/6	18EN	8 th'rds
4 1/6		14 1/3	18 1/2	20EN	14 thr'ds
7 1/6	20	5 2/3	32 5/6	24EN	24 thr'ds
11 1/6	27 1/2	17 1/6	55 5/6	28EN	40 thr'ds

A—from bottom end to 1st tap.
B—from 1st tap to 2nd tap.
C—from last tap to top end coil.

THE NATIONAL COMPANY
A.G.S. COMMERCIAL SET

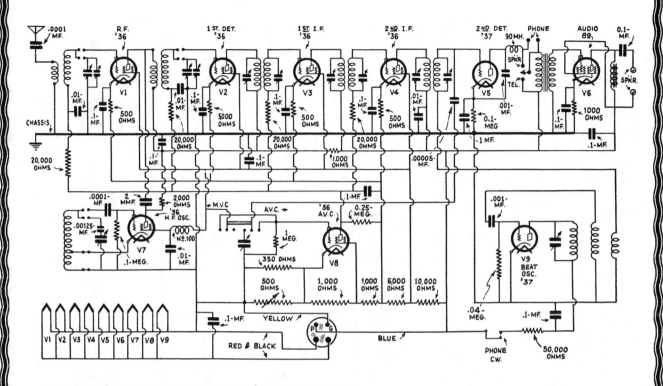

Front view of new "AGS" commercial or "pro" type short wave receiver, with "single dial" control. Yes, it's a super-het!

"AGS" S-W Receiver

Nine tubes are used in a circuit comprising a stage of tuned R.F. amplication and first detector, employing screen grid tubes, a high frequency oscillator; two stages of extremely selective high-gain screen grid "I.F." amplification; "I.F." power detector; automatic volume control, working in conjunction with both R.F. and I.F. amplifiers; beat frequency oscillator, and pentode output with provision for either phones or loudspeaker. Tubes used: 4 '236s, 4 '237s, 1 '238.

Outstanding Features: Tuned R.F. stage preceding first detector. (Image suppression — improved signal-to-noise ratio—improved "weak signal" response.) Single dial *straight frequency line* tuning (270°). Calibration curves and Station Chart on panel. Coil change from front of panel. Automatic volume control or manual volume control, as desired. Extremely rigid mechanical construction from very heavy aluminum plate. Relay rack mounting (size 8¾x19"). Frequency range 2400 to 15,000 kc. Additional coils to extend the range to 20,000 kc. Heterodyne oscillator for c.w. reception.

The receiver is absolutely "*single control.*" There are no trimmers, antenna coupling devices, or other secondary adjustments; merely one single, accurately calibrated, frequency control, plus the volume control. The oscillator is of the electron-coupled type as modified for use in high frequency superheterodynes, resulting in an extreme degree of frequency stability and steadiness of frequency

THE NATIONAL COMPANY
3 TUBE S.W. SET

This reliable 3-tube set can be used very nicely with a pair of 2000 ohm or higher resistance headphones, or the output may be fed into a power amplifier of one or two stages. This receiver may be operated from "A" and "B" batteries or also from a good "A" and "B" eliminator.

Data on the coils and tuning condenser in the detector circuit are given herewith. The antenna is choke-coupled to the grid of the R.F. tube and may have an inductance of 60 to 85 millihenries. The regeneration is controlled by a 50,000 ohm potentiometer which accurately varies the voltage applied to the screen grid of the pentode detector.

The radio frequency choke, R.F.C. in the tickler plate circuit of the detector, may have a value of about 28 to 30 millihenries, or it may consist of about 700 turns of No. 36 insulated magnet wire (silk covered enamel) wound on a ½" diameter dowel stick or cork. Wind coil in slot or between separators 3/16" wide, random fashion. This choke is of extreme importance and it is therefore strongly recommended that the finest type possible, such as the new National type 100, wound on an Isolantite core, be utilized. The coupling resistors, joining the output of the detector to the input of the A.F. stage, have values of .25 megohm and 1 megohm, respectively, the two resistors being coupled through a .01 mf. fixed condenser.

The 6-pin coil socket is mounted about 1" above the metal subpanel and the coils should be kept a distance of at least ¾" from the metal cabinet, to avoid all undue losses and also broadening of the tuning or lack of selectivity. The forms used are made of National R39 material, which ensures the minimum loss at these high frequencies; this material is far superior to ordinary bakelite. The tickler coil "T" may be wound in the slot at the bottom of the National form; the primary winding can be wound in between the turns of the secondary.

The R.F. choke coupling the antenna to the ground and to the grid of the R.F. tube may comprise 350 turns of No. 36 (silk covered enamel) magnet wire wound on a ½" diameter dowel stick, the coil being 3/16" thick and random wound (helter skelter style). The main variable tuning condenser connected across the secondary, S, of the regenerative coupler has a capacity of 90 mmf. or .00009 mf. You can substitute other values of tuning condenser such as 140 mmf. if you already have coils or data for that value of tuning condenser and wish to use them instead.

Data on National "Short Wave" Coils

The secondary winding of the coils is shunted by 90 mmf. (.00009 mf.) variable condensers. Diameter of coil forms 1½ inches:

No. 10 coils, covering from 9 to 15 meters:
  Secondary 2 5/6 turns of No. 16 Enamel
  Primary 1 5/6 turns of No. 34 Enamel
  Tickler 3 turns of No. 32 Double Silk.
No. 11 coils, covering from 14.5 to 25 meters:
  Secondary 6¼ turns of No. 16 Enamel
  Primary 3 5/6 turns of No. 34 Enamel

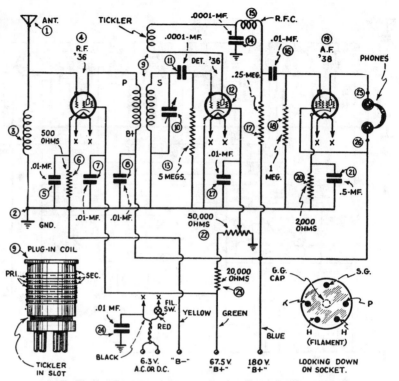

The schematic wiring diagram for the 3-tube receiver

Tickler 3 turns of No. 32 Double Silk.
No. 12 coils, covering from 23 to 41 meters:
  Secondary 11 5/6 turns of No. 18 Enamel
  Primary 7 5/6 turns of No. 34 Enamel
  Tickler 3 turns of No. 32 Double Silk.
No. 13 coils, covering from 40 to 70 meters:
  Secondary 19 5/6 turns of No. 18 Enamel
  Primary 12 5/6 turns of No. 34 Double Silk
  Tickler 4 turns of No. 32 Double Silk.

No. 14 coils, covering from 65 to 115 meters:
  Secondary 34 5/6 turns of No. 24 Enamel
  Primary 21 5/6 turns of No. 34 Double Cotton
  Tickler 4 turns of No. 32 Double Silk.
No. 15 coils, covering from 115 to 200 meters:
  Secondary 62 5/6 turns of No. 28 Enamel
  Primary 38 5/6 turns of No. 32 Double Silk
  Tickler 5 turns of No. 32 Double Silk.

THE NATIONAL COMPANY
NC-5 S.W. CONVERTER

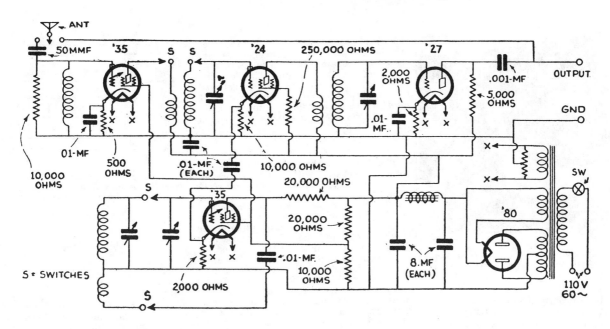

The I. F. Amplifier

In order that the converter might be universal in its application to any type of broadcast receiver, a stage of high-gain I. F. has been incorporated as an integral part of the converter.

In addition to furnishing a voltage gain of approximately 100, this stage of amplification also serves as a low-impedance output, or coupling-network; so that the output lead from the converter may be connected directly to the antenna post of the broadcast receiver, without either making the set oscillate, or having the antenna coupling system of the set act as such an extremely low-impedance "load" on the output circuit of the first detector as to prevent an appreciable signal getting into the broadcast receiver.

The output lead, from the plate circuit of this I. F. tube to the antenna post of the broadcast receiver, is shielded in order to prevent pick-up of any local interference or low-frequency transmitting station. To further minimize the possibility of any such pick-up troubles, where the converter is used with an unshielded R. F. chassis, an I. F. frequency of 575 kc. has been selected.

The Oscillator of New Design

A rather unusual oscillator, or frequency-changing arrangement, will be noticed upon examination of the circuit diagram. In the first place, a '35 tube is employed for this purpose, rather than the more conventional '27 and, in addition, coupling is *not* obtained through a small condenser between the grids, as has been common practice in the past. Instead, a new and unique coupling arrangement has been developed whereby the cathode return circuits of the detec-

tor and the oscillator are coupled together. Such a method automatically provides just the right amount of coupling, at all times, to produce a minimum of "hiss" and other such noise; which has been so troublesome with converters in the past.

In fact it is rather surprising that more engineers, working on the short-wave superhet and superhet converter design problems, have failed to realize to what an extent improper "mixing" is responsible for the bad reputation of "converters" and "superhets," for high noise-to-signal ratio. The use of a properly designed pre-amplifier, in conjunction with the variable-mu oscillator, and the cathode-return "mixing circuits," results in a signal-to-noise ratio closely approaching the very favorable one obtainable from a good T.R.F. set, such as the "SW5 Thrill Box."

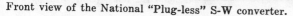

Front view of the National "Plug-less" S-W converter.   Rear view of converter, which has inductance change switches

THE NATIONAL COMPANY
H.F.C. 5 METER CONVERTER

above—Rear top view of U.S.W.
converter with shield covers removed.

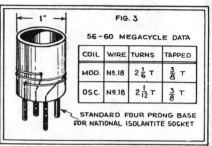

FIG. 3

56–60 MEGACYCLE DATA

COIL	WIRE	TURNS	TAPPED
MOD.	Nº 18	$2\frac{1}{6}$ T	$\frac{3}{8}$ T
OSC.	Nº 18	$2\frac{1}{12}$ T	$\frac{3}{8}$ T

STANDARD FOUR PRONG BASE
FOR NATIONAL ISOLANTITE SOCKET

Winding data for the coils used in the
5-meter band converter.

1550 kc. (about 200 meters), the frequency at which the converter is designed to operate. If it so happens that a powerful station is operating on this frequency, the receiver should be detuned sufficiently to avoid the possibility of interference. Detuning as much as 30 or 40 kc. has no appreciable effect upon the ganging.

Tubes

The design of this unit is such that it may be operated with either the 6-volt D.C. heater type tubes or 2½-volt A.C. tubes. In the first case, two '36's are employed for detector and oscillator and a '37 for the output coupling tube. For A.C. operation, the corresponding tubes are two '24's and one '27; '35 tubes may be substituted for the '24's if desired. A certain amount of care must be exercised in the selection of tubes or trouble will be experienced from *microphonics* or noise resulting from leakage between heater and cathode. This latter trouble appears as a loud grating or scratchy hum. As a general rule, tubes of recognized quality having standard characteristics will prove entirely satisfactory. No special matching is required, since ample provision for balancing tube capacities, etc., is incorporated in the various circuits.

Antenna

The antenna requirements are not in any way critical, although as a general rule a single wire as high as possible will give best results. The directional effects of various types of antenna are often very pronounced at high frequencies, so that the use of a vertical antenna located well away from any surrounding objects usually gives best results. The length may be between 5 and 50 feet over-all. A longer wire is not recommended, as it tends to increase the noise-to-signal ratio.

Intermediate Frequency Amplifier

The circuit of the type HFC converter is such that almost any broadcast receiver will be quite satisfactory for use as the I.F. and audio amplifier. For best results, the receiver should have a fair degree of sensitivity and should be stable. If the receiver has a tendency to oscillate, it will be somewhat emphasized when the converter is connected, which may make it impossible to fully advance the volume control without causing overall oscillation. Extreme sensitivity and selectivity are not required, since the converter employs a high gain I.F. stage and is in itself quite selective. As a matter of fact, the use of an extremely selective broadcast receiver is something of a disadvantage, especially when hunting for signals or when receiving signals having a large degree of frequency modulation. The broadcast receiver should be capable of tuning to a frequency of

Power Supply

The filament or "A" supply may be either a 6-volt storage battery or 2½-volt transformer, depending upon the type of operation desired. In most installations, no connection between the storage battery and B-minus is required, although under certain conditions it may be advisable to ground one side. When a 2½-volt filament transformer is employed, the center of the winding should be grounded by means of a tap on the secondary or a center-tapped resistor having a total resistance of 10 or 20 ohms. The "B" supply may consist of either "B" batteries or a "B" eliminator, the batteries being preferable where fluctuating line voltages are encountered. The voltages are not critical and may be between 67 and 75 for the screen circuits and 135 to 180 for the plate circuits. Reference to the circuit diagram will show that the "B" batteries are subjected to a certain amount of current drain when the converter is not in use. The B-minus should, therefore, be disconnected during idle periods.

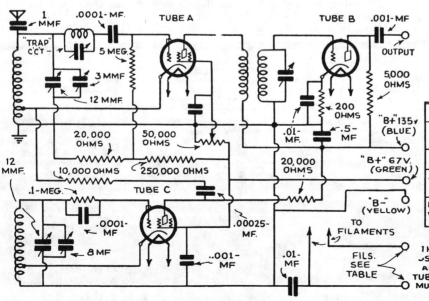

NATIONAL
ULTRA-HIGH FREQUENCY
CONVERTER
TYPE H.F.C

	A.C	D.C
TUBE A	'24 OR '35	'36
TUBE B	'27	'37
TUBE C	'24 OR '35	'36
FILAMENT VOLTAGE	2½V, A.C	6 V., D.C

NOTE –
THIS CONVERTER MAY BE USED ON EITHER DIRECT OR ALTERNATING CURRENT. PROPER TUBES AND FILAMENT VOLTAGE MUST BE USED. (SEE TABLE.)

THE NATIONAL COMPANY
TYPE H.F.R. ULTRA S.W. SET

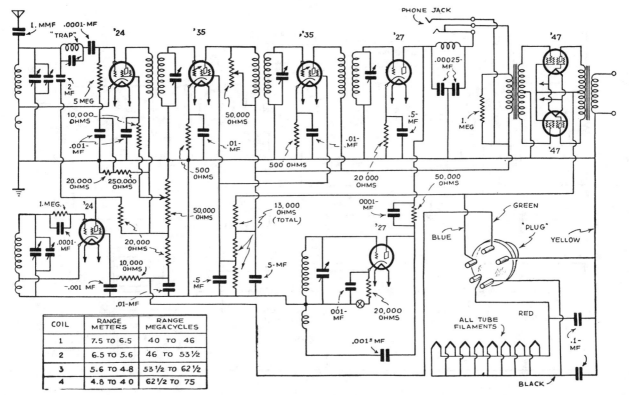

COIL	RANGE METERS	RANGE MEGACYCLES
1	7.5 TO 6.5	40 TO 46
2	6.5 TO 5.6	46 TO 53½
3	5.6 TO 4.8	53½ TO 62½
4	4.8 TO 4.0	62½ TO 75

Diagram of connections used in the National 9-tube superheterodyne for ultra-short-wave reception, covering 4 to 7½ meters or 75 to 40 megacycles.

Coils

The standard coils accompanying the receiver cover a wave length range of approx 4 to 7½ meters subject to variation, since slight differences in tubes, trimmer and padding condensers, setting, wiring, etc., may alter the range considerably. The design is such, however, that adequate overlap is always provided between coils. The coils are numbered as shown, two coils of corresponding numbers being used in the oscillator and detector circuits. The coils having the red mark on the base should be placed in the detector coil's socket (front compartment) which is also marked red. The oscillator coils and socket are marked black. The coils must be placed firmly down in their sockets or trouble will be experienced in obtaining correct ganging and maintaining calibration. It will be noticed that the connecting leads between the ends of certain coils and the pins in the coil form are bent. These leads must not be straightened or altered in any way, since the coils are individually calibrated by carefully adjusting the leads in the laboratory.

Intermediate Frequency Amplifier

The I.F. amplifier of this receiver is tuned to a frequency of approximately 1550 kc. The circuits employed in the amplifier are entirely conventional. Reference to the circuit diagram will show an I.F. trap in the first detector grid circuit. This is also tuned to 1550 kc. and may best be checked after the receiver is put in operation by setting the condenser with a bakelite screwdriver at the point which gives maximum background noise. If it so happens that a powerful local station is operating on the intermediate frequency, the amplifier should be detuned sufficiently to avoid the possibility of interference. Detuning as much as 30 or 40 kc. has no appreciable effect upon the ganging.

Controls

From left to right, the controls are: first detector regeneration control, first detector trimmer, the main tuning dial, I.F. amplifier or volume control and the beat frequency oscillator switch. The beat frequency oscillator is tuned to the intermediate frequency and is coupled to the second detector.

COIL NUMBER	1		2		3		4	
	DET.	OSC.	DET.	OSC.	DET.	OSC.	DET.	OSC.
TOTAL TURNS	3⅛	2¾	2⅜	2⅛	1⅞	1⅝	1¼	1⅛
FRACTIONAL TURN BETWEEN TOP AND BOTTOM OF COIL	⅔	⅔	⅜	⅜	⅓	⅓	¼	¼
RANGE Mc	40 TO 46		46 TO 53½		53½ TO 62½		62½ TO 75	

TUNING CONDENSER, CAP. (DET. AND OSC.) = 18 MMF EACH. (TYPE, NEW 270° STRAIGHT FREQUENCY LINE

I.F. COIL DATA :-
FORM 1¼" DIA 2½" LONG
PRI. = 50 TURNS No. 32 D.S
SEC. = 100 TURNS No. 28 ENAM
(SAME DIRECTION)
64 TURNS PER INCH
TUNED WITH A 4-70 MMF
COMPRESSION TYPE I.F
TUNING CONDENSER

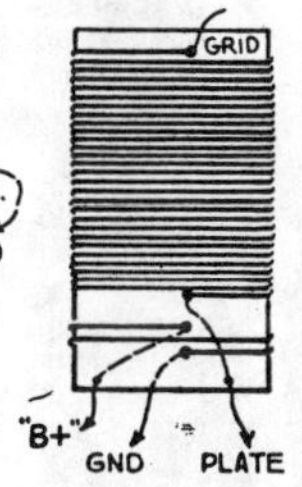

I.F. TRAP CCT TUNED TO 1550 KC.
COIL L2 = 100T 10-68 LITZ WIRE ON A ⅝" FORM.
CAP C2 = 4-70 MMF

THE NATIONAL COMPANY
TYPE S.W. 58 SET

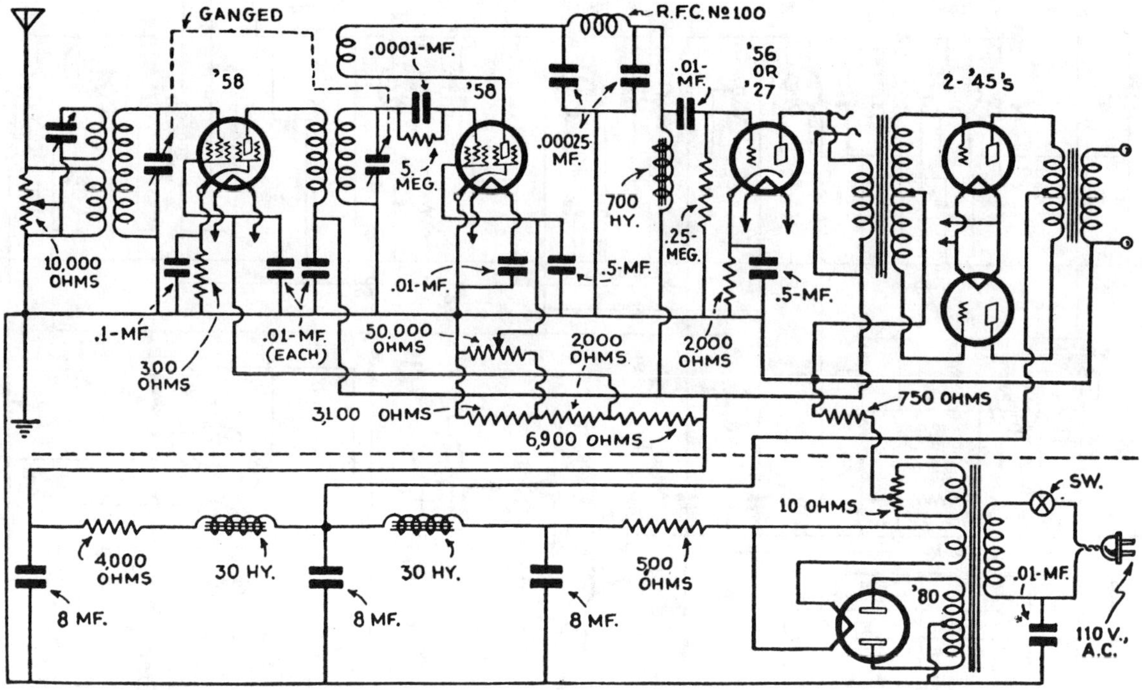

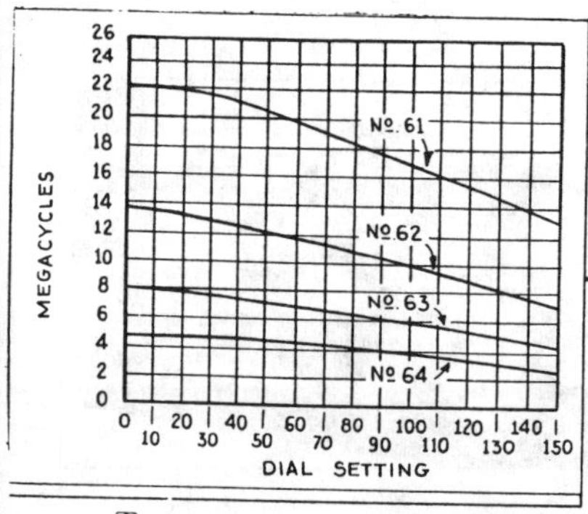

Tuning curves showing the bands covered by each of the plug-in coils.

IN past years the scientific design of short-wave receivers has been retarded by the lack of laboratory equipment which would enable the short-wave engineer to predetermine and check performance with the technique and precision that has long been possible on the lower and conventional broadcast frequencies. It has been difficult to generate accurately known radio frequency potentials at frequencies above 6,000 kc., and sensitivity and gain measurements at wavelengths under fifty meters have necessarily been subject to an elastic interpretation, to say the least. However, the engineer now has a tool in his hands which enables him to produce controlled potentials at very high frequencies, and this apparatus has contributed greatly to the design of the T.R.F. short-wave receiver. The equipment functions on the principle of a balanced detector circuit generating high frequency harmonics, the voltages of which bear a close relationship to the strength of the plate current. Once this relationship has been established, it is only necessary to determine the plate current to ascertain the value of the R.F. potential.

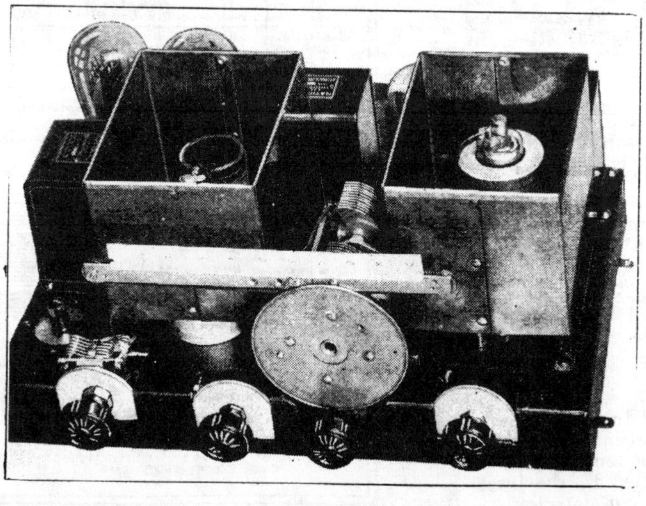

The R.F. Pentode

The type '58 tube has been made to order for a circuit of this design. Its high amplification factor, trans-conductance, and above all its high plate impedance, enable the engineer to obtain a degree of sensitivity and selectivity in the radio frequency circuits which has previously been impossible.

These tubes, having twice the impedance of the '24, of course necessitated the design of special coils, data on which are contained in the following coil table:

Coil Winding Data

Coil	*Primary*	*Secondary*	*Tickler*
No. 61	6¼ turns	6¼ turns	2 turns
No. 62	10⅝ turns	11⅝ turns	2½ turns
No. 63	15⅝ turns	19⅝ turns	3 turns
No. 64	28⅝ turns	34⅝ turns	3 turns

The tuning curves in Fig. 5 show the manner in which the various bands are covered with these coils. The inductors are wound on the low loss R-39 material, which, in conjunction with Isolantite, is the only insulating material employed in the SW-58. The two tuning condensers each have a capacity of 90 mmf. and the R.F. choke has an inductance of 2.5 m.h.

THE NATIONAL COMPANY
TYPE SW 5 D.C. "THRILL BOX"

Problems of Circuit Alignment

It has been customary in short-wave receivers to use two types of coils, the antenna-coupler or R.F. input coils being without ticklers and having provision for coupling in the antenna through an antenna winding or a small condenser. It was decided in this case to avoid the condenser-coupling method; for the double reason that it tends to cause severe mistuning and that noises, especially power-line noises, appear to be somewhat more severe with such an input than with an antenna coil. A variable condenser in series with the antenna does not provide proper compensation when the antenna is inductively coupled, the series condenser having to be reset with almost every change in tuning. It therefore amounts to another tuning control and spoils the single-control feature entirely. A "vernier" condenser connected across the tuning condenser will give fairly good compensation but will *not quite* maintain alignment across the tuning scale.

The regeneration control is by varying the detector's screen-grid voltage; as this has both a positive action and a wide range which will produce the desired action with any ordinary tube, and with "B" and "A" voltages within about 20% either way from the proper values.

Special Coil Design

The use of a 4-prong coil form to fit the standard UX socket, or a 5-prong socket like the UY standard, was very attractive but was abandoned in favor of a special 6-prong socket, which permits complete independence of the three windings, which appear on each coil form.

The method of ganging with these coils depends on their winding arrangement. The heavy-wire winding is spaced to occupy a length about equal to its diameter. This is not quite the best theoretical form but is one which is conveniently maintained for all the coils of the series, while at the same time it is not far from the ideal. This heavy-wire winding is used as the secondary, regardless of the position in which the coil is used. Putting it differently, the heavy-wire winding is always the tuned winding which feeds the next grid—whether that be the R.F. tube or the detector.

"Inter-wound", with this heavy-wire winding, is a primary winding of fine wire, which has 66% as many turns as the secondary. This ratio gives almost all of the R.F. gain which would be obtained with a primary having as many turns as the secondary. At the same time it does not transfer as much capacity from the plate circuit into the next tuned circuit (grid circuit), where it is not wanted.

The third winding on the form is close-wound of small silk-covered wire in a narrow groove at the lower end of the spool. It is normally used as the detector tickler.

The coil ranges are so located that exchanges are not necessary in any band,

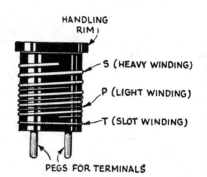

This drawing shows how the secondary (S), primary (P), and tickler or slot winding (T) are placed on the coil form.

except the standard broadcast range where rapid traverse is never necessary.

The coils marked ** are standard equipment with the set, either A.C. or D.C.

Cat. No.	Wave Range Meters	Purpose
10	9—15	International broadcasts; commercial signals; amateur 28-megacycle band.
11**	14—25	International broadcasts; commercial signals; amateur 14-megacycle band.
12**	23—41	International broadcasts; commercial signals.
13**	38—70	International broadcasts; commercial signals; amateur 7-megacycle or 7,000-kc. band.
14**	65—115	International broadcasts; amateur 3.5-megacycle or 3,500-kc. band.
15	115—200	Amateur 1.75-megacycle or 1,750-kc. band.
16	200—360	Regular broadcasts.
17	350—550	Regular broadcasts.

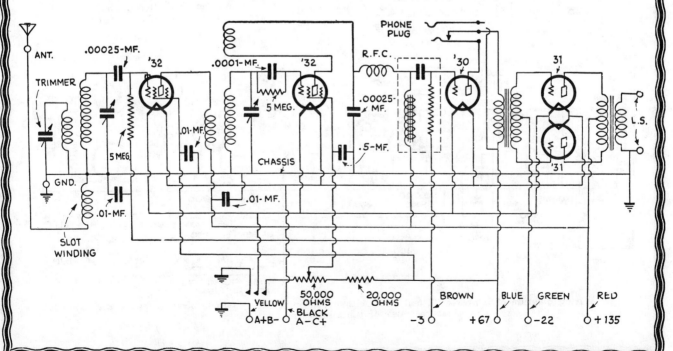

THE NATIONAL COMPANY
TYPE SW5 A.C. "THRILL BOX"
(IMPROVED)

The Variable MU Detector

While designed primarily for use in R.F. amplifiers in order to eliminate cross talk due to undesired rectification, it has been found that the '35 variable mu tube makes an ideal regenerative short-wave detector. True enough, the data sheets supplied with the tubes of this type state that they are unsatisfactory as detectors; this reference, however, is to their use as plate rectifiers in broadcast receivers. As a grid-leak condenser regenerative detector, especially at the higher frequencies, numerous investigations made in the "National" laboratories during recent months have shown quite the reverse to be the case.

From past experience in designing high-frequency receivers employing the '24 type of screen-grid tube as a grid leak-condenser regenerative detector, it had been found that the most satisfactory of the various methods of regeneration control was the variation of screen voltage by means of a potentiometer. How would the action of the '35 tube as a grid leak-condenser regen-

erative detector differ from that of the '24 when its screen voltage was shifted? For some unknown reason, the tube manufacturers in their data sheets and their so-called engineering and specification reports, as supplied to the radio set manufacturers, seem to be surprisingly consistent in at least one respect; namely, the complete omission of any curves that might throw some light on the subject.

It was soon found, however, that this relation is of an *inverse exponential* nature. Thus, when the screen voltage of the '35 is increased, the tube rapidly approaches an oscillating condition. The nearer the tube approaches the "spill over" point, however, the less effect the increasing of screen voltage has upon its tendency to oscillate. Consequently, this affords a regeneration control that permits of readily obtaining and maintaining a higher degree of regeneration, with the attendant smooth sliding into oscillation so much sought after in S.W. receivers of the past—and obtained in them to only a fair degree by the care-

ful selection of tubes and the juggling of grid leak and condenser values.

This same characteristic of the '35, that permits of this higher order of regeneration, also results in a more stable condition with regard to the holding of the regeneration adjustment when once set. There seems to be entirely lacking that tendency, of regenerative detector tubes of the past, to "pop" suddenly into oscillation on the slightest provocation.

Of course the '35 tube is also used in place of the original '24 in the specific manner for which the '35 was developed; so that anyone having one of the original "SW5" models, who wishes to use the variable mu tube in the R.F. and detector circuits, will find that but one change is necessary: namely, the substitution of a 500-ohm R.F. biasing resistor for the 350-ohm value employed in the former set. No change in the detector circuit is required.

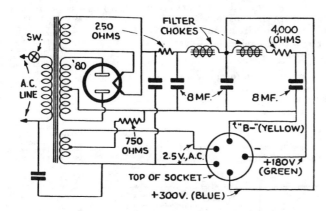

R.F. Transformer Coils				
"Brown"	No. 10	Range	9 to 15	meters
"Black"	No. 11	Range	14.5 to 25	"
"Red"	No. 12	Range	23 to 41	"
"White"	No. 13	Range	40 to 70	"
"Green"	No. 14	Range	65 to 115	"
"Blue"	No. 15	Range	115 to 200	"
"Orange"	No. 16	Range	200 to 360	"
"Yellow"	No. 17	Range	350 to 550	"
"Purple"	No. 18	Range	500 to 850	"

Band-Spread Coils		
No. 11A	20 meter band	
No. 13A	40 meter band	
No. 14A	80 meter band	

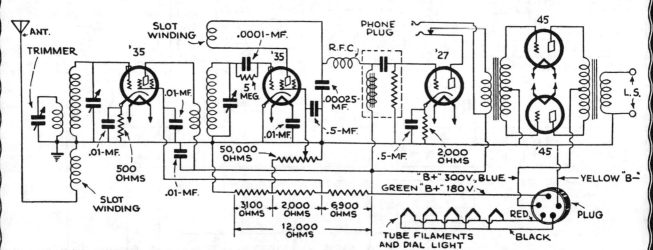

Wiring diagram of improved "Thrill Box" with provision for new vari-mu '35 tubes in the R.F. and detector stages; also '45 tubes in second audio stage.

NEWARK RADIO CORP.

NEWARK SHORT-WAVE CONVERTER

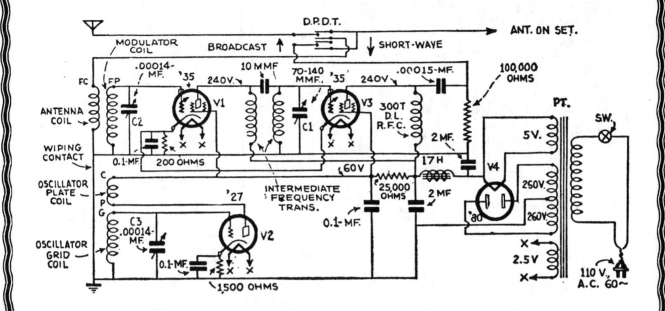

Diagram of the connections used in the Newark Short Wave Converter which includes an I.F. stage.

The converter contains, in addition, a stage of I.F. amplification. The chassis utilizes four tubes in all: a modulator, an oscillator, an intermediate frequency amplifier and a rectifier.

The use of the rectifier tube in conjunction with the built-in power unit makes the converter independent of the broadcast receiver supply.

Three plug-in coils are used to cover the short-wave range. Only one coil is changed when shifting from one band to another; there are six contacts on each coil form. Five of the contacts are made through a five-prong UY base on the coil forms, while the sixth contact is made through a special wiping contact on the side of the coil form.

The forms are tubular, with an outside diameter of 1½ inches and a length of 2¾ inches.

Coil Winding Data

Coil	Turns	Type Winding	Wire Size	Range meter.
1	Ant. 10, Mod. 34, Plate 13, Grid 20	Tight wound	No. 24	80-200
2	Ant. 9, Mod. 20, Plate 11, Grid 13	Tight wound	No. 22	45-85
3	Ant. 8, Mod. 7, Plate 8, Grid 7	Spacing of 1 wire thickness	No. 22	15-49

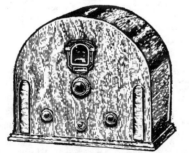

The converter is housed in a very neat cabinet.

Coupling between the oscillator and the modulator circuits is effected by introducing the oscillator output into the modulator circuit through the screen grid of the modulator tube. The stage of I.F. amplification incorporated in the converter is sharply tuned and has a high gain. It should not be thought of merely as a coupling stage, as it materially adds to the sensitivity and selectivity of the whole system. The resonant frequency of the I.F. stage can be varied between 900 kc. and 1,100 kc. by adjusting the trimmer condenser C1, located at the back of the chassis. This operation is performed by means of a wooden screw driver. For best results, the circuit should be resonated in the vicinity of 1,000 kc., as highest amplification will be obtained from the I.F. amplifier at this frequency. However, if there is a strong broadcast signal operating on this frequency, any other frequency between 900 and 1,100 kc. may be selected by properly adjusting condenser C1.

Coil Winding Data for I. F. Transformer and R. F. Choke

I.F. Transformer: Form 1⅛" x 3". Primary, 700 turns No. 36 single silk enameled wire. Secondary, 130 turns No. 28 enameled wire.

There is no inductive coupling between primary and secondary; all coupling between these coils is effected by a 10 to 15 mmf. condenser.

R.F. Choke: 300 turn duo-lateral. No. 36 single silk enameled wire.

Controls

The main tuning control is located at the center of the front panel, at the left is the modulator tuning knob used for vernier adjustments. On the right side is another knob controlling a D.P.D.T. switch. Under the main tuning control is the power supply switch.

The function of the D.P.D.T. switch is to connect the aerial to the converter's input and the converter's output to the broadcast receiver when the switch is at one position, or to connect the aerial directly to the broadcast receiver and to disconnect the converter when the switch is in its alternate position. The use of this switch makes it unnecessary to change the aerial wire from the converter to the broadcast receiver everytime the converter is turned off.

NORDEN-HAUCK LABS.

ADMIRALTY SUPER - 15

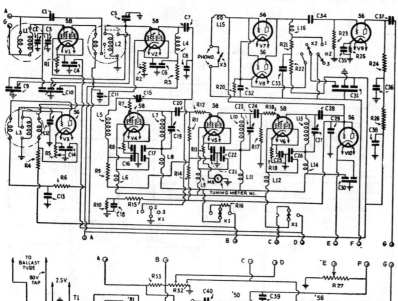

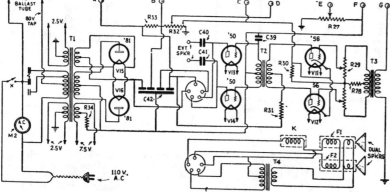

Broadcast Coil data: The antenna coupler L-1 consists of an outside coil 1¼" in diameter with 88 turns of No. 30 enameled wire, space wound. The primary consists of a small spool ½" center, ¼" slot, wound with 200 turns of No. 36 D. S. C., scrambled. The detector coupler L-2 has a secondary winding of 88 turns No. 30 enameled wire, space wound, 1¼" tube. The primary consists of a spool with a ½" center, ¼" slot, 75 turns, scrambled wound No. 36 D. S. C. The primary spool of both coils L-1 and L-2 are centered in the secondary coil. The oscillator coupler L-3 has the grid coil wound on a 1¼" tube and consists of 52 turns No. 28 enameled wire, space wound. The plate and pick-up coils are close wound with No. 30 D. S. C. wire, on a 1" tube, centered inside of the grid coil. The middle separation between the plate and pick-up coil windings is ¼". L-5 and L-13 coils consist of 245 turns, close wound No. 36 D. S. C. wire, on a 1½" tube. L-4, 7, and 10 consist of 240 turns each, close wound with No. 36 D. S. C. wire on a 1½" tube. All of these coils and couplers are shielded. The condenser unit C-12 consists of two section 500 mmf. each, straight line wavelength capacity which tune the antenna coupler and detector coil; and the 3rd section is 300 mmf. capacity with special curve plates to track with the other sections. This last condenser of course, is to tune the oscillator coupler.

Short Wave Coil Data: The antenna coupler L-1 is not changed for short wave reception. The detector coupler L-2 and the oscillator coupler L-3 are changed. Both the short wave coils L-2 and L-3 have 5-point selector switches which connect to taps on the coil windings. The amplification obtained in the Super-15 is so great that any small loss incurred by using a tapped coil is negligible. The short wave detector coupler L-2 consists of a single winding of 16 turns No. 24 bare copper wire, space wound ⅛" and tapped at 11 turns, 7½, 4½, and 3 turns; size of the tubing used for the coil form is 2" O. D. Wavelength range approximately 16-80 meters. The short wave oscillator coupler L-3 consists of an outside coil form 2" O. D. with the grid winding of 14 turns, tapped at 13, 3¼, 8¾, 5¾ and 2 turns. The inside coil winding is on a 1¾" form, with a plate coil and the pickup coil at six turns each, close wound in the center and spread from each other ¼". The outside winding of the coupler L-3 is a No. 24 bare copper wire, and the inside windings No. 24 D. C. C. wire. The outside grid winding is spaced ⅛". These coils are listed as type "A"

There is another set of coils type "B" covering a wavelength range between 80-200 meters approximately. The windings are on the same sized forms as the type "A" coils. The detector coupler L-2 has 38 turns, No. 24 bare copper wire, tapped at 26, 17 and 14 turns, by a 3-point selector switch, spaced ⅛". The oscillator coupler L-3 has an outside grid winding of 32 turns, spaced ⅛", No. 24 bare copper wire, and the inside windings consisting of 11 turns each No. 24 D. C. C. wire, close wound in the center of the form and spread from each other ¼". There is no selector switch on this coil, but the small knife switch shunts a 29 mmf. condenser across the grid winding in order to cover the entire frequency band.

General List of Parts

L1—Antenna Coupler, shielded.
L2—Tuned R. F. Transformer, shielded.
L3—Oscillator Coupler, shielded.
L4—1st Detector Plate Coil (band pass).
L5—1st Int. Grid Coil (band pass).
L6—1st Int. S.G. Choke.
L7—1st Int. Plate Coil.
L8—1st Int. Plate Choke.
L9—2nd Int. S.G. Choke.
L10—2nd Int. Plate Coil.
L11—2nd Int. Plate Choke.
L12—3rd Int. S.G. Choke.
L13—3rd Int. Plate Coil.
L14—3rd Int. Plate Choke.
L15—Second Detector Grid Choke.
L16—Second Detector Plate Choke.
K—1st Filter Choke, 300 ohms, 175 mils.
F1, F2—Speaker Field Windings, 2500 ohms each, carrying 70 mils.; used as second choke for filter.
C1—15 mf. Mica Condenser (used on short-wavelengths).
C2, C3—Ant. Circuit Trimmer Condensers.
C4—.25 mf. 2-sec. By-pass Condenser (1st R.F., S.G. and Cathode).
C5—1st Detector Trimmer Condenser (panel control).
C6—.25 mf. 2-sec. By-pass Condenser (1st det. S.G. and Cathode).
C7—100 mmf. Air Condenser, 1st Detector Plate Tuning circuit.
C8—.03 Mica, 1st det. to 1st Int. Coupling Condenser.
C9—2-sec. Air Condenser R.F. and 1st Det. Tuning Compensator.
C10—3-sec. R.F., 1st Det. and Oscillator Tuning Condensers.
C11—.25 mfd. R.F. Plate By-pass Condenser.
C12—65 mmf. Oscillator Trimmer Condenser.
C13—.1 mf. Paper Condenser, osc. plate.
C14—.25 mf. 2-sec. By-pass Condenser, osc. cathode and 1st det. S.G.
C15—100 mmf. 1st Int. Grid Tuning Condenser.
C16—1st Int. Grid and Cathode By-pass Condenser, 2-sec. .25 mf. ea.
C17—1st Int. Plate and S.G. By-pass Condenser, 2-sec. .25 mf. ea.
C18—.25 mf. By-pass Condenser.
C19—100 mmf. Air Condenser tuning 1st. int. plate.
C20—100 mmf. Mica Condenser, 1st int. to 2nd int. coupling.
C21—2nd Int. Grid and Cathode By-pass Cond., .25 mf. ea.
C22—2-sec. .25 Mfd. Cond., 2nd Int. Plate and S.G. By-pass.
C23—100 mmf. Air Condenser, 2nd int. Plate tuning.
C24—100 mmf. Mica Condenser, 2nd Int. to 3rd Int. Coupling.
C25—.25 mf. Paper Cond., 3rd Int. Cathode By-pass.
C26—2-sec. .25 Mfd. Cond., 3rd Int. Plate and S.G. By-pass.
C27—100 mmf. Air Condenser tuning 3rd Int. Plate.
C28—100 mmf. Mica Condenser coupling 3rd Int. and Detector.
C29—1. mf. Paper Condenser, A.V.C. Plate By-pass.
C30—2-sec. .25 Mfd. Paper Condenser, A.V.C. Cathode and Plate voltage.
C31—3-sec. 1. Mfd. 2nd Det. Plate, Cathode and Grid By-pass Cond.
C32—.03 mf. 2nd Det. Cathode By-pass Condenser.
C33—1. mf. Paper Cond. Detector Plate By-pass.
C34—1. mf. Paper Cond. Det. Plate to Audio Grid Coupling.
C35—.25 mf. Paper Cond. 1st Audio Cathode By-pass.
C36—.25 mf. Paper Cond. 1st Audio Plate By-pass.
C37—1. mf. Paper Cond. 1st Audio Plate to PP Transformer Primary.
C38—.25 mf. Paper Condenser, 1st Audio Plate Voltage By-pass.
C39—500 mmf. Mica Condenser, 3rd Audio By-pass.
C40, C41—.1 mf. Paper Condenser to External Speaker Connections.
C42—4-sec. 2 Mfd. Ea. Filter Condensers, rated 1000 volts.
R1—3000 Ohms, R.F. Cathode Resistor.
R2—50,000 Ohms Variable, 1st Det. Cathode Resistor.
R3—150,000 Ohms, 1st Det. Plate.
R4—4,000 Ohms, 1st Det. S.G.
R5—10,000 Ohms, Oscillator Cathode Resistor.
R6—50,000 Ohms, Oscillator Plate.
R7—10,000 Ohms, Grid Suppressor 1st Int.
R8—2400 Ohms, 1st. Int. Cathode Bias.
R9—100,000 Ohms, 1st Int. Grid Return.
R10—100,000 Ohms, Auto. Vol. Control Plate.
R11—2 Meg., 2nd Int. Grid Return.
XR13—3000 Ohms, 2nd Int. Cathode Bias Resistor.
XR12—10,000 Ohms, Grid Suppressor 2nd Int.
R14—100,000 Ohms, 2nd Int. Grid Return.
R15—100,000 Ohms, Auto. Vol. Control Plate.
R16—30,000 Ohms, Compensating Resistor, Int. Plates.
R17—2 Meg., 3rd Int. Grid.
R18—10,000 Ohms, Grid Suppressor, 3rd Int.
R19—3000 Ohms, 3rd Int. Cathode Bias.
R20—50,000 Ohms, 2nd Det. Cathode Bias.
R21—50,000 Ohms, 2nd Det. Plate.
R22—25,000 Ohms, 2nd Det. Plate feed.
R23—150,000 Ohms, 1st Audio Grid.
R24—50,000 Ohms, 1st Audio Plate.
R25—2700 Ohms, 1st Audio Cathode Bias.
R26—25,000 Ohms, 1st Audio Plate Feed.
R27—1,500 Ohms, Variable Cathode Resistor for A.V.C.
R28—50,000 Ohms, 2nd Audio Grid Return.
R29—250,000 Ohms, Dual Section Resistor, Audio Volume Control.
R30—1400 Ohms, Voltage Divider.
R31—15,000 Ohms, 2nd Audio Plates.
R32—11,000 Ohms, Voltage Divider.
R33—6,000 Ohms, Voltage Divider.
R34—775 Ohms, 3rd Audio Bias.
X—Power Switch.
X1—Radio Amplifier Gain Switch, 3-pole, 3-position.
X2—Audio Tone Switch, 2-pole, 3-position.
X3—Phono Switch, S.P., S.T.
M1—O-5 M.A. Plate Tuning Meter.
M2—0-150 V. A.C. Voltmeter for recording line voltage.
T1—750 Volt Power Transformer.
T2—Type 250 PP Input Transformer.
T3—Interstage PP Transformer coupling 1st and 2nd stages.
T4—Type 250 Output Audio Transformer for Dual Speakers.

Tubes Used

VT1, VT2, VT4, VT5, VT6—Type '58 R. F. Pentodes.
VT3, VT7, VT8, VT9, VT10, VT11, VT12—Type '56 Tubes.
VT13, VT14—Type '50 Super Power Tubes.
VT15, VT16—Type '81 Heavy Duty Rectifiers.

NORDEN-HAUCK LABS.
MODEL 34 – NAVY

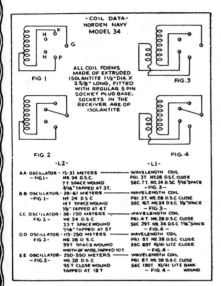

-COIL DATA-
NORDEN NAVY
MODEL 34

ALL COIL FORMS
MADE OF EXTRUDED
ISOLANTITE 1½" DIA. X
2 5/8" LONG, FITTED
WITH REGULAR 5 PIN
SOCKET PLUG BASE.
SOCKETS IN THE
RECEIVER ARE OF
ISOLANTITE

-L2-

AA OSCILLATOR - 15-31 METERS
-FIG.1-
№ 24 D.S.C.
7 T SPACE WOUND
3/16" TAPPED AT 3 T.

BB OSCILLATOR - 28-61 METERS
-FIG.1-
№ 24 D.S.C.
14 T SPACE WOUND
⅛" TAPPED AT 4 T

CC OSCILLATOR - 56-120 METERS
-FIG.2-
№ 24 D.S.C.
23 T SPACE WOUND
7/16" TAPPED AT 5 T

DD OSCILLATOR - 115-250 METERS
-FIG.2-
№ 28 D.S.C
39 T SPACE WOUND
WIDTH OF WIRE, TAPPED 10 T.

EE OSCILLATOR - 250-550 METERS
-FIG.2-
№ 28 D.S.C
95 T CLOSE WOUND
TAPPED AT 18 T

-L1-

WAVELENGTH COIL
PRI. 3 T. №28 D.S.C. CLOSE
SEC. 7 T. №24 D.S.C 3/16"SPACE
-FIG. 3-

WAVELENGTH COIL
PRI. 3 T. №28 D.S.C CLOSE
SEC. 16 T. №24 D.S.C ⅛"SPACE
-FIG. 3-

WAVELENGTH COIL
PRI. 4 T. №28 D.S.C. CLOSE
SEC 29 T. №24 D.S.C 1/16"SPACE
-FIG. 4-

WAVELENGTH COIL
PRI. 5 T. №28 D.S.C CLOSE
SEC. 65 T. 10/41 LITZ CLOSE
-FIG. 4-

WAVELENGTH COIL
PRI. 6 T. №28 D.S.C. CLOSE
SEC. 130 T. 10/41 LITZ BANK
-FIG. 4- WOUND

High Sensitivity and Selectivity

The new receiver has a sensitivity of less than ¼ micro-volt per meter throughout the entire frequency range, which permits the reception of the very weakest signals. The sensitivity of any receiver is limited by the proportion of *background noise* in relation to *signal strength*. Great care has been exercised in the design of this set to insure the highest possible ratio of signal-to-noise, so that satisfactory reception is possible even under adverse conditions. Actually this ratio is about 4:1 which is higher than any other receiver now available, the designers of this set state, and signals that are inaudible to other receivers are easily picked up with the Navy Model 34 with satisfactory quality and volume.

By employing variable condensers of different capacities on different frequency bands, the Model 34 achieves great ease of tuning, even at the highest frequencies (shortest wave lengths). Thus the usually congested short-wave broadcast bands no longer present a problem with this tuning system. The continuous *band-spread* tuning arrangement provides ample separation of all stations, regardless of frequency and is so arranged, for example, that stations on the 25 meter band will occupy as much as three degrees on the band-spread tuning dial, thus illustrating the unusual ease of tuning of this instrument.

This set uses one 2B7 tube in a special circuit, which actually represents a stage of intermediate amplification as well as the control of the radio frequency amplification of the set. The control grid current for the automatic volume control tube is obtained through means of a separate winding in the last intermediate transformer, thereby eliminating drain on the detector grid. A switch cuts the "AVC" in or out of circuit, a desirable feature for CW reception.

Bill of Material for Modern Navy Model 34

L-1 Antenna Inductance specially wound on threaded isolantite forms
L-2 Oscillator Coil specially wound on threaded isolantite forms
L-3, 4, 5, 6, 7, 8, 9, 14 8 M.H. Chokes
L-10, 11 85 M.H. Chokes
L-12, 13 Power Filter Chokes
L-15 C.W. Oscillator Coil
C-1 W.L. Tank Condenser .000138 mf. cap.
C-2 W.L. Band-Spread Tuning Condensers .0000125 mf. cap.
C-3 Auxiliary W.L. Band-Spread Tuning

Condenser .000029 mf. cap.
C-4 Oscillator Tank Condenser .000138 mf. cap.
C-5 Oscillator Band-Spread Tuning Condenser .0000125 mf. cap.
C-6 Auxiliary Oscillator B a n d - Spread Tuning Condenser .000029 mf. cap.
C-7, 8, 9 Three-Section .1; .1; .1 mf. By-Pass Condenser Block
C-10, 14, 15 .003 mf. By-Pass Condenser
C-11, 12, 13, 16, 17, 19 .005 mf. By-Pass Condenser
C-18 .1 mf. By-Pass Condenser
C-20 8. mf. By-Pass Condenser
C-21 .000006 mf. Coupling Condenser
C-22 .0001 mf. Condenser
C-23, 24, 25 8. mf. Filter Condensers
C-26 1. mf. Audio By-Pass Condenser
C-27, 28, 29 .00025 mf. Filter Condenser
C-30 .02 mf. Blocking Condenser
C-31 .05 mf. Audio By-Pass Condenser
C-32 .0001 mf. Grid Condenser
C-33 .5 mf. Condenser
R-1, 4, 9, 13 25,000 ohms Fixed
R-2, 3, 18 15,000 ohms Fixed
R-5 1. Megohm Fixed
R-6 500,000 ohms Fixed
R-7, 8 200 ohms Fixed
R-10, 16 100,000 ohms Fixed
R-11 250,000 ohms variable Potentiometer
R-12 3,000 ohms Fixed
R-14, 17, 22 50,000 ohms Fixed
R-19, 20 125,000 ohms Fixed
R-21 6,950 ohms Fixed
J-1 Headphone Jack
P.T.-1 Power Transformer
A.T.-1 Audio Output Transformer
R.F.T.-1, 2, 3 Intermediate Frequency Transformers 465 K.C.
R.F.T.-4 A.V.C. Transformer
S.W.-1 Audio Tone Switch
S.W.-2 A.V.C. Switch
S.W.-3 Power Switch
S.W.-4 C.W. Oscillator Switch
V.T.-1, 5 Type 57 Tube, RCA Radiotron
V.T.-2, 3, 4, 6 Type 58 Tubes, RCA Radiotron
V.T.-7 Type 2B7 Tube, RCA Radiotron
V.T.-8 Type 2A5 Tube, RCA Radiotron
V.T.-9 Type 80 Tube, RCA Radiotron

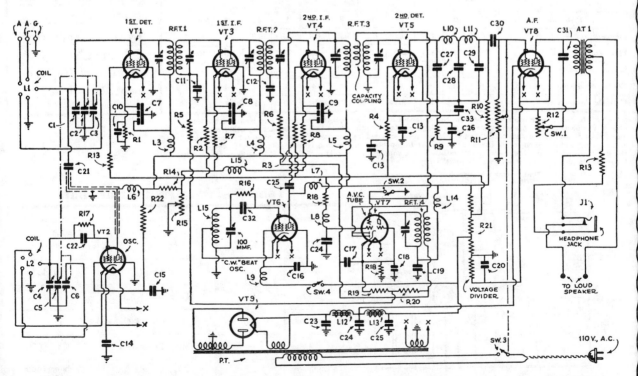

Wiring diagram of the new Navy model 34 multi-wave superhet receiver. This set represents one of the highest class receivers thus far offered to short-wave "fans" and "hams."

PHILCO RADIO & TELEVISION CO.
MODEL - 16

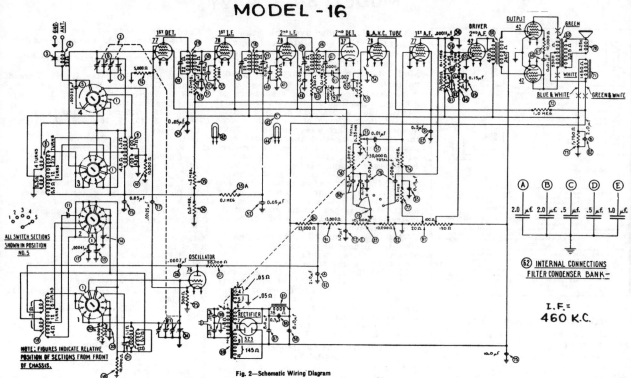

Fig. 2—Schematic Wiring Diagram

520 KC. to 23 M.C.

Table 1—Tube Socket Data*—A. C. Line Voltage 115 Volts

Circuit	1st Det.	Osc.	1st I. F.	2nd I. F.	2nd Det.	Inter-Station Noise Supr. Circuit	1st A. F.	2nd A. F. (Driver)	Output		Rectifier
Type Tube	77	76	78	78	37	78	77	42	42	42	5-Z-3
Filament Volts—F to F........	6.3	6.3	6.3	6.3	6.3	6.3	6.3	6.3	6.3	6.3	4.7
Plate Volts—P to K..........	220	53	225	230	0	1.8	130	220	340	340	400
Screen Grid Volts—SG to K....	80	—	80	80	—	1.8	1.8	220	340	340	—
Control Grid Volts—CG to K...	1.6	6.4	0	0	.2	1.6	.4	.6	34	34	—
Cathode Volts—K to F........	4.2	1.9	2.2	2.5	0	0	0	0	0	0	—

NOTE—These values are for Model 16-122. Model 16-121 uses a Type 80 Rectifier Tube. See Note, page 4, at end of Replacement Parts List.

* All of the above readings were taken from the underside of the chassis, using test prods and leads, with a suitable A. C. voltmeter for filament voltages, and a high-resistance multi-range D. C. voltmeter for other readings. The *Philco Model 048 All-Purpose Set Tester* is highly recommended for this use. Volume control set at maximum and station selector turned to low frequency end; interstation noise suppression circuit potentiometer turned all the way to the right; and toggle switch (interstation noise suppression circuit) in "ON" ("S") position. Readings taken with a plug-in adapter will **NOT** be satisfactory.

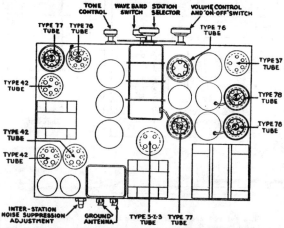

Fig. 1—Top View of Chassis, Showing Tube Locations and Major Parts

Table 2—Power Transformer Data

Terminal	A. C. Volts	Circuit	Color
1—2	105—125	Primary	White
3—5	6.3	Filament	Black
6—7	5.0	Filament of 5-Z-3	Blue
8—10	800	Plates of 5-Z-3	Yellow
4	—	Center Tap of 3—5	Black—Yellow Tracer
9	—	Center Tap of 8—10	Yellow—Green Tracer

PHILCO RADIO & TELEVISION CO.

MODEL 4

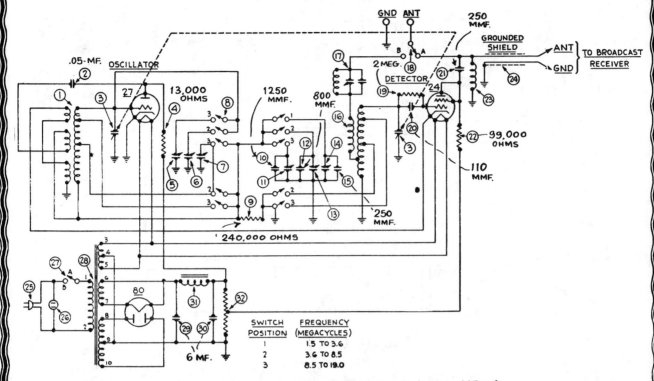

SWITCH POSITION	FREQUENCY (MEGACYCLES)
1	1.5 TO 3.6
2	3.6 TO 8.5
3	8.5 TO 19.0

Table 1—Tube Socket Readings—Line Voltage—115 volts

Tube		Filament Volts	Plate Volts	Screen Grid Volts	Control Grid Volts	Cathode Volts
Type	Circuit					
27	Oscillator	2.4	110	..	.1	0
24	Detector	2.4	25	25	.3	0
80	Rectifier	5.0	170/170	..	..	..

NOTE: The above voltage readings were taken from the socket terminals on the underside of the chassis, using a Weston multi-range voltmeter, 1000 ohms per volt. The radio set tester cannot be used either for voltage or plate current readings because of the effect of the long leads through the set tester cord.

Table 2—Power Transformer Voltages

Terminals	A. C. Volts		Color
1—2	105—125	Primary	White
3—5	2.5	Filament of 24 and 27	Black
6—7	5.0	Filament of 80	Light Blue
8—10	340	Plates of 80	Yellow
4	...	Center Tap of 3—5	Black with Yellow Tracer
9	...	Center Tap of 8—10	Yellow with Green Tracer

Table 3—Condenser Data

Nos. on Figs. 1 and 2	Capacity Mfd.	Container
20	.00011	Blue and Golden Yellow
12	.0008	Green and Orange
10	.00125	Blue and Orange
2	.05	Black Bakelite Container
29 30	6.	Electrolytic

Table 4—Resistor Data

Nos. on Figs. 1 and 2	Power (Watts)	Resistance (Ohms)	COLOR		
			Body	Tip	Dot
32		4750 4750	Long Tubular		
4	1.	13000	Brown	Orange	Orange
22	1.	99000	White	White	Orange
9	.5	240,000	Red	Yellow	Yellow
19	.5	2 Megohms	Red	Black	Green

PHILCO RADIO & TELEVISION CO.

MODEL 14

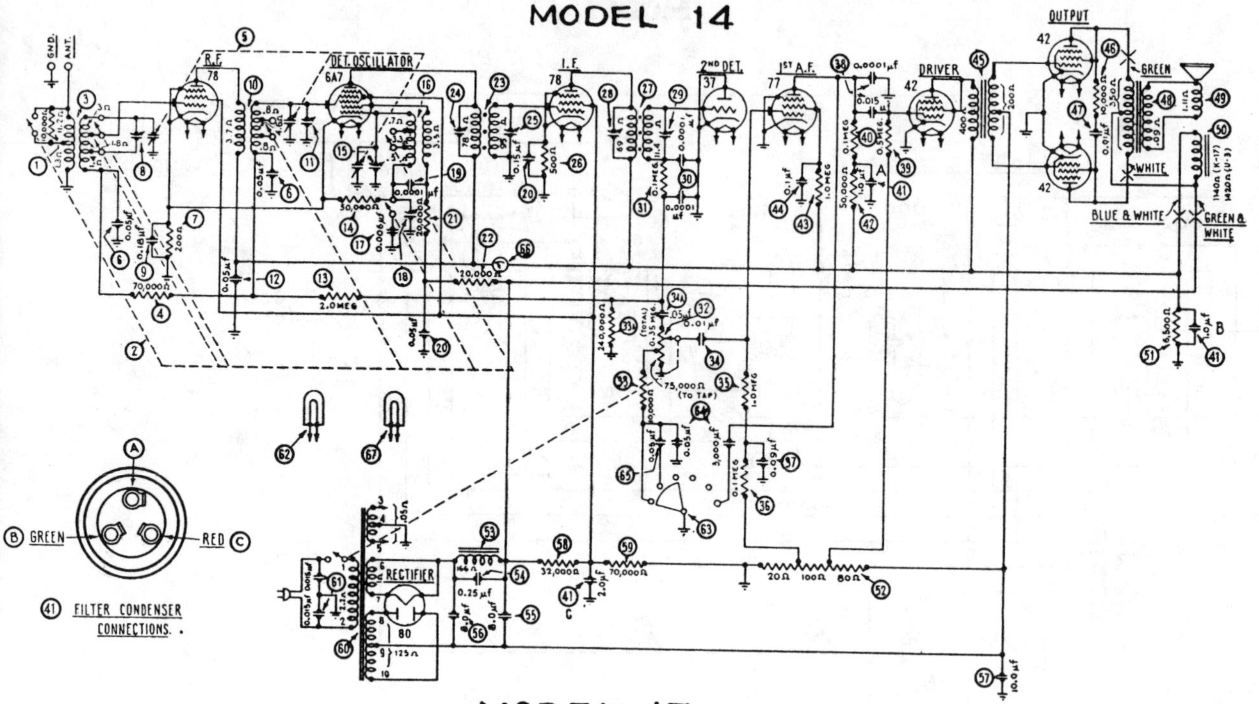

(A) GREEN — RED (C)

(41) FILTER CONDENSER CONNECTIONS.

MODEL 17

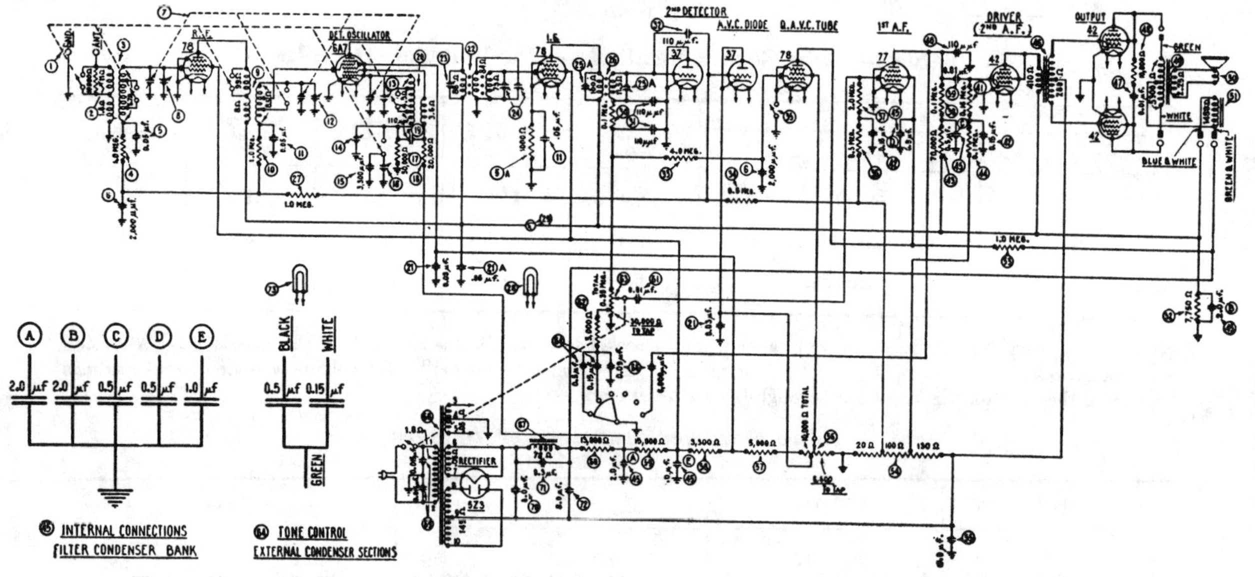

(A) (B) (C) (D) (E)

2.0 µf   2.0 µf   0.5 µf   0.5 µf   1.0 µf

BLACK   WHITE

0.5 µf   0.15 µf

GREEN

(A) INTERNAL CONNECTIONS FILTER CONDENSER BANK

(B) TONE CONTROL EXTERNAL CONDENSER SECTIONS

Table 1—Tube Socket Data*—A. C. Line Voltage 115 Volts

Circuit	R.F.	Det. Osc.	I.F.	2nd Det.	A.V.C.	Inter-Station Noise Supr. Crt.	1st A.F.	Driver	Output		Rectifier
Type Tube	78	6A7	78	37	37	78	77	42	42	42	5Z3
Filament Volts—F to F..	6.3	6.3	6.3	6.3	6.3	6.3	6.3	6.3	6.3	6.3	4.7
Plate Volts—P to K.....	220	220	225	0	0	45	45	230	340	340	400
Screen Grid Volts—Sg to K... (6A7-G3-5 to K)	75	58	75	—	—	50	50	230	340	340	...
Control Grid Volts—CG to K.. (6A7-G4 to K)	Negligible	Negligible	3.7	.25	.25	.24	.24	.24	34.	34.	...
Cathode Volts—K to F..	0	0	3.7	0	11.	0	0	0	0	0	...

6A7-G1 to K = 22.0 Volts
6A7-G2 to K = 140.0 Volts

NOTE: These values are for Model 17-122. Model 17-121 uses a Type 80 Rectifier.

Table 2—Power Transformer Data

Terminal	A. C. Volts	Circuit	Color
1–2	105–125	Primary	White
3–5	6.3	Filament	Black
6–7	5.0	Filament of 5Z3	Blue
8–10	800	Plates of 5Z3	Yellow
4	...	Center Tap of 3-5	Black—Yellow Tracer
9	...	Center Tap of 8-10	Yellow—Green Tracer

PHILCO RADIO & TELEVISION CO.

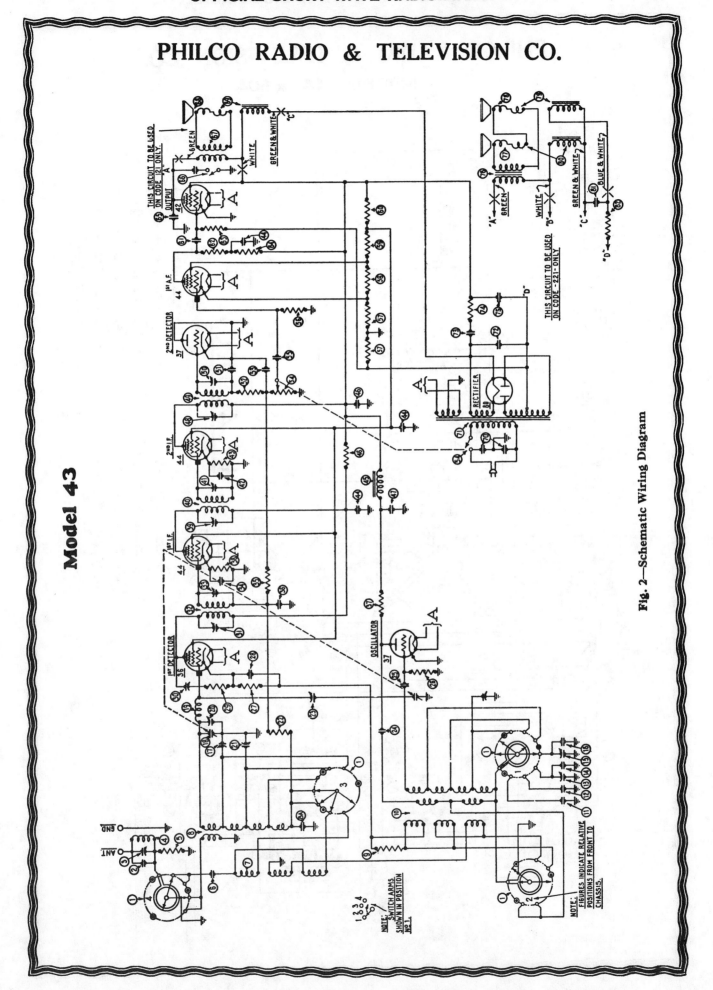

Model 43

Fig. 2—Schematic Wiring Diagram

PHILCO RADIO & TELEVISION CO.
MODELS 44 & 504

I.F. 460 K.C.

NOTE: ALL SWITCH SECTIONS SHOWN IN POSITION Nº4

NOTE: FIGURES INDICATE RELATIVE POSITION OF SWITCH SECTIONS FROM FRONT OF CHASSIS

PHILCO RADIO & TELEVISION CO.
MODEL 60

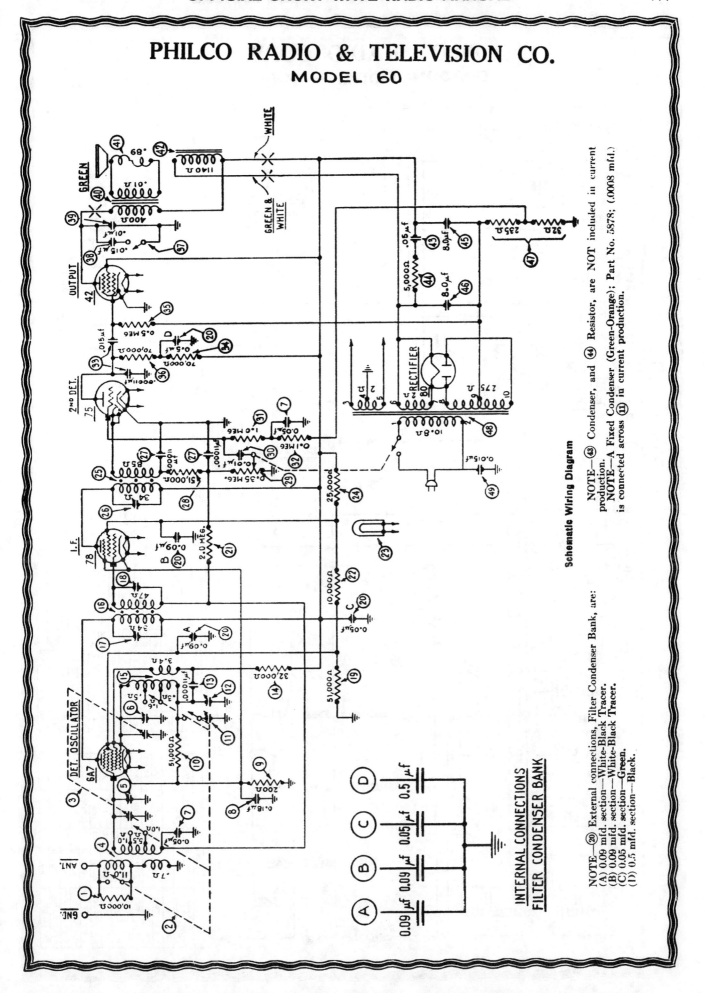

Schematic Wiring Diagram

NOTE—㊸ Condenser, and ㊹ Resistor, are NOT included in current production.
NOTE—A Fixed Condenser (Green-Orange); Part No. 5878; (.0008 mfd.) is connected across ⑪ in current production.

NOTE—⑳ External connections, Filter Condenser Bank, are:
(A) 0.09 mfd. section—White-Black Tracer.
(B) 0.09 mfd. section—White-Black Tracer.
(C) 0.05 mfd. section—Green.
(D) 0.5 mfd. section—Black.

INTERNAL CONNECTIONS
FILTER CONDENSER BANK

PILOT RADIO CORP.
DRAGON MODEL ~ 84

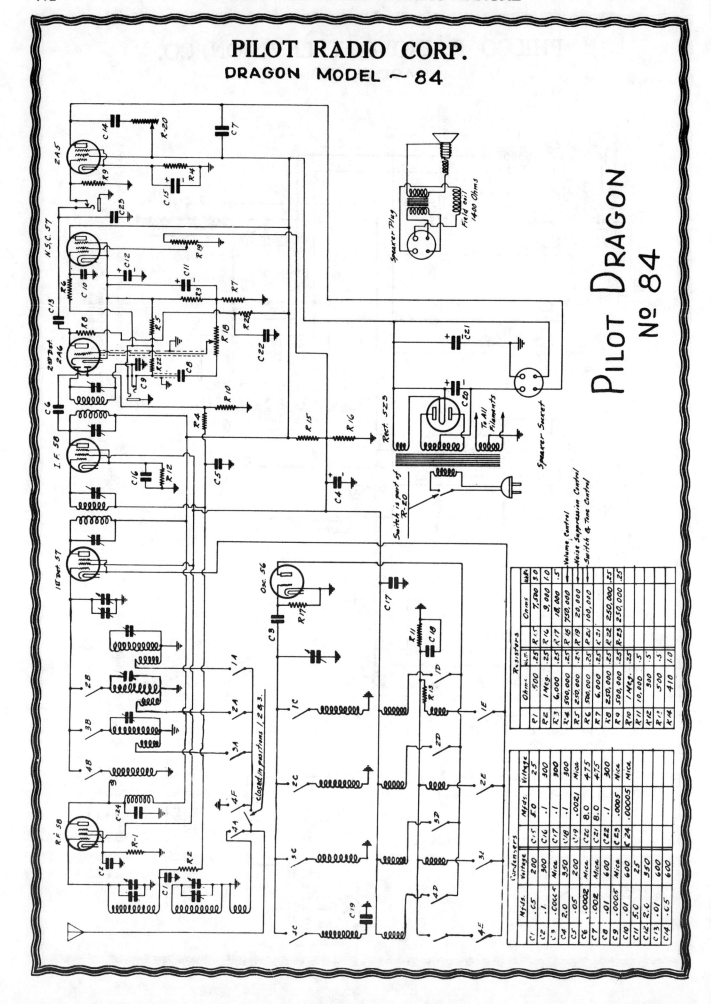

PILOT RADIO & TUBE CORP.
UNIVERSAL SUPER-WASP

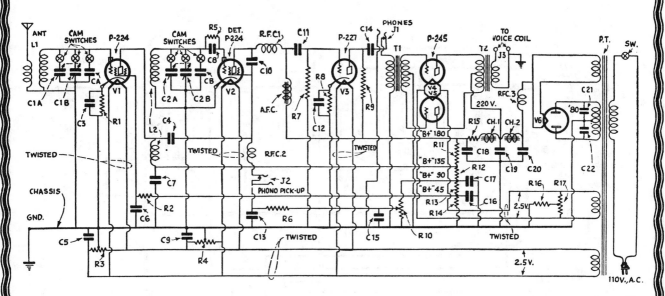

Schematic diagram of new "Universal Super-Wasp", the four sets of R.F. coils being represented as one for simplicity. The capacities CA and CB are fixed .0004 m.f. condensers used only to tune the highest waveband, above 470 meters. Note the usual regeneration method. All plug-in coils are eliminated by a simple yet effective switching scheme.

The standard Universal receiver now uses a total of six tubes, including the rectifier. Its great advantage lies in its wavelength changing switch, which eliminates the plug-in coils that heretofore have been the great nuisance in short-wave work. The coils are fixed inside the set, and are thrown in and out of the circuit by means of a very ingenious pair of rotary cam switches contained in molded bakelite housings. This switch, which is controlled by a simple little knob on the front panel, has seven positions, and covers seven wavelength ranges as follows: (1) 15 to 23 meters; (2) 22 to 41; (3) 40 to 75; (4) 70 to 147; (5) 146 to 270; (6) 240 to 500; and (7) 470 to 650.

Wiring Diagram

At a first glance this appears to be rather complicated, but a closer study will reveal it to be quite easily understandable. The set uses one stage of screen-grid tuned-radio-frequency amplification, a regenerative screen-grid detector, one impedance-coupled audio stage using a '27 tube, and a push-pull output stage using two '45's. For the sake of simplicity, the four antenna couplers fixed inside the set are represented as a single coil, L-1, and the four detector coils as L-2; each of these coils has two windings. L-1 has a primary and a secondary, and L-2 a combination primary-tickler and a secondary. One end of each winding is permanently grounded, either directly to the chassis or through a non-inductive condenser, as in the case of the primary-ticker. The other ends of the respective coils are

brought to contacts on the cam switches, and are connected in the proper sequence as the switches are turned.

The antenna and detector tuning condensers, marked C-1 and C-2 in the diagram, are actually double units; one section has a maximum capacity of 130 mmf. and the other 415 mmf. They have a common rotor connection but separate stators; the latter are also brought out to contacts on the cam switches, there being 15 contacts altogether. For the sake of convenience we will refer to the small condenser section as A and the large section as B, and to a separate fixed loading condenser of .0004-mf., capacity as C. We will also refer to the pairs of coils as 1, 2, 3 and 4.

How Coils and Condensers Are Switched

When the wavelength switch is set to range one, coils 1 and condensers A are connected together; range two, coils 2 and condensers A; range 3, coils 3 and condensers A; range 4, coils 3 and condensers A and B; range 5, coils 4 and condensers A; range 6, coils 4 and condensers A and B; range 7, coils 4 and condensers A, B and C. The shift from one range to another is made in an instant, and it is not necessary to open the set or disturb anything in it.

The primary and the tickler windings of the No. 3 antenna and detector coils are tapped in one place; part of the windings being used for wave-range 3 and the whole thing for range 4. The primaries and the ticklers of the No. 4 coils are tapped in two places, for use on ranges 5, 6 and 7.

The method of coupling the radio-frequency stage to the detector, and the system of regenerating through the combination primary and tickler, were adopted after exhaustive investigation by David Grimes and Edgar Messing. It is the logical method of coupling for screen-grid operation on the short-waves and provides very smooth regeneration, the control of which does not affect the tuning circuits. Thus it is possible to log stations very definitely and to duplicate the dial settings at any time.

If you will follow the circuit carefully, you will see that the radio-frequency current from both the plate and the screen of the detector tube is led back to the tickler winding T, through the .00004-mf. condenser between the screen and plate and the .0005-mf. condenser at the lower junction of this circuit. The R.F. choke coils in the plate and screen leads prevent the R.F. current from going any place else. The actual control of regeneration is provided by a 50,000-ohm potentiometer regulating the screen voltage.

Incidentally, a phonograph pick-up jack is connected directly in the screen lead as shown, and the regeneration control potentiometer then acts as a volume control on the phonograph music.

The plate voltage for the detector tube is fed through a high-inductance choke coil, rather than through a fixed resistor. The choke coil allows the plate voltage to assume the value necessary for efficient operation and, at the same time, prevents the audio-frequency component of the plate current from leaking off through the "B" circuit.

PILOT RADIO & TUBE CORP.
A.C. SUPER – WASP

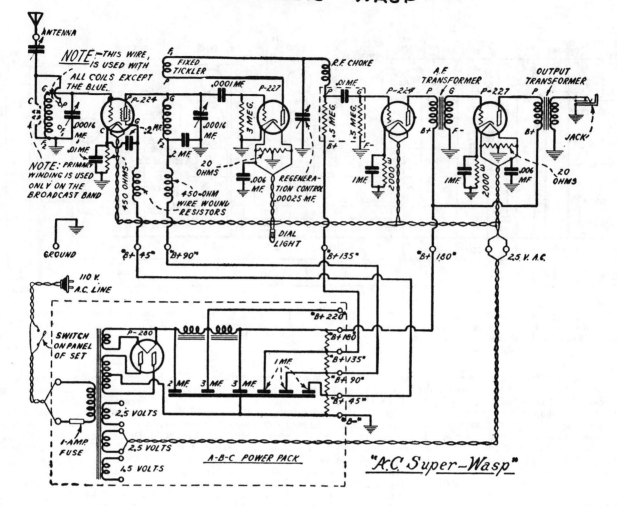

COMBINED S.W. AND B.C. SET

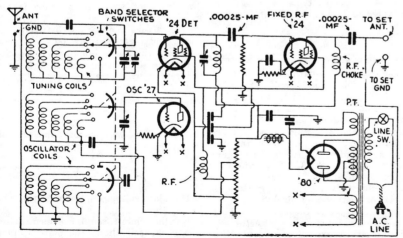

Above: Hook-up of Pilot short wave "converter," which tunes in S-W's when connected to "broadcast" receiver. Note that wire joining lower ends of tuning coils should be "grounded". With gang switch in 6th position, aerial is connected direct to broadcast set; also detector grid is "grounded".

The first intermediate frequency is 550 kc. This value, however, is not critical and, if there is a strong local station broadcasting on that frequency, the broadcast section tuning control can be set either slightly above or slightly below 550 without impairing the sensitivity of the set. This should be done because the broadcast section is so sensitive that there may be some direct pick-up that will cause interference with short wave reception.

The second intermediate frequency is that of the standard broadcast section, 175 kc. This means that the oscillator in the broadcast section will be operating at approximately 550 plus 175 kc., or 725 kc. It is impossible to shield this oscillator absolutely and its harmonics will therefore be picked up by the short-wave section. The third harmonic, for example, is 2175 kc., which is slightly under 150 meters. At about the middle of the fifth band, therefore, the short-wave section will appear to have tuned in on a very strong carrier that will have no modulation.

POSTAL RADIO CO.
9 TUBE INTERNATIONAL

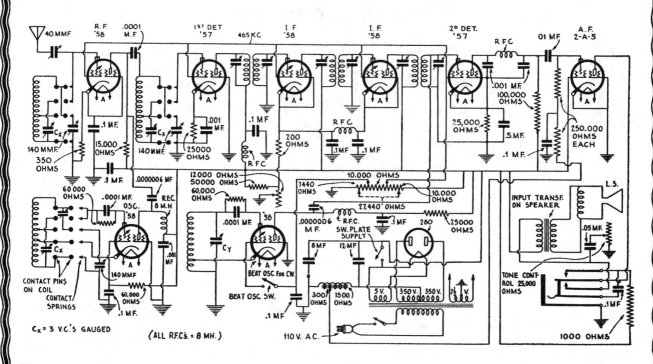

Diagram of connections used in the Postal International 9-Tube Superhet. All three coils, R.F., detector, and oscillator, are simultaneously changed in one single operation from the front panel, by a simple one-half inch movement of the complete coil shield box or "drawer"; the set being converted from ordinary tuning to band-spread tuning in a jiffy. The bands are spread over 65 degrees of the dial.

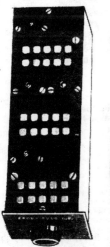

Close up of one set of coils in sliding shield box; in one position of the "drawer" "normal" tuning results—moved one half inch, "band-spread" is provided.

The main circuit of the 9-tube super consists of a 58 tuned R.F. stage, a 57 high sensitivity first detector, a 58 electron-coupled oscillator, a 58 first I.F. amplifier, a 58 second I.F. amplifier, 57 second detector, 58 electron-coupled audio "beat" oscillator for "CW" reception, a 2A5 output power tube in the A.F. stage, delivering three watts of undistorted audio signal energy; the rectifier tube being a type 280.

The I.F. amplifier stages are tuned to 465 kilocycles, with dual tuned I.F. transform-

ers, which are wound with Litz wire, a voltage gain of approximately 100 times being thus obtained.

This "pro" type receiver while especially designed for commercial and amateur short-wave communication purposes, is simultaneously an excellent short-wave receiver for the general short-wave "fan" who is interested in hearing the "foreign" DX stations, due to the high sensitivity and selectivity of the set.

List of Parts—Postal Superhet

4—Postal Multiformers
1—Special Postal socket, for Multiformer
1—3 gang 140 mmf. Postal condenser
1—40 mmf. Ant. comp. condenser
3—465 K.C. I.F. transformers
1—Audio beat oscillator coil 456 K.C.
1—Power transformer, to handle 9 tubes
1—12 mf. condenser 450 volt working v.
1—8 mf. condenser 450 v. working v.
1—12,000 ohm. volume control and switch
1—75,000 ohm tone control
1—Single circuit jack, with single pole double throw switch
1—Toggle switch for "B" supply
1—Rotor switch for audio "beat" oscillator
1—dial and front plate
5—58 sockets
2—57 sockets
1—2A5 sockets
1—280 sockets
1—Speaker 5 prong socket
5—8 millhenry R.F. chokes
1—Ant. Gnd. binding post.
8—.1 mf. tubular condensers
1—.05 mf. tubular condenser
1—.01 mf. tubular condenser
4—.001 mica fixed condensers
3—.0001 mica fixed condensers
1—.0000006 mmfd. condenser

1—25 watt wire-wound resistor 27,440 ohm; tapped 10,000 ohm, 10,000 ohm, and 7,440 ohm.
1—10 watt wire-wound resistor 1,000 ohm
2—60,000 ohm, 113 watt, pigtail resistors
1—15,000 ohm, 1 watt pigtail resistor
1—60,000 ohm, 1 watt pigtail resistor
3—25,000 ohm, 1 watt pigtail resistors
2—250,000 ohm, 1 watt pigtail resistors
1—350 ohm, 1 watt pigtail resistor
1—200 ohm, 1 watt pigtail resistor
1—100,000 ohm, 1 watt pigtail resistor
1—Cord and plug
1—Chassis 11"x19"x3"
1—Steel front panel 9¼"x20½"
6—Knobs.

To change from one band to another all the operator has to do is to pull out one "coil drawer" and insert another, the front panel of each coil unit being engraved with the frequency range in K.C. to which it responds. By simply pulling the drawer out ½ inch the advantage of "band-spread" tuning may be enjoyed, the band being spread over 65 degrees of the dial for example, in all of the four amateur bands.

A tuned radio frequency amplifier stage is used ahead of the first detector, which helps to eliminate repeat spots or image-frequency interference found in many superhet receivers. The variable condenser connected in series with the antenna circuit, coupled to the grid of the R.F. amplifier stage, makes it possible to match the R.F. and first detector tuning circuits, so that they will track accurately.

RADIO TRADING CO.
DOERLE 3 TUBE
"SIGNAL GRIPPER"

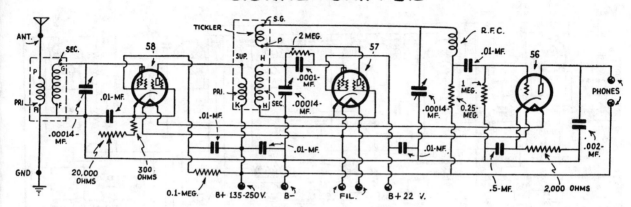

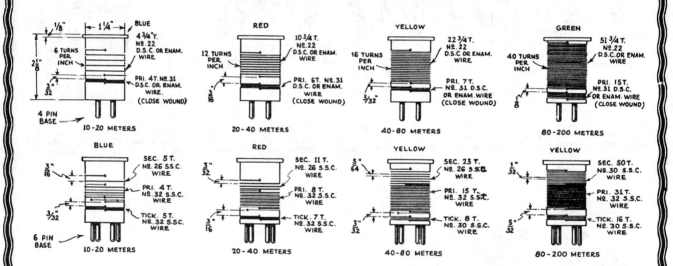

The old system of coupling between the R.F. stage and the detector, which is tuned impedance, is done away with and inductive coupling is used. The coils for this arrangement were obtained from the Radio Trading Co., and have three windings, one for the grid, one for the tickler, and a winding inter-woven with the grid coil for the plate circuit of the R.F. tube. Four of these coils will be needed and it is cheaper to buy them than to wind them by hand, because this is quite a difficult task.

Various Sources of Power Usable

This receiver can be operated from various sorts of power supply arrangements and is adaptable to any location whether A.C. power is available or not. For those having A.C. power it is suggested that this set be run from a regular power supply, delivering from 180 to 250 volts with a 2.5 volt filament winding. A 22 volt tap will be required for the screen of the detector tube, of course. It might be well to state here that the voltage applied to the screen should not exceed 22 volts under any consideration, because the sensitivity of the receiver will be very much affected by running the screen at a higher potential. Also the regeneration control will

not operate smoothly if the voltage is not of this value. If one wishes to operate this set from batteries it can be done very nicely with no change in the circuit. It's just a matter of changing the tubes to the automobile type and running them from a six-volt storage battery and using "B" batteries for the plate supply. 135 volts will work very nicely, although higher voltage is recommended if full signal strength is to be had. For operating on a regular power supply from 110 volts A.C., a 58 will be needed for the tuned R.F. stage, a 57 for the detector, and a 56 as the output tube. When operating from a storage battery with "B" batteries for the plate supply, a 78 will be used for the R.F. tube, a 77 for the detector and a 37 for the audio tube.

Operation

The operation of this receiver is exactly the same as before it was changed, as far as tuning is concerned. The two tuning condensers will have to be tuned at the same time, and the stations formerly received on this set will be received on practically the same dial settings, because the new coils tune exactly the same as the old ones. Tuning of the R.F. stage, however, will be much sharper than before; in fact the selectivity of the whole set is far greater than when it used the 2 volt type tubes.

List of Parts for the New "Doerle" 3-Tube A.C. Receiver

1—Drilled Metal Chassis, Radio Trading Co.
1—R.F. Choke Coil, Radio Trading Co.
1—Set of 4 Special Three-Winding Coils, Radio Trading Co.
1—Set of 4 Regular Doerle Coils, Radio Trading Co.
5—.01 mf. Fixed Condensers, Flechtheim.
1—.002 mf. Fixed Condensers, Flechtheim.
1—.5 Bypass Condenser, Flechtheim.
1—300 Ohm Resistor.
1—100,000 Ohm Resistor, Lynch.
1—250,000 Ohm Resistor, Lynch.
1—1 Megohm Resistor, Lynch.
1—2 Megohm Resistor, Lynch.
1—2,000 Ohm Resistor, Lynch.
1—2,000 Ohm Resistor, Variable.
3—Six Prong Sockets, Eby (National; Hammarlund; Na-ald).
1—Five Prong Socket, Eby (National; Hammarlund; Na-ald).
1—Four Prong Socket, Eby (National; Hammarlund; Na-ald).
2—Triple-Grid Tube Shields, Hammarlund (National).
1—.0001 Fixed Condenser, Flechtheim.
3—Hammarlund .00014 mf. Tuning Condensers.
2—Tuning Dials, National or other make.
1—Antenna Ground Terminal Strip, Eby.
1—Phone Terminal Strip, Eby.
1—Five Wire Cable.

RADIO TRADING CO.

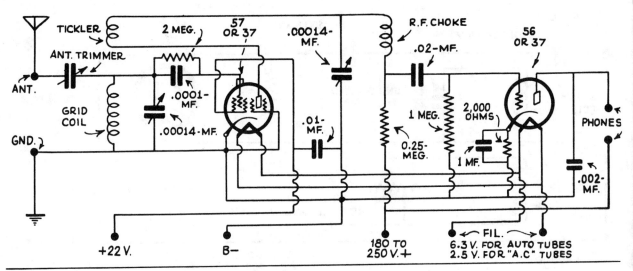

TICKLER · ANT. TRIMMER · 2 MEG. · 57 OR 37 · .00014-MF. · R.F. CHOKE · .02-MF. · 56 OR 37

ANT. · GRID COIL · .0001-MF. · .00014-MF. · .01-MF. · 1 MEG. · 2,000 OHMS · PHONES

GND. · 0.25-MEG. · 1 MF. · .002-MF.

+22 V. · B− · 180 TO 250 V.+ · FIL. 6.3 V. FOR AUTO TUBES 2.5 V. FOR "A.C" TUBES

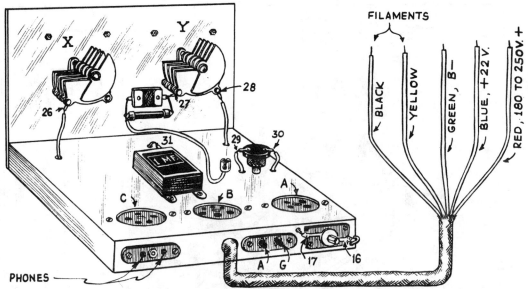

FILAMENTS

X · Y · 28

26 · 27

31 · 29 · 30

C · B · A

PHONES · A G · 17 · 16

BLACK · YELLOW · GREEN, B− · BLUE, +22 V. · RED, 180 TO 250 V.+

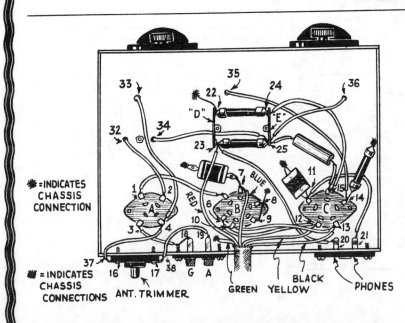

ELECTRIFIED DOERLE 2 TUBE ~ A.C. RECEIVER

33 · 35 · 22 · 24 · 36 · "D" · "E" · 32 · 34 · 23 · 25

✳ = INDICATES CHASSIS CONNECTION

1 · 2 · 7 BLUE · 11 · RED · A · 6 · B · 8 · 15 · 14 · 10 · 9 · 12 · C · 3 · 4 · 18 · 19 · 13 · 20 · 21

✳ = INDICATES CHASSIS CONNECTIONS

37 · 16 · 17 · 38 · G A · ANT. TRIMMER · GREEN · YELLOW · BLACK · PHONES

RADIO TRADING CO., New York, N.Y.

RADIO TRADING CO.

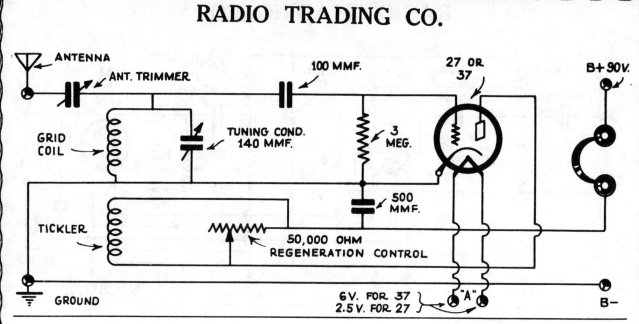

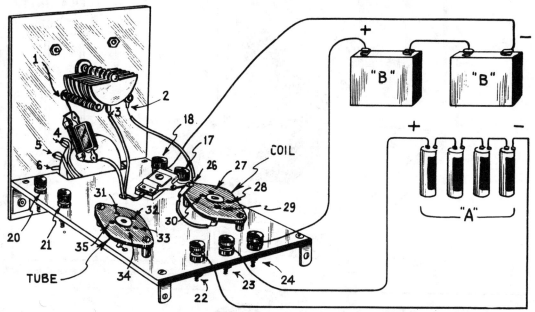

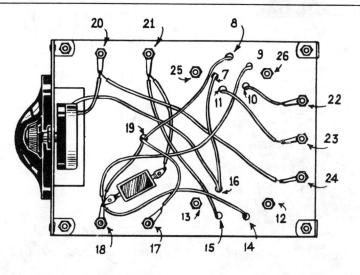

OSCILLODYNE 1-TUBE WONDER SET

RADIO TRADING CO.
New York, N.Y.

RADIO TRADING CO.

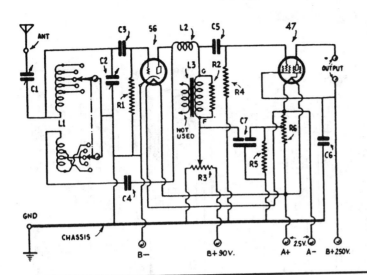

2 TUBE A.C. OSCILLODYNE RECEIVER

RADIO
TRADING Co.
New York, N.Y.

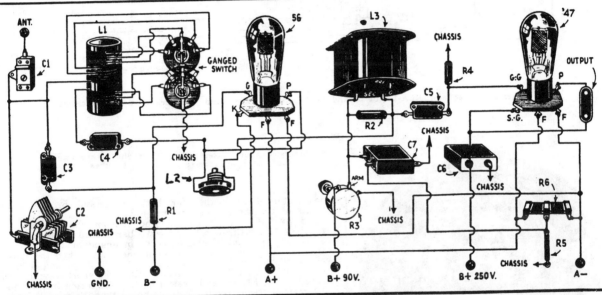

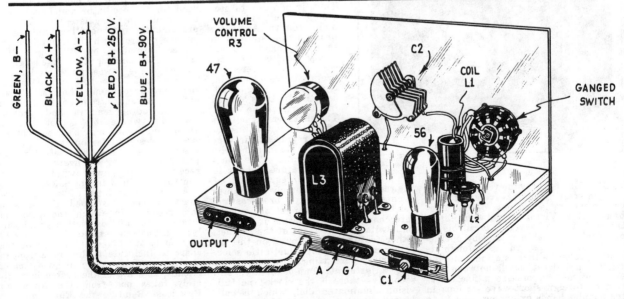

RCA VICTOR, INC.
MODEL K-64

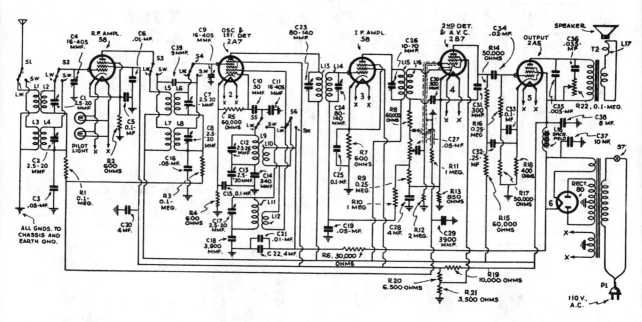

Diagram for the new RCA-Victor 2-band model, "short-wave" and "broadcast" 6-tube receiver.

Photo above shows latest RCA-Victor short-wave and broadcast band receiver, utilizing 6 tubes.

● ONE of the very newest and highly interesting short-wave and broadcast receivers is the new 6-tube, two-band, RCA Victor model here illustrated. This receiver is available in different style cabinets and by operating a switch it reproduces through its loud speaker either stations on the regular broadcast band, between 200 and 550 meters, or else a fine selection of foreign short-wave stations in the popular bands extending from 19 to 55.5 meters. Between the limits of the short-wave band available in this receiver at the throw of a switch are included four of the internationally assigned short-wave broadcast bands, located at 49, 31, 25, and 19 meters, respectively. Thus, in addition to providing fine entertainment from the American broadcasting stations in the usual band, this receiver permits direct reception of interesting programs from the principal short-wave broadcast transmitters located in all parts of the world. The short-wave facilities afforded by this instrument represent the very newest engineering developments. The short-wave feature is built in as an integral part of the radio chassis, not simply an adapter connected to an old-style broadcast receiver. Both tuning ranges are quickly interchangeable by means of a push-pull switch on the front of the cabinet. Other features to be found on this receiver are the vernier dual-ratio selector drive, permitting either rapid or fine adjustments independently, and secondly—there is the clock-type full-vision illuminated dial, which is calibrated directly in terms of frequency for both short and broadcast band ranges.

As the diagram shows, each of the two wave bands made available in this set have independent tuned couplers or inductances and when the two-way switch S-1, 2, 3, 4, 5, and 6 knob is turned, either the short wave or broadcast coils are connected into circuit. It will be noted that the new 2A7 tube is used for the oscillator and first detector, while the very latest circuit improvements incorporating automatic volume control with the second detector are provided, by utilizing a 2B7 tube. The loud speaker is energized by one of the newest power audio frequency tubes, the 2A5. The manufacturers recommend the use of an outdoor antenna from 25 to 75 feet long, including lead in and ground wire, and where this is not possible an inside antenna may be used. These sets are designed for operation on 110 volts, 60 cycle A.C. and sets for 220 volt A.C. circuits are available. This receiver has its power switch and tone control combined in one knob.

A brand new form of tuning time-chart has been developed for use with this receiver in which the program "time on the air" is plotted graphically; also Eastern Standard Time, as well as G.M.T. are given at the top of the chart to facilitate tuning in European stations. This set will win many friends, as a flip of the switch immediately takes one from the American Broadcast Band to the European and South American circuit.

RCA VICTOR, INC.

MODEL K-80, K-80-X; & K-85

This all wave super-heterodyne receiver is of the continuous tuning type utilizing a straight super-heterodyne circuit in all bands. The bands are as follows:

Selector Switch Position	Frequency Range (Kilocycles)	Wave Length Range (Meters)
X	150–410	2000–732
A	540–1500	555–200
B	1500–3900	200–77.0
C	3900–10000	77.0–30
D	8000–18000	37.5–16.7

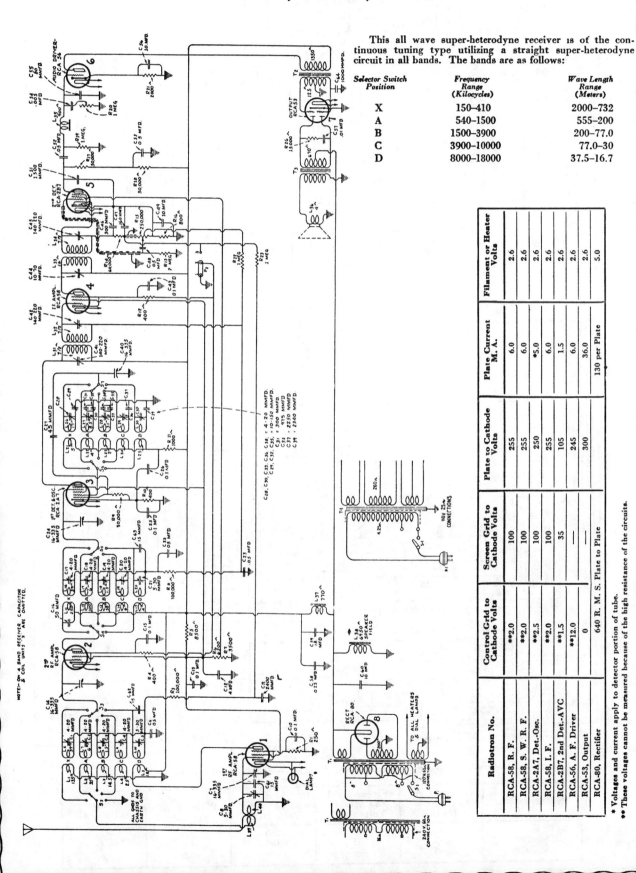

Radiotron No.	Control Grid to Cathode Volts	Screen Grid to Cathode Volts	Plate to Cathode Volts	Plate Current M. A.	Filament or Heater Volts
RCA-58, R. F.	**2.0	100	255	6.0	2.6
RCA-58, S. W. R. F.	**2.0	100	255	6.0	2.6
RCA-2A7, Det.-Osc.	**2.5	100	250	*5.0	2.6
RCA-58, I. F.	**2.0	100	255	6.0	2.6
RCA-2B7, 2nd Det.-AVC	**1.5	35	105	1.5	2.6
RCA-56, A. F. Driver	**12.0	—	245	6.0	2.6
RCA-53, Output	0	—	300	36.0	2.6
RCA-80, Rectifier	640 R. M. S. Plate to Plate			130 per Plate	5.0

* Voltages and current apply to detector portion of tube.
** These voltages cannot be measured because of the high resistance of the circuit.

RCA VICTOR, INC.
MODEL RO-23

SHORT WAVE CONVERTER CHASSIS

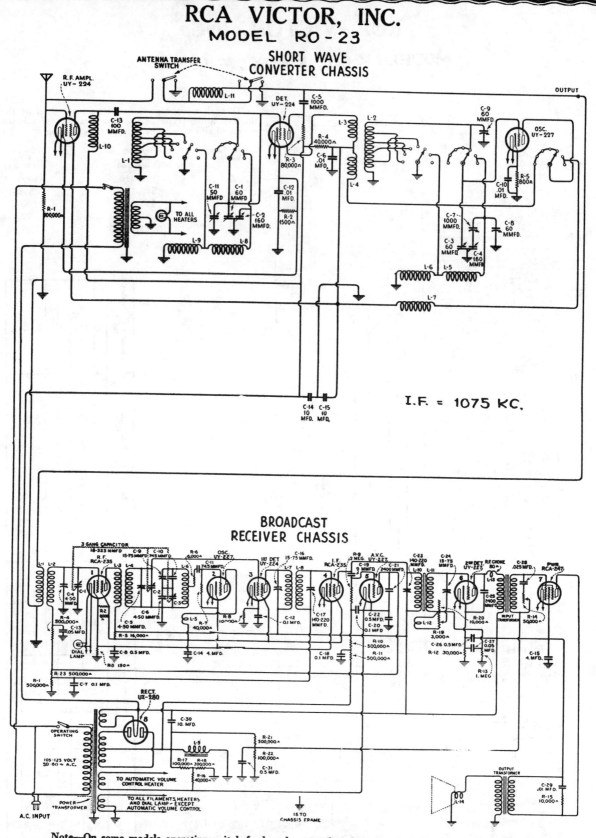

I.F. = 1075 KC.

BROADCAST RECEIVER CHASSIS

Note—On some models operating switch for broadcast receiver is in circuit to Converter.

Radiotron No.	Control Grid to Cathode Volts D. C.	Screen Grid to Cathode Volts D. C.	Plate to Cathode Volts D. C.	Plate M. A.	Heater Volts A. C.
R. F.	—3	50	260	1.0	2.66
Detector	—3	50	180	1.0	2.66
Oscillator	—5	—	50	5.0	2.66

RCA VICTOR, INC.

SWA 2   SW. CONVERTER

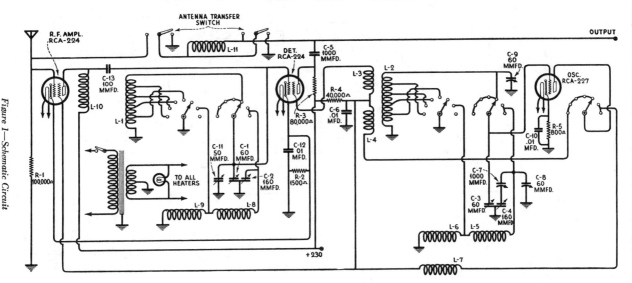

Figure 1—Schematic Circuit

RCA Victor Short Wave Converter SWA-2 is a three tube, single control short wave unit designed to convert all short wave signals from 13.8–200 meters to a single frequency so that they may then be amplified by means of the usual broadcast receiver.

One Radiotron UY-224 is used as an R. F. Amplifying stage, one UY-224 as the detector and one UY-227 as the oscillator.   Heater current for these Radiotrons is obtained from a small transformer incorporated in the unit.   Plate supply is obtained from the broadcasting receiver.

A wafer connector is supplied that may be inserted under the tube socket when a receiver using a UX-280 rectifier and a filter in the negative side of the line is used.   Under these conditions—most modern receivers are so designed that this is true—the plate supply to the converter is obtained through the contact on the wafer connector to the UX-280 filament.   On receivers where this condition does not exist, but where Pentode output tubes are used, the wafer connector can be used to make connection to the screen grid of the Pentode.   On receivers where neither condition exist any connection that gives a filtered D. C. output of from 180 to 260 volts between the contact and ground will be suitable.

Due to the SWA-2 being identical with the converter chassis used in the RO-23, reference to the RO-23 Service Notes should be made for data pertaining to Service work.

REPLACEMENT PARTS

Stock No.	DESCRIPTION	List Price	Stock No.	DESCRIPTION	List Price
2747	Cap—Grid contactor cap—Package of 5............	$0.50	6109	Knob—Knob with pointer—Package of 5...........	$1.75
2977	Knob—Station selector, or Resonator knob—Package of 5	2.50	6110	Dial lamp shield and indicator.....................	.50
3058	Resistor—100,000 ohms—Carbon type—1 watt—Package of 5	2.50	6111	Escutcheon—Range switch knob escutcheon—Package of 5	1.80
3153	Resistor—1500 ohms—Carbon type—1 watt—Package of 5	2.75	6112	Cushion—Receiver chassis rubber cushions—Package of 4	.50
3285	Cord—Drive cord—Package of 5...............	1.00	7062	Capacitor—Adjustable capacitor—15–70 mmfd......	1.00
3286	Spring—Drive cord tension spring—Package of 5....	1.40	7298	Capacitor—0.01 mfd.	.80
3288	Socket—UY Radiotron socket—Complete with insulation strip	.50	7406	Capacitor—Double adjustable capacitor—One section 10–70 mmfd.—One section 800–1000 mmfd...	1.10
3289	Contact lug—Complete with mounting rivets—Package of 10	.50	7407	Coil—High frequency detector coil...............	1.05
3290	Switch—Antenna—"Off and On"—Toggle type—2 used—Complete with mounting nut	1.00	7408	Coil—Low frequency detector and oscillator coil..	1.45
3291	Board—Terminal board with two soldering terminals complete with mounting rivets—Located on switch bracket—Package of 5	.50	7409	Coil—High frequency oscillator coil.............	1.85
3292	Drive shaft with pulley—Package of 5.............	2.35	7410	Capacitor—Variable capacitor—7 plate—Complete with mounting nut and washers...............	1.75
3293	Coil—For resistor board assembly.............	.65	8806	Transformer—Filament power transformer.........	3.25
6100	Coil—Coil assembly with mounting eyelet—For switch and bracket assembly	.75	8807	Transformer—Filament power transformer—110 volts—25 cycle.....................	5.75
6101	Socket—Dial lamp socket and bracket with mounting rivets	.50	8808	Transformer—Filament power transformer—220 volts—60 cycle.................	3.40
6102	Capacitor—1000 mmfd.—Package of 5...........	2.50	8809	Board—Resistor board less resistors, capacitors and coil..................	1.00
6103	Resistor—800 ohms—Carbon type—1 watt—Package of 5	2.00	8810	Lever—Switch lever assembly—Comprising shaft, 3 switch levers and coupling bushing.............	.70
6104	Resistor—80,000 ohms—Carbon type—1 watt—Package of 5	2.00	8811	Switch—Range switch complete with mounting washer and nut..................	6.60
6105	Resistor—40,000 ohms—Carbon type—3 watt—Package of 5	2.00	8812	Capacitor—Variable tuning capacitor assembly.....	5.10
6106	Coupling—Switch lever shaft coupling with 2 taper pins—Package of 5	.50	8813	Dial drum and scale..................	1.20
6107	Switch—Toggle type—Power switch............	1.00	10820	Capacitor—100 mmfd..............	.50
6108	Binding post—Complete with terminal lug, mounting washer and mounting nut—Package of 5.....	1.75		**CABINET**	
			3229	Escutcheon—Tuning dial escutcheon with mounting screws	.70
			6113	Foot—Cabinet felt foot—Package of 15...........	.50
			9399	Cabinet—Complete less equipment..............	12.00

RCA VICTOR, INC.
MODELS 140, 141, 141-E, 240 & AVR 1

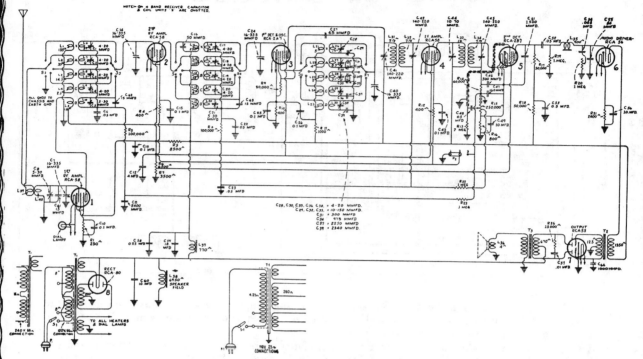

Figure A—Schematic Circuit Diagram

Radiotron No.	Control Grid to Cathode Volts	Screen Grid to Cathode Volts	Plate to Cathode Volts	Plate Current M. A.	Filament or Heater Volts
RCA-58, R. F.	**2.0	100	255	6.0	2.6
RCA-58, S. W. R. F.	**2.0	100	255	6.0	2.6
RCA-2A7, Det.-Osc.	**2.5	100	250	*5.0	2.6
RCA-58, I. F.	**2.0	100	255	6.0	2.6
RCA-2B7, 2nd Det.-AVC	**1.5	35	105	1.5	2.6
RCA-56, A. F. Driver	**12.0	—	245	6.0	2.6
RCA-53, Output	0	—	300	36.0	2.6
RCA-80, Rectifier	640 R. M. S. Plate to Plate			130 per Plate	5.0

* Voltages and current apply to detector portion of tube.

** These voltages cannot be measured because of the high resistance of the circuits.

External Oscillator Frequency	Dial Setting	Location of Line-Up Capacitors	Position of Selector Switch	Adjust for	Number of Adjustments To Be Made
445 K. C.	Any setting that does not bring in station.	At rear of chassis	Any position that does not bring in station.	Maximum output.	4
370 K. C.	370 K. C.	Bottom of chassis	X	Maximum output.	3
175 K. C.	Set for signal.	Top of chassis.	X	Maximum output while rocking dial back and forth.	1
1400 K. C.	1400 K. C.	Bottom of chassis.	A	Maximum output.	3
600 K. C.	Set for signal.	Top of chassis.	A	Maximum output while rocking dial back and forth.	1
3900 K. C.	3900 K. C.	Bottom of chassis.	B	Maximum output.	3
1710 K. C.	Set for signal.	Top of chassis.	B	Maximum output while rocking dial back and forth.	1
10 M. C.	10 M. C.	Bottom of chassis.	C	Maximum output.	3
15 or 18 M. C.	15 or 18 M. C.	Bottom and top.	D	Maximum output. Adjust oscillator trimmer until two points are noted where signal is heard. Use for adjustment the higher frequency of these two points. This will be the point lying counter-clockwise from the other point.	4

RCA VICTOR, INC.
AR-1145

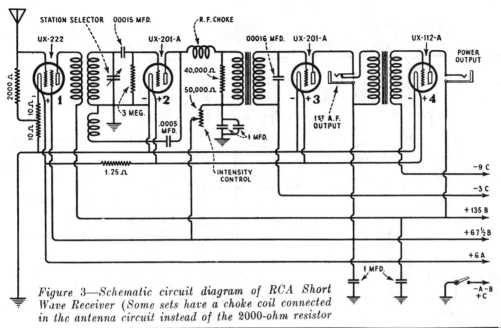

Figure 3—Schematic circuit diagram of RCA Short Wave Receiver (Some sets have a choke coil connected in the antenna circuit instead of the 2000-ohm resistor

Coil No.	Frequency Range		Wavelength Range Meters
	Megacycles	Kilocycles	
1	20—12	20,000—12,000	15— 25
2	12—7.2	12,000— 7,200	25— 42
3	7.2—4	7,200— 4,000	42— 75
6	......	1,500— 940	200—320
7	......	940— 550	320—545

VOLTAGE READINGS AT RADIOTRON SOCKETS

Intensity Control Near Zero. Operating Switch "On." All Batteries Connected. (See Figure 10.) Radiotrons in Sockets, or Test Set. Loudspeaker Plugged in Second Audio Stage Jack.

Radiotron	Fil. Volt.	Grid Volt.	Plate Volt.	Plate Current
Coupling UX-222	3.2	*Control grid 1.5 *Screen grid 67.5	130.0	Plate 3.5 mil. amp. *Screen 0.5 mil. amp.
Detector UX-201A	5.0	....	30-60 (Depending on position of intensity control)	0.65 to 1.5 mil. amp.
1st Audio Amp. UX-201A	5.0	3.0	65	1.1 mil. amp.
2d Audio Amp. (Power) UX-112A	5.0	9.0	130.0	4.0 mil. amp.

* These readings cannot be measured by ordinary methods as with the Weston Model 537 test set.

RCA VICTOR, INC.

EARLY SHORT WAVE CONVERTER

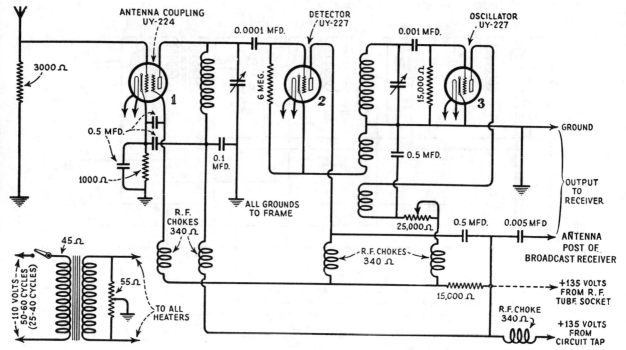

Schematic diagram of Short Wave Adaptor

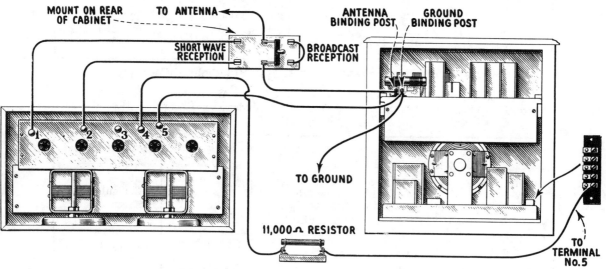

Connections to RCA Radiolas 80, 82, 86 and RAE-68

Set the dial on the Receiver at about the middle of the scale (approximately 1000 K.C.) and at a position no powerful or local station comes in. With, for example, the No. 2 coils in place, vary the two tuning dials keeping both readings approximately alike. In doing this, proceed slowly as the right hand dial settings are extremely sharp and the stations may be missed if the motion is too rapid. Many code stations will be heard, and having found one, adjust both dials carefully for maximum signal. The setting of the antenna dial is much broader and may not read exactly like the oscillator dial for maximum signal strength.

E. H. SCOTT RADIO LABS.
12 TUBE ALL-WAVE SET

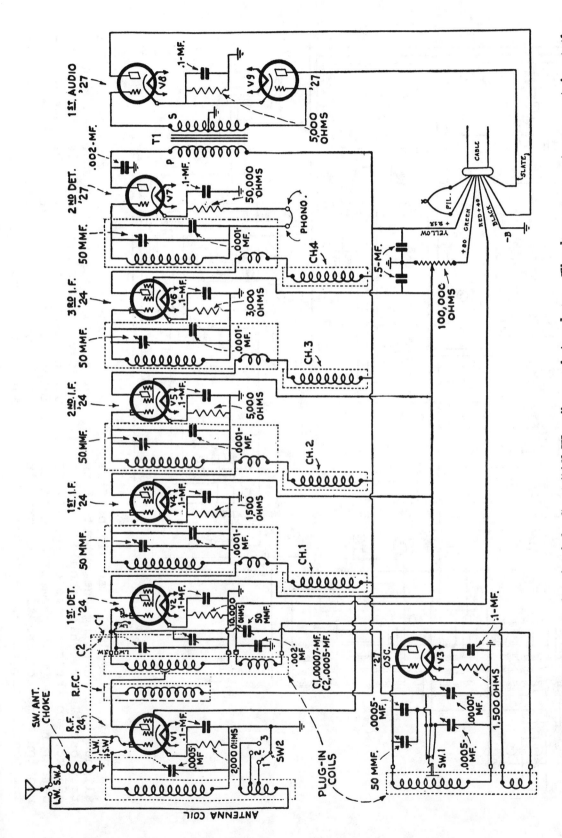

Schematic circuit of the receiver chassis of the Scott "All-Wave" superheterodyne. The long-wave antenna post is at the rear of the chassis; and the short-wave post on the shield over the tuning gang. The output feeds a push-pull '45 pack; a push-pull '50 unit is also obtainable.

McMURDO SILVER, INC.
MASTERPIECE ~ I

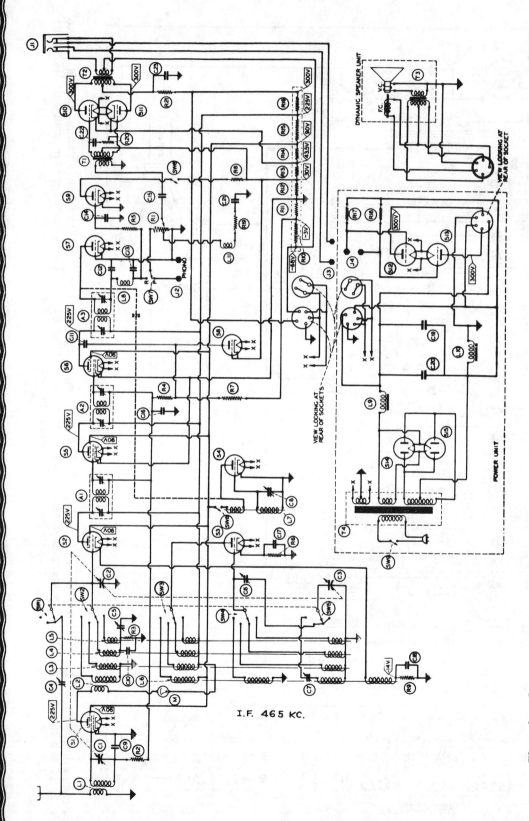

I.F. 465 KC.

The i.f. amplifier should be first aligned by connecting the output of the test oscillator, which should be set at 465 kc., to the grid cap of the '57 first detector tube (with normal set grid connection removed) and to the ground binding post of the tuner. With the bottom plate removed, the six trimmer screws found beneath each i.f. transformer can should be adjusted for maximum deflection of the tuning meter, taking care that the oscillator output is kept low enough so that the volume control of the set can be well advanced during this adjustment.

McMURDO SILVER, INC.
MASTERPIECE — II

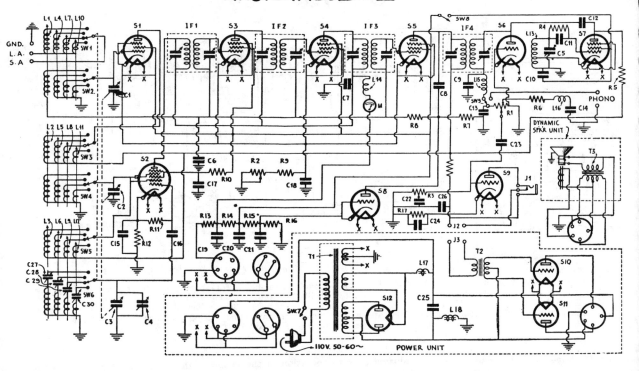

Here we have the schematic circuit diagram for the new McMurdo Silver "Masterpiece ·II" All-Wave Superheterodyne Receiver.

A New Kind of Sensitivity Control

A special squelch tube was used in the first MASTERPIECE connected to function as a valve refusing all signals and noise below a certain level when it was in use. By this means the set could be tuned from station to station with dead silence between stations when desired. This arrangement had two disadvantages. It required an extra tube, and its cut-off level had to be set at some arbitrary point—it could not easily be set to agree with the different local noise conditions found in different locations. Also, it was found that many stations constituting good noise-free entertainment, would, in the course of their normal and continuous slight fading, fade across any arbitrarily established cut-off level, resulting in a periodic cut-off of reception, or if fading was rapid, in choppy, distorted reception when the squelch circuit was in use.

An r.f. (radio-frequency) sensitivity control has been substituted for the squelch circuit, which was a switch like valve and has been eliminated, therefore. This sensitivity control can be adjusted to suppress any prevailing condition or level of local noise, which is obviously advantageous, and eliminates entirely the possibility of choppy reception of stations fading slightly across the cut-off level of any automatic squelch or valve circuit. It permits of adjustment when desired, of the r.f.—i.f. (i.f.=intermediate frequency) gain to the exact degree desired, almost wholly independently of the actual loud speaker volume control, and over all ordinary operating ranges, has no effect on the automatic volume control action, or on the operation of the audio volume level control.

The use of the better r.f.-i.f. sensitivity control eliminates the arbitrarily adjusted squelch tube, and together with three new and meritorious tubes recently introduced, permits of the elimination of a total of three tubes net, allowing somewhat better results to be obtained than with the original fifteen at first employed.

Band-Spread Tuning Arrangement

The use of two dials instead of a single tuning dial is the second change referred to. The receiver is completely tuned by the right-hand dial and its single knob, as was the first MASTERPIECE. The second dial is simply a vernier, or band-spread tuning dial to permit the short wave bands such as the 6000, 9500 and 12,000 kc. short-wave broadcast bands being spread out over a whole full dial scale for easy tuning. It may likewise be used to spread the five amateur bands for easy tuning—or even small segments of the broadcast band. It is purely a vernier, not a second tuning control. It need not be used at all in operating the set, yet its use makes for much easier tuning of the short-wave bands.

Tuning of the short-wave bands in the first receiver was made easy by a 28:1 dial ratio. This was necessarily slow to tune with over any range, but even more important, did not permit of spreading the different short wave stations far enough apart on the dial scale itself to make for easy reading for the eye. With this new band-spread dial, the main tuning dial need only be set at, say 6.2 for the 6000 kc. or 50 meter short wave broadcast band, and all the stations in this band will be found spread out nicely on the vernier dial—actually making short-wave tuning easier than is broadcast band tuning on ordinary receivers.

Tuning is not rendered easy by any high ratio tuning dial, which will necessarily be mechanically stiff. By the use of 6:1 automatic take-up gear drives with opposed gears (an equivalent of the beautifully smooth helical gear control) the mechanical operation not only is smooth and entirely free of slippage, wear or backlash, but the control knobs turn with extreme ease and absence of effort. Tuning is thus made faster, easier, simpler, and easier to read. This simple mechanical change is invaluable, and in the hands of a novice can make all the difference between skipping over foreign short wave stations or having them actually easier to find than broadcast band stations.

The Complete Circuit

Thus the revised tube circuit line-up is·

'58 r.f. stage, 2A7 combined first detector and electron-coupled oscillator (the first combination tube so far introduced which gives actually better results than separate tubes performing the same functions), three '58 i.f. stages (the third stage used for selectivity, not for gain—its additional gain cannot be used), '56 first audio stage, 2A3 pushpull Class A fifteen watt power output stage and 5Z3 rectifier.

It can be seen that from the original tube complement the '56 oscillator and '57 first detector have been replaced by the even better 2A7 tube, which performs both functions with higher gain and the desirable frequency stability and uniform output of the electron-coupled oscillator. Thus one tube is eliminated, and the performance is improved a bit. The next tube eliminated is the squelch tube, referred to previously.

Two '80 rectifiers were originally used. The new 5Z3 thermionic high vacuum rectifier, having the same power capacity as two '80s, allows one of the original two rectifiers to be dropped, thus effecting further simplification.

Actually, however, the improved and simplified receiver uses twelve tubes, the twelfth tube being a '58 in the added third dual tuned i.f. stage. The entire i.f. (intermediate-frequency) amplifier is air-tuned, making for permanency of setting in all climates, for the first time at no loss of selectivity. This tube is added only because it is the simple and obvious means of coupling the two extra tuned circuits added to the i.f. amplifier to set the selectivity up to absolute 9 kc.

A tuned or transposed antenna or lead-in system may be used when desired by virtue of separate antenna coupling coils for each of the four bands.

A tuned r.f. stage is used on both broadcast and short wave; the additional gain of this stage cuts down oscillator hiss and results in a very fine signal-to-noise ratio. It also eliminates the repeat spot or image interference so common in many improperly designed superhet receivers.

SILVER-MARSHALL, INC.
S-M - 726

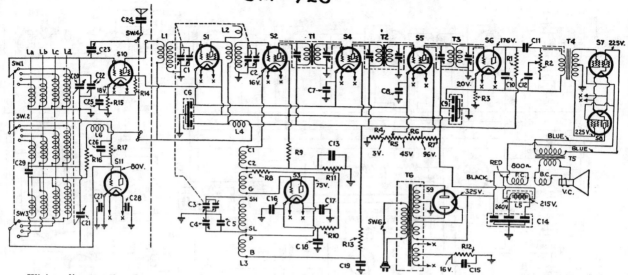

Wiring diagram for the latest "S-M"—726 SW., "All-Wave" Super-het. For short wave reception the aerial feeds into a S-W detector S-10 and oscillator S-11; in all 2 oscillators and 3 detectors are used. Tubes—S2, S-10—'24's; S3, S6, S11—'27's; S7, S8—'47's; S1, S4, S5—'51's (R. C. A. 235); S9—'80 rectifier. This set is a combination of a 9 tube vari-mu pentode broadcast super-het, and an 11 tube short wave super, using switches instead of plug-in coils. Pentodes are used at S7 and S8.

"Double-Super-het" Principle Used

It is obviously not practical to build a superheterodyne receiver for both short and broadcast wavelengths with two different intermediate-frequency amplifiers; for the cost of the equipment would be very considerable. This problem has been nicely solved in the "726 SW" by designing the main I.F. frequency amplifier for 175 kc.; this being preceded by the oscillator, first detector, and R.F. tube for broadcast band reception. As soon, however, as the receiver is shifted over to operation in the range of 10 to 200 meters, a scheme popularly known as "double suping" is resorted to —that is, the use of two intermediate frequencies with two oscillators, one fixed and one variable.

Setting the Dial on Broadcast Receiver

Specifically, the broadcast tuning dial is set to some clear channel in the neighborhood of 650 kc. (it may actually be anywhere between 600 and 700 kc.) and this done, the broadcast band R.F. amplifier tube and first detector, together with their tuned circuits, become the first level of intermediate-frequency amplification; which takes place obviously at the setting of the broadcast dial, or at 650 kc. approximately. A short-wave first detector is then placed ahead of the R.F. amplifier tube, which has now become an I.F. amplifier tube; and to this tube is coupled a short-wave oscillator which is arranged to track away from the short-wave first detector by approximately 650 kc., in order to produce the first intermediate frequency.

S-M CONVERTER

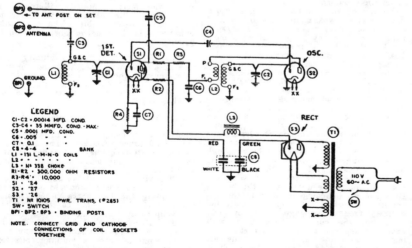

Wiring diagram of the Silver-Marshall S-W converter. It plugs into any A.C. socket.

Parts Necessary

The converter consists of a '24 first detector, with plug-in coils which are tuned by a .00014-mf. vernier or midget type of condenser. It was not thought desirable to bring the condenser control out to a vernier dial; since it is not particularly critical in setting, and operation may be more easily mastered when the first detector's tuning is regarded as a vernier or trimmer adjustment rather than as a regular tuning control. The oscillator circuit, however, is extremely sharp; employing a somewhat similar coil to that of the first detector, it is tuned by a .00014-mf. condenser controlled by a vernier dial. A '27 tube is the oscillator, and power supply for both tubes of the converter is obtained from a small, self-contained power unit.

A '26 tube is used as a rectifier, being fed by a small transformer which also supplies filament current for the '24 detector and '27 oscillator. Filtration is provided by one high-inductance choke and two 4-mf. dry-electrolytic, semi-self-healing condensers.

SILVER-MARSHALL, INC.
S-M - 727

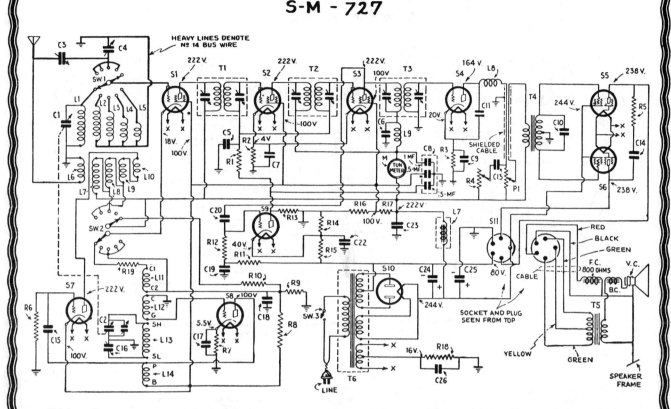

Wiring diagram for new All-wave Silver-Marshall Super-heterodyne designed for operation on 110 volt, 60 cycle A.C. circuit.

Model S.M. 727 Shortwave-Broadcast Receiver

C1-C2—2 Gang Variable Condenser — 365 Mmf. Max. ± 5 Mmf. 0° — 90° ± 1 Mmf. 90° — 180° ± ½ of 1% ; use with dial scale.
C3—25 Mmf. Trimmer Condenser.
C4—200 Mmf. Variable Midget Condenser.
C5—0.1 Mf. Condenser—Sprague.
C6—0.1 Mf. Condenser—Sprague.
C7—0.1 Mf. Condenser—Sprague.
C8—1.0, .5, .5 Mf. Condenser.
C9—1.0 Mf. Condenser 150 V. (Dual with C-22).
C10—.001 Mf. Condenser Mica.
C11—.001 Mf. Condenser Mica.
C13—.025 Mf. Condenser—Sprague.
C14—.006 Mf. 700 V.—Sprague.
C15—.00015 Mf. Mica.
C16—Oscillator Trimmer Condenser.
C17—0.1 Mf. Condenser—Sprague.
C18—0.1 Mf. Condenser—Sprague.
C19—0.1 Mf. Condenser—Sprague.
C20—.0005 Mf. Condenser Mica.
C22—1.0 Mf. Condenser, 150 V. (See C-9).
C23—4 Mf. Dry Electrolytic Cond. 450 V.
C24—8 Mf. Dry Electrolytic Cond. 450 V.
C25—4 Mf. Dry Electrolytic Cond. 450 V.
C26—0.1 Mf. Condenser—Sprague.
L1—197 Broadcast Antenna Coil (550-1,500 K.C.).
L2—202 Short Wave Antenna Coil (1.56-3.46 megacycles).
L3—201 Short Wave Antenna Coil (3.51-5.36 megacycles).
L4—200 Short Wave Antenna Coil (5.54-10.29 megacycles).
L5—199 Short Wave Antenna Coil (9.6-18.15 megacycles).
L6—198 Oscillator Coil.
L7—10145 Choke. (Iron Core Plate Filter.)
L8-L9—281 R.F. Choke. (IF and 2d Det. C'cts.)
M—Tuning Meter—15 M.A.
P1—100,000 Ohm Vol. Control (Com. with A.C. Switch).
R1—100,000 Ohm Resistor—1 Watt Carbon.

R2—400 Ohm Resistor—Wire Wound.
R3—100,000 Ohm Resistor—1 Watt Carbon.
R4—½ Megohm Tapered Variable Resistance.
R5—25,000 Ohm Resistor—1 Watt Carbon.
R6—300,000 Ohm Resistor—1 Watt Carbon.
R7—400 Ohm Resistor—Wire Wound.
R8—60,000 Ohm Resistor—1 Watt Carbon.
R9—3,500 Ohm Resistor—1 Watt Carbon.
R10—300,000 Ohm Resistor—1 Watt Carbon.
R11—1 Megohm Resistor—1 Watt Carbon.
R12—1 Megohm Resistor—1 Watt Carbon.
R13—300,000 Ohm Resistor—1 Watt Carbon.
R14— 10,000 Ohm Resistor—1 Watt Carbon.
R15— 10,000 Ohm Resistor—1 Watt Carbon.
R16— 10,000 Ohm Resistor—2 Watt Carbon.
R17— 6,500 Ohm Resistor Ohmite Red Devil—3 Watt.
R18—220 Ohm Resistor Ohmite Red Devil—2 Watt.
R19—400 Ohm Resistor—Wire Wound.
S1—'24 Tube.
S4-S7-S8-S9—'27 Tubes.
S5-S6—'47 Tubes.
S2-S3—'51 Tubes.
S10—'80 Tube.
S11—Speaker Socket.
SW1-SW2—Tandem Change-over Switch.
SW3—A.C. Switch (Combination with volume control).
T1—Q-1 I.F. Transformer.
T2—Q-2 I.F. Transformer.
T3—Q-3 I.F. Transformer.
T4—10159 Audio Transformer.
T5—10200 Output Transformer.
T6—10202 Power Transformer.

Silver-Marshall All-Wave Super-Het. Coil Data

Referring to the I.F. transformers, T-1, 2 and 3 in the diagram, these have 125-mmf. (max.) variable condensers as trimmers across primaries and secondaries as shown. Each I.F. transformer has two coils in an aluminum shield, and each coil consists of 209 turns (10

strands) No. 41 S.S.E. Litz. (inductance 1 m.h. at 1,000 cycles and resistance 7.6 ohms). The coils are bank-wound and have an internal diameter of ¾ inch, and each coil is ⅜ inch thick. The 2 coils are spaced about 1¼ inches apart and should be not less than ¼ inch from the aluminum shield can. Both coils are wound in the same direction.

Harmonic Generator Coil Data

1.56 to 3.46 megacycles. Primary L-2—48¾ turns, No. 27 enameled wire, space wound. Secondary L-7—18½ turns No. 36 D.S.C. close wound.

3.51 to 6.36 mc. primary L-3—23¾ turns No. 21 enameled wire, space wound. Secondary L-8—12½ turns No. 36 D.S.C. close wound (these coils so far specified wound on 1¼ inch outside diameter bakelite tubing).

5.54 to 10.29 mc. primary L-4—16¾ turns No. 19 enameled wire, space wound ; secondary L-9—12½ turns No. 36 D.S.C. close wound.

9.6 to 18.15 mc. primary L-5—6¾ turns, No. 17 enameled wire, space wound. Secondary L-10—10½ turns No. 36 D.S.C. close wound.

Broadcast Coil Data

Primary 450 turns No. 34 D.C.C. wire, bunch wound .5 inch in length on a form 2 inches in diameter. Secondary 83 turns No. 24 enameled wire wound on the same tube, which is 3⅜ inches long.

Oscillator Coil: L-6, L-11—11 turns No. 36 enameled wire, close wound.

L-12—30 turns No. 36 enameled wire, close wound.

L-13—84 turns No. 28 enameled wire, wound 68 turns per inch.

Coils L-11, 12 and 13 are wound on 1¼ inch outside diameter tube; these three coils being placed in a row with ⅜ inch space between coils, the order being L-11, L-13, L-12.

Over one end of L-13, near coil L-11, is wound coil L-14, comprising 43 turns No. 36 enameled wire, close wound, on a tube 1½ inches outside diameter.

SILVER-MARSHALL, INC.
S-M-728 S.W.

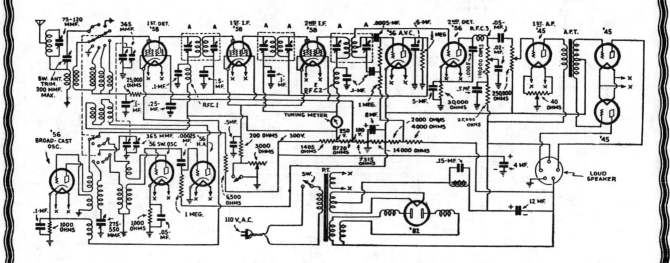

Wiring diagram of the new 728 S-W Silver-Marshall, 13 to 550 meter, 12-tube superheterodyne. Capacity A-75-120 mmf., R.F. chokes have a value of 7 millihenries.

The set is a twelve tube superheterodyne using the new "Q" system developed in the S-M laboratories which for the first time permits of accurate short wave dial calibration at the factory. As a broadcast set it uses ten tubes, and as a short wave set twelve. These two extra tubes are in no sense a conventional converter, for both essentially function as the short wave oscillator and replace the broadcast oscillator entirely on short waves.

The circuit involves a '58 detector and '56 oscillator for the broadcast band, or for the three short wave bands from 1,500 to 22,700 kc., a '58 first detector and '56 oscillator with its '56 harmonic generator tube, which allows one oscillator circuit to cover this wide frequency range without changes. Following this is a two stage 465 kc. i.f. amplifier using two '58 tubes, a '56 A.V.C. tube, '56 second detector, a '45 first audio or driver stage and a "Class A Prime" push-pull '45 output stage, with, of course, an '82 mercury vapor rectifier tube. In changing from one short-wave band to another, only the antenna circuit is changed, the broadcast oscillator being dropped completely out of circuit for the short wave bands.

The audio system is totally new, as it uses only a pair of '45s in the output stage, driven by a third '45, yet turns out over eight watts of undistorted power output, with the ability to handle strong signal peaks of up to twelve and sixteen or more watts—just as much to the ear as will '46 in Class B audio, but without any of the '46s' serious harmonic distortion at home volume levels.

The 728SW has a noise suppressor system adjustable for each and every specific location, so that all noise can be cut out at the throw of a switch for local and medium distance reception. This is accomplished by a switch and a semi-variable resistor (on the rear of the chassis) that allows the first i.f. grid to be biased negative enough to cut sensitivity to a point where only signals stronger than local noise can be heard in the silent position of the noise-suppressor switch.

The radio frequency chokes shown in the plate circuits of the first detector, second intermediate frequency amplifier and second detector, each have a value of 7 millihenries. The coils used in the new 728SW super-het are wound with Litz wire in many cases, and are difficult to wind accurately, let alone trying to calibrate them. However, for the benefit of the short-wave "intelligentsia," the coil winding data is given in the text that follows:

INTERMEDIATE TRANS.
1st and 2nd—
Pri.—211 turns No. 41 S.S.E. Litz wire
Sec.—211 turns No. 41 S.S.E. Litz wire
Wound on $\frac{3}{4}$" form—spaced $1\frac{1}{16}$" with loss ring between coils
3rd—
Pri.—211 turns No. 41 S.S.E. Litz wire
Sec.—211 turns No. 41 S.S.E. Litz wire—tapped at 150 turns.
Wound on $\frac{3}{4}$" form—spaced $\frac{7}{16}$"

BROADCAST OSC. COIL:
Tank—95 turns No. 27 P.E. wire—space wound 60 T.P.I. on $1\frac{1}{4}$" dia. form.
Grid—35 turns No. 36 P.E.—close wound
Plate—45 turns No. 36 P.E.—close wond
Pickup—13 turns No. 36 P.E. space wound
Grid, plate, pickup wound on $\frac{7}{8}$" dia. form.

SHORT WAVE OSC. COIL:
Tank—29 turns No. 20 P.E.—space wound—winding length $1\frac{1}{8}$" wound on $1\frac{1}{4}$" dia. form.
Plate—26 turns No. 36 S.E.—space wound—winding length $1\frac{1}{8}$" wound on $\frac{7}{8}$" dia. form.
Grid—9 turns No. 36 S.E.—wound in spacing of tank

ANTENNA COIL—BROADCAST:
Secondary—Approx. 121 turns No. 30 P.E. wire—space wound—83 T.P.I.
Primary—450 turns No. 34 D.C.C. wire—random wound.

SHORT WAVE COILS
1st COIL
Secondary—34 turns No. 27 P.E. wire—space wound
Pickup—20 turns No. 36 S.E. wire—close wound spaced $\frac{1}{8}$" from sec.
Wound on $1\frac{1}{4}$" dia. form
2nd COIL:
Secondary—$19\frac{1}{2}$ turns No. 21 P.E. wire—space wound
Pickup—20 turns No. 36 S.E. wire—close wound spaced $\frac{1}{8}$" from sec.
Wound on $\frac{7}{8}$" dia. form.
3rd COIL:
Secondary—$6\frac{1}{2}$ turns No. 19 P.E. wire—space wound—length, $\frac{7}{8}$"
Pickup—20 turns No. 36 S.E. wire—close wound spaced $\frac{1}{8}$" from sec.
Wound on $\frac{7}{8}$" dia. form.

SILVER-MARSHALL, INC.

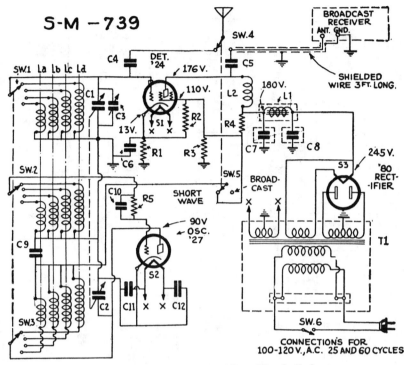

Wiring diagram of the new Silver-Marshall short wave converter here described.

The wave-changing switch has four positions and provides tuning in the following wave-bands: 10 to 20 meters; 20 to 40 meters; 40 to 80 meters: 80 to 200 meters.

The small knob, at the right of the main tuning dial, changes the antenna lead from the "converter" to the "antenna post" on the "broadcast" receiver with which the converter is operating.

Converter Uses Only Three Tubes

The "S-M 739" Super converter is a full A.C.-operated screen-grid one, with "self-contained" power supply. (This converter is supplied in kit form; and also completely wired, at small additional cost.

The converter consists of a tuned detector, using a '24 tube, in conjunction with a '27 tube in a specially-designed oscillator circuit, and an '80 type rectifying tube. The power supply operates from any 100-120 volt, 25-60 cycle circuit or (by quick change of two power transformer leads) from any 200-240 volt, 25-60 cycle alternating current lighting circuit; and provides all "A," "B," and "C" power for the converter. The converter comes wired for operation on 100-120 volt 25-60 cycle current.

"BEARCAT" RECEIVER

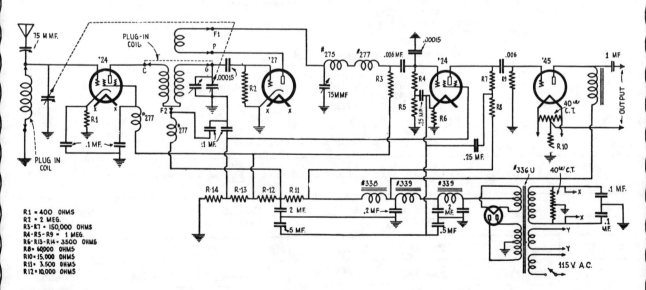

Wiring diagram of S-M "Bearcat" receiver for short waves. It employs a stage of shield grid radio frequency, regenerative detector and two stages of special audio frequency amplification.

With this receiver it is possible to spread the crowded ham bands and secure close tuning on foreign phone stations, without dismantling the set, by using a little midget condenser that is built right into the circuit. To that unique feature has been added the increased amplification of the tuned radio frequency screen-grid stage and also the screen-grid first audio stage; still maintaining the convenience of single-dial operation. It uses, as well, a '27 type detector, '45 output, and '80 rectifier.

The "Bearcat" is built on a rugged steel chassis. It has one tuning dial with series antenna condenser and regenerative control, switch, and vernier.

The receiver can be purchased in kit form or completely wired. Plug-in coils are furnished, giving a range from 16.6 up to 200 meters.

SPARKS-WITHINGTON CO.
MODEL 61 – 62

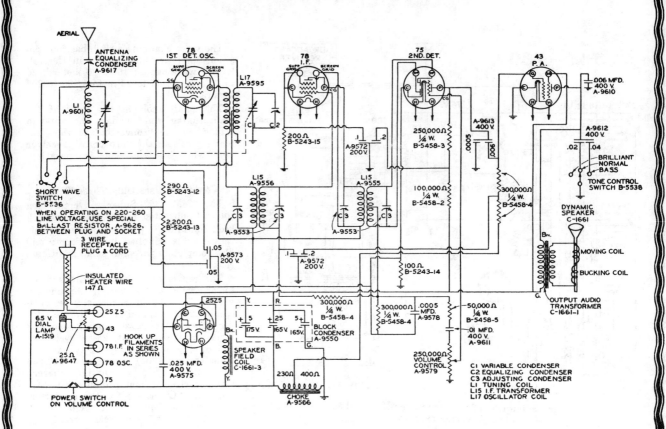

Operates on any A. C. or D. C. current, **100 to 130** volts and **200 to 260** volts—ANY cycle.

Tunes all stations from 540 to 1500 on regular broadcast band, and from 1500 to 5000 kilocycles on short wave band (200 meters to 60 meters).

Antenna tuning that permits exact matching of receiver to any antenna, which makes for maximum sensitivity.

Sparton latest type precision-built high efficiency tubes:
1 78 detector oscillator.
1 78 intermediate amplifier.
1 75 second detector-automatic volume control.
1 43 power pentode.
1 25Z5 dual rectifier.

VOLTAGE ANALYSIS

Line Voltage <u>120</u>   Position of Volume Control <u>Full</u> with Antenna Disconnected

Tube	Location	Heater or Filament	Plate	Control Grid —	Screen Grid +	Plate Current M. A.
25Z5	Rectifier	22 - 28*	-------	-------	-------	-------
43	Power	22 - 28*	92 - 98	14 - 18	98 - 108	18 - 25
75	Detector-A.V.C.	5.9 - 7*	35 - 45	.4 - .65	-------	.19 - .22
78	I. F. Amplifier	5.9 - 7*	98 - 108	1.8 - 2.5	98 - 108	7 - 10
78	Det.-Oscillator	5.9 - 7*	98 - 108	14 - 18	98 - 108	4 - 6

Voltage across speaker field is 100 - 120 volts.
*Readings slightly less when taken with a low resistance voltmeter.

SPARKS-WITHINGTON CO.
MODEL AR-50

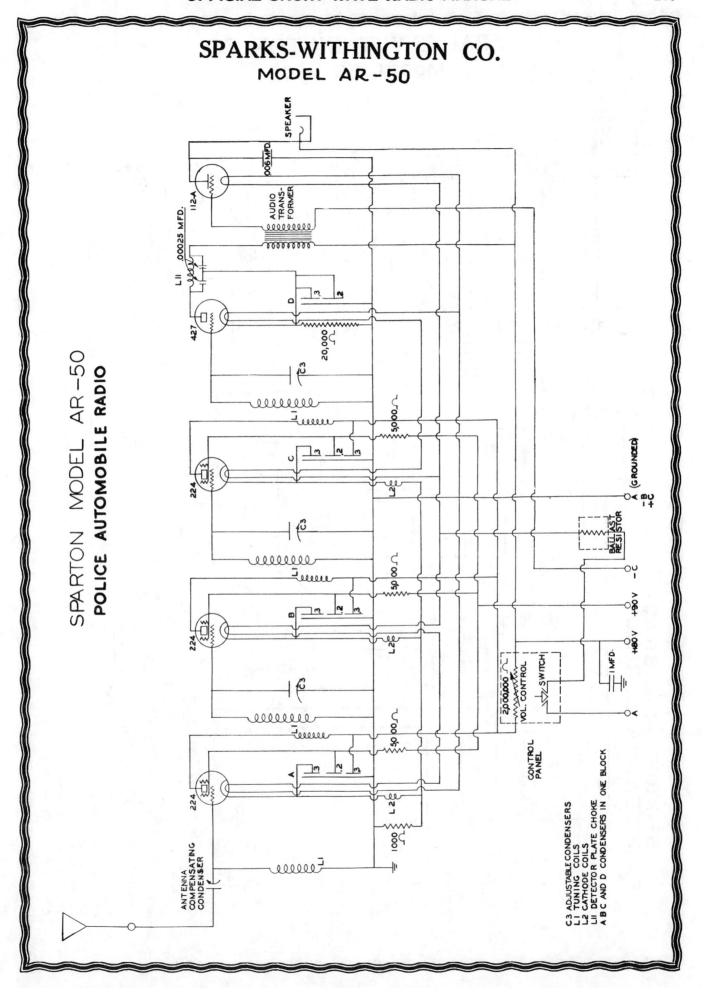

SPARTON MODEL AR-50
POLICE AUTOMOBILE RADIO

C3 ADJUSTABLE CONDENSERS
L1 TUNING COILS
L2 CATHODE COILS
L11 DETECTOR PLATE CHOKE
A B C AND D CONDENSERS IN ONE BLOCK

SPARKS-WITHINGTON CO.
MODEL 55 A.C.

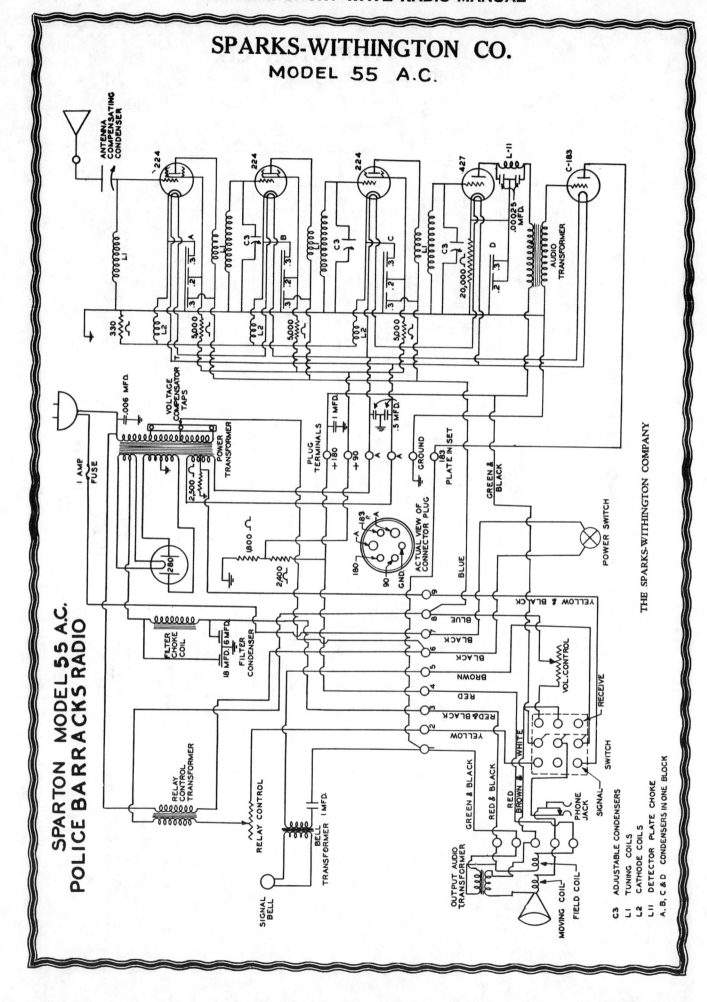

SPARTON MODEL 55 A.C.
POLICE BARRACKS RADIO

THE SPARKS-WITHINGTON COMPANY

C3　ADJUSTABLE CONDENSERS
L1　TUNING COILS
L2　CATHODE COILS
L11　DETECTOR PLATE CHOKE
A, B, C & D CONDENSERS IN ONE BLOCK

STEWART-WARNER CORP.

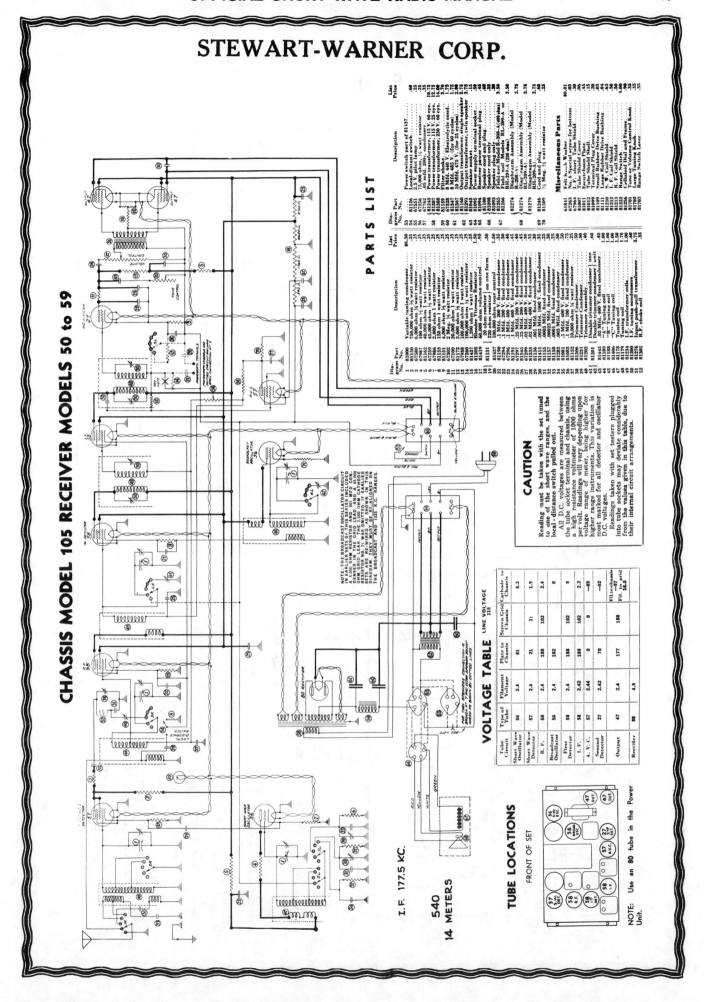

CHASSIS MODEL 105 RECEIVER MODELS 50 to 59

I.F. 177.5 KC.

540

14 METERS

PARTS LIST

Diagram Part No.	Description	List Price
1	Variable tuning condenser	$6.50
2	750,000 ohm ½ watt resistor	.35
3	6,000 ohm ½ watt resistor	.35
4	40,000 ohm 1 watt resistor	.40
5	40,000 ohm 1 watt resistor	.40
6	150 ohm 1 watt resistor	.40
7	4,000 ohm ½ watt resistor	.35
8	2,000 ohm ½ watt resistor	.35
9	20,000 ohm ½ watt resistor	.35
10	20,000 ohm ½ watt resistor	.35
11	100,000 ohm ½ watt resistor	.35
12	1,200 ohm 1 watt resistor	.40
13	60,000 ohm volume control	1.95
14	20 ohm resistor	1.50
15	230 ohm Vitreous resistor	.40
16	100,000 ohm tone control	1.50
17	100,000 ohm ½ watt resistor	.35
18	1 Mfd. 200 V. fixed condenser	.45
19	1 Mfd. 200 V. fixed condenser	.45
20	1 Mfd. 400 V. fixed condenser	.45
21	.1 Mfd. 400 V. fixed condenser	.45
22	.25 Mfd. 400 V. fixed condenser	.45
23	.01 Mfd. fixed condenser	.30
24	.001 Mfd. fixed condenser	.30
25	.01 Mfd. fixed condenser	.30
26	.0005 Mfd. fixed condenser	.30
27	.003 Mfd. fixed condenser	.30
28	.003 Mfd. fixed condenser	.30
29	10,000 ohm ½ watt resistor	.45
30	Trimmer Condenser	.50
31	Trimmer Assembly	
32	Double trimmer condenser } one	.60
33	.02 Mfd. 400 V. fixed condenser	.60
34	"A" Tuning coil	1.00
35	"B" Tuning coil	1.00
36	"C" Tuning coil	1.50
37	Tuning coil	1.70
38	I.F. tuning condensers	.60
39	I.F. tuning transformer	5.75
40	R.F. choke coil	.33

Diagram Part No.	Description	List Price
53	Power switch part of 81457	
54	Local-distant switch	.60
55	2.5 V. pilot lamp	.25
56	500 ohm ½ watt resistor	.35
57	.05 mfd. 200v. condenser	.25
58	Power transformer, 115 V. 60 cyc.	10.75
59	Power transformer, 115 V. 25 cyc.	16.75
60	Power transformer, 230 V. 60 cyc.	14.00
61	Filter choke	1.75
62	8 Mfd. Electrolytic cond.	2.70
63	8 Mfd. 475 V. (for 60 cycles)	1.75
64	10 Mfd. 475 V. (for 25 cycles)	2.00
65	Output transformer, twin speaker	2.75
66	Speaker socket	.45
67	Field coil Model RH-209-A	3.50
	Field coil Model RH-209-A (400 ohm)	
68	Diaphragm Assembly (Model RH-209-A, 230 ohm)	2.75
	Diaphragm Assembly (Model RH-209-A)	
69	Cord and plug	2.75
70	Speaker plug only	.25

Miscellaneous Parts

Description	List Price
Felt Washer	$0.01
No. 6 Special screw for bottom	.03
Tube Shield	.30
Tube Shield Cover	.15
Escutcheon Plate	.45
Tube Shield Base	.15
Terminal Plug Screw	.03
Small Rubber Drive Bushing	.04
Large Rubber Drive Bushing	.65
S. W. Coil Shield	.45
R.F. Coil Shield	.50
Range Switch	.90
Celluloid Dial and Frame	4.00
Tone and Volume Control Knob	.35
Large Tuning Knob	.35
Range Switch Lever	.35

NOTE—THE BROADCAST OSCILLATOR CIRCUIT IN EARLIER SETS OF THIS SERIES INCLUDED A 1,000 OHM RESISTOR AND .001 MFD CONDENSER IN THE GRID LEAD, AND A 45,000 OHM GRID LEAK IN THE 2,000 OHM CATHODE RESISTOR. NO LONGER USED. EARLIER SETS ARE REWIRED AS SHOWN IN THIS DIAGRAM. THEY MUST BE RE-ALIGNED ON THE BROADCAST AND 1325 K.C. RANGES.

CAUTION

Reading must be taken with the set tuned to one of the short wave ranges, and the local - distance switch pulled out.

All D.C. voltages are measured between the tube socket terminal and chassis, using a high resistance voltmeter of 1,000 ohms per volt. Readings will vary depending upon voltage range of meter, being higher for higher range instruments. This variation is most marked for all detector and oscillator D.C. voltages.

Readings taken with set testers plugged into tube sockets may deviate considerably from the values given in this table, due to their internal circuit arrangements.

VOLTAGE TABLE

LINE VOLTAGE 115

Tube Circuit	Type of Tube	Filament Voltage	Plate to Chassis	Screen Grid to Chassis	Cathode to Chassis
Short Wave Oscillator	56	2.4	81	5.2	
Short Wave Detector	57	2.4	21	21	1.9
R. F.	58	2.4	188	102	2.4
Broadcast Detector	56	2.4	102		0
First Detector	58	2.4	188	102	9
I. F.	58	2.42	188	102	2.2
A. V. C.	57	2.44	0	0	−89
Second Detector	27	2.42	70		−92
Output	47	2.4	177	188	Fil.-chassis −97, Fil. to grid 16.5
Rectifier	80	4.9			

TUBE LOCATIONS

FRONT OF SET

NOTE: Use an 80 tube in the Power Unit.

STEWART-WARNER CORP.
R-111

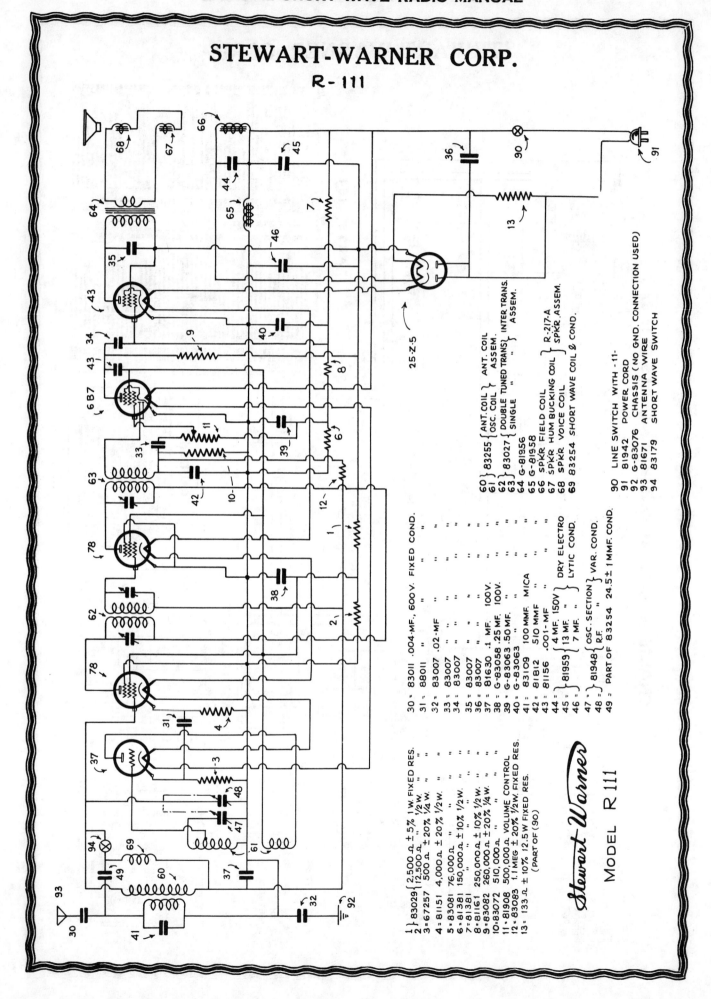

Stewart-Warner

MODEL R 111

1 = 83029 { 2,500 Ω ± 5% 1 W. FIXED RES.			
2 = { 12,500 Ω. " " ½ W.			
3 = 67257 500 Ω ± 20% ¼ W. "			
4 = 81151 4,000 Ω ± 20% ½ W. "			
5 = 83081 76,000 Ω. "			
6 = 81381 150,000 Ω ± 10% ½ W. "			
7 = 81381 " "			
8 = 81161 "			
9 = 83082 250,000 Ω ± 10% ½ W. "			
10 = 83072 260,000 Ω ± 20% ¼ W. "			
11 = 81908 500,000 Ω. VOLUME CONTROL			
12 = 83083 1.1 MEG ± 20% ½ W. FIXED RES.			
13 = 133 Ω ± 10% 12.5 W FIXED RES. (PART OF (90)			

30 = 83011 .004 MF, 600 V. FIXED COND.			
31 = 88011 " "			
32 = 83007 .02 MF " "			
33 = 83007 " "			
34 = 83007 " "			
35 = 83007 " "			
36 = 83007 " "			
37 = 81630 .1 MF, 100 V. "			
38 = G-83058 .25 MF. 100 V. "			
39 = G-83063 .50 MF. " "			
40 = G-83063 " "			
41 = 83109 100 MMF. MICA "			
42 = 81812 510 MMF " "			
43 = 81156 .001 MF " "			
44 = 81959 { 4 MF. 150 V } DRY ELECTRO			
45 = { 13 MF. } LYTIC COND.			
46 = { 7 MF. }			
47 = 81948 { OSC. SECTION } VAR. COND.			
48 = { R.F. }			
49 = PART OF 83254 24.5 ± 1 MMF. COND.			

60 = 83255 { ANT. COIL } ANT. COIL		
61 = { OSC. COIL } ASSEM.		
62 = 83027 { DOUBLE TUNED TRANS } INTER. TRANS.		
63 = { SINGLE " " } ASSEM.		
64 = G-81956		
65 = G-81958		
66 = SPKR. FIELD COIL		
67 = SPKR. HUM BUCKING COIL } R-217-A		
68 = SPKR. VOICE COIL } SPKR. ASSEM.		
69 = 83254 SHORT WAVE COIL & COND.		

90 = LINE SWITCH WITH -11-	
91 = 81942 POWER CORD	
92 = G-83076 CHASSIS (NO GND. CONNECTION USED)	
93 = 81671 ANTENNA WIRE	
94 = 83179 SHORT WAVE SWITCH	

STEWART-WARNER CORP.

R-115

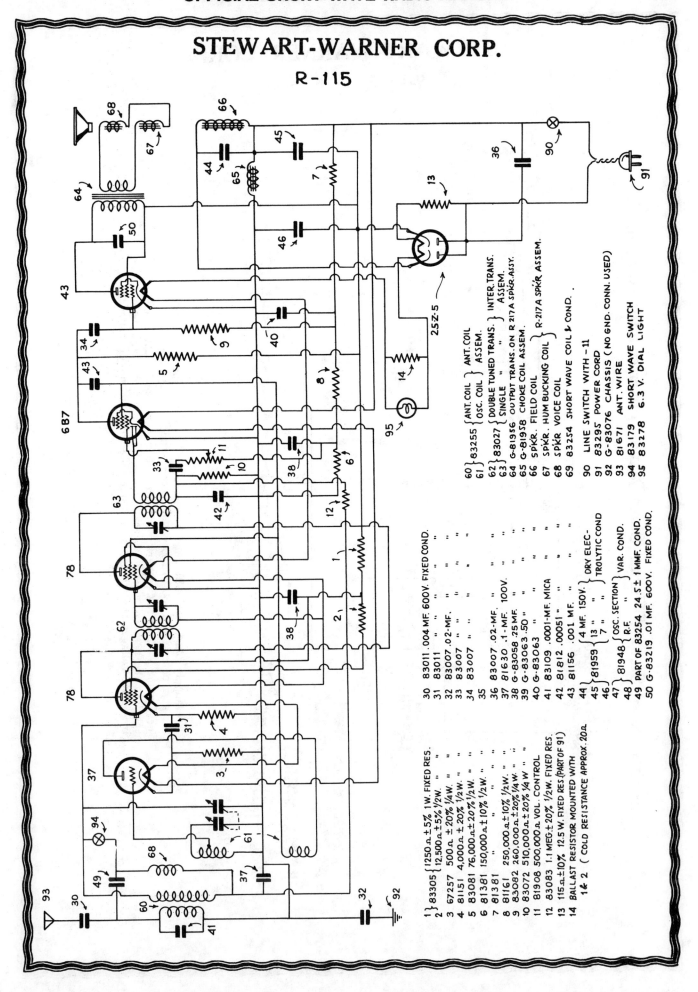

STEWART-WARNER CORP.
301-A, B & E

Circuit Data of Stewart-Warner Short Wave Converter R301-A, B, and E

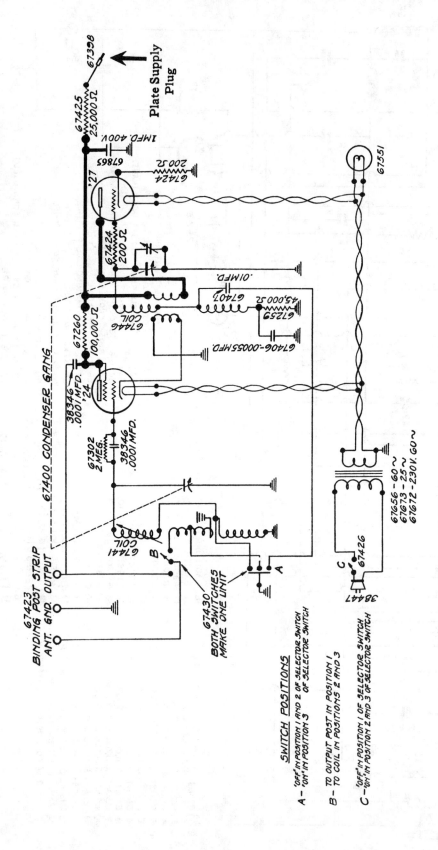

Plate supply plug (#67398) must be connected to some source of filtered D.C. at a potential of 180 to 280 volts. **Recommended voltage is 250.** The Ground Post of the converter is the negative return and **must** be connected to the negative side of the external plate supply. The table below gives plate voltages at both tubes for three different plate supply voltages.

Plate Supply Voltage	'24 Plate	'27 Plate
180	26	70
250	34	93
280	37	102

SWITCH POSITIONS

A — "OFF" IN POSITION 1 AND 2 OF SELECTOR SWITCH
"ON" IN POSITION 3 OF SELECTOR SWITCH

B — TO OUTPUT POST IN POSITION 1
— TO COIL IN POSITIONS 2 AND 3

C — "OFF" IN POSITION 1 OF SELECTOR SWITCH
"ON" IN POSITION 2 AND 3 OF SELECTOR SWITCH

SUPERTONE PRODUCTS CO.

4 TUBE A.C.

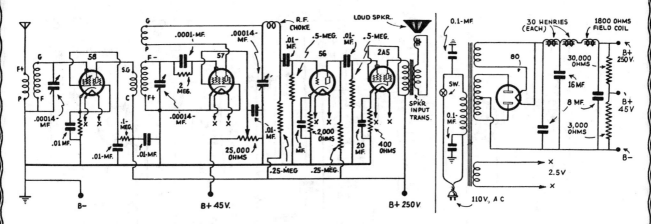

Interesting diagram showing the hook-up of the Supertone 4-Tube A.C. operated short-wave receiver, utilizing the new 2A5 output tube; also connections of power supply unit.

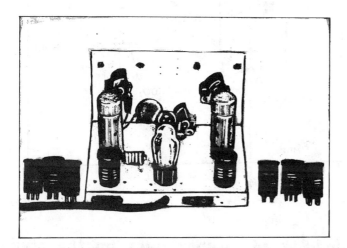

A glance at the diagram herewith shows that a tuned stage of radio frequency amplification, employing a 58 tube, is used to build up the signal strength before it is detected. Next we come to the tuned three coil coupler which links the R.F. stage with the regenerative detector, employing a 57 tube. Dual regeneration control is provided through the medium of the variable condenser connected between the R.F. choke and ground, the second regeneration control being the 25,000 ohm potentiometer, the arm of which connects to the shield grid of the 57 detector tube.

The output from the detector stage is resistance coupled as shown into the first audio stage which uses a 56 tube. This tube is biased by a 2,000 ohm resistor shunted by a 1 mf. condenser. Out of this first audio tube we pass into a resistance-coupled network, and once more into the grid of the output stage tube, which is a 2A5. 250 volts B plus plate supply is applied to the screen-grid and plate of the 2A5, through the loud speaker transformer as shown in the diagram.

Both the R.F. and detector stages are tuned by means of .00014 mf. variable condensers and standard plug-in coils, which have been described many times in this journal, as well as the present number, can be used with them. The antenna stage

utilizes a 4-pin base coil having two windings on it; the detector stage employs a 6-pin base coil, having three windings on it, a primary, secondary, and tickler winding. A wavelength range from 15 to 200 meters is thus made available by the use of these coils, which can be easily purchased on the market or wound from data given in this as well as past numbers of SHORT WAVE CRAFT. The potentiometer used to regulate the voltage applied to the screen grid of the detector besides acting as a regeneration auxiliary control, also serves the useful purpose of a *volume leveler*. The R.F. choke used in the plate circuit of the detector may be one of the Hammarlund type, the new 10 MH., size being all right. If *motor-boating* should result a lower value of grid resistors in the audio circuit may be used than those shown. If *motor-boating* occurs one may also try using lower value plate resistors in the resistance-loaded plate circuits. Note that the 400 ohm biasing resistor for the 2A5 output tube is shunted by a high capacity electrolytic condenser, having 20 mf.

Power Supply Unit Details

A great deal of experimenting was done on not only the placement but also the make-up of the plate supply unit, and as the diagram shows it is very simple in its make-up. As a protection against *tun-*

able hums being transferred through the power transformer from the 110 volt A.C. 60 cycle supply circuit, 0.1 mf. condensers are connected from either side of the 110 volt primary circuit to ground.

One of the most usual sources of hum, whenever an A.C. operated plate supply unit is employed, lies in the filter circuit; in the particular filter network two 30 henry iron-core chokes are used together with the field winding on the loud speaker as an additional inductance to smooth out the rectified current. As but two plate voltages, 45 and 250 volts respectively, are necessary for operating this set, two resistors, one of 3000 and one of 30,000 ohms, only are required to form the voltage divider. Three liberal sized electrolytic condensers are used in the high voltage filter, two of 8 mf. each and one of 16 mf. No condenser is used between the B minus side of the filter and the juncture between the speaker field winding and the 30 henry choke.

List of Parts

COILS:
One set of UX-base short-wave plug-in coils, four coils to a set. (See page 213.)
One set of six-pin, three-winding, plug-in coils, four coils to a set. (See page 213.)
One Hammarlund radio-frequency choke coil.

CONDENSERS:
Three Hammarlund 0.00014 mf. tuning condensers.
Six 0.01 mf. condensers.
One 1 mf. bypass condenser.
One 20 mf. electrolytic bypass condenser.
One 0.0001 mf. grid condenser.

RESISTORS:
Two 400-ohm pigtail resistors.
One 0.1 meg. pigtail resistor.
One 2 meg. pigtail resistor.
Two 0.25 meg. pigtail resistors.
Two 0.5 meg. pigtail resistors.
All above resistors are 1 watt.
One 25,000 ohm potentiometer.

OTHER REQUIREMENTS:
Four six-pin and two UY sockets (extra UY is for voltage cable, extra six-pin per coil; one UX socket, for other coil).
One dynamic speaker for 2A5 output, with output transformer "built in;" field coil, 1,800 ohms. Cone diameter is 6 inches.
One chassis.
Two vernier dials.
Two knobs.
Two tube shields and bases.

SUPERTONE PRODUCTS CO.
4 TUBE   D.C.

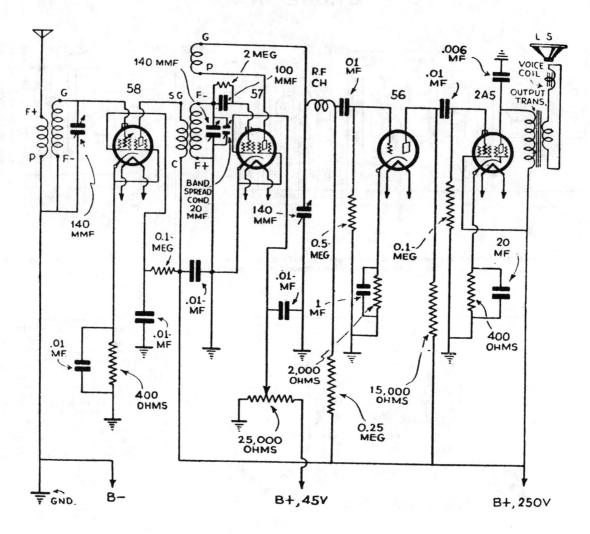

How the Supertone Band-Spread 4-tuber lines up the stages.

Here's the new Supertone 4-tube Band-Spread short-wave receiver.

As can be seen from the wiring diagram, the receiver has two main tuning controls, one controls the tuning of the R.F. stage and the other the detector. As most of us know, the R.F. stage on a tuned R. F. short-wave receiver is usually rather broad in comparison with the detector tuned circuit and the most critical adjustment, of course, is the detector tuning condenser. It can be readily appreciated that if a small condenser around 20 mmf. were shunted across a 140 mmf. tuning condenser, the small condenser could be tuned over a considerable range without getting entirely out of resonance with the R.F. stage grid circuit and in this manner provide a very efficient and economical means of obtaining band spread. The method of tuning this receiver would be to tune the two 140 mmf. condensers together to a definite short-wave broadcast band, and then do all other tuning with the added 20 mmf. condenser.

TRANSFORMER CORP. OF AMERICA
CLARION A.C.–240 & 25-240
S.W. SUPERHET

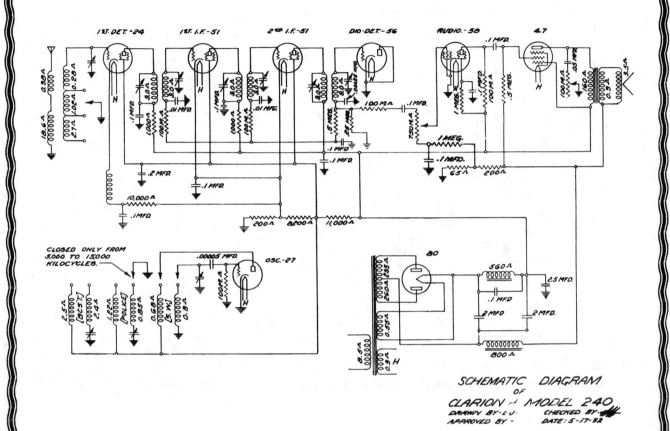

SCHEMATIC DIAGRAM
OF
CLARION ~ MODEL 240
DRAWN BY- L.U.   CHECKED BY-
APPROVED BY -   DATE: 5-17-32

CIRCUIT RESISTANCE ANALYSIS

Model 240 Socket to ground

Stage	Grid	Cath-ode	Heater	Plate	Screen G	Suppr G	Space G
Autodyne	4.0	10,300	0.15	20,400	8,400	.....	.....
1st.I.F.	850,000	200	0.2	20,400	8,400	.....	.....
2nd.I.F.	850,000	200	0.25	19,400	8,400	.....	.....
Dio Det.	250,000	0.17	0.33	0.11	.....	.....	.....
Oscilla-tor	100,000	0.25	0.15	8,400	.....	.....	.....
Audio	Infini-ty	0.1	0.12	120,000	Infin-ity	0.1	.....
Pentode	500,000	.....	0.25	20,000	.....	.....	19,400
Recti-fier.	.....	.....	19,900	1,320 / 1,360	.....	.....	.....

Note: Readings of one megohm and over are given as "infinity". The first three significant figures, only are interpreted from the ohm meter in each reading; the individual resistance in the circuit can be readily checked upon removal of chassis.

VOLTAGE ANALYSIS

Model 240

No.	Stage	A	B	C	K	Sc.G	Ip.	Su.g
1	Autodyne	2.C	230	8.0	10	75	.6	.....
2	Oscillator	2.0	100	0	0	.....	7.0	.....
3	1st. I.F.	2.0	250	.0	3.0	75	2.0	.....
4	2nd. I.F.	2.0	250	0	3.0	75	2.0	.....
5	Audio	2.0	190	.4	0	25	1.0	.4
6	Dio Det	2.0	0	0	0	0	0	.....
7	Output.	2.25	250	16.0	.....	250	30.	.....
8	Rectifier	4.8	300	.....	.....	.....	.....	.....

Vol. control "full on".
Band switch "broadcast".
Tested with Weston model 565 analyzer.
Line: 115 Volts.

TRY-MO RADIO CO.
POWERTONE — WALLACE

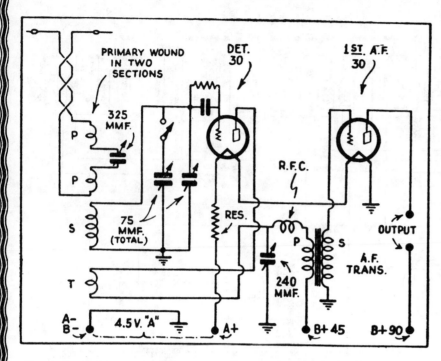

Connections used in the new receiver here described; note split antenna coil.

The two-tube Wallace receiver, illustrated herewith, can be operated with type 30 tubes and dry cells. The extremely low filament drain of the type 30 tubes enables the user to operate the receiver from an ordinary 4½ volt "C" battery for filament lighting and a single 22½-45 volt "B" battery for plate current. A reasonable amount of volume is secured when this method is used but the receiver will still bring in all stations that are otherwise heard when a larger number of tubes is used. In fact, a one-tube receiver will bring in as many stations as a 3- or 4-tube receiver, but the signals are not received with sufficient volume to make reception enjoyable.

The coil winding data are as follows:
15 to 40 meter coil, antenna coupler—16 turns (2 windings of 8 turns each). Secondary coil, 6½ turns, double space wound, No. 14 DCC wire. Regeneration coil 6½ turns, closely wound, spaced ½" from secondary winding. Use No. 14 DCC wire.
40 to 75 meter coil. Antenna coupler—Same for all coils. Secondary coil—16 turns, closely wound. Use No. 22DCC wire. Regeneration coil—15 turns, closely wound, spaced about 1¼" from secondary coil. Use No. 22 DCC wire.
75 to 150 meter coil. Antenna coupler—Same for all coils. Secondary coil—28 turns, closely wound. Use No. 22 DCC wire. Regeneration coil—26 turns, closely wound. Use No. 22 DCC wire. Space as far as possible from secondary coil (about 1").
150 to 200 meter coil—Antenna coupler—same for all coils. Secondary coil—70 turns No. 26 DCC wire, closely wound. Regeneration coil—20 turns, No. 26 DCC wire, spaced about ½" from secondary coil. Coil forms 1⅝" diameter.

The Circuit

As an examination of the diagram will show, the set uses a standard regenerative detector fed into a power pentode audio stage, coupled by the resistance-condenser method. One variation from the standard type of circuit is in the aerial series condenser. Instead of placing a semi-variable capacity in the set, a variable condenser of 60 mmf. is mounted directly on the panel. The adjustment of this condenser is extremely important in a short-wave receiver, as many of you know.

By correctly adjusting this condenser, the natural wavelength of the aerial is shifted and this often results in an unconscious selection of half-wave and quarter-wave aerial conditions. Whatever the merits of the "measured" aerials may be, at least it is true that the regenerative effect in the detector is increased by decreasing the series capacity which assures satisfactory oscillation on even the highest frequencies.

The flexibility of this little receiver is of course due to the use of the 25Z5 as a rectifier when A.C. is the supply; the tube being floated otherwise on the D.C. line. For D.C. operation the only use of the tube is for the resistance of its heater, as you can see by referring to the circuit diagram. The filament of this tube in addition to the 200 ohm, 20 watt series resistor serves to reduce the line voltage to the correct value for the two other tubes.

UNIVERSAL A.C.-D.C. PORTABLE

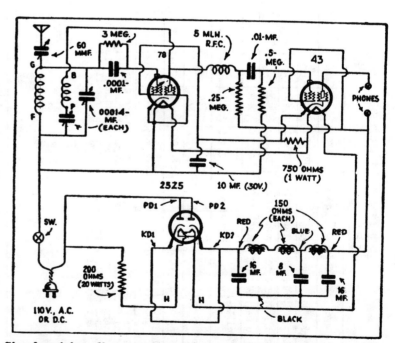

Simple wiring diagram used in the 110-volt A.C.-D.C. portable illustrated above. The plate current is rectified by a 25Z5 tube. The whole outfit is very light.

TRY-MO RADIO CO.
REGENT-FOUR S.W. SET

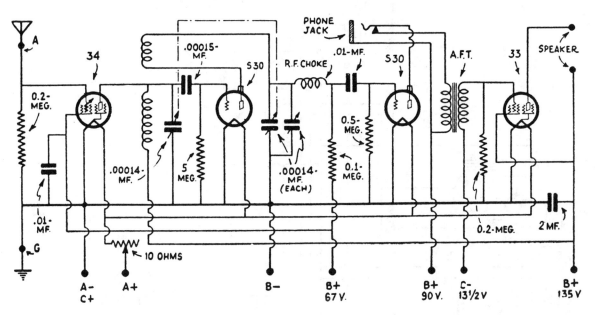

Here is the way the various components of the "Regent-Four" are hooked up.

Parts List for "Regent-Four."

1—Try-Mo "Regent-Four" Foundation Kit, (including drilled panel and base)
1—Powertest special 2-gang condenser .00014 mf.
1—Powertest .00014 mf. variable condenser
1—Powertest set of 4 plug-in coils (Alden, or other make coils suited to operation with a .00014 mf. tuning condenser may be used. (See page 236.)
1—Powertest R.F. Choke, 60 mh.
2—Powertest .01 mf. condensers
1—Powertest 2 mf. condenser

2—Powertest 200,000 ohm resistor
1—Powertest 100,000 ohm resistor
1—Powertest 500,000 ohm resistor
1—Rheostat, 10 ohms
2—5-prong sockets (Eby, Na-ald, National or Hammarlund.)
2—4-prong sockets (Eby, Na-ald, National or Hammarlund.)
1—Powertest special phone jack
1—Audio transformer
4—Eby binding posts
1—Powertest .00015 mf. condenser
1—Powertest 5 megohm grid-leak

1—Powertest 7 wire battery cable
1—Powertest Regent dial with escutcheon plate
3—Matched knobs
3—Screen-grid caps
2—Triad S-30 tubes
1—Triad 33 power pentode tube
1—Triad 34 screen-grid tube
2—No. 6 dry cell batteries
3—45 volt "B" batteries
1—22½ volt "C" battery
2—tube shields

POWER TONE – 4

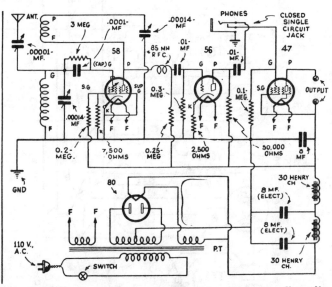

Above—we see the handsome appearance of the new A.C. operated short wave 4-tube receiver here described, as well as diagram showing the efficient arrangement of the circuit.

UNITED AMERICAN BOSCH CORP.
MODEL 108 POLICE CAR RADIO

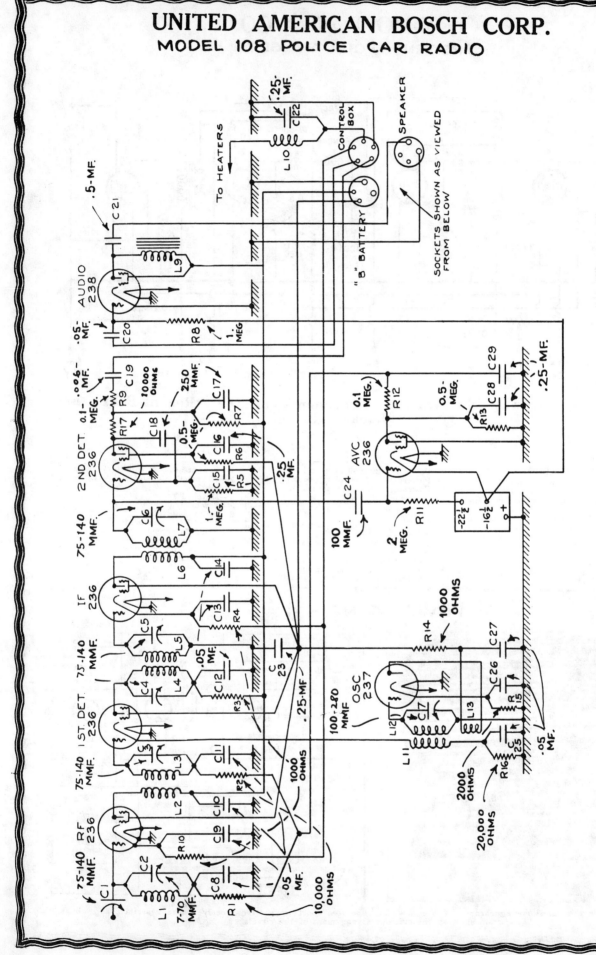

WIRING DIAGRAM—MODEL 108 POLICE CAR RADIO

UNITED AMERICAN BOSCH CORP.
MODEL 260-C
WORLD CRUISER

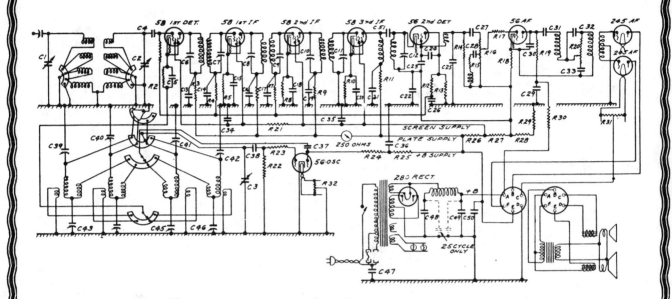

The oscillator and antenna coils are changed for each band, thus providing practically four separate and complete short-wave receivers in one instrument.

● ONE of the latest and finely engineered multi-wave receivers is that here illustrated. Whenever one of the various short-wave bands are selected, the operator is automatically notified of the fact by a change in color of the illuminated full-vision scale. A tuning meter, placed just above the main tuning scale, facilitates and simplifies the tuning operation. Whenever a different frequency band is selected, a different set of antenna and oscillator coils are automatically switched into circuit, so that practically four separate and complete 10-tube receivers are provided in one set. Automatic volume control is provided on all bands; other features include—silent tuning control; true-pitch tone control; properly blended, dual loud speakers; push-pull power-output stage using two 45 type tubes. Following are numerous constants or values of the condensers and resistors used in this new Bosch Multi-Wave Superheterodyne Receiver. Those interested in the values of the coils and tuning condensers in the oscillator and antenna circuits may refer to back numbers of SHORT WAVE CRAFT in which details or constants of such circuits have been repeatedly given.

"Close-up" of the 10-tube multi-wave superhet chassis.

R1—500,000 ohms	R5—1,500 ohms		
R2—5,000 ohms	R6—1,000 ohms		
R3—1,000 ohms	R7—100,000 ohms		
R4—100,000 ohms	R8—1,500 ohms		
R9—1,000 ohms	C15—.05 mfd.		
R10—350 ohms	C16—.005 mfd.		
R11—1,000 ohms	C17—.005 mfd.		
R12—500,000 ohms	C18—.05 mfd.		
R13—1 megohm	C19—.005 mfd.		
R14—100,000 ohms	C20—.05 mfd.		
R15—10,000 ohms	C21—.005 mfd.		

R16—500,000 ohms	C22—.05 mfd.	C1—Tuning	C40—alignment
R17—100,000 ohms	C23—100 mfd.	C2—Tuning	C41—alignment
R18—1,500 ohms	C24—.05 mfd.	C3—Tuning	C42—alignment
R19—25,000 ohms	C25—100 mmf.	C4—100 mmf.	C43—alignment
R20—100,000 ohms	C26—.25 mfd.	C5—.05 mfd.	C44—alignment
R21—10,000 ohms	C27—.05 mfd.	C6—alignment	C45—alignment
R22—50,000 ohms	C28—.025 mfd.	C7—alignment	C46—alignment
R23—150 ohms	C29—4. mfd.	C8—alignment	C47—.01 mfd.
R24—10,000 ohms	C30—.001 mfd.	C9—alignment	C48—8 mfd.
R25—10,000 ohms	C31—.5 mfd.	C10—alignment	C49—8 mfd.
R26—3700 ohms	C32—.05 mfd.	C11—alignment	C50—4 mfd.
R27—2270 ohms	C33—.05 mfd.	C12—alignment	C51—1,000 mmf.
R28—230 ohms	C34—.05 mfd.	C13—.005 mfd.	C52—8 mfd.
R29—1280 ohms	C35—.05 mfd.	C14—.005 mfd.	C53—4 mfd.
R30—10,000 ohms	C36—.25 mfd.		
R31—Mid tap	C37—.05 mfd.		
R32—Mid tap	C38—100 mmf.		
	C39—alignment		

Note: R26, R27, R28, R29—tapped unit; C13, C14, C16, C17. C19, C21—single unit; C29. C48. C49, C50—single unit.

UNITED AMERICAN BOSCH CORP.
MODEL 360 ALL-WAVE SET

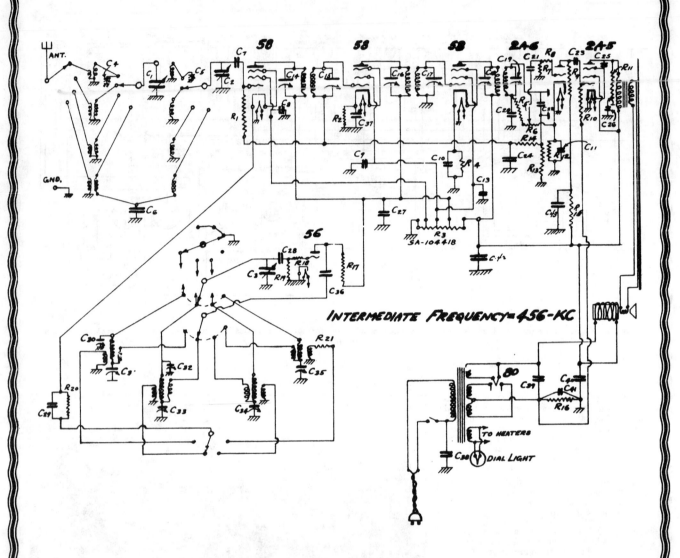

SCHEMATIC WIRING DIAGRAM OF THE MODEL 360 RECEIVER

R-1	1/2 meg. 1/4 watt	R-17 2500 ohms 1/2 watt	C-10 .05 mfd.	C-25 .005 mfd. 3-ply
R-2	300 ohms 1/4 watt	R-18 50 ohms 1/2 watt	C-11 .25 mfd.	C-26 .05 2-ply
R-3	Multiple	R-19 20,000 ohms 1/4 watt	C-12 ----	C-27 .05 mfd. 3-ply
R-4	300 ohms 1/4 watt	R-20 2000 ohms 1/4 watt	C-13 .05 mfd.	C-28 100 mmf mica
R-5	5000 ohms 1/4 watt	R-21 200 ohms 1/4 watt	C-14 30 - 100 mmf mica	C-29 .05 mfd. 2-ply
R-6	500,000 ohms vol.		C-15 " " " "	C-30 7 - 70 mmf
R-7	2 meg. 1/4 watt		C-16 " " " "	C-31 300 mmf variable
R-8	100,000 ohms 1/4 watt	C-1 ----	C-17 " " " "	C-32 7 - 70 mmf
R-9	250,000 ohms 1/4 watt	C-2 Variable condenser	C-18 " " " "	C-33 1200 mmf variable
R-10	1/2 meg. 1/4 watt	C-3 ----	C-19 " " " "	C-34 1200 mmf variable
R-11	Variable	C-4 Trim condenser		C-35 1500 mmf variable
R-12	5000 ohms 1/4 watt	C-5 Trim condenser	C-20 100 mmf mica	C-36 .05 mfd. 2-ply
R-13	1 meg. 1/4 watt	C-6 .05 mfd. 2-ply	C-21 .005 mfd. 3-ply	C-37 .05 mfd.
R-14	1/2 meg. 1/4 watt	C-7 100 mmf mica	C-22 100 mmf mica	C-38 .01 mfd. 4-ply
R-15	-----	C-8 .05 mfd.	C-23 .005 mfd. 3-ply	C-39 8 mfd. electro
R-16	400 ohms 1 watt	C-9 .05 mfd.	C-24 .05 mfd. 2-ply	C-40 4 mfd. electro
				C-41 20 mfd. electro

WHOLESALE RADIO SERVICE CO.
MASTER — 6 SHORT WAVE SET

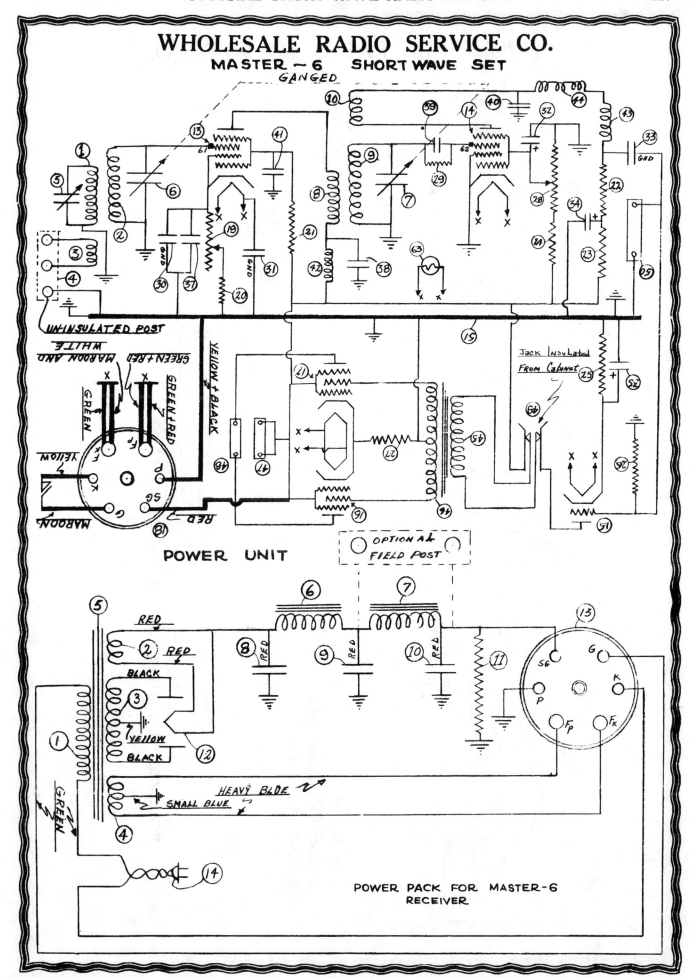

POWER UNIT

OPTIONAL FIELD POST

POWER PACK FOR MASTER-6 RECEIVER

WHOLESALE RADIO SERVICE CO.

MODEL
10-12

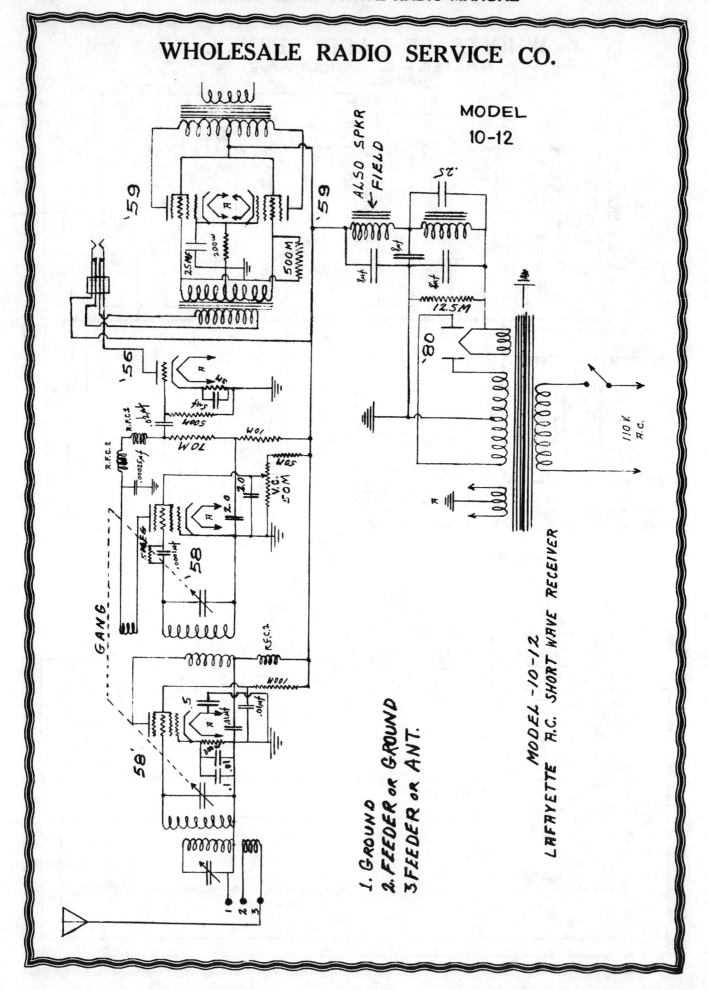

1. GROUND
2. FEEDER or GROUND
3. FEEDER or ANT.

MODEL-10-12
LAFAYETTE A.C. SHORT WAVE RECEIVER

WHOLESALE RADIO SERVICE CO.

MODEL
M-15

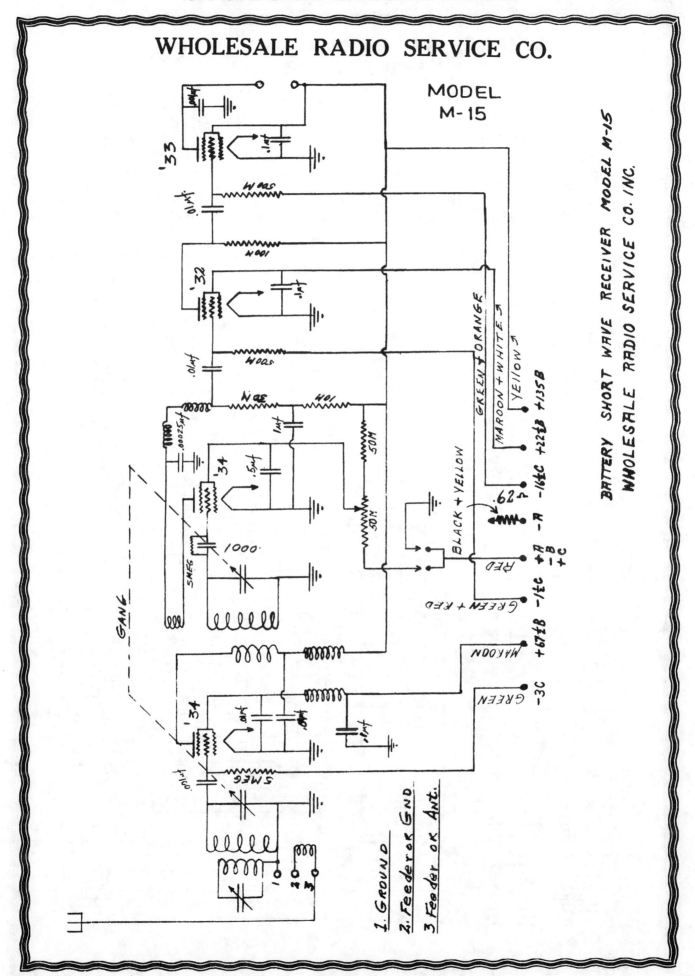

BATTERY SHORT WAVE RECEIVER MODEL M-15
WHOLESALE RADIO SERVICE CO. INC.

1. GROUND
2. Feeder & Gnd.
3. Feeder or Ant.

WHOLESALE RADIO SERVICE CO.
E-20 & E-204

Lafayette
"Receiver"
Models E-20 and E-204

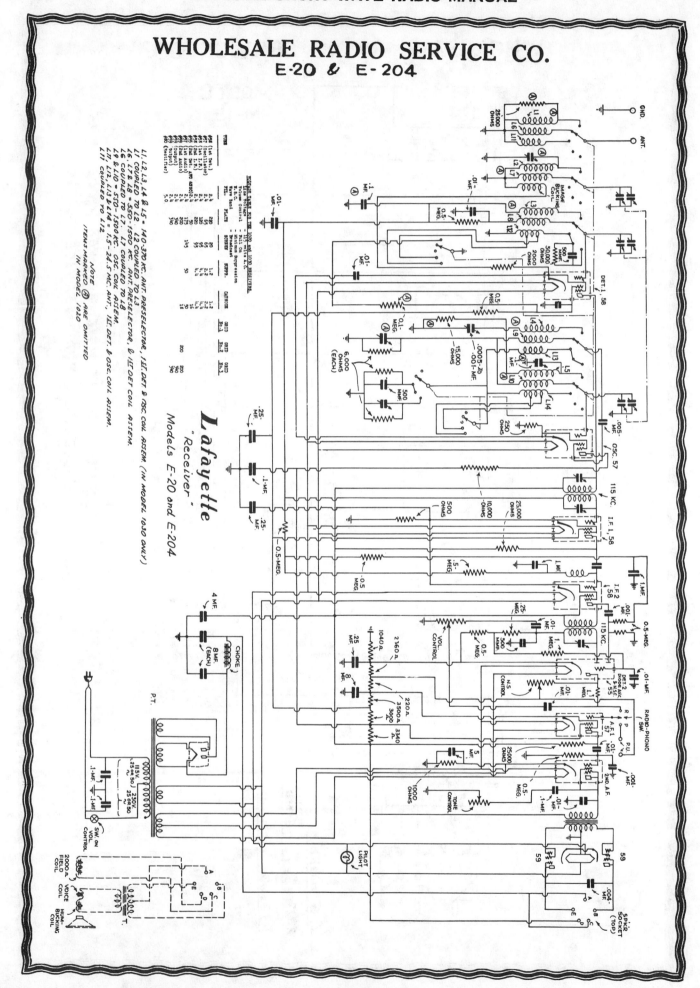

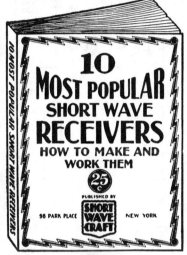

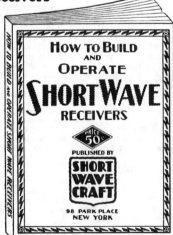
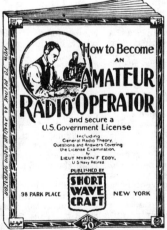

WHOLESALE RADIO SERVICE CO.
MODEL N-39 CIRCUIT № N-41 SET

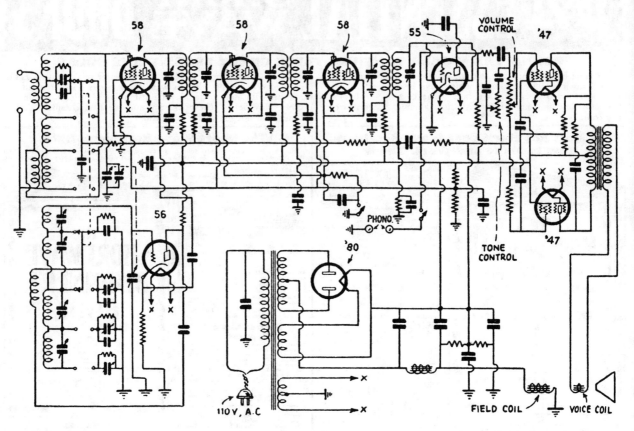

Diagram of Lafayette all wave, non "plug-in" receiver, using latest tubes and A.V.C.

I.F. = 508 KC.

● ONE of the things that retarded the development and commercialization of combination short-wave and broadcast receivers for a long time was a commonly held prejudice against the use of anything but removable plug-in coils for the various wave ranges. Radio men who were brought up on copper-tubing inductances, bus-bar wiring and open-face baseboards threw up their hands in horror at the mere mention of wave-changing switches or tapped coils.

Now we all admit that plug-in coils, if properly made and employed, possess certain low-loss characteristics. Dyed-in-the-wool "hams" of the old school still swear by 'em, but the new and uprising generation, perhaps spoiled a little by some of the other conveniences of modern life, swear *at* them. People nowadays don't want to wrestle with delicate coils that stick in their receptacles as if bolted down. They have "free-wheeling" in their cars; they want the equivalent of free-wheeling in their radio receivers.

This new set is a superheterodyne, consisting of a 56 oscillator, 58 first detector, two stages of intermediate amplification using 58's, (tuned to 508 kc.) 55 second detector, push-pull 47 audios, and 80 rectifier. The 55, known as the "duplex diode triode", provides automatic volume control, a desirable feature for broadcast operation and a high-

ly important one for short-wave reception, in which fading effects are often very pronounced. The peculiar circuit connections of this interesting tube, which has just recently appeared on the market, merit considerable study. The two diodes and the triode are independent of each other except for the common cathode sleeve, which has one emitting surface for the diodes and another for the triode. In this particular circuit the two diode plates are tied together to form a single diode, which performs at the same time the functions of perfect half-wave rectification and automatic volume control; in addition, the triode unit works independently as an audio amplifier under its own optimum conditions.

Manual volume control is provided by a potentiometer in the audio grid circuit. Tone control is effected by a filter between the triode section of the 55 and the 47 output tubes. These controls are marked in the schematic diagram.

Instead of the usual conglomeration of coils usually associated with all-wave receivers, there are only two coils in this outfit, one for the oscillator (L1) and the other for the first detector input circuit, L2. The top end of each is connected, respectively, to the grids of the oscillator and first detector tubes. Each coil is tapped in four places, the taps being brought out to contacts on a

simple rotating switch. A wavelength range of 12 to 555 meters is covered by this arrangement, in four steps as follows: 12 to 33 meters; 32.3 to 96.2 meters; 72.4 to 216 meters; and 195 to 555 meters. Wave changing is accomplished by turning the lower left knob on the front of the set.

For the convenience of the operator, the indicator scale, which is illuminated, is divided into four sections. The lowest one, covering the broadcast band, is calibrated in kilocycles; the upper three, for the high-frequency bands, in megacycles. Different units are used here merely for the sake of convenience, as kilocycle figures would occupy too much space on the scale. In addition, the vernier dial which controls the dual tuning condensers has two "speeds" or driving ratios: a medium ratio for the broadcast band, and a very high ratio for the short wave bands. This little feature will undoubtedly be applauded by people who have had experience with short-wave superhets and know how critically fine the tuning can sometimes be!

An incidental provision in the set is made for a *phonograph pick-up* or *microphone*, with a switch on the back of the chassis to turn the instrument on or off. The chassis, of course, is of modern all-metal construction with all sensitive units completely shielded. Overall the receiver stands 19 inches high, 16 inches wide and 11½ deep.

With the tubes once installed, tuning and wave changing are accomplished from the front of the set. There are four knobs, as follows: lower left, wave changing; upper center, tuning; lower center, volume control; lower right, tone control. The wave-changing switch, being of light and simple construction, turns easily, and saves the temper of the operator.

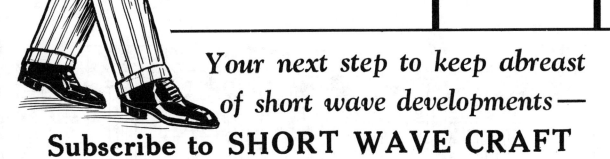

WIRELESS EGERT ENG. CO.
SWS-9 SET

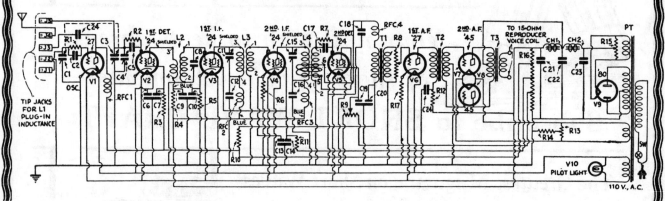

Schematic circuit of the "SWS-9" short-wave super; the volume control is potentiometer R8, in the first audio input. R9 is a regeneration control for the second detector. The oscillator condenser C1 and antenna tuning condenser C3 are ganged.

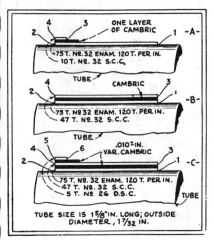

TUBE SIZE IS 1 5/8 IN. LONG; OUTSIDE DIAMETER, 1 7/32 IN.

Specifications of I.F. transformers, in the following order: A, L2; B, L3; C, L4—second-detector, with feed-back coil having terminals 5 and 6.

Assembly of the Receiver

Let us begin with the chassis, which consists of an inverted tray measuring 10 x 20 x 2 inches; it is made of 3/32-inch aluminum, bent over on all edges.

In the specifications which follow, both in the figures and in the text, dimensions for holes for audio transformers and chokes are not included; since it is felt that most constructors will prefer to use their own transformers. The placement of the transformers, in the factory model, however, is shown in the photograph reproduced here.

The shield can for the oscillator and first detector tubes and tuning condensers is made of 1/16-in. aluminum and measures 4 7/8 x 5 1/5 x 8 1/2 inches; it is provided with a cover.

The shield can for the inductances of

the oscillator and first detector circuits is made of 12-ounce copper and measures 2 x 3 7/8 x 2 1/2 inches deep; into its rectangular opening fits a bakelite plate 2 x 3 7/8 x 1/4-inch thick, which is drilled for five General Radio pin-plugs (four of these being spaced 7/8-in., and the last one, to "polarize" the construction, 1 in.).

Since both oscillator and detector coils are wound on the same tube forms, the coupling between them is rather high. It is therefore necessary to use a high intermediate frequency (1,600 kc.) in order to prevent the detector from being blocked by the oscillator.

List of Parts

Two *Hammarlund* "Type ML-7" 140-mf. variable condensers (C1, C3);
One *Tobe Deutschmann* .00015-mf. fixed condenser (C2—to be mounted directly on cap of screen-grid tube);
One *Pilot* "Type J-23" .0001-mf., variable condenser, (C4);
One *Dubilier* .00015-mf. fixed condenser, (C5);
Four *Polymet* "siamese" 0.25-mf., by-pass condensers (C6, C7, C9, C10, C13, C14, C19, C20);
Three *Hammarlund* "Type EC-80" 80-mmf. equalizing condensers (C8, C11, C15);
Two *Sangamo* .01-mf. fixed condensers, (C12, C16);
One *Polymet* .0002-mf. fixed condenser (C17);
Two *Polymet* .002-mf. fixed condensers, (C18, C24);
One *Dubilier* 1,000-volt 1-mf. fixed condenser (C21);
Two *Polymet* electrolytic 8-mf. fixed condensers (C22, C23);
One *Polymet* 0.5-mf. fixed condenser, (C25);
One *Durham* 25,000-ohm fixed resistor (R1);
Two *Durham* 2-meg. resistors (R2, R7);
Three *Lynch* 10,000-ohm fixed resistors (R3, R4, R10);
Two *Durham* 500-ohm fixed resistors (R5, R6);
One *Clarostat* 500,000-ohm potentiometer (R8);
One *Clarostat* 50,000-ohm potentiometer (R9);
One *Durham* 5,000-ohm fixed resistor (R11);
One *Lynch* 10,000-ohm heavy duty limiting resistor (R12);
Two *Durham* 20-ohm center-tapped resistors (R13, R15);
One *Cresradio* 780-ohm 25-watt resistor (R14);
One *Electrad* 20,000-ohm tapped voltage divider (R16);
One *Durham* 2,000-ohm resistor (R17);
Four *W.E.E.Co.* 115-millihenry R.F. chokes (RFC1, RFC2, RFC3, RFC4);
Five *General Radio* plug-in jacks and plugs (J1, J2, J3, J4, J5; these must be insulated from the metal front panel.—*Technical Editor*):

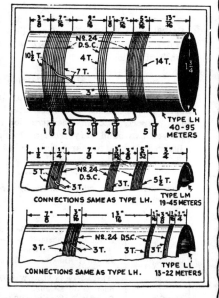

CONNECTIONS SAME AS TYPE LH.

Winding data for the short-wave oscillator and first-detector inductances, that are combined in one shielded coil unit, L1; this plugs into a front-of-panel receptacle.

One *National* "Type VHCC" drum dial;
One *Arrow-H.&H.* toggle switch, Sw.;
One Metal front panel;
Three *W.E.E.Co.* "Types LL, LM, and LH" shielded plug-in inductances, one each (L1: see text);
Three *W.E.E.Co.* "Types IF1, IF2, IF3" shielded I.F. transformers (one each L2, L3, L4; see text);
One *Amertran* first stage 'DeLuxe' A.F. transformer (T1);
One *Amertran* "Type 151" input push-pull A.F. transformer (T2);
One *Amertran* "Type 443" (for dynamic reproducer), or "Type 442" (magnetic reproducer), output push-pull A.F. transformer (T3);
Two *Thordarson* 30-henry, 75-ma. filter chokes (Ch. 1, Ch. 2);
One *W.E.E.Co.* "Type PT 116" power transformer (PT);
Nine *Cinch* tube sockets, three UX and six UY.

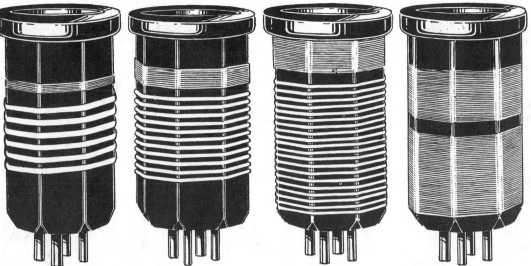
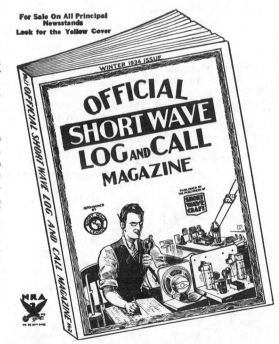

WORKRITE RADIO CORP.
WALKER SUPER CONVERTER

THE super-heterodyne converter gives every indication of providing the ultimate in efficient short-wave reception. Every short-wave radio fan already possesses a broadcast band receiver of some sort. Now remains the means of converting the broadcast band receiver, so that short-wave stations throughout the world may be tuned in, with simplicity and sensitivity the prime factors.

The Walker Super-Converter—a handy and efficient means of converting your present broadcast band receiver to tune in short-wave programs and police calls, over the band from 15 to 200 meters. This converter, which depends upon the electric-light socket for its source of power, may be used with all types of receivers, whether A.C., or battery type, and including superheterodynes.

The principle of converting your present T.R.F. or neutrodyne receiver into a short-wave superheterodyne is quite simple. All receivers have an R.F. "tuning circuit", which covers the broadcast band of 200-550 meters (equivalent to 1,500-500 kilocycles). In addition, there is the audio amplifier circuit, which greatly increases the volume of the signal detected and amplified at radio frequency by the tuning circuit. In all superheterodyne receivers there are, in addition to these circuits possessed by T.R.F. or neutrodyne receivers, the oscillator and first-detector circuits. The Walker Super-Converter contains an oscillator and a first detector, as well as a stage of screen-grid radio-frequency

amplification ahead of the detector to further boost the volume. The signal is picked up by the converter and passed on to the receiver for additional amplification; thereby utilizing each circuit of the receiver and making unnecessary the purchase of an extra speaker or the erection of a special antenna. *All tuning is done with the simplified controls of the converter.*

The receiver is not disturbed, since there is no need to "plug in" to furnish the power necessary to operate a converter. A connection is made from the converter directly to the antenna post of the receiver. There is no overloading of the receiver's power supply; since this new converter draws its power directly from the light socket. There can be no possible damage to the receiver.

Regeneration provides a degree of sensitivity and selectivity otherwise unattainable; it has been credited with the efficiency of an additional tube. Where is there a "dyed in the wool" short-wave fan who does not insist upon regeneration? A stage of screen-grid, radio-frequency amplification, ahead of the regenerative detector, not only insures greater volume, but also provides for the use of any length of antenna; in addition, it prevents radiation, which might disturb your neighbor's short-wave set. The oscillator and the detector tuning condensers are ganged and provided with a fine vernier dial for sharp tuning. A small midget condenser, connected in parallel with the oscillator condenser, permits obtaining exact resonance at all wavelengths.

MODEL 4X CONVERTER

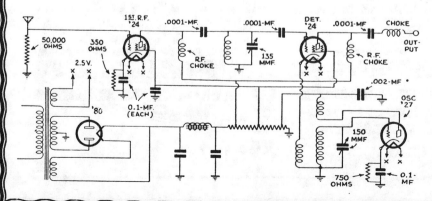

The Model 4X super-converter here illustrated was designed by Mr. George W. Walker, who has carried on a tremendous amount of experimenting and research on converters of every description; if anyone knows how to design and build a converter it is Mr. Walker. The converter here shown employs four tubes —2-'24 screen grid, 1-'27, and 1-'80 tubes. The wavelength range covered is 14 to 200 meters; the popular band from 23 to 50 meters is tuned without changing the coils.

RUDOLPH WURLITZER MFG. CO.
MODEL SW-80

550 – 1500 - KC.
1450 - 3700 - KC
3500 - 9000 - KC.
8500 - 22,000 - KC.

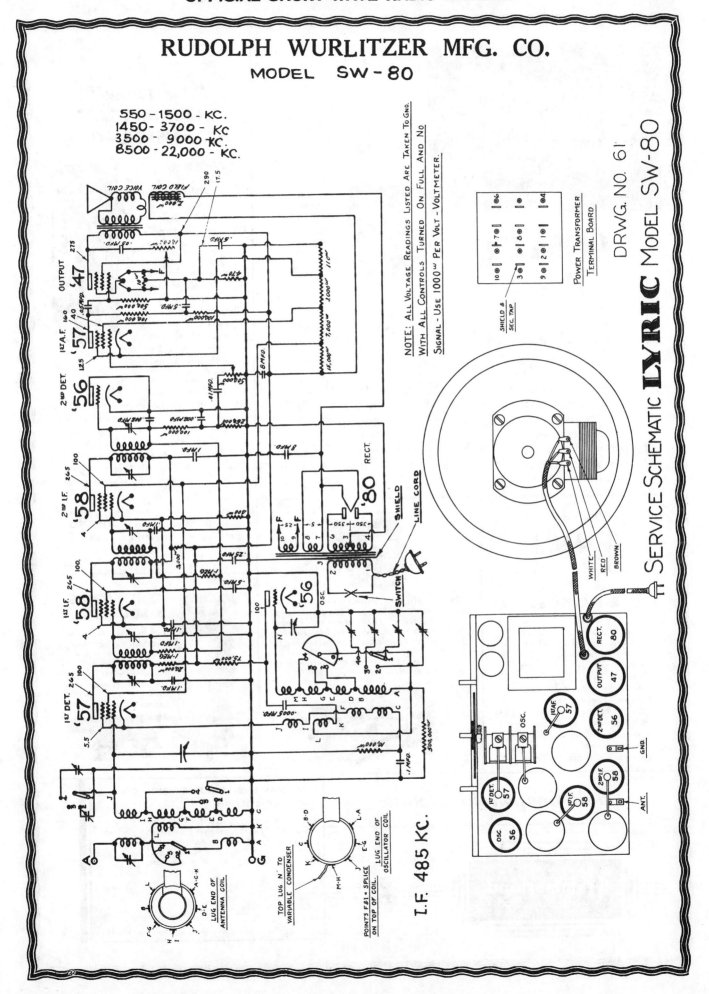

NOTE: ALL VOLTAGE READINGS LISTED ARE TAKEN TO GND. WITH ALL CONTROLS TURNED ON FULL AND NO SIGNAL - USE 1000ᵂ PER VOLT - VOLTMETER.

POWER TRANSFORMER TERMINAL BOARD

DRWG. NO. 61

Service Schematic **LYRIC** MODEL SW-80

I.F. 485 KC.

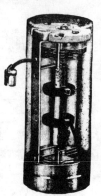
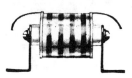

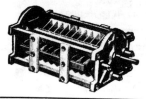

Building Old-Time Radios

by T. J. Lindsay

Back in the late 20's and early 30's when newly discovered short-wave frequencies were coming into use, building a receiver was the quickest and lowest cost way of listening in.

As you've seen, the simplest and most often described receiver in this book, has been the regenerative receiver. The performance possible with just a handful of components is incredible. And although I've built receivers of more modern design, I keep coming back to the lowly regenerative. It never fails to amaze me.

Getting parts

Getting the components you need to build the receivers described in this book will be difficult at best. There are a few dealers who can supply old vacuum tubes, but when you finally do find a tube you need, be prepared to pay. With everything being solid-state these days, the vacuum tube factories have shut down, and prices have increased, in some cases, dramatically. It's like trying to buy parts for antique automobiles.

Suppose you do find the tube you need. Just finding a socket can be a problem. And then finding capacitors, dial drives, and all the rest can be difficult.

One place to look for old parts is at radio amateur flea markets called hamfests. You'll find upcoming hamfests listed in amateur radio magazines. Over the past ten years, I've been grabbing antique parts whenever I've seen them. As a result I have a carton filled with old time capacitors, vernier dial drives, plug in coil forms and a lot of the rare components described in the first part of this book.

But you don't need old parts, necessarily. You can use the circuits described here and in other old books, and adapt them to modern components. A solid state regenerative receiver will actually perform much better than the old tube set and will be much easier to build.

Transistors replace vacuum tubes

The heart of the regenerative receiver is the vacuum tube. The tube has a very high input impedance which simply means that the large current flowing through the tube is controlled by a small voltage on the control grid. Practially no current is needed by the grid — just voltage. Usually, you must apply a couple of hundred volts to the plate to get current to flow while applying about a -5 volts to the grid as bias.

A 50c junction field effect transistor (JFET or FET) will replace the tube. It, too, is voltage controlled, but here you need only apply 9 volts or so to the drain which corresponds to the tube's plate, and bias the gate (similar to the grid) with -2 volts.

Consider this for a minute. Instead of paying several dollars for a tube, finding the appropriate socket, coming up with 6 or 12 volts of filament power, and high voltage for the plate which is no fun if you accidentally tie yourself into it, you merely solder the FET to the circuit board and attach a standard 9-volt transistor radio battery. Working with transistors is so much easier and safer. Who would ever want to go back to tubes?

A favorite story of mine is one concerning the building of a make-shift receiver. One evening, wanting to try something different, I dug through the junk box and came out with a handful of components. With alligator clips and bits of wire I clipped a messy looking radio together on the table top. I wound a tank coil with one of my wife's old plastic curlers and strung a 20 foot antenna across the basement ceiling. When I attached a ordinary 9-volt battery, signals started popping in. Because I didn't have a vernier dial drive on the tuning capacitor. The best I could do was roughly set it, and "fine tune" it by moving my hand closer or further from the set (body capacitance put to good use). After about three minutes I heard a call from the captain of a ship near Seattle asking for emergency help in navigating shallow water. Not bad. . . Considering I'm sitting here in my basement near Chicago! Incredible considering how extremely simple the receiver is!

With the FET you get the advantage of the very high input impedance found in vacuum tubes. In a simple regen receiver the frequency is selected by the tank circuit, that is, the resonant coil and capacitor circuit. A low input impedance amplifying device would suck too much energy out of the tank circuit. The "Q" drops way off. The result is that your receiver won't be able to separate signals very well, or may not even operate at all. A high input impedance device like the vacuum tube or FET extracts very little energy from the tank circuit, resulting in very high "Q" and excellent performance.

Biasing

Tubes and transistors must have the right voltage applied to each of their leads before they can amplify. That's the job of the resistors you see in radio diagrams. They're there to "bias" the tube or transistor.

You can replace the tubes in old radio diagrams with transistors. About all you'll have to change are the bias resistors. In some cases the gain of a transistor is so much greater than the tube it replaces, that you may have to eliminate some of the amplification stages that follow.

I've built a number of regenerative receivers using old tube designs, but substituting transistors for the tubes. It works! It's cheaper, simpler, and even works better. You can do it, too.

You'll see that most of the old circuits in this book use a triode or a screen-grid tube as the regenerating detector. To bias these tubes, the cathode was most often connected directly to ground or, better yet, through a resistance of a couple of hundred ohms. A very high value resistance from 1 to 10 megohms allows direct current to get to the grid and bias it, but keeps the radio frequency AC from being short-circuited to ground and lost.

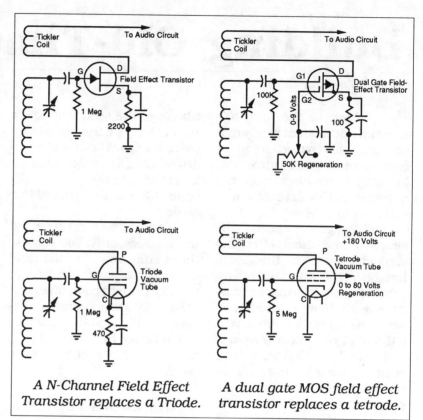

A N-Channel Field Effect Transistor replaces a Triode.

A dual gate MOS field effect transistor replaces a tetrode.

How do you accomplish the same with a FET? Much the same way. The low-cost MPF102 n-channel JFETs I've used work quite well with 2 to 3 thousand ohms in the source leg (same as the cathode) and a gate (same as the grid) bias resistor of 1/2 to 5 megohms. Increasing the gate resistance may make regeneration harder to control, while too low a value decreases sensitivity of the detector to incoming signals.

Decreasing the source bias resistance increases the amount of current flowing through the FET which increases heating which in turn can upset receiver stability or melt the FET. Increasing the resistance too far can cause the detector to stop working.

In early regenerative designs, regeneration was controlled by varying the plate voltage of the triode. When screen-grid tubes were developed, it was found that better performance could be obtained by varying the screen grid voltage while leaving the plate voltage fixed.

Replacing a screen-grid tube with an FET is not much more difficult if you use a double gate MOSFET transistor such as the 40673. MOS means metal-oxide-semiconductor, another way of fabricating an FET. The 40673 has two gates. Use one in the normal way, but hook the other leg to 0-9 volts variable. The variable leg changes the overall gain of the transistor and will affect the regeneration. I haven't tried this arrangement in regen receivers, but others have used it with success in other applications. Many of the screen-grid tube hookups described earlier should work with a 40673 if biased correctly.

Sample Circuits

Here are a few designs from my notebook that have worked well. I don't claim to be an expert in radio design. I'm just an experimenter like you. That means that although I had good luck with them, they're not guaranteed to be fool-proof circuits.

Circuit A uses an MPF102 JFET with tickler feedback. You have seen this basic design a dozen times already in the preceding pages. The 2700 ohm source bias resistor could probably be anything from 1000 to 4700 ohms, and the gate bias resistor can vary from 100k to 5 meg. You'll have to experiment to see what gives you the best results. The 10 mfd capacitor near the regeneration control could probably be anything from 1 to 50 mfd, and is not at all critical. The .001 mfd cap in the tickler lead is a little more touchy. It has to be big enough to allow radio frequencies to go through to ground, but not so big that audio frequencies will be shorted out.

With a properly wound and adjusted tickler coil, this circuit can really perform. This is the "table top circuit" that I used together with a one-transistor audio amp to pull in the ship at sea that one night.

Circuit B uses cathode feedback. I worked out this variation to avoid having to wind and rewind tickler coils before getting it right. In my tests, the detector worked nicely when the tap was anywhere from 25% to 50% of the coil. If you use this circuit, be sure to turn the power off when changing coils. Should the source lead become disconnected, the JFET will be destroyed.

Circuit C is a variation

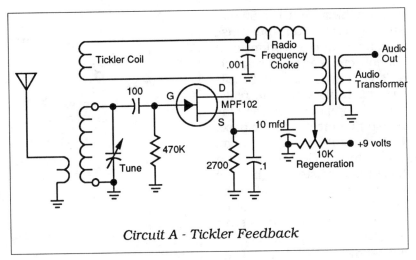

Circuit A - Tickler Feedback

of circuit B. Here some of the radio frequency energy flowing through the coil tap and source leg is shorted to ground by a 500 ohm pot in series with a .01 mfd capacitor. The 4 mHz test model worked quite well.

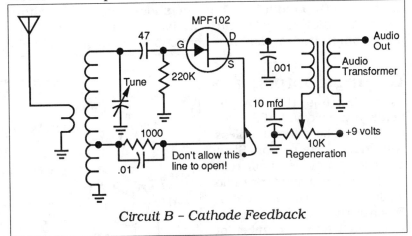

Circuit B – Cathode Feedback

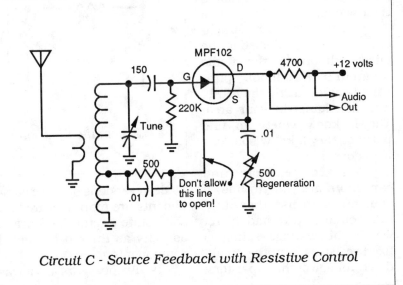

Circuit C - Source Feedback with Resistive Control

Once you have audio flowing out of the detector, you need to amplify it. The simplest audio amp I use is a single high-gain NPN bipolar transistor shown in Circuit D. The 10 mfd input capacitor is not criti-

243

cal. Anything from 1 to 50 mfd should work. You must use high impedance earphone(s) in the collector lead to get respectable gain and to prevent melting the transistor. The really critical component is the bias resistor connected to the base. Put a voltmeter on the transistor's collector. Install a base bias resistor so that the meter shows about half of the supply voltage. In other words, if you're powering your receiver with 9 volt battery, the collector will show about 4 1/2 volts when you have the correct bias resistor installed. The value can be anywhere from 50k to 2 meg depending on the gain of the transistor. A pot might be handier to install than a fixed resistor. This circuit was used in the "table top radio."

Circuit E is my favorite audio amp. It provides ear-

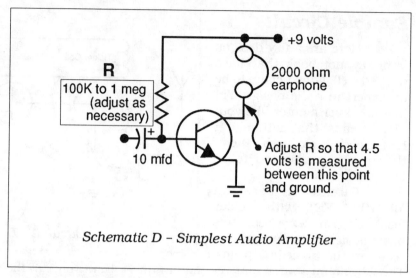

Schematic D – Simplest Audio Amplifier

splitting volume when I want it, yet does not oscillate. It's something I stole from a Radio Amateur's Handbook a few years ago. If you don't already have a copy of the handbook, you should. It's a great source for basic information and constuction techniques despite the fact that you won't find regen's discussed anymore. Use the resistor values shown in the schematic. The capacitors can be almost anything close. Run-of-the-mill NPN transistors are used and give good results.

Getting Components

Getting the components you need to build these radios can be difficult unless you know where to look. Start by picking up several different electronics and amateur radio magazines off the newsstand. Towards the end of each magazine you'll find a number of parts suppliers advertising their wares.

Another list of suppliers is usually published in the latest edition of the Radio Amateur's Handbook published each year by the ARRL. And ask hams in your area. They'll know where to find parts or they'll know someone who does.

Fixed resistors and capacitors are easy to find. If you intend to use them in a tube circuit, you had better watch the wattage rating of fixed resistors, and the breakdown voltage of the capacitors.

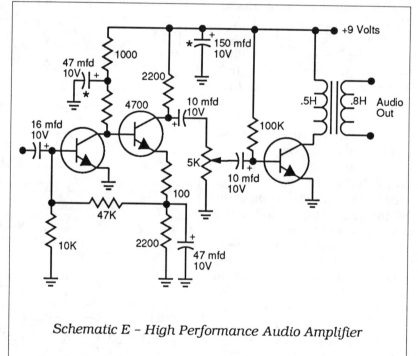

Schematic E – High Performance Audio Amplifier

Modern versions of these components are manufactured for solid state circuits and aren't as hefty as the old-fashioned components.

Twenty years ago it was easy to find junk TV's sitting in alleys waiting to be picked up by trashman. Stripping a chassis would often yield almost all of the components needed for a regen receiver. The same can be done today, but most chassis are solid state. Garage sales, flea markets, and hamfests are great places to pick up parts in the form of transistor radios, tape recorders, CB's and so on.

The hardest components to find are variable capacitors and drives. The old multi-plate capacitors manufactured by Hammarlund, EF Johnson, and others are hard to find. More than once I've bought a junk, homebuilt amateur transmitter for $3, and carried the 40 pound monster half a mile just to get my hands on five beautiful variable capacitors. Variable capacitors are out there waiting to be found, but if you can't get to a hamfest, you may have to pay quite a bit from a dealer.

Some surplus component dealers have offered a plastic bag of 10 variable capacitors intended for AM-FM radios for $6 or $7. The AM jobs are usually 365pfd variables while the FM variables are some smaller value. Some of the units even have a built in 3-to-1 gear reduction.

Variables that have too high a value can be modified by pulling out a few of the moving plates mounted on the shaft. You have to be careful not to bend the remaining plates, otherwise they'll touch the stator plates, short out, and make the capacitor useless. You might want to put two 365 pfd variables in parallel, leaving one in its original state for use as a bandset capacitor, while removing all but two moving plates in the other capacitor to use it for fine tuning, or bandspread. It's not at all complicated, and it works very well.

Much more difficult to find are vernier dial drives. You can still get some made in the Orient, and they're pretty good. They're 5 to 1 planetary gear re-

duction, if I remember right, meaning that you will have to turn the tuning knob 2 1/2 turns to get the capacitor to revolve just one-half turn from full close to full open (maximum to minimum capacitance). Some type of vernier dial drive is necessary if you are to be able to smoothly and carefully zero in on the station you want to hear.

I prefer the old-fashioned drives like National Radio's Velvet Verniers which were made for years with beau-

tiful huge bakelite knobs, and later in stainless steel. I will have to admit that they're appearance may be why I like them, since they're 5 to 1 drives just like smaller new drives. Finding them is very difficult, but if you see one, grab it.

Plug in coil forms are like dial drives, hard to find, at best. The best forms, so the old timers say, were made by Hammarlund and Insuline Corp. They were 1 1/2" in diameter, had 4, 5, or 6 pins, and were ribbed for easy winding. As you can see in the preceding

pages, coils could be wound for any desired band, and plugged in at will. These forms were plentiful at hamfests 10 years ago, but now they're hard to find. Millen Corp was making new forms until just a couple of years ago. Some beautiful porcelain sockets that worked perfectly with these coil forms turned up in a scrap yard as part of obsolete telephone microwave gear. You must always be on the lookout for parts. A dyed-in-the wool scavenger will accumulate odds and ends for

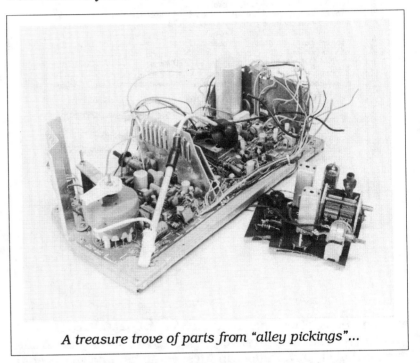

A treasure trove of parts from "alley pickings"...

that day when construction begins. Start looking now, so that you have what you need when you need it.

Coil forms can be made from all kinds of things ranging from PVC drain pipe, plastic pill bottles, paper towel tubes, solid wood, and just about anything non-metallic approximately circular in shape. Some coils I've used have slipped inside collars permanently mounted on the radio chassis. The coils were held with a friction fit and electrical connections were made with

245

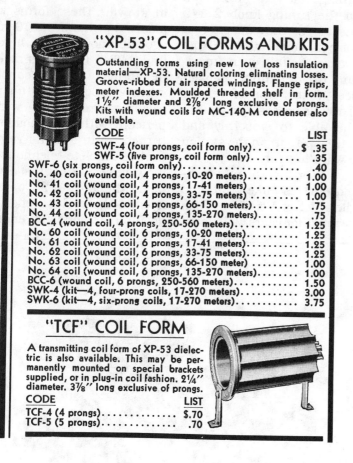

Hammarlund advertisement from 1939 ARRL "Radio Amateur's Handbook"

alligator clips. The only real problem with such an arrangement is that slight vibrations in the connecting wires can produce noise in the receiver audio output called "microphonics". Modern radio coils are wound on slug-tuned plastic forms and on toroids. Refer to the Radio Amateur's Handbook for complete details.

Recommendations and precautions for use can be found in the construction articles in this book. On pages 79 through 83 you'll find ideas for making your own components.

Test Equipment

What kind of electronic test equipment do you need to build these radios? Very simple equipment.

First, you need a volt-ohm meter, something to measure voltage and resistance. You can spend a $100 on a digital voltmeter (DVM) which are very accurate and very convenient. But you can get by with a simple old-fashioned analog volt-ohm meter. The hardware store here in town offers them for about $12. And they're every bit as accurate as you're going to need.

If you are serious about building radios, then by all means get a grid-dip meter. This device is an oscillator that produces radio signals that can be used to check the natural frequency of the coil-capacitors circuits you're building. It has a large calibrated dial with frequencies etched on it that correspond to the ranges of each of the special test coils that plug into the meter.

Suppose you're trying to wind a coil that you're going to use with a capacitor you found in an old radio, and you want it to tune the 6 mHz shipping band. Look through tables or make some calculations and wind a test coil. Hook it up to the capacitor. Turn on the grid-dip meter and bring the grid-dip meter coil near the coil and capacitor you're testing. As you turn the grid-dip knob with the calibrated dial, you'll see the meter on the dip meter drop suddenly, and then bounce up. If you tune back slowly, you'll be able to find a place where the meter drops to a very low value, the "dip". Once you find that point, the natural frequency can be read off the calibrated dial. If you see 8.2 mHz on the calibrated dial, you'll know the test coil you wound does not have enough turns. Add a few turns and try again, until the meter dips at 6.5 mHz or so. When you use this coil and capacitor in the receiver you're building, you'll know that it will be close to the 6 mHz band you want.

A grid-dip meter will tell you the natural resonant frequency of an L-C circuit. If you use you have a capacitor of unknown value, you can determine its value by hooking it up to a coil of a known value and measuring its natural frequency. If you plug these values into the formula for natural frequency, you can determine the value of the capacitor. The

Set 1 — Front View
If you can scavange enough old components, or those that look old, you can build a simple, yet hot-performing receiver that looks like something out of World War I. This set uses a National Radio "Velvet Vernier" manufactured in the 30's and 40's together with early radio knobs. The case is made from stained dimension lumber with masonite front panel which was sprayed with several coats of black lacquer, each layer being wet sanded. A coat of wax made the masonite look so much like Bakelite that one oldtimer asked me where I found it. Controls — upper left: switch for future crystal calibrator; left center: volume, on-off; lower left: earphone jack; center: tuning; right upper: antenna tuning capacitor; right lower: regeneration.

same procedure works to determine the inductance of unknown coils.

A dip meter is an oscillator. If you set the calibrated dial on 8 mHz, you can be sure it's putting out signal near 8 mHz. You can tune your receiver to see if you can hear it. In other words, the dip meter is also a small transmitter you can set to almost any frequency.

Grid-dip meters get their name from the fact that they use vacuum tubes and that the L-C circuit being tested "sucks" energy out of the grid which registers on the meter. Like everything else these days, transistors and even tunnel diode are now used. Now they're called just "dip meters". Heathkit and Millen have offered such units. You'll find old Allied Radio dip meters (like mine) that use an honest-to-goodness vacuum tube at hamfests for a few dollars. They're the nuts for building regen receivers.

Another handy device is the RLC bridge. You can clip a resistor, inductor or capacitor to the terminals and twirl the calibrated dial until the meter dips just like the grid dip meter. Then you read the value off the dial.

A bridge will tell you the value of variable capacitors salvaged from old receivers, the value of unmarked resistors and inductors, or can give you an idea of the turns ratio of audio transformers by measuring the inductance. For the person who wants to scrounge parts from old electronic equipment, they're a necessity. Don't panic, though, because I built radios for years without one.

If you have money burning a hole in your pocket,

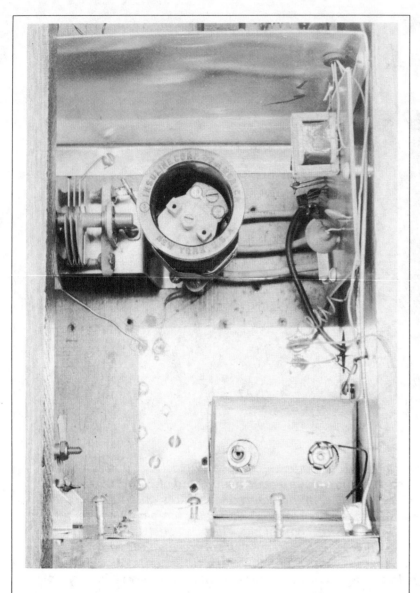

Set 1 — Detector Compartment

At the top is a plug-in style "Insuline" coil form mounted behind a 15 pfd variable capacitor mounted on a bracket above the wooden floor of the set. Three stiff 12 gauge wires connect the tuning tank circuit with the detector printed circuit card mounted on the back wall to the right.

At the bottom is a large 9 volt battery. This radio was in storage when photographed, so the battery had been disconnected. Tinned brass sheet and aluminum is used for shielding behind the front masonite panel and between the audio and detector sections. The sheet beneath the battery provides a low-inductance ground for the detector circuits. The device inside the coil form is a variable capacitor that is used as a bandset capacitor. The coil was originally used in a 1950 amateur transmitter obtained for $4 several years ago.

buy a low-cost frequency meter. You can stick a small coil near the receiver coil, and the frequency meter will tell you the exact frequency you're receiving. In a sense, that's kind of crazy — you're using a high-tech piece of test equipment to keep track of a low-tech bare-bones receiver. But if you have a counter, use it.

In the end, a regen receiver builder should have a voltmeter and a dip meter. That's all you need to have fun building receivers.

Common Problems

Howling

One of the earliest problems I ever remember encountering in building regen receivers was howling. You would have the earphones and the volume up, trying to "dig out" a weak signal. You'd advance the regeneration just a bit and all of a sudden the receiver would give out the loudest God-awful howl you had ever heard. Your eyes would cross in pain as you ripped the earphones off your head.

My grandfather built regens in the 20's, but fifty years later, still didn't know what caused the howling. The cause is simple. The output of the audio amplifier is getting back into the regenerative stage. Remember that the regenerative stage is amplifying both at radio and audio frequencies. The only signal that should be put into the regen stage is a weak signal from the antenna.

The output of the regenerative circuit should be audio which is amplified hundreds of times by the audio amp whose output is supposed to go directly to earphones or speaker. If something goes wrong, and some of the audio finds its way into the input side of the regenerative stage, both the regen detector and audio amp act as one big audio amplifier, and the whole receiver goes into audio regeneration, or oscillation.

The most common way for this unwanted feedback to occur is through the power supply. The audio amplifier draws quite a bit of power. On a strong signal, enough power is drawn to actually lower the battery voltage. The regenerative detector sees this change in voltage, and it changes the gain. Although the concept might be a bit confusing, the change in battery voltage actually allows audio to get into the regen stage.

The cure is not too difficult. What you need to do is provide simple voltage regulation. Circuit F shows a simple regulator. The audio amplifier sucks power through resistor R which drops the voltage a small amount. The capacitor C is quite large, and acts as a storage device to filter out changes in the voltage, especially changes at audio frequency.

Suppose we have an amplifier which needs about 10mA of current and that we have a 9 volt battery supply. A one volt drop across resistor R would require:

1 volt / 10 mA = 100 ohms

by Ohm's law. Although you can calculate the minimum size capacitor needed to filter a particular frequency, I usually just use a very large capacitor, say 150 mfd or more.

In schematic E, the audio amp, you'll see two voltage regulating capacitors marked with an "*".

There are other meth-

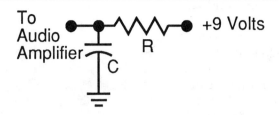

Schematic F - Simple Voltage Regulator to Cure Howling

ods, one of the best being the use of a dropping resistor as above, but using a Zener regulating diode in place of the capacitor. If this doesn't solve the regulation problem, nothing will!

Another wise choice is to use a "stiff" power supply. What that means is that the battery is large and can deliver so much power, that even when the audio amp is demanding enormous current when amplifying a large signal, the battery voltage doesn't fluctuate. This

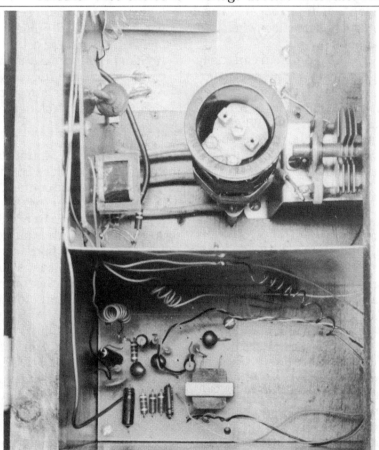

Set 1 — Audio Section

In a compartment of its own is the audio amplifier shown earlier in Circuit E. The tangle of wires bring audio in from the detector, shuttle audio back and forth to the volume control, and send it to the earphone jack on the front panel, bring power in from the battery, take switched power to the audio board, and variable DC to the detector board. This simple receiver is exceptionally stable and easy-to-use. It is one of the best receivers I've ever built.

is a brute-force-and-ignorance method, but it works.

Although we're talking about solid state circuits here, the same general concepts were to blame for the howling in my grandfather's tube-design regenerative receiver. A well-built regen receiver does not howl.

Spurious Frequencies

Another problem I've encountered is "crazy tuning". You'll be tuning up through the 6 mHz band when all of a sudden you get a lot of noise and "hash". Often regenerative control has no effect when this happens.

The problem can easily be seen on a frequency meter. You'll see the frequency smoothly increase: 6, 6.1, 6.2, 6.3, and then 42.7! What happened? Simple. The regenerative stage is now oscillating at a new frequency so strongly that it has taken over the receiver.

These parasitic oscillations are often a result of the physical layout of printed circuit boards, shielding plates, coils, leads, and so on. Some of my receivers have had nasty parasitics, while others have had none.

You should have a solidly grounded metal front plate to eliminate hand capacity problems, and probably another plate to shield the audio amp from radio frequencies being generated in the regenerative detector stage. But too much metal, too close to a coil can can act as a capacitor, or can upset the inductance of the coil. At certain settings of the tuning capacitor, the regenerative will want to use these unwanted effects to oscillate at unwanted frequencies.

Every receiver acts differently. Different coils have

Set 2 — Front View
A blend of old and new dial drives mounted on a polished aluminum panel attached to a stained dimension lumber cabinet hide the workings of another regen receiver. The modern Millen drive at the far left is the bandset control. The circular middle drive is probably fifty years old and provides bandspread. The two smaller controls are for volume and regeneration. This large cabinet houses two small printed circuit boards which make up the radio.

different stabilities. Mount the main tuning coil close to the tuning capacitor and connect it with large diameter wire, or better yet, flat ribbon or braids to keep the unwanted inductance very low. Make sure shield plates are at least a coil diameter away from the tuning coil — more if you can provide it.

These instabilities are much more of a problem with transistors than with tubes, because tube gain decreases as frequency increases. Modern transistors have no such decreases until frequencies get very, very high. The net result is that transistors love to be unstable.

It is wise when building regen receivers to use the fabrication techniques that radio amateurs use when building solid-state transmitters, such as double-clad printed circuit board, ferrite beads, and toroids. Often a ferrite bead

slipped over the gate or base lead of the transistor before you solder it into the circuit will help. The bead acts as an inductor which attenuates higher frequency signals flowing into the device more than lower frequencies. In a sense, the bead lowers the higher frequency gain of the device, making it act more like a vacuum tube and making it more immune to parasitic oscillations.

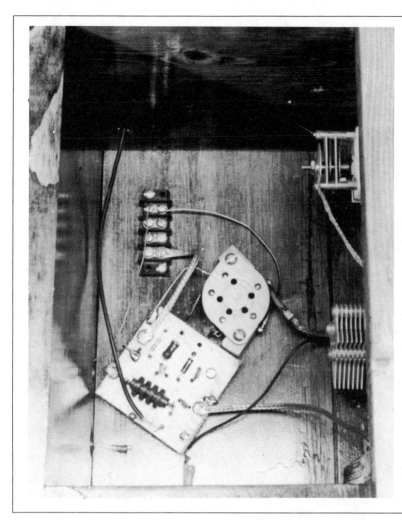

Set 2 —
Detector Compartment

Here we see the detector section with the coil removed from the socket. The capacitor at the lower right is driven by the new Millen drive. The bandspread capacitor at the top is a high-voltage military surplus 15 pfd variable. The detector is constructed on double-sided printed circuit board using techniques often described in amateur radio literature and the Amateur Radio Handbook. The flat braids were stripped from an old coaxial cable and were stiffened by flooding with molten solder. Their wide flat shape provides very low inductance connections between components operating at radio frequency. The audio compartment is at the top.

Coil Adjustment

Adjusting tickler coils can be a nuisance since it is often a trial-and error-process of adding and removing tickler turns until regeneration is smooth. But once you have exactly the right number of turns, you'll find it was worth the trouble. You can advance the regeneration slowly and smoothly. If you've done it right, you can tune in on top of a single-sideband (SSB) signal and make it intelligible by slowly adjusting regeneration. In fact, the mark of solid, stable, well-built regen receiver is its ability to quickly and easily tune in a SSB signal and hold it for a length of time without readjustment.

Set 2 — Audio Section

Built on perf board, this audio amplifier provides plenty of volume. The cabinet was built oversized intentionally to allow for experimentation. The messy wiring is proof of the continual changes this set undergoes.

Set 3 — Front View

    A $7 World War II frequency meter provided a beautiful National Radio stainless steel "Velvet Vernier" dial drive tied to a small variable capacitor mounted inside a steel chassis and case. Most of the controls were removed and replaced with modern standard components. The original meter was left in rather than have a gaping hole in the front panel. Being well-shielded and "built like a tank", this set is stable and solid. In the audio section was inserted an audio active filter to single out one particular tone to make copying Morse code in crowded bands much easier.

Tickler coils are a pain because the coil must be wound and rewound, soldered and resoldered, removed and installed until it is right. If you use the tapped coil type of feed back (a Hartley oscillator), you can avoid most of the hassle. To adjust regeneration, you merely move the tap up and down the coil as needed, usually just by unsoldering and resoldering to another turn. It's fast and easy.

The oldtimer's say that the tickler coils are better because you can get a more precise fix on the feedback, and hence, smoother regeneration. The tapped coil method may give regeneration that is harder to control. But I haven't experimented with Hartley-type detectors enough to know whether this is true or not.

Other Ideas

You might consider using varactor diodes to replace tuning capacitors. These unusual devices develop a capacitance across their diode junction that is proportional to the DC voltage being applied. If you change the voltage, you change the capacitance. Instead of using hard-to-find and/or expensive variable capacitors and dial drives, you can use a varactor to replace the variable capacitor and a low-cost and readily available potentiometer to supply a variable voltage. These methods are used in receivers, TV tuners and so on. Check modern radio publications to see what types are available and how to use them.

Look into toroids if you want to build really tiny receivers. Toroids can replace the large tuning inductors with donut-shaped coil only a half inch in diameter. And there are no coupling effects with shielding or other coils!

There have been a great many old receiver diagrams published over the years, and you could spend a lifetime collecting them. If you just want to build receivers, don't spend too much time looking, because the some of the very best diagrams are here in this book.

If you want diagrams for solid state regen receivers, you won't find many. Solid

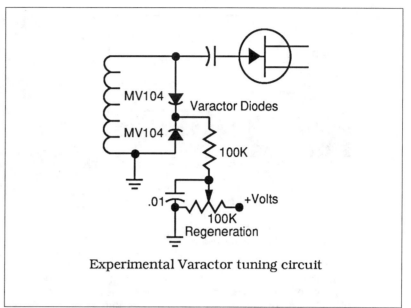

Experimental Varactor tuning circuit

state devices make the construction of complicated receivers easy. Not many people have developed antique circuits with modern devices. Yet if you study diagrams for oscillators, receivers, and even radio amateur transmitters, you'll get many ideas suitable for inclusion in regen circuits.

Simple receivers can in

no way compete against modern high-tech designs in overall performance. But the lowly regenerative receiver can provide more listening and construction enjoyment, dollar-for-dollar and component-for-component than just about any other design.

Build, experiment, and enjoy.

References:

CQ Magazine, June 1972 details construction of a two tube DX regenerative receiver from 1931. A search through "Reader's Guide" at your local library should turn up other articles as well. For subscription information write:

CQ Magazine
76 N. Broadway
Hicksville NY 11801

You should have a copy of "Radio Amateur's Handbook" if you intend to build radios and other electronic equipment. Write for information on the handbook and other excellent publications from

American Radio Relay League
Newington CT 06111

One of My Favorite Magazine Articles

The following four pages have been reproduced from the February 1966 issue of CQ magazine. The late Walt Burdine reported on the regenerative receiver built by Hijame Suzuki of Japan. Reprinted by permission of the publisher.

NOVICE

WALTER G. BURDINE,* W8ZCV

ABOUT a year and a half ago I received a letter from Hijame Suzuki of Tokyo, Japan and printed it in our column. The results of that letter has brought many letters about his receiver and his record of stations confirmed. I have received 37 letters about his letter and many of them wanted to know more about Hijame's receivers and location. I wrote to Hijame for more dope on his wonderful little receivers and about amateur radio operators and the license examinations in Japan. The following is part of a condensation of the letters from Hijame. I'm sure you will enjoy them as much as I did. By the way I might also do some other columns like this if you like to know more of amateur radio in other countries.

Right after the story of Hijame's appeared in *CQ* (Dec. 1964) he sent me a diagram of his receivers and a nice long 24 page letter. Here is part of one paragraph. "I'm very happy to see my letter and photo to you has been printed in your and "our" *CQ*. I have received 5 letters concerning the receiver asking for circuit diagram and one of them is asking me to sell him my receiver to him. But I don't want to sell my radio to anyone, I want to confirm 200 countries, using with my peanut sets. I will send to you

*R.F.D. 3, Waynesville, Ohio 45068.

Front view of receiver, dials may be bought from any radio supply house.

a diagram of my receiving sets so that your friends may build them if they want.

"Some people ask me for a circuit diagram of my receiver, they almost never ask me nothing other than a circuit diagram. But could they make a good regenerative receiver as I did? They can not make a good one just by looking at the diagram, I hold. The simpler a receiver, the more difficult to build it. The number of turns of my coils are not theoretical ones but experimental ones with or without a signal generator (200 kc to 30 mc). I made the 3.5, 7.0 and 14.0 mc coils first without an r.f. generator. From the first time I intended to make surely solid-state coils, so I used 1 mm d.c.c. wire for the 7.0 and 14.0 mc coils. This is very important because the coils are the heart of the simple short-wave receivers. So I have 6 or 7 hearts for my receiver. I tried but 50 mc is not so sensitive. 50 mc is a v.h.f. band and we can soon realize that it is difficult to make a sensitive and stable circuit for v.h.f. with low frequency techniques. The 50 mc coils are rather sensitive for a simple circuit.

"Thinking simply I can get a voltage amplification of 140 times (47.6 db) with the 7.0 mc coil so I could hear 127 countries on this band. I have now heard 202 countries. I made the 202 country mark last month with 176 confirmations. I think this is a new world's record with a simple receiver according to one of the top Japanese DXers. I am very happy to report to you that I have now finished my main aim as an s.w.l. I used 50 months for this goal and spent about

The receiving position of Hijmae Suzuki, Tokyo Japan and some of his QSL collection. All pictures were taken by Hijame with a simple box camera.

Top view of receiver.

254

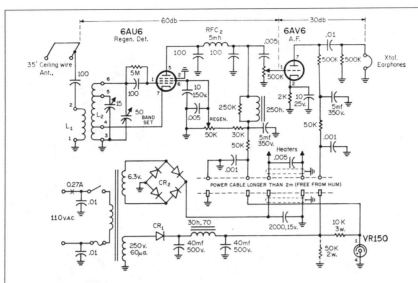

Fig. 1—Circuit of the regenerative receiver.

$42.00 to collect QSL cards from 176 countries. If I will not stop s.w.l.ing, I'll probably be able to confirm 200 countries with my simple 2 tube regenerative receiver fed by a 10 meter antenna. I need not a big thousand dollar receiver at present, my simple receiver will do.

"My pen-pal G3LHJ had worked 234 countries with 201 confirmations with 50 watts and a quad as of last December. JA6PA worked 130 countries with only 8 watts and JA2LC worked about 120 countries with an 807 and a three tube regenerative set (one rectifying tube) five or six years ago, getting a DXCC certificate. This JA2LC's attractive results with a homemade regenerative receiver was one of the main inspirations for me and I got started to listen to the DX with a simple regenerative receiver with my own modifications and ideas.

"I have heard 40 zones with 38 zones confirmed. I need zones 2 and 33 for completion of confirmed all zones. I have heard 66 countries on s.s.b. with 31 confirmed. I do not have a QSL card from W1. I have heard 2 countries on 160 meters, 25 on 80, 127 on 40 meters, 170 on 20,

41 on 15, 2 on 10, 2 on 6 and 1 on 2 meters. We do not have the 220 mc band and I have heard no one on 432. I have heard 158 stations on 6 meters and have been awarded the '50 MC 100' award of the JARL. The sunspot activity is at its worst at this time and I'm sure that I could have done better in 1957-58. I am doing very well with my simple receiver and am glad to know that my results are believed by you and many others.

"I have learned much of my English through s.w.l.ing and made many friends. Walt, I have learned another thing by "s.w.l.ing". Of course the technical approach of radio engineering is very important, but what amateur radio stations have as their goal is the world-wide friendship. I have a theory that the so-called DX country type of ham is not a real ham. Of course there are time limitations but the local QSOs are important, in other words, the DXCC is not so important as the deep philosophical point of view. I believe that friendship is more important than DXCC."

In his February letter, Hijame tells me that

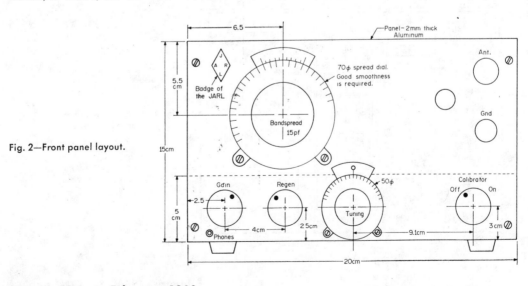

Fig. 2—Front panel layout.

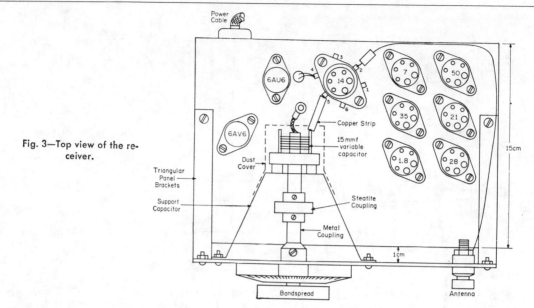

Fig. 3—Top view of the receiver.

he is 24½ years of age and is studying to be a radio engineer. He says, "I am going to finish my student time with one more studying year at my institute till March 31, 1966. I would like to remain a student always in my heart till death.

"I live in a small house made of wood and have never tried the receiver in a concrete or steel house. My location is bad, noise-wise. My closest neighbors are only 50 feet away from my room and my house is surrounding many other houses. My family has only a very small backyard here. The antenna is a 10 meter wire under the roof of my house, at ceiling high. I use this antenna because the antenna vibration caused by the wind affects the stability of my receiver. I have a ground plane for 2 meters, 30 feet high and even this affects the indoor antenna. With this ceiling antenna and my simple receiver I heard and confirmed UA4KED on

7 mc s.s.b. and heard the rare M1ZG on 3.5 mc A1. The receiver is very sensitive but is bothered by local man made noises caused by motors, fluorescent lamps, T.V. sets, radio sets, automobiles and other electrical appliances. I always unplug the BC set when I want to really get the DX. I use d.c. on the filaments. The radiation from TV sets is the worst."

I am building one of these receivers and expect to be listening in on you quite a lot soon. I will also be on 40 c.w. looking for you as soon as the antenna is up at my home. I will be running low power, using a long-wire. When I first started, I used a receiver nearly like this. My receiver used a 6D6 regenerative detector with a 6C5 and a 41 audio amplifier.

To use this receiver with a speaker, this unit will need another stage of audio, such as a 6AQ5 or similar amplifier tube. I will give you

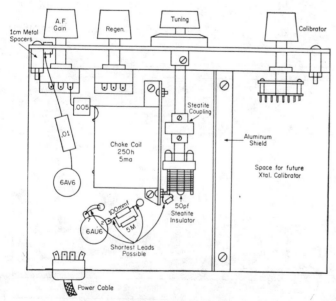

Fig. 4—Bottom view of the receiver.

Coils for the receiver. The three higher frequency coils may be carefully wired in the socket or the bottom cut off the form, the coils installed and the form glued back in place. Adjust all cathode taps so to get the set to regenerate at about ½ setting of the regeneration potentiometer.

Rear View of the receiver. 20 meter coil in socket.

more information on the use of this receiver. By the way almost all of the parts for this receiver including the power supply and extra audio can be built from the parts in any old television set that you have laying around.

Don't spoof at the little set, you might be like the fellow that I was listening to the other night. He was running 30 watts and the fellow that he was talking to was kidding him about his low power. The QRP fellow asked him if he had ever tried low-power, and said he was having a bushel of fun working the bands with his peanut whistle. I have always believed that if they can get here with low power I would be ashamed to admit that I couldn't do the same. Try low-power and a good antenna, see what you are missing.

I must extend my thanks to Hijame Suzuki, JARL-JA1-3477, Inokashira 2-33-12 Mitaka-Shi, Japan for the information for this column. He tells me that *CQ* cost him 300 yen before the price increase and it is always two months later than we get it. He is copying code at 20 w.p.m. and the Japanese code at 18 w.p.m. Student activities keeps him too busy but he will be on the air this summer. He graduates March 31, 1966. I'll be looking for you on the air, Hijame.

As an afterthought: the choke used in the plate circuit of the 6AU6 is an article that may be hard to get so in an emergency use the primary of an audio transformer. A single plate to grid unit may be used.

Next month we will show the Japanese licensing procedures & regulations along with Hijame's plans for 6, 2, 432 mc super regenerative detectors. 73, Walt, W8ZCV

TABLE I—COIL DATA FOR THE HIJAME SUZUKI RECEIVER

Freq. (Mc)	L_1	L_2	Cathode Tap†	Band Spread Tap†
1.8 (1.74-3.25)	4¼ t. #26e. 5mm from L_2	72½ t. #24e. close wound	1 t.	None
3.5 (3.05-4.95)	4¼ t. #26e. 1-10 mm from L_2	36⅝ t. #24e. 1mm pitch	¼ t.	None
7.0 (6.81-12.2)	4 t. #24e. 6mm from L_2	20¼ t. #18e. 2mm pitch	⅜ t.	12½ t.
14 (12.0-22.4)	4 t. #23e. 1mm pitch. 9mm from L_2	9½ t. #18e. pitch 3mm	⅜ t.	5 t.
21 (17.4-30.1)	2½ t. #20 tinned, 1.5mm pitch, 4mm from L_2 *	10½ t. #20 tinned, 1.5mm pitch *	½ t.	6¼ t.
28 (21.6-40.5)	2-1/12 t. #20 tinned, 1.5mm pitch, 4mm from L_2 *	7½ t. #20 tinned, 1.5mm pitch *	½ t.	4¼ t.
50 (36-66)	2 t. #20 tinned, 2mm pitch, 2.5mm from L_2 *	6 t. #20 tinned, 2mm pitch, 10mm dia. *	1½ t.	3¼ t.

*Can be Airdux or equiv.
†Turns from cold (ground) end of coil. 1 inch = 22.4 mm (approx 3/64″ = 1 mm)
All coils wound on 1¼″ dia. 6 prong coil forms except where coil stock is used.

Receiver Construction

● A ONE-TUBE REGENERATIVE RECEIVER

THE SIMPLEST receiver capable of giving at all satisfactory results in everyday operation is one consisting of a regenerative detector followed by an audio amplifier. This type of receiver is sufficient for headphone reception, and is quite easy to build and adjust. A dual tube may be used for both stages, thereby reducing cost.

Figs. 1101 to 1105 show such a receiver, using a 6C8G twin-triode tube, one triode section being the regenerative detector and the other the audio amplifier. The circuit diagram is given in Fig. 1103. The grid coil, L_1, is tuned to the frequency of the incoming signal by means of condensers C_1 and C_3, C_1 being the bandsetting or general coverage condenser and C_3 the bandspread condenser. Regeneration is supplied by means of the tickler coil L_2; the variable plate by-pass condenser, C_2, is the regeneration control. The receiver is coupled to the antenna through C_5, a low-capacity trimmer condenser. R_1 and C_4 are the grid leak and grid condenser.

The audio amplifier section of the tube is coupled to the detector by the audio transformer T_1. Bias for the audio stage is supplied by a midget flashlight cell, this type of bias being quite convenient as well as cheaper than other methods. The choke, *RFC*, is necessary to prevent r.f. current from flowing in the primary winding of the audio transformer; without the choke the regeneration control condenser C_2 may be ineffective. A switch, S_1, is provided for turning off the "B" supply when transmitting.

This receiver is laid out so that it can be converted into the two-tube superhet described in the next section, using most of the same parts over again and utilizing the same chassis and panel. The superhet will give improved performance, but is a little more difficult to build and adjust. By building the one-tube receiver first, the beginner will acquire experience in the operation of regenerative

Fig. 1102 — A rear view of the one-tube regenerative receiver. The grid condenser and grid leak are supported by their wire leads between the stator plates of the tuning condenser and the grip cap on the tube.

Fig. 1101 — A one-tube regenerative receiver, using a double triode as a regenerative detector and audio amplifier.

196 CHAPTER ELEVEN

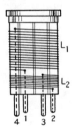

Fig. 1103 — Circuit diagram of the one-tube regenerative receiver.
C₁, C₂ — 100-μμfd. variable (Hammarlund SM-100).
C₃ — 15-μμfd. variable (Hammarlund SM-15).
C₄ — 100-μμfd. mica.
C₅ — 3-30-μμfd. mica trimmer (National M-30).
R₁ — 1 megohm, ½ watt.
L₁, L₂ — See coil table.
T₁ — Audio transformer, interstage type, 3:1 ratio (Thordarson T13A34).
S₁ — S.p.s.t. toggle switch.
RFC — 2.5-mh. r.f. choke.

circuits which will be helpful in building and using the two-tube receiver.

The construction of the receiver is shown in the photographs. The chassis measures 5½ by 9½ by 1½ inches. The three variable condensers are mounted on the panel three inches from the bottom edge, with C_3 in the center, C_1 at the right and C_2 at the left. The condensers are 3½ inches apart, center to center. The tube socket is directly behind C_3, its center being 2¼ inches from the panel; the coil socket is 2½ inches to the right. The audio

BOTTOM OF SOCKET
OR COIL FORM

Fig. 1105 — Method of winding coils for the one-tube regenerative receiver.

Fig. 1104 — Bottom of chassis view of the one-tube regenerative receiver. Construction and wiring are extremely simple.

transformer is mounted along the rear chassis edge as shown. All ground connections may be made directly to the chassis, making sure that the paint is scraped away and that good contact is secured.

The wiring underneath the chassis is very simple. In the photograph, Fig. 1104, the antenna connection strip is at the left, with C_5 supported by the wiring to the antenna post. The ground connection is soldered to a lug under the nut holding the connection strip in place. The choke RFC also is supported by the wiring. The bias battery (the zinc can is the negative terminal) is soldered to a lug strip as shown. The headphone connections are made by means of tip jacks mounted on the rear edge of the chassis. Filament and plate power are brought in through a four-wire cable which enters the chassis through the rear edge.

The coils are made as shown in Fig. 1105 and the coil table. Both windings should be in the same direction. Using the standard pin numbering for four prong sockets, pin 1 connects to ground, pin 2 to the plate of the detector, pin 3 to RFC and the stator plates of C_2, and pin 4 to the stator plates of C_1 and C_3. L_1 for the B, C and D coils should have its turns evenly spaced to occupy the specified length; the wire may be held in place when the coil is finished by running some Duco cement along the ridges of the coil forms. Be sure to clean any excess soldering flux from the coil pins after the wires are soldered in place; rosin flux in particular will form a thin insulating film over the pins and prevent contact when the coil form is inserted in the socket.

The heater supply for the receiver may be either a 6.3-volt filament transformer (the 1-ampere size will be ample) or a 6-volt battery. A 45-volt "B" battery should be used for the plate supply. The "B" current drain is only a few milliamperes, and a medium- or small-size "B" battery will give excellent service.

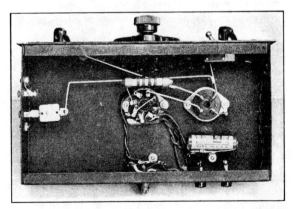

CHAPTER ELEVEN *197*

ONE-TUBE REGENERATIVE RECEIVER COIL DATA

Coil	Grid Winding (L_1)	Tickler (L_2)
A	56 turns No. 22 enamelled	15 turns No. 24 enamelled
B	32 " " " "	8 " " " "
C	18 " " " "	5 " " " "
D	10 " " " "	5 " " " "

All coils wound on 1½-inch diameter forms (Hammarlund SWF-4). Grid windings on coils B, C and D spaced to occupy a length of 1½ inches; grid winding on coil A close-wound. Tickler coils all close-wound, spaced ⅛ inch from bottom of grid winding. See Fig. 1105

Frequency range: Coil A — 1700 to 3200 kc.
B — 3000 to 5700 kc.
C — 5400 to 10,000 kc.
D — 9500 to 18,000 kc.

After the set is completed and the wiring checked to make sure that it is exactly as shown, insert the C coil (selected because signals can usually be heard in this range at any time of the day or night) in the coil socket and connect the headphones, antenna and ground, and the heater supply. After the heater supply has been connected for a few minutes, the tube should feel warm to the touch and there should be a visible glow from the heater. The "B" battery can now be connected and the switch, S_1, closed.

Now turn the regeneration condenser, C_2, starting from minimum capacity (plates all out) until the set goes into oscillation. This phenomenon is easily recognizable by a distinct click, thud or hissing sound. The point where oscillation just begins is the most sensitive operating point at that particular dial setting.

The tuning dial may now be slowly turned, the regeneration control knob being varied simultaneously (if necessary) to keep the set just oscillating. A number of stations will probably be heard. A little practice will make tuning easy.

If the set refuses to oscillate, the sensitivity will be poor and no code signals will be heard on the frequencies at which such signals should be expected. It should oscillate easily, however, if the coils are made exactly as shown. It sometimes happens that the antenna takes so much energy from the set that it cannot oscillate, this usually resulting in "holes" in the range where no signals can be picked up (and where the hissing sound cannot be obtained). This can be cured by reducing the capacity of C_5 (unscrewing the adjusting screw) until the detector again oscillates. If it still refuses to oscillate, the coil L_2 must be moved nearer to L_1 or, in extreme cases, a turn or two must be added to L_2. This is best done by rewinding with more turns rather than by trying to add a turn or two to the already-wound coil. For any given band of frequencies, adjust C_5 so that the detector oscillates over the whole range, using as much capacity at C_5 as is possible. This will give the best compromise between dead spots and signal strength. It will be found that less advancing of the regeneration control, C_2, is required at the high-frequency end of a coil range (C_1 at or near minimum capacity) than at the low-frequency end. The best adjustment of the antenna condenser, C_5, and the feedback coil, L_2, is that which requires almost a maximum setting of the regeneration control at the low-frequency end (maximum capacity of C_1) of any coil range.

Coil A misses the high-frequency end of the broadcast band, but it is possible to hear police stations and the 160-meter amateur band with it, as well as other services. The amateur band is most easily located by listening at night (when there is the most activity), setting C_3 at maximum and slowly tuning with C_1 until some of the police stations are heard. These stations operate on 1712 kc., so that once found they become "markers" for the low-frequency end of the band. Further tuning then should be done with the main tuning dial, and many amateur stations should be heard.

Locating the amateur bands on the other coils is done in much the same manner, by searching carefully with C_1. The 3.5–4.0-Mc. amateur band will be found on coil B at about 80% setting of C_1; it will be easiest to locate this band by setting C_3 at minimum capacity (plates unmeshed) and adjusting C_1 until amateur 'phone stations are heard. Again this is best done at night, when the activity is heaviest on this band. On coil C, the 7-Mc. amateur band will be found with C_1 meshed about 60%; the 14-Mc. band (coil D) is found with C_1 meshed about 20%.

A suitable antenna for the receiver would be 50 to 75 feet long, and as high and clear of surrounding objects as possible. The ground lead should preferably be short; a ground to a heating radiator or water piping is usually good.

198 CHAPTER ELEVEN

Reprinted from the 1942 "Radio Amateur's Handbook" 19th edition